RECRYSTALLIZATION
AND RELATED ANNEALING PHENOMENA

Pergamon Titles of Related Interest

Books

BEVER
Encyclopedia of Materials Science and Engineering
8 Volume Set

BLOOR, BROOK, FLEMINGS & MAHAJAN
The Encyclopedia of Advanced Materials
4 Volume Set

CAHN
Encyclopedia of Materials Science and Engineering
Supplementary Volumes 1, 2 & 3

Journals

Acta Metallurgica et Materialia

Calphad

Journal of Physics and Chemistry of Solids

Materials Research Bulletin

Progress in Crystal Growth and Characterization of Materials

Scripta Metallurgica et Materialia

RECRYSTALLIZATION
AND RELATED ANNEALING PHENOMENA

by

F.J. HUMPHREYS

University of Manchester Institute of Science and Technology, U.K.

and

M. HATHERLY

University of New South Wales, Australia

PERGAMON

ELSEVIER SCIENCE Ltd
The Boulevard, Langford Lane
Kidlington, Oxford OX5 1GB, UK

First edition 1995
Second impression 2002

Library of Congress Cataloging in Publication Data

Recrystallization and related annealing phenomena/
F.J. Humphreys and M. Hatherly
p. cm.
Includes index.
1. Recrystallization (Metallurgy). 1. Hatherly, M. II. Title.
TN690.H969 1995
671.3'6 – dc20
94-44667

British Library Cataloguing in Publication Data
A catalogue record for this book is available from the British Library

ISBN: 0 08 041884 8 (Cased version)
ISBN: 0 08 042685 9 (Flexi version)

Printed in The Netherlands.

CONTENTS

CHAPTER 3
THE STRUCTURE AND ENERGY OF GRAIN BOUNDARIES 57

CHAPTER 4   85
THE MOBILITY AND MIGRATION OF BOUNDARIES

CHAPTER 5   127
RECOVERY AFTER DEFORMATION

CHAPTER 6   173
RECRYSTALLIZATION OF SINGLE-PHASE ALLOYS

CHAPTER 11
RECOVERY AND RECRYSTALLIZATION DURING AND AFTER HOT DEFORMATION

363

CHAPTER 12
CONTROL OF RECRYSTALLIZATION.

CHAPTER 13
COMPUTER MODELLING AND SIMULATION
OF ANNEALING

APPENDIX   435

SYMBOLS

The following notation is generally used in the text. The subscripts i or n indicate the use of letters or numbers for particular symbols. On rare occasions where the letters or symbols are used for other purposes, this is specifically stated.

b	Burgers vector of a dislocation
c, c_n, C, C_n, K_n	These denote "local" constants which are defined in the text.
d	Diameter of second-phase particle
D	Grain or subgrain diameter
D_i	Diffusivity (s=bulk diffusion, b=boundary diffusion, c=core diffusion)
F_v	Volume fraction of second-phase particles
G	Shear modulus
E_i	Energy. e.g. stored energy of deformation E_D
k	Boltzmann contant
M	Boundary mobility
N_v	Number of grains or second-phase particles per unit volume
N_s	Number of particles per unit area
P or P_i	Pressure on a boundary
Q or Q_i	Activation energy, (for diffusion:- s=bulk, b=boundary, c=core)
R	Grain or subgrain radius
s	Shear strain
t	Time
T, T_m	Temperature, melting temperature (K)
v	Velocity of dislocation or boundary
V	Volume
α, β	Constants
γ	Energy of an interface or boundary
$\gamma_{SFE}, \gamma_{RSFE}$	Stacking fault energy, reduced stacking fault energy
γ_b	Energy of a high angle boundary
γ_s	Energy of a low angle boundary
ϵ	True strain
$\dot{\epsilon}$	True strain rate

θ	Misorientation across a boundary
λ	Interparticle spacing as defined in equation 8.3
ν	Poisson ratio
ν_0	Atomic vibrational frequency
ρ	Dislocation density
$\Sigma = n$	Coincidence grain boundaries. 1/n is the fraction of sites common to both grains
σ	True stress
τ	Shear stress
Φ, ϕ_1, ϕ_2	Euler angles (defined in the appendix)
Ω	Orientation gradient

ABBREVIATIONS

BKD	Backscattered Kikuchi diffraction
CSL	Coincidence site lattice
EBSP(D)	Electron backscatter patterns (diffraction)
HAGB	High angle grain boundary
HVEM	High voltage transmission electron microscope
JMAK	Johnson-Mehl-Avrami-Kolmogorov kinetic model
LAGB	Low angle grain boundary
ND,RD,TD	Normal, rolling and transverse directions in a rolled product
ODF	Orientation distribution function
PSN	Particle stimulated nucleation of recrystallization
SEM	Scanning electron microscope
SFE	Stacking fault energy
SIBM	Strain induced boundary migration
STEM	Scanning transmission electron microscope
TEM	Transmission electron microscope

PREFACE

Recrystallization and the related annealing phenomena which occur during the thermomechanical processing of materials have long been recognised as being both of technological importance and scientific interest. These phenomena are known to occur in all types of crystalline materials; they occur during the natural geological deformation of rocks and minerals, and during the processing of technical ceramics. However, the phenomena have been most widely studied in metals, and as this is the only class of material for which a coherent body of work is available, this book inevitably concentrates on metallic materials.

Although there is a vast body of literature going back 150 years, and a large collection of reviews which are detailed in chapter 1, there have only been two monographs published in recent times on the subject of recrystallization, the latest nearly 20 years ago. Since that time, considerable advances have been made, both in our understanding of the subject and in the techniques available to the researcher.

Metallurgical research in this field is mainly driven by the requirements of industry, and currently, a major need is for quantitative, physically-based models which can be applied to metal-forming processes so as to control, improve and optimise the microstructure and texture of the finished products. Such models require a more detailed understanding of both the deformation and annealing processes than we have at present. The development of the underlying science to a level sufficient for the construction of the required models from first principles provides a goal for perhaps the next 10 to 20 years.

The book was written to provide a treatment of the subject for researchers or students who need a more detailed coverage than is found in textbooks on physical metallurgy, and a more coherent treatment than will be found in the many conference proceedings. We have chosen to emphasise the scientific principles and physical insight underlying annealing rather than produce a comprehensive bibliography or handbook.

Unfortunately the generic term **annealing** is used widely to describe two metallurgical processes. Both have a common result in that a hardened material is made softer, but the mechanisms involved are quite different. In one case, associated with the heat treatment of ferrous materials, the softening process involves the $\gamma \rightarrow \alpha$ phase transformation. In the second case, which is the one relevant to this book, the softening is a direct result of the loss via recovery and recrystallization, of the dislocations introduced by work hardening.

It is not easy to write a book on recrystallization, because although it is a clearly defined subject, many aspects are not well understood and the experimental evidence is often poor and conflicting. It would have been desirable to quantify all aspects of the phenomena and to derive the theories from first principles. However, this is not yet possible, and the reader will find within this book a mixture of relatively sound theory, reasonable assumptions and conjecture. There are two main reasons for our lack of progress. First, we cannot expect to understand recovery and recrystallization in depth unless we understand the nature of the deformed state which is the precursor, and that is still a distant goal. Second, although some annealing processes, such as recovery and grain growth are reasonably homogeneous, others, such as recrystallization and abnormal grain growth are heterogeneous, relying on local instabilities and evoking parallels with apparently chaotic events such as weather.

It must be recognised that we are writing about a live and evolving subject. Very little is finished and the book should therefore be seen as a snapshot of the subject at this particular time as seen by two scientists who are undoubtedly biased in various ways. We hope that when a second edition of this volume is produced in perhaps 10 years time, or a new treatment is attempted, many aspects of the subject will have become clearer.

Recovery and recrystallization depend on the nature of the deformed state and involve the formation, removal and movement of grain boundaries. For these reasons we have included treatments of the deformed state in chapter 2, and the nature of grain boundaries in chapter 3. These are both large topics which merit complete books in themselves, and we have not attempted a comprehensive coverage but have merely aimed to provide what we regard as essential background information in order to make the volume reasonably self-contained. Chapter 4 is concerned with the migration and mobility of grain boundaries, and this contains some background information.

The main topics of the book - recovery, recrystallization and grain growth are covered in chapters 5 to 11 and include specific chapters on ordered materials, two-phase alloys and annealing textures. In order to illustrate some of the applications of the principles discussed in the book we have selected a very few technologically important case studies in chapter 12. The final chapter outlines the ways in which computer simulation and modelling are being applied to annealing phenomena, and in the Appendix we provide an introduction to the measurement and representation of textures for the benefit of readers who are not specialists in this area.

<table>
<tr><td>John Humphreys</td><td style="text-align:right">Max Hatherly</td></tr>
<tr><td>Manchester Materials Science Centre</td><td style="text-align:right">School of Materials</td></tr>
<tr><td>University of Manchester Institute of</td><td style="text-align:right">University of New South Wales</td></tr>
<tr><td>Science and Technology</td><td style="text-align:right">Australia</td></tr>
<tr><td>U.K.</td><td></td></tr>
</table>

August 1994

The need to reprint the book has provided an opportunity for us to correct some of the errors and to carry out minor modifications to the text.

May 1996

ACKNOWLEDGEMENTS

The authors would like to acknowledge a great debt to those with whom they have discussed and argued over the subjects covered by this book over a period of very many years. During the writing of the book we have had particularly useful discussions and correspondence with Brian Duggan, Bevis Hutchinson and Erik Nes. A large number of others have helped by providing advice, material and in many other ways. They include Sreeramamurthy Ankem, Mahmoud Ardakani, Christine Carmichael, Michael Ferry, Brian Gleeson, Gunther Gottstein, Brigitte Hammer, Alan Humphreys, Claire Humphreys, Peter Krauklis, Lasar Shvinderman, Tony Malin, Paul Munroe, Nigel Owen, Phil Prangnell, Fred Scott, Karen Vernon-Parry and David Willis. Needless to say, the authors accept all blame for the mistakes in the book.

Problems of communication during the preparation of this book were eased by the hospitality afforded by the University of New South Wales to FJH and by UMIST to MH.

Finally, we must acknowledge the great patience and understanding shown by our wives Anna and Lorna who gave us the freedom to write this book.

Chapter 1

INTRODUCTION

1.1 THE ANNEALING OF A DEFORMED MATERIAL

1.1.1 Outline and terminology

The free energy of a crystalline material is raised during deformation by the presence of dislocations and interfaces, and a material containing these defects is thermodynamically unstable. Although thermodynamics would suggest that the defects should spontaneously disappear, in practice the necessary atomistic mechanisms are often very slow at low homologous temperatures, with the result that unstable defect structures are retained after deformation (fig 1.1a).

If the material is subsequently heated to a high temperature (**annealed**), thermally activated processes such as solid state diffusion provide mechanisms whereby the defects may be removed or alternatively arranged in configurations of lower energy.

The defects may be introduced in a variety of ways. However, in this book we will mainly be concerned with **those defects, and in particular dislocations, which are introduced during plastic deformation**. The point defects introduced during deformation anneal out at low temperatures and generally have little effect on the mechanical properties of the material. In considering only materials which have undergone substantial plastic deformation, we necessarily limit the range of materials with which we will be concerned. Metals are the only major class of crystalline material to undergo substantial plastic deformation at low homologous temperatures, and much of this book will be concerned with the annealing of deformed metals. However at high temperatures, many minerals and ceramics readily deform plastically, and the annealing of these is of great interest. In addition, some annealing processes such as **grain growth** are relevant to cast or vapour deposited materials as well as to deformed materials.

On annealing a cold worked metal at an elevated temperature, the microstructure and also the properties may be partially restored to their original values by **recovery** in which annihilation and rearrangement of the dislocations occurs. The microstructural changes during recovery are relatively homogeneous and do not usually affect the boundaries between the deformed grains. The changes in microstructure due to recovery are shown schematically in figure 1. Similar recovery processes may also occur during deformation, particularly at high temperatures, and this **dynamic recovery** plays an important role in the **creep** and **hot working** of materials.

Recovery generally involves only a partial restoration of properties because the dislocation structure is not completely removed, but reaches a metastable state (fig 1.1b). A further restoration process called **recrystallization** may occur in which new dislocation-free grains are formed within the deformed or recovered structure (fig 1.1c). These then grow and consume the old grains, resulting in a new grain structure with a low dislocation density. (fig 1.1d). Recrystallization may take place during deformation at elevated temperatures and this is then termed **dynamic recrystallization**.

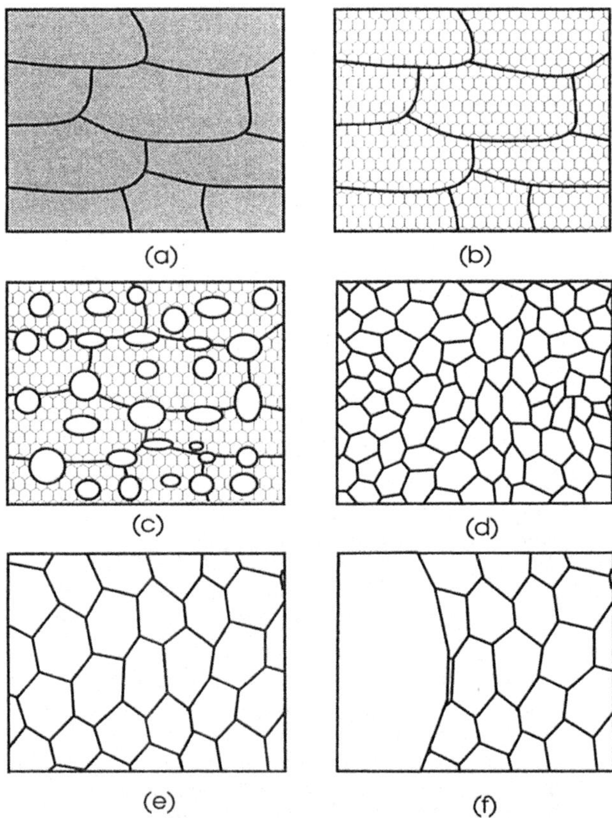

Fig. 1.1. Schematic diagram of the main annealing processes a) Deformed state, b) Recovered, c) Partially recrystallized, d) Fully recrystallized, e) Grain growth, f) Abnormal grain growth.

Although recrystallization removes the dislocations, the material still contains grain boundaries, which are thermodynamically unstable. Further annealing may result in **grain growth**, in which the smaller grains are eliminated, the larger grains grow, and the grain boundaries assume a lower energy configuration (fig 1.1e). In certain circumstances this **normal grain growth** may give way to the selective growth of a few large grains (fig 1.1f), a process known as **abnormal grain growth or secondary recrystallization**.

Some annealing processes such as recovery and normal grain growth occur relatively homogeneously throughout the microstructure, whereas others such as recrystallization and abnormal grain growth are very heterogenous. It is sometimes convenient to distinguish between these as **continuous** or **discontinuous** processes. It is now known that these distinctions are not always clear and that a recrystallized microstructure may sometimes evolve in a homogeneous manner.

1.1.2 The importance of annealing

Many metallic materials are produced initially as large castings which are then further processed in the solid state by forging, rolling, extrusion etc to an intermediate or final product. These procedures, which may be carried out hot or cold, and which may involve intermediate anneals, are collectively termed **thermomechanical processing**. Recovery, recrystallization and grain growth are core elements of this processing.

To a large extent the mechanical properties and behaviour of a metal depend on the dislocation content and structure, the size of the grains and the orientation or **texture** of the grains. Of these, the dislocation content and structure are the most important. The mechanical properties depend primarily on the number of dislocations introduced during cold working and their distribution. As this increases from $\sim 10^{11} \text{m}^{-2}$, typical of the annealed state, to $\sim 10^{16} \text{m}^{-2}$, typical of heavily deformed metals, the yield strength is increased by up to 5-6 times and the ductility decreased. If the strain hardened material is subsequently heated to $\sim T_m/3$ dislocation loss and rearrangement occur and this is manifested by a decrease in strength and increased ductility. There is an enormous literature on the magnitude of these changes and any adequate treatment is beyond the scope of this book. For details of these the reader is referred to the appropriate volumes of **Metals Handbook**.

The grain size and texture are determined mainly by the recrystallization process. There are numerous examples of the need to control grain size. For example, a small grain size increases the strength of a steel and may also make it tougher. However, a large grain size may be required in order to reduce creep rates in a nickel-based superalloy for use at high temperatures. Superplastic forming, in which alloys are deformed to large strains at low stresses, is becoming an important technological process for the shaping of advanced materials. Great ingenuity must be exercised in producing the required grain size and preventing its growth during high temperature deformation. The control of texture is vital for the successful cold forming of metals, a particularly important example being the deep drawing of aluminium or steel beverage cans.

1.2 HISTORICAL PERSPECTIVE

1.2.1 The early development of the subject

Although the art of metalworking including the procedures of deformation and heating has been practised for thousands of years, it is only comparatively recently that some understanding has been gained of the structural changes which accompany these processes. The early history of the annealing of deformed metals has been chronicled very elegantly by Beck (1963), and it is clear that the pace of scientific understanding was largely dictated by the development of techniques for materials characterisation. It should be noted that this constraint still applies.

1.2.1.1 Crystallinity and crystallization

In 1829, the French physicist Felix Savart found that specimens from cast ingots of various metals exhibited acoustic anisotropy, and concluded that cast ingots consisted of crystals of different orientations. He also found that although the anisotropy was changed by plastic deformation and subsequent annealing, heating alone produced no change. This is the first recorded evidence for a structural change occurring during the annealing of a cold worked metal.

In the mid 19th century, the concept of the crystallization of metals was extensively discussed, and it was widely thought that plastic deformation rendered metals amorphous. This belief arose from the inability to observe grain structures in the deformed metals using visual inspection. On reheating the deformed metal however, the grain structure could sometimes be seen (Percy 1864, Kalischer 1881), and this was then interpreted as **crystallization** of the metal from its amorphous state.

The introduction of metallographic techniques by Sorby, culminating in his paper of 1887 took the subject a step forward. He was able to study the elongated grains in deformed iron and note that on heating, a new equiaxed grain structure was produced, a process which he termed **recrystallization**. Furthermore, Sorby recognised that the distorted grains must be unstable, and that recrystallization allowed a return to a stable condition. Despite Sorby's work, the idea that cold worked metals were amorphous persisted for some years and was not finally abandoned until the Bakerian lecture by Ewing and Rosenhain in 1900 in which it was clearly shown that plastic deformation took place by slip or twinning, and that in both of these processes the crystal structure was preserved.

1.2.1.2 Recrystallization and grain growth

Although recrystallization had been identified by the beginning of the 20th century, recrystallization and grain growth had not clearly been distinguished as separate processes. The outstanding work of Carpenter and Elam (1920) and Altherthum (1922) established that stored energy provided the driving force for recrystallization and grain boundary energy that for grain growth. This is shown by the terminology for these processes used by Altherthum - Bearbeitungsrekristallisation (cold-work recrystallization) and Oberflächenrekristallisation (surface tension recrystallization).

In 1898 Stead had proposed that grain growth occurred by grain rotation and coalescence, and although Ewing and Rosenhain presented convincing evidence that the mechanism was one of boundary migration, Stead's idea was periodically revived until the work of Carpenter and Elam finally settled the matter in favour of boundary migration.

1.2.1.3 Parameters affecting recrystallization

By around 1920, many of the parameters which affected the recrystallization process and the resultant microstructure had been identified.

Kinetics - The relationship of the recrystallization temperature to the melting temperature had been noted by Ewing and Rosenhain (1900), and Humfrey (1902) showed that the rate of recrystallization increased with an increase in annealing temperature.

Strain - Sauveur (1912) found that there was a critical strain for recrystallization, and a relationship between grain size and prior strain was reported by Charpy (1910). Carpenter and Elam (1920) later quantified both of these effects.

Grain growth - In a very early paper on the control of microstructure during annealing, Jeffries (1916) showed that **abnormal grain growth** in thoriated tungsten was promoted in specimens in which normal grain growth had been inhibited.

Further developments in the understanding of recrystallization were not possible without a more detailed knowledge of the deformed state. This was provided by the development of dislocation theory in 1934, and a notable early review of the subject following the advent of dislocation theory is that of Burgers (1941).

From about this period it becomes difficult to distinguish the papers of historical interest from the early key papers which are still relevant to current thinking, and the latter are cited as appropriate within the various chapters of this book. However, it may be helpful to the reader to have a source list of books, reviews and conferences on the subject from within the last 40 years, and this is given below.

1.2.2 Key literature (1952-1993)

Monographs on Recrystallization

Byrne, J.G. (1965), **Recovery, Recrystallization and Grain Growth**. McMillan, New York.

Cotterill, P. and Mould, P.R. (1976), **Recrystallization and Grain Growth in Metals**. Surrey Univ. Press. London.

Multi-author, edited compilations on Recrystallization

Himmel, L. (ed), (1963), **Recovery and Recrystallization of Metals**. Interscience, N.York. (1963)

Margolin, H. (ed), (1966), **Recrystallization, Grain growth and Textures**. ASM, Ohio, USA.

Haessner, F. (ed), (1978), **Recrystallization of Metallic Materials**. Dr. Riederer-Verlag. GMBH Stuttgart.

<u>Review articles and books containing chapters on Recrystallization.</u>

Burke, J.E. and Turnbull, D. (1952), **"Recrystallization and Grain Growth"**. Progress in Metal Phys. <u>3</u>, 220.

Beck, P.A. (1954), **"Annealing of Cold-worked Metals"**. Adv. Phys. <u>3</u>, 245.

Leslie, W.C., Michalak, J.T. and Aul. F.W. (1963), **"The annealing of cold-worked iron"**. In. **Iron and its Dilute Solid Solutions**. Ed. Spencer and Werner. Interscience. New York. 119.

Christian, J.W. (1965), **The Theory of Transformations in Metals and Alloys**. Pergamon, Oxford.

Jonas, J.J., Sellars, C.M. and Tegart, W.J. McG. (1969), **"Strength and Structure Under Hot Working conditions"**. Met. Revs. <u>130</u>, 1.

Martin, J.W. and Doherty, R.D. (1976), **The Stability of Microstructure in Metals**. Cambridge University Press.

Cahn, R.W. (1996), in **Physical Metallurgy**. Eds. Cahn and Haasen. North-Holland, Amsterdam. 4th ddition.

Hutchinson, W.B. (1984), **"Development and Control of Annealing Textures in Low-Carbon Steels"**. Int. Met. Rev., <u>29</u>, 25.

Honeycombe, R.W.K. (1985), **The Plastic Deformation of Metals**. Edward Arnold.

Humphreys, F.J. (1991), **"Recrystallization and Recovery"**, in. **Processing of Metals and Alloys**. 371, ed. R.W. Cahn. VCH, Germany.

<u>Proceedings of Conferences</u>

<u>International Recrystallization Conference Series</u>
Chandra, T. (ed). (1991), **Recrystallization'90**. TMS, Warrendale, USA.

Fuentes, M. and Gil Sevillano, J. (eds), (1992), **Recrystallization'92**. Trans Tech Pubs. Switzerland.

<u>International Texture Conference (ICOTOM) Series</u>
Conferences are held every 3 years.
The 11th Conference was held in Xi'an, China in 1996.

<u>International Risø Symposia</u>
Held annually in Risø, Denmark, the proceedings in 1980, 1983, 1986, 1991 and 1995 are of particular relevance.

<u>International Grain Growth Conferences</u>
Abbruzzese, G. and Brozzo, P. (eds). (1991). **Grain Growth in Polycrystalline Materials**. Trans Tech Publications, Switzerland.

<u>Other International Conferences</u>

Institute of Metals (1973), **Recrystallization in the Control of Microstructure**. London. Keynote papers published in Metal Science J. <u>8</u>. (1974).

Institute of Metals (1978), **Recrystallization in the Development of Microstructure.** Leeds. Published in Metal Science, <u>13</u>. (1979).

Institute of Materials (1990), **Microstructure and Mechanical Processing**. Cambridge 1990. Keynote papers published in Materials Science and Tech. <u>6</u>. (1990).

Fundamental of Recrystallization - Zeltingen, Germany. A number of short papers from this meeting are published in Scripta Metall. Mater. <u>27</u>, (1992)

1.3 FORCES, PRESSURES AND UNITS

The annealing processes discussed in this book mainly involve the migration of internal boundaries within the material. These boundaries move in response to thermodynamic driving forces, and specific quantitative relationships will be discussed in the appropriate chapter. It is however useful at this stage to set out some of the terminology used and also to compare the energy changes which occur during the various annealing processes with those which drive phase transformations.

1.3.1. Pressure on a boundary

The processes of recovery, recrystallization and grain growth are all driven by the defect content of the material. Consider a small part of the microstructure of a single-phase crystalline material as shown in figure 1.2, which consists of two regions **A** and **B** separated by a boundary at position **x**. Let us assume that the two regions contain different defect concentrations and that the free energies of these regions per unit volume are G^A and G^B respectively.

The boundary will move if the Gibbs free energy of the system is thereby lowered, and if an area **a** of the boundary moves a distance **dx**, then the change in free energy of the system is

$$dG = dx(G^A - G^B).a \qquad\qquad (1.1)$$

The **force F** on the boundary is given by **dG/dx**, and the **pressure P** on the boundary, is given by **F/a**, and thus

$$P = -\frac{1}{a}\frac{dG}{dx} = G^A - G^B = \Delta G \qquad\qquad (1.2)$$

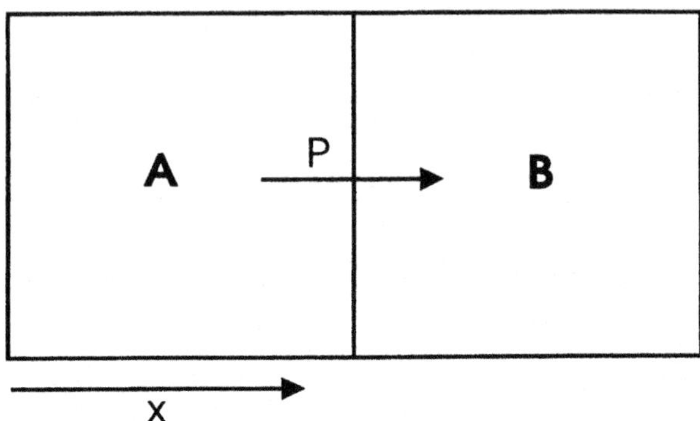

Fig. 1.2. The pressure on a boundary

If ΔG in equation 1.2 is given in units of Jm^{-3}, then the pressure on the boundary (**P**) is in Nm^{-2}. There is some confusion in the literature regarding terminology, and the terms **force on a boundary** and **pressure on a boundary** are both used for the parameter which we have defined above by **P**. As **P** has units of Nm^{-2} which are those of pressure, then there is some logic in using the term **pressure**, and we will adopt this terminology.

1.3.2 Units and the magnitude of P

Although we will be discussing the forces and pressures acting on boundaries in some detail in later chapters, it is useful at this stage to examine, with examples, some of the forces involved in annealing. This will serve to demonstrate the **units** used in the book and also to give some idea of the relative magnitudes of the forces involved in annealing. A good discussion of forces arising from a variety of sources is given by Stüwe (1978).

Recrystallization - driving pressure due to stored dislocations
The driving force for recrystallization arises from the elimination of the dislocations introduced during deformation. The stored energy due to a dislocation density ρ is $\sim 0.5\rho Gb^2$, where **G** is the shear modulus and **b** the Burgers vector of the dislocations. A dislocation density of 10^{15} - 10^{16} m^{-2}, which is typical of the cold worked state, in copper ($G = 4.2 \times 10^{10}$ Nm^{-2}, $b=0.26nm$) therefore represents a stored energy of $\sim 2\times10^6$ - 2×10^7 Jm^{-3} (~ 10-100 J/mol) and gives rise to a driving pressure for recrystallization of $\sim$ **2-20MPa**.

Recovery and grain growth - driving pressure due to boundary energy
Recovery by subgrain coarsening and grain growth following recrystallization are both driven by the elimination of boundary area. If the boundary energy is γ per unit area and the boundaries form a three-dimensional network of spacing D, then the driving pressure for growth is given approximately as $3\gamma/D$. If the energy of a low angle grain boundary (γ_s) is

0.2 Jm^{-2}, and that for a high angle boundary (γ_b) is 0.5Jm^{-2}, we find that **P $\sim$ 0.6MPa** for the growth of 1μm subgrains during recovery, and that **P $\sim$ 10^{-2} MPa** for the growth of 100μm grains.

Comparison with the driving forces for phase transformations

It is of interest to compare the energy changes which occur during annealing, as discussed above, with those which occur during phase transformations. For example a typical value of the latent heat of fusion for a metal is $\sim$ **10kJ/mol** and that for a solid state transformation is $\sim$ **1kJ/mol**. We therefore see that the energies involved in annealing of a cold worked metal are very much smaller than those for phase transformations.

Chapter 2

THE DEFORMED STATE

2.1. INTRODUCTION.

The emphasis in this chapter is quite different to that of the remaining chapters of the book. The most significant of the many changes associated with recrystallization and other annealing phenomena is the decrease in the density of dislocations. In this chapter we are concerned with dislocation **accumulation** rather than dislocation **loss** and with the increases in stored energy that are a result of deformation. The dislocations provide the driving force for the annealing phenomena dealt with in the remaining chapters.

A comparison of this chapter with those dealing with annealing will reveal discrepancies between our current knowledge of the deformed state and our requirements for the understanding of annealing. For example, we currently have an incomplete understanding of the rates of dislocation accumulation during deformation and of the large scale deformation heterogeneities which are important in nucleating recrystallization, and this impedes the formulation of quantitative models of recrystallization. On the other hand, we now have a great deal of information about the formation of microbands during deformation, but have not yet formulated annealing theories which take these into account. Such discrepancies provide useful pointers to areas which need further research.

During deformation the microstructure of a metal changes in several ways. First, and most obvious, the grains change their shape and there is a surprisingly large increase in the total grain boundary area. The new grain boundary area has to be created during deformation and this is done by the incorporation of some of the dislocations that are continuously created during the deformation process. A second obvious feature, particularly at the electron microscope level, is the appearance of an internal structure within the grains. This too, results from the accumulation of dislocations. Except for the small contribution of any vacancies and interstitials that may have survived, the sum of the energy of all of the dislocations and new interfaces represents the stored energy of deformation. There is one other consequence of deformation that is relevant to the study of annealing processes. During deformation the orientations of single crystals and of the individual grains of a polycrystalline metal change relative to the direction(s) of the applied stress(es). These changes are not random and involve rotations which are directly related to the crystallography of the deformation. As a consequence the grains acquire a preferred orientation, or texture, which becomes stronger as deformation proceeds.

Every stage of the annealing process involves loss of some of the stored energy and a corresponding change in microstructure. The release of stored energy provides the driving force for recovery and recrystallization, but it is the nature of the microstructure that controls both the development and growth of the nuclei that will become recrystallized grains and their orientation. If these changes are to be understood it is essential that we begin by examining the nature of the deformed state, the generation of microstructure and particularly the development of inhomogeneities in that microstructure. Unfortunately our knowledge of these matters is still imperfect and Cottrell's assessment of the situation more than 40 years ago is still valid.

*"Few problems of crystal plasticity have proved more challenging than work hardening. It is a spectacular effect, for example enabling the yield strength of pure copper and aluminium crystals to be raised a hundredfold. Also, it occupies a central place in the subject, being related both to the nature of the slip process and to processes such as **recrystallization** and creep. It was the first problem to be attempted by the dislocation theory of slip and may well prove the last to be solved."*

Cottrell (1953)

2.2 THE STORED ENERGY OF COLD WORK.

2.2.1 Origin of the stored energy

Most of the work expended in deforming a metal is given out as heat and only a very small amount ($\sim 1\%$) remains as stored energy in the metal. This stored energy, which provides the source of all the property changes that are typical of deformed metals, is derived from the point defects and dislocations generated during deformation. However the mobility of vacancies and interstitials is so high that except in the special case of deformation at very low temperatures, point defects do not contribute significantly to the stored energy of deformation. In the common case of deformation at ambient temperatures almost all of the stored energy is derived from the accumulation of dislocations and the essential difference between the deformed and the annealed states lies in the dislocation content and arrangement. Because of this, discussion of the significance of the deformation microstructure during recovery and recrystallization must be based on the density, distribution and arrangement of dislocations.

The increase in dislocation density is due to the continued trapping of newly created mobile dislocations by existing dislocations and their incorporation into the various microstructural features that are characteristic of the deformed state. One of the simplest of these is the grain shape. During deformation, the grains of a polycrystalline metal change their shape in a manner that corresponds to the macroscopic shape change. As a result there is an increase in grain boundary area. Consider the case of a cube-shaped grain during rolling. After 50% reduction the surface area of this grain is increased by $\sim 16\%$; after 90% reduction the increase is 270% and after 99% reduction it is 3267%. The retention of contiguity requires that this new grain boundary area be continuously created during deformation and this is done by the incorporation of some of the dislocations generated during deformation.

The energy associated with this increase in area represents a significant part of the stored energy of cold working and obviously it will be greater for small grain sizes and large strains. Gil Sevillano et al. (1980) have considered the case of a severely compressed metal ($\epsilon = 5$, i.e. reduction in thickness $\sim 99.3\%$), with grain size 10μm and cubic grains. It was assumed that the grain boundary energy remained constant at 0.7Jm^{-2}. The energy stored in the grain boundaries under these conditions was predicted to be $\sim 10^{6}$Jm^{-3}, i.e. ~ 71J/mol for copper.

The rate of increase of grain boundary area per unit volume depends very much on the mode of deformation. The grains of a rolled sheet become laths, those of a drawn wire become needles and those of a compressed specimen are disc shaped. Figure 2.1 shows the calculated increase as a function of strain for several deformation modes.

A second obvious feature of the deformation microstructure is the appearance of an internal structure within the grains. This may take several forms but all of these involve the creation

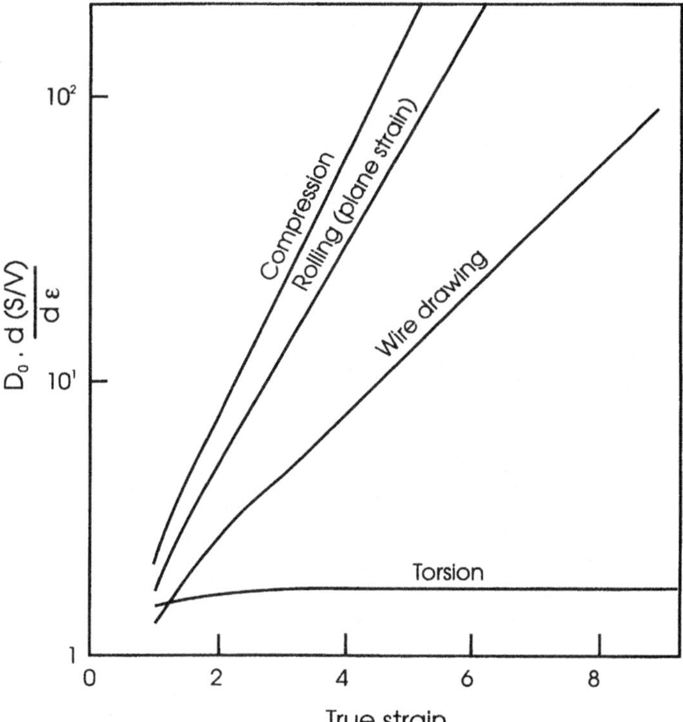

Fig. 2.1. Rate of growth of grain boundary area per unit volume (S/V) for different modes of deformation assuming an initial cubic grain of size D_0, (after Gil Sevillano et al. 1980).

of boundaries of some sort. Many of the newly created dislocations are subsequently located in these internal boundaries. A further source of dislocations and, therefore, of stored energy is associated with the presence in the metal of second-phase particles that may deform less readily or not at all. This incompatibility results in the formation of additional dislocations. The deformation of multi-phase materials will be considered in chapter 8.

In a typical lightly deformed metal the stored energy is about $10^5 Jm^{-3}$. This is a surprisingly low value. It represents only about 0.1% of the latent heat of fusion ($\sim 13kJ/mol$ for the solidification of copper) and is very much smaller than the energy changes associated with phase transformations ($0.92kJ/mol$ for the α-γ transformation in iron). As a consequence of this any phase transformations that may occur at the recrystallization temperature, e.g. precipitation of a second phase or an ordering reaction, may have a profound effect on the recrystallization behaviour. Nevertheless this small amount of stored energy is the source of all the strengthening that occurs during deformation and its loss leads to all of the property changes that occur during annealing.

The increase in dislocation density during deformation arises from both the trapping of existing dislocations and the generation of new dislocations. During deformation, the dislocations (of Burgers vector **b**) move an average distance L, and the **dislocation density** (ρ) is related to the true strain (ϵ) by

$$\epsilon = \rho \, b \, L \tag{2.1}$$

The value to be attributed to L and its variation with strain are responsible for much of the uncertainty in current theories of work hardening. In some cases it is possible to define limiting values, e.g. grain size or interparticle spacing. Consideration of these has given rise to what Ashby (1970) has called geometrically stored and statistically necessary dislocations. Further consideration of the work hardening of metals is however, outside the scope of this volume.

2.2.2 Measurements of stored energy.

The measurement of stored energy is not easy. It may be measured directly by calorimetry or determined indirectly from the change in some physical or mechanical property of the material.

2.2.2.1 Calorimetry

Early calorimetric methods involved the use of large and elaborate calorimeters in which the heat flux emitted by a specimen during heating was measured, usually with reference to a standard. A good account of the results of the extensive early work can be found in Bever et al. (1973). However the magnitude of the energy change involved is so low that even the best calorimetric methods require great attention to detail. The values obtained depend on such factors as composition (particularly impurity levels), grain size, and the extent and temperature of deformation. A brief summary follows:

(i) Stored energy increases with strain. Typical values for copper deformed in tension at room temperature are shown in figure 8.7a.

(ii) Variations in redundant work may produce large differences in the energy stored if the method of deformation is changed. In the case of copper, values of 3.2-5.7 J/mol for tensile deformation should be compared to 3.8-8.3J/mol for comparable compression and ~95 J/mol for wire drawing.

(iii) At low and medium strain levels ($\epsilon < 0.5$) more energy is stored in fine grained material than in coarse grained. At high strain levels the energy stored is usually found to be independent of grain size, although Ryde et al. (1990) have reported a higher stored energy in copper with a coarse grain size.

The development of differential scanning calorimeters (e.g. Schmidt 1989, Haessner 1990) has resulted in a renewed interest in the calorimetric measurement of stored energy. Schmidt (1989) and Haessner (1990) have examined the energy release from several metals after deformation by torsion at -196°C. Their results are shown in figure 2.2. The low temperature peaks are associated with the loss of point defects and can be ignored for our purpose but the high temperature peaks are due to dislocation loss and recrystallization. The magnitude of the stored energy ranges from 21.5 to 220J/mol. Analysis of the recrystallization peaks, produced the results shown in table 2.1. The values of stored energy for lead and aluminium are very much lower than those for copper and silver and are a direct consequence of the differences in **stacking fault energy** (γ_{SFE}). The stacking fault energy, which is a function of the material, determines the extent to which unit dislocations dissociate

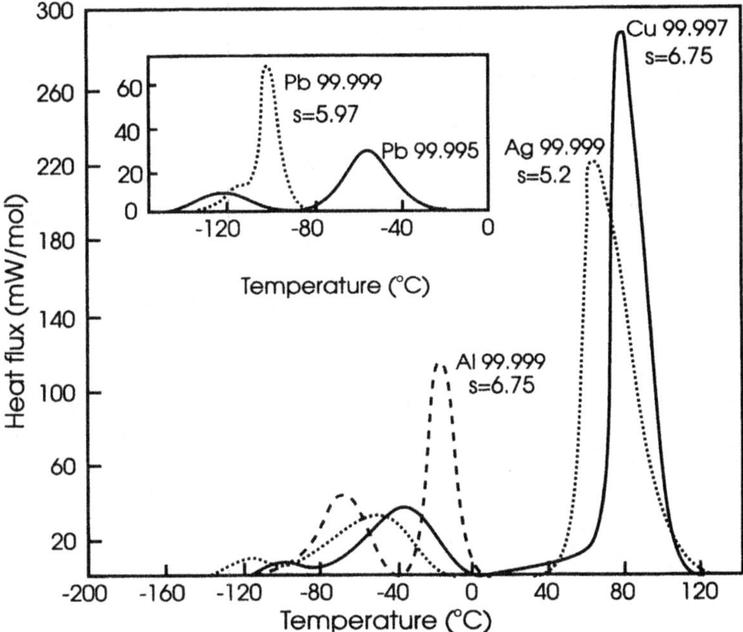

Fig. 2.2. Calorimetric readings during heating of aluminium, lead, copper and silver specimens deformed in torsion at -196°C. to several different values of surface shear strain (s), (unpublished results of J. Schmidt, quoted by Haessner 1990).

Table 2.1
Recrystallization data for metals deformed at 77K.
(Schmidt 1989).

	Al 99.999at%	Pb 99.999at%	Cu 99.997at%	Ag 99.999at%
Reduced stacking fault energy (x1000)	26	15	4.7	2.6
Shear strain	6.75	5.97	6.75	5.2
Stored energy, J/mol	69.6	21.5	216	220
Expended energy, J/mol	3151	1400	5592	4914
E(stored)/E(expended)	0.022	0.015	0.039	0.045
Dislocation density[*], m^{-2}	3.1×10^{15}	1.7×10^{15}	10×10^{15}	8.7×10^{15}

* calculated from $[E_{stored})/Gb^2]$

into partial dislocations. Such dissociation, which is promoted by a low value of γ_{SFE}, hinders the climb and cross slip of dislocations, which are the basic mechanisms responsible for recovery. Dislocation theory therefore predicts that high values of γ_{SFE} should promote dynamic dislocation recovery and table 2.1 shows that γ_{SFE} for the two groups is consistent with this expectation. Attention is also drawn to the small fraction of the energy expended during deformation which is stored in the materials.

2.2.2.2 X-ray line broadening
A measure of the stored energy of a deformed metal can also be found from analysis of X-ray line broadening. This technique measures only the inhomogeneous lattice strain energy and the difference between this and the calorimetric results provides a striking indication of the significance of the dislocation content to stored energy. The X-ray values are typically of the order of 8-80J/mol for heavily cold worked metals in which the calorimetric value is 250-800J/mol.

2.2.3 Relationship between stored energy and microstructure

2.2.3.1 Stored energy and dislocation density
In materials in which the dislocation density is low, the dislocation content may be measured directly by transmission electron microscopy. However, the density of dislocations in even moderately deformed metals is such that they cannot be counted accurately, and inhomogeneous distribution of the dislocations, for example into a cell structure, makes measurement even more difficult. An estimate of the dislocation density may also be obtained from the mechanical properties of the material. For example, a relationship between flow stress (σ) and dislocation density of the form

$$\sigma = c_1 G b \rho^{1/2} \tag{2.2}$$

where c_1 is a constant of the order of 0.5, and **G** is the shear modulus, has been shown to hold for a wide variety of materials (e.g. McElroy and Szkopiak 1972).

If the energy of the dislocation core is neglected and if isotropic elasticity is assumed, then the energy $\mathbf{E_{dis}}$ per unit length of dislocation line is given approximately by

$$E_{dis} = \frac{Gb^2 f(v)}{4\pi} \ln(\frac{R}{R_0}) \qquad (2.3)$$

where:
$\mathbf{R}$ is the upper cut-off radius (usually taken to be the separation of dislocations ($\rho^{-1/2}$),
$\mathbf{R_0}$ is the inner cut-off radius (usually taken as between b and 5b),
$\mathbf{f(\nu)}$ is a function of Poisson's ratio (ν), which, for an average population of edge and screw dislocations is $\sim (1-\nu/2)/(1-\nu)$.

For a dislocation density ρ the stored energy is then

$$E_D = \rho E_{dis} \qquad (2.4)$$

Although this relationship is appropriate if the dislocations are arranged in such a way that the stress fields of other dislocations are screened, the energies of the dislocations present in real materials are not wholly represented by such simple considerations. Dislocations in even moderately worked metals are kinked and jogged and are found in pile-ups and in intricate tangles. Dislocation theory shows that the energy of a dislocation depends on its **environment** and is for example highest in a pile-up and lowest when in a cell or subgrain wall.

In most cases where only very approximate values of dislocation energy are needed then equation 2.3 can be simplified to

$$E_{dis} = c_2 \, G \, b^2 \qquad (2.5)$$

where c_2 is a constant of the order of 0.5.

The stored energy is then given as

$$E_D = c_2 \, \rho \, Gb^2 \qquad (2.6)$$

2.2.3.2 Stored energy and cell/subgrain structure

If the deformation microstructure consists of **subgrains** (§2.3.2.1), then the stored energy may be estimated from the subgrain diameter (D) and the specific energy (γ_s) of the low angle grain boundaries which comprise the subgrain walls. The area of low angle boundary per unit volume is $\sim 3/D$ and hence the energy per unit volume (E_D) is given approximately by

$$E_D \approx \frac{3\gamma_s}{D} \approx \frac{\alpha\gamma_s}{R} \qquad (2.7)$$

where α is a constant of ~ 1.5

As discussed in §3.3, the boundary energy (γ_s) is directly related to the misorientation (θ) across the boundary (equations 3.5 or 3.6) and therefore equation 2.7 may be expressed in terms of the parameters D and θ, both of which may be measured experimentally as:

$$E_D = \frac{3\gamma_0\theta(A - \ln\theta)}{D} \approx \frac{K\theta}{D} \qquad (2.8)$$

where A and γ_0 are defined with equation 3.5 and K is a constant.

It has been found experimentally that both the cell/subgrain size and the misorientation may be dependent on the grain orientation, and therefore, from equation 2.7, we expect the stored energy to vary in the different texture components of the material. As will be discussed later, this may have important implications for the recrystallization behaviour. The pioneering work in this area was a study of 70% cold rolled iron by Dillamore et al. (1972). At this level of rolling the microstructure consists basically of a cell-type structure and it was found that the cell size and the misorientation between neighbouring cells were orientation dependent (fig 2.3). For the $\{hkl\} < 110 >$ components of the rolling texture, small cells and large misorientations were associated with rolling plane orientations near $\{110\}$, while larger cells and small misorientations occurred for $\{001\}$ orientations. From equation 2.8 it is seen that if the dislocations are concentrated mainly in cell or subgrain walls then the stored energy should be greatest for small cells and large misorientations, and Dillamore et al. concluded that

$$E_{110} > E_{111} > E_{112} > E_{100}.$$

This result has proved to be very useful in providing understanding of the recrystallization textures found in low carbon steels. However only rolling plane sections were examined in

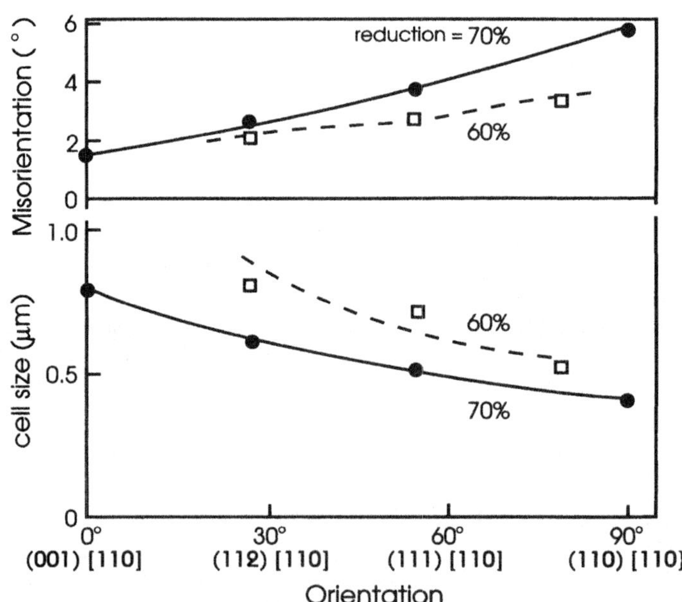

Fig. 2.3. Variation of cell size and cell boundary misorientation in rolled iron as a function of local orientation, (after Dillamore et al. 1972).

this early work and it was assumed that the subgrains were hexagonal prisms with height equal to the diameter; it will be shown later in this chapter that the actual microstructure of deformed metals is more complicated. Every and Hatherly (1974), using the X-ray line broadening method found an orientation dependence of the stored energy similar to that calculated by Dillamore et al. (1972). In their analysis of the substructure they reported the following values for a 70% rolled, killed steel:

$$\{110\} = 13.5; \{111\} = 8.7; \{211\} = 6.8; \{100\} = 4.8 \text{J/mol}.$$

These workers also examined the microstructure of their specimens using sections normal to both the rolling and transverse directions. It was found that the predominant, high energy components ($\{110\} <uvw>$ and $\{111\} <uvw>$) consisted of cells elongated in both the transverse and rolling directions. These cells were 0.15-0.20μm thick, 2-3 times longer, and had orientations within $10°$ of a $<110>$ zone between $\{110\}$ and $\{111\}$. The other, low energy components ($\{211\} <uvw>$ and $\{100\} <uvw>$) were associated with larger equiaxed cell structures of diameter 0.30 - 0.45μm, and orientations lying in zones within $30°$ of $\{100\}$. Similar results were obtained by Willis and Hatherly (1978) from an interstitial free steel.

2.3. THE MICROSTRUCTURE OF DEFORMED METALS.

2.3.1. The characteristics of the microstructure

This introductory section provides a simple description of the most significant features observed in the microstructure of deformed, polycrystalline metals. In subsequent sections the formation of these features will be discussed in more detail.

In cubic metals the two basic methods of deformation are slip and twinning and the most significant material parameter with respect to the choice of method is the value of γ_{SFE}. The planes and directions involved are a function of the crystal structure and tables 2.2 and 2.3 show these data for cubic metals.

As shown by table 2.2 the crystallography of slip in fcc metals is simple. In general slip takes place on the most densely packed planes and in the most densely packed directions. Together these factors define the slip system, $\{111\} <110>$. It has been known for many years, however, that other systems sometimes operate at high temperatures and particularly in metals with high values of γ_{SFE}. The systems observed involve slip on $\{100\}$, $\{110\}$, $\{112\}$ and $\{122\}$ planes. More recently, unusual, low-temperature slip has been reported on $\{111\}$, $\{110\}$ and $\{122\}$ planes and Yeung (1990) has observed slip on a number of non-octahedral planes in the low stacking fault energy alloy 70:30 brass at high values of strain. For summaries of recent work in this area the reader is referred to the papers of Bacroix and Jonas (1988), Yeung (1990) and Maurice and Driver (1993).

In bcc metals slip occurs in the close packed $<111>$ directions but the slip plane may be any of the planes $\{110\}$, $\{112\}$ or $\{123\}$; each of these planes contains the close packed slip direction, $<111>$. The choice of slip plane is influenced by the temperature of deformation. At temperatures below $T_m/4$ $\{112\}$ slip occurs; between $T_m/4$ and $T_m/2$ $\{110\}$ slip is favoured and at temperatures above $T_m/2$ $\{123\}$ is preferred. At room temperature iron slips on all three planes in a common $<111>$ direction and the term pencil glide is

Table 2.2
Crystallography of slip in cubic metals.

Structure	Slip System	
	Plane	Direction
fcc	{111}	<110>
bcc	{110}	<111>
	{112}	<111>
	{123}	<111>

Table 2.3
Crystallography of twinning in cubic metals.

Structure	Twinning shear	Twinning Plane	Twinning Direction
fcc	0.707	{111}	<112>
bcc	0.707	{112}	<111>

used to describe the nature of the slip process in this case. One consequence of such slip is the wavy nature of the slip lines seen on pre-polished surfaces of deformed specimens.

The actual planes and directions associated with slip and twinning correspond to the system with the greatest resolved shear stress and are differently oriented from grain to grain in polycrystalline metals. In general, slip or twinning processes initiated in one grain are confined to that grain and can be readily distinguished from those occurring in neighbouring grains. It should not be thought however, that deformation is homogeneous in any grain of the aggregate. When a single crystal specimen is deformed it is usually free to change its shape subject only to the need to comply with any external constraints (e.g. grip alignment in a simple tension test). This freedom does not exist for the individual grains of an aggregate which are subjected to the constraints exercised by every one of several neighbours each of which is deforming in a unique manner. It will be obvious that contiguity must be maintained if deformation is to continue and it is a direct consequence of this need that the deformation processes will be different in various parts of any particular grain. This is discussed further in §2.3.5 and §2.3.6, but for the moment it is sufficient to understand that a polycrystalline metal responds to deformation by developing orientations that are different from grain to grain and different from region to region within an individual grain. This development of microstructural inhomogeneity begins at a very early stage of deformation and is easily seen by several methods; (i) on pre-polished surfaces, (ii) on suitably etched specimens, (iii) during TEM examination of thin foil specimens and (iv) by back-scattered electron images and diffraction patterns (EBSD) in the SEM.

Inhomogeneity of deformation has been recognized for a long time and the differently oriented regions within a grain have been identified in various ways. However major attention was first focused on the matter by Barrett (1939) who argued that inhomogeneities of this type contributed to the inability to predict the strain hardening behaviour and the orientation changes taking place during deformation. Following Barrett the term **deformation band** will be used to describe a volume of constant orientation that is significantly different to the orientation(s) present elsewhere in that grain. Other workers have introduced terms such as kink bands, bands of secondary slip etc. to describe deformation bands of various types and occasional use will be made of the former term.

The different features are illustrated schematically in figure 2.4 which shows a region B, which has a different orientation to that in the grain proper A. The region (T) at the edge of the deformation band where the orientation changes from B to A is not a grain boundary. It has a finite width and will be called a **transition band.** In many cases deformation bands occur with approximately parallel sides and involve a double orientation change A to C and then C to A. A deformation band of this special type will be called a **kink band** following the nomenclature of Orowan (1942). The micrographs of figure 2.5 show the two types of inhomogeneity and figure 2.5a shows a grain which has fragmented into regions of different orientation. The magnitude of the orientation changes involved is well illustrated in figure 2.5b, while the complexity of the deformation process is clear from figure 2.5c, which shows part of the same field after etching in a strain sensitive etching reagent. The development of deformation bands is an inevitable consequence of the deformation of polycrystals (and also of constrained single crystals), but further details of the microstructure are determined by the crystallographic nature of the deformation process.

The inhomogeneity referred to above arises independently of whether or not slip or twinning is the significant deformation mode. These basic methods of deformation must now be considered in more detail. As mentioned earlier the most significant material parameter with respect to the choice of deformation method is the value of γ_{SFE}. Table 2.4 shows values of γ_{SFE} for a number of metals.

Values of γ_{SFE} in the literature vary considerably and depend particularly on the method of measurement. For this reason the data given in Table 2.4 should be regarded generally as

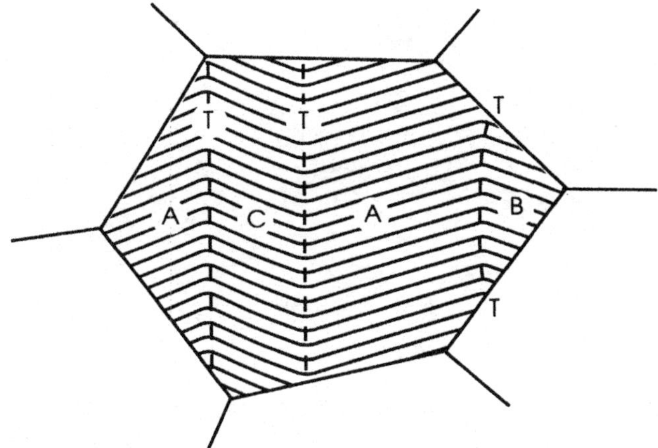

Fig. 2.4. Schematic diagram of deformation bands, transition bands and kink bands.

approximations. There has been an increasing use in recent years of a reduced (or normalized)) stacking fault energy defined as

$$\gamma_{RSFE} = \frac{\gamma_{SFE}}{Gb} \qquad (2.9)$$

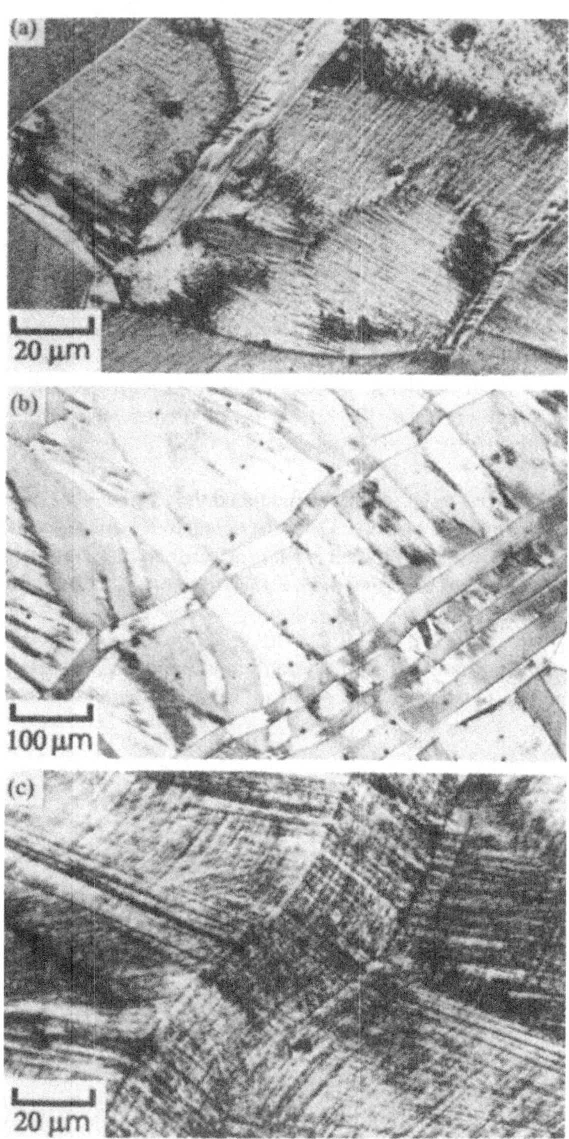

Fig. 2.5. Microstructure of compressed 70:30 brass: (a) compressed 16%, showing division of grain into deformation bands, (etched in cupric ammonium chloride); (b) compressed 12%, showing kink bands, (etched in ferric chloride); (c) central portion of figure 2.5b, (etched in Jacquet's sensitive thiosulphate reagent).

Table 2.4
Stacking fault energy of metals
(Murr 1975).

Metal	γ_{SFE} (mJm^{-2})	Metal	γ_{SFE} (mJm^{-2})
Aluminium	166	Zinc	140
Copper	78	Magnesium	125
Silver	22	91Cu:9Si	5
Gold	45	Zirconium	240
Nickel	128	304 stainless steel	21
Cobalt (fcc)	15	70Cu:30Zn	20

In general bcc metals deform by slip, as do fcc metals with medium to high values of γ_{SFE} such as copper (~ 80mJm^{-2}) and aluminium (~ 170mJm^{-2}). In metals with low values of γ_{SFE} such as silver or in alloys like 70:30 brass and austenitic stainless steels with $\gamma_{SFE} \sim 20$mJm^{-2} the dislocations dissociate to form stacking faults and twinning is the preferred mode of deformation. The tendency to deform by twinning is increased if the deformation temperature is lowered or the strain rate increased.

In cph metals deformation begins by slip but, because of the lack of sufficient slip systems to accommodate the imposed strain, slip is soon accompanied by twinning as an important deformation mode.

Irrespective of whether slip or twinning is the major deformation mode the microstructure is complicated by the presence of inhomogeneities that are smaller than deformation bands. Some examples of these are given in figure 2.6 for materials with different values of γ_{SFE} that have been deformed to low and high levels of strain. In moderately deformed aluminium, copper, nickel etc. (fig 2.6a) the dislocations are arranged in a cellular structure that is approximately equiaxed. Superimposed on this are long, thin, plate-like features that form initially on the {111} planes (fig 2.6b). These are called **microbands**. At higher levels of strain, typical of those that can be generated by rolling, somewhat similar long features are present that are aligned parallel to the rolling plane (fig 2.6c). These bands, which are then profuse, are often clustered and the clusters are separated by cells similar to those present at low strains. At these higher levels of strain ($\epsilon > 1$) a macroscopic inhomogeneity known as a **shear band** also occurs. These have a morphology that is related explicitly to the deformation geometry and are easily seen in the optical microscope (fig 2.6d).

As discussed above, the microstructure in a metal deforming by slip typically comprises a three-dimensional structure of regions of low dislocation density, bounded by walls of high dislocation density. Such structures have been referred to as both **cells** and **subgrains**. The distinction lies in the nature of the boundary. If the boundary is diffuse, consisting of a tangled array of dislocations, it is a cell structure. If the boundary is sharp and consists of a well ordered dislocation array (the low angle grain boundaries discussed in §3.3) then it is more properly described as a subgrain structure.

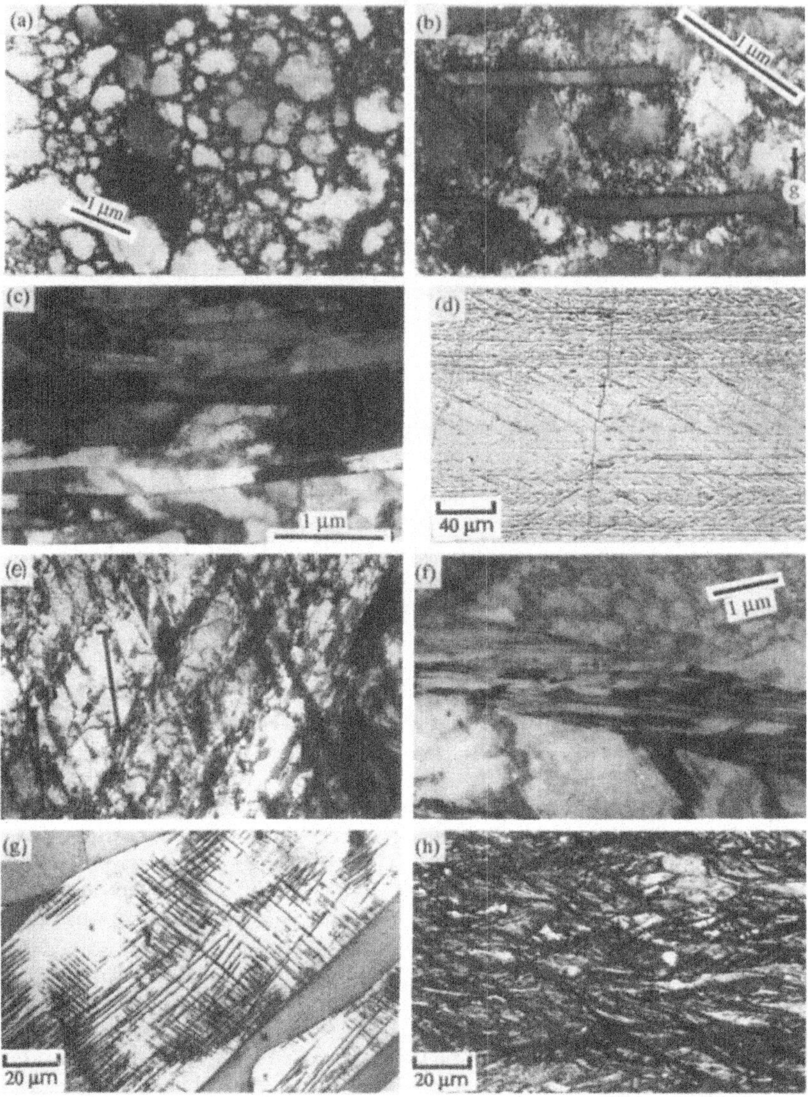

Fig. 2.6. Some characteristic features of the microstructure of deformed metals: (a) cell structure in 25% cold rolled copper; (b) microbands in 18% cold rolled copper; (Malin and Hatherly 1979); (c) microbands in 98% cold rolled copper (Malin and Hatherly 1979), (d) shear bands developed by light rolling on pre-polished, and lightly scratched, surface of 83% cold rolled copper, (Malin and Hatherly 1979); (e) stacking faults in 15% cold rolled, copper-13at.% aluminium alloy (Malin 1978); (f) deformation twins in 30% cold rolled 70;30 brass (Duggan et al. 1978b); (g) Strain markings in 70:30 brass after 14% compression, (etched in ferric chloride); (h) shear bands in 75% cold rolled, 70:30 brass, (etched in ferric chloride), (Hatherly and Malin 1979). (All electron micrographs taken from TD-plane sections; $1\mu m$ marker parallel to rolling direction).

In materials with low values of γ_{SFE} (silver, austenitic stainless steels, many copper-rich alloys, etc.) a dislocation cell structure does not form and the dislocations dissociate to form arrays of stacking faults on the twin planes (fig 2.6e). At a comparatively early stage of deformation, thin bands of very fine deformation twins develop on the {111} planes (fig 2.6f). These bands are easily seen on polished and etched surfaces and are the classical **strain markings** of the early literature (fig 2.6g). At higher levels of strain ($\epsilon > 1$) shear bands develop (fig 2.6h). These occur much more profusely than in metals of higher SFE and the morphology is also quite different.

The above classification of microstructure in metals that deform by slip is based on a large volume of work from a number of laboratories. While differences of interpretation are common there has been general agreement about the major features and particularly about nomenclature. The terminology has evolved gradually over more than 60 years. In the last few years Hansen and his collaborators at Risø National Laboratory in Denmark have described the dislocation structure seen in copper, aluminium, nickel and iron in terms of a different and more complete set of observed features. In this nomenclature the cell block corresponds to the present deformation band and the microband is of either first or second generation. The former refers to a non-crystallographic, double walled feature that separates neighbouring cell blocks and is therefore, somewhat akin to the transition bands described here. Second generation microbands are crystallographic in nature and identical to the microbands of this text. The Risø description is based on the use of the low energy dislocation structure concept (Kuhlmann-Wilsdorf and Wilsdorf 1989) which attributes the increase in dislocation density during straining to the trapping of dislocations in low energy configurations. Full details of this work are given in Hansen (1990) and Bay et al. (1992). The Risø study was confined initially to relatively low strain levels ($\epsilon < 1$) but more recent work has extended this to $\epsilon > 2$ (Bay et al. 1992).

2.3.2. The microstructures formed by slip.

Deformation begins with glide on the most favourably oriented slip system and this results in the formation of slip lines and slip bands on polished surfaces. The term **slip line** is used here to define the individual steps of the elementary structure described by Wilsdorf and Kuhlmann-Wilsdorf (1953). The term **slip band** refers to the clusters of slip lines seen by Brown (1952); these are the common slip manifestations seen in the optical microscope at low strain levels. Slip lines correspond therefore to discrete steps some 30nm apart and 1-5nm high; the step height associated with a slip band is 100-200nm. As deformation proceeds slip is not confined to a single system and at a relatively low strain the array of intersecting slip lines and slip bands becomes so confused that a clear description is no longer possible.

2.3.2.1 Cells and subgrains.
The dislocation structures seen in thin foil TEM specimens bear little resemblance to the obviously crystallographic slip line/slip band patterns seen on polished surfaces. At very low strains corresponding to those at which the first slip bands have already appeared, tangles of dislocations are observed and these are sometimes seen to be associated with the traces of slip planes. However this association does not persist and as the tangles become connected the non-crystallographic cell structure of figure 2.6a develops. In most grains this is already clear at a strain level $\epsilon < 0.05$. The cell structure is equiaxed and the cell diameter is usually 0.5-2μm. The interiors of the cells are relatively free of dislocations and at low

strains the tangled dislocations that make up the cell walls can still be identified as such, so that the cell wall has a finite width. In general the orientation difference between neighbouring cells is small ($<2°$).

In some respects the development of the cell structure is surprising. The cell size is an order of magnitude larger than the spacing of the surface slip lines and about one fifth of the slip line length. If the slip lines accurately reflect the deformation processes occurring within the metal, many slip planes must have operated within the volume of each cell before it was fully developed. There is no evidence of this activity in the microstructure and for this and other reasons it seems likely that the cell structure is a relaxation configuration that develops in the bulk material during or after the dislocation movement induced by the applied stress. Support for this view is provided by the cell walls. As the strain level is increased the walls become quite sharp and at the same time the interiors become even more free of dislocations indicating that substantial recovery has occurred (these might now be called subgrain boundaries). In some respects these changes are similar to those occurring during static recovery (see chapter 5). The strain level associated with this sharpening during deformation at room temperature is related to the melting point and is ~ 0.1-0.2 for 99.998% aluminium, ~ 1.0-1.2 for copper and ~ 1.6 for iron.

The effects of further deformation on a newly developed cell structure are still uncertain and the literature is confused. To some extent this is due to the difficulties associated with specimen preparation from heavily strained materials. One of the simplest methods of deformation to high strain levels is rolling, but specimen preparation is then difficult. Foils can be prepared from rolled sheet with any of the three axes of principal strain parallel to the foil surface. Preparation is obviously simplest if rolling plane sections are used (rolling plane or ND foils) and this was the technique used for much of the early work (prior to 1966). Unfortunately the structures in such foils are poorly defined and very little information was gained at strain levels >0.2. The difficulty arises partly from the fact that the grains of a polycrystalline specimen change their shape in a way that corresponds approximately to the aggregate itself. The grains of rolled sheets become flat ribbons and the typical planar features of the microstructure cannot be distinguished from each other. The directional properties of the various, microstructural heterogeneities described above make them largely invisible in such foils and the cell structure is confused. It is not surprising that microbands were not reported in the early literature based on rolling plane foils and their discovery had to wait until foils were prepared with other geometries. Most of our present knowledge comes from the examination of longitudinal (TD) foils in which the foil surface is normal to the transverse direction of rolled sheet.

Although an equiaxed cell structure is well developed at low strain levels in all metals that deform by slip, equiaxed cells are still present at much higher strains. It is clear therefore that the cells formed at low strains are not subject to the shape changes that occur in the grains and it follows that cell creation is a continuous process. As pointed out earlier, relaxation in the bulk material must be involved and the cell structure may not be directly related to the deformation process. Despite this, there has been much discussion of the cell size. It has been generally believed that the size decreases as γ_{SFE} decreases to 40mJm^{-2} and with increasing strain; the rate of change becoming less as the strain increases. Some collected results for a number of metals are given in figure 2.7. It has been pointed out (Gil Sevillano et al. 1980), that the initial rapid decrease in cell size shown in figure 2.7 occurs at a rate that is greater than would be expected if the cells were to change shape in accordance with the macroscopic shape changes. Such behaviour suggests that most of the

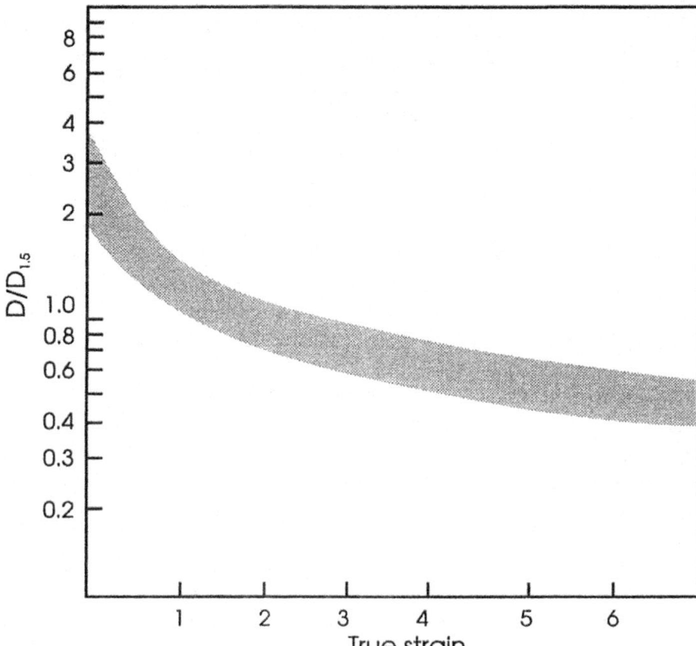

Fig. 2.7. Average cell size as a function of strain for Al, Cu, Fe, Ni, Cr, Nb, (from data assembled by Gil Sevillano et al. 1980 from different authors). The cell size is expressed as a fraction of $D_{1.5}$, the cell size at a strain of 1.5. Values of $D_{1.5}$ are Al $\sim 0.5\mu$m, Cu $\sim 0.3\mu$m, Ni $\sim 0.3\mu$m, Fe $\sim 0.3\mu$m, Cr $\sim 0.4\mu$m, Nb $\sim .02\mu$m.

dislocations generated at this stage are involved in the creation of new cell walls. At higher strain levels ($\epsilon > 1$) the dimensions change more slowly than macroscopic considerations would suggest and this has been attributed to the continuous annihilation of cell walls.

Measurement of cell dimensions is difficult and the reliability of much of the early work in this area is doubtful. In assessing the results of figure 2.7 (and many other "quantitative" measurements of deformed metal microstructures) a number of factors must be borne in mind (Malin and Hatherly 1979). These include:

(i) The temperature of deformation. At $0.31T_m$ the room temperature deformation of pure aluminium must involve substantial recovery effects at high strain levels.
(ii) In the investigations shown, the cell size was measured by using a linear intercept method and longitudinal sections. This technique averages the results from equiaxed cells of diameter ~ 1 μm and long elongated microband type features of thickness $\sim 0.2\mu$m.
(iii) There was an early, general belief that the cells formed at low strains became elongated as deformation proceeded. The long microbands shown in figure 2.6b could not have originated in this way.

A comparison of figures 2.6a-c will give some idea of the difficulties involved in the measurement of cell size.

There have been relatively few studies of the effect of strain on the orientation change across cell or subgrain boundaries. Once again much of the early work which was based on the analysis of TEM spot diffraction patterns, is of limited accuracy (Duggan and Jones 1977). The average misorientation between cells and subgrains is generally found to increase with strain (Gil Sevillano et al. 1980), although in polycrystalline aluminium, as shown in figure 2.8, there is evidence that the misorientation saturates at a strain of ~ 1 at a value of $\sim 2°$.

The mean subgrain misorientation is however only one of the parameters describing the variation of orientation within a grain. Of particular importance for the **nucleation** of recrystallization (§6.6) are the **long range orientation gradients** in the material. Figure 2.9a shows the misorientation in deformed aluminium relative to an arbitrary reference subgrain, as a function of distance from the starting subgrain. It may be seen that although the nearest-neighbour misorientations are generally low, there are some significant orientation gradients in the grain and these are different in the rolling and normal directions.

We therefore need a minimum of two further parameters to describe the variation of orientation. These are the orientation gradient, which may be defined (Ørsund et al. 1989) as $\Omega = \mathbf{d\theta/dx}$, the rate of accumulation of misorientation measured from a trough to a peak in the data or vice versa (fig 2.9b), and a further parameter representing the wavelength of the fluctuations, e.g. the mean peak-to-peak distance in figure 2.9a. The data of figure 2.9 yield values of $\Omega=1.6$ deg/μm and wavelengths of $\sim 8\mu$m in the normal direction and $>20\mu$m in the rolling direction. Ørsund et al. (1989) also found that these three parameters varied with grain orientation. We therefore see that even a crude representation of orientation variation within a grain requires three parameters, and if misorientations are to be more fully defined, then misorientation axes as well as angles are required. There are alternative methods of representing this type of data (e.g. Weiland 1992), and the measurement and representation of the spatial distribution of local orientations, which is a very complex subject, will be briefly discussed in the Appendix. However, we should note here that although our present knowledge is extremely limited, a detailed knowledge of local orientations is a necessary precursor to a complete description of the recrystallization process.

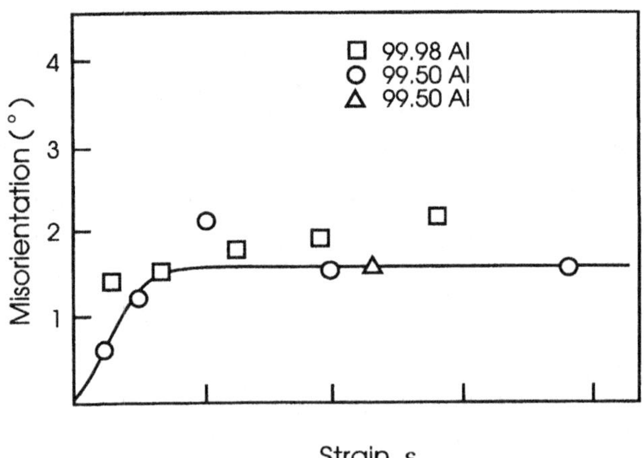

Fig. 2.8. Average subgrain misorientation as a function of rolling strain for commercial purity aluminium, (after Furu 1992).

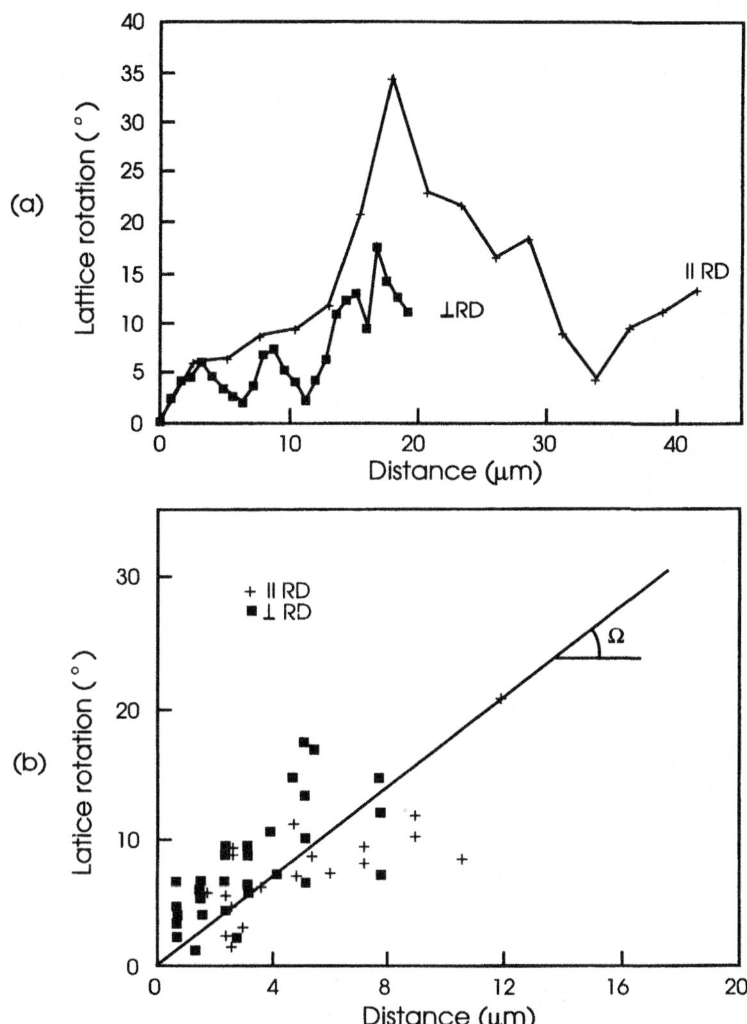

Fig. 2.9. Misorientation in the $\{112\}<111>$ component of commercial purity aluminium cold rolled 90%. a) Misorientation angle relative to a reference subgrain as a function of distance. b) Orientation gradient (Ω), (after Ørsund et al. 1989).

Although the misorientations have been discussed above in terms of cells and subgrains, it is likely that data such as that of figure 2.9 includes substantial contributions from some of the other microstructural features discussed below.

2.3.2.2 Microbands.
A feature of the microstructure in metals that deform primarily by slip, which has generated much controversy, is the microband (fig 2.6b). These long thin plate-like features are typically 0.1-0.3μm thick and are confined to single grains. The walls consist of dislocation

assemblies that are similar to those that make up the cell walls but unlike the cells the dislocation content inside a microband is relatively high. In general the orientation within a microband is only slightly different to that in the adjoining cells and most reports suggest that the orientation is relatively constant along the bands. Microbands were first described and named by Bourelier and Le Héricy (1963) in copper that had been rolled to a strain of 0.2. They reported that the cell structure had superimposed, long, thin plate-like regions that were 0.1-1μm thick and relatively free of dislocations. Many reports have confirmed that microbands form first on the {111} slip planes (e.g. Hatherly 1982). Although they have been found in single crystals of copper at strains as low as 0.02 (Malin et al. 1981), microbands are not common in polycrystals at $\epsilon < 0.1$. At higher levels of strain ($\epsilon = 0.2$) they are sometimes seen on more than one slip plane and the intersections are then marked by prominent shear displacements; similar shear off-sets are also observed whenever a microband is terminated at a grain boundary. The shear associated with these intersections is ~ 1 (Malin and Hatherly 1979). As might be expected the first-formed microbands are affected by the crystallographic rotations and shape changes that accompany further deformation and the slip plane association is lost. With further deformation the microbands cluster together and at high strains ($\epsilon > 1.5$) the dominant features of the microstructure are the visually similar long aligned features shown in figure 2.6c. Although these are usually referred to as microbands it is difficult to see how microbands that form initially on the slip planes achieve the bodily rotation required for alignment at high strain levels. The arrays of dislocations that define their boundaries sharpen in a fashion similar to that observed for the cell structure and at a similar strain level.

Long thin banded features with all the characteristics of microbands are sometimes observed as clusters (somewhat similar to figure 2.6c) across which a large and cumulative orientation difference occurs. These clusters are transition bands and the orientation difference across the cluster may be as high as 60°.

2.3.2.3 Shear bands.

At high strain levels, $\epsilon > 1.2$ for copper, a new mode of deformation occurs and a new microstructural feature appears. Shear bands were described in detail by Adcock (1922), but until the work of Brown (1972) with aluminium and Mathur and Backofen (1973) with iron, were largely forgotten. These bands correspond to narrow regions of intense shear that occur independently of the grain structure and independently also of normal crystallographic considerations. In rolled material they occur at $\sim 35°$ to the rolling plane and parallel to the transverse direction. In metals that deform by slip, shear bands form in colonies in each of which only one set of parallel bands develops (fig 2.6d). The colonies are usually several grains thick and the bands in alternate colonies are in opposite sense, so that a herringbone pattern develops (Malin and Hatherly 1979). Any grain boundaries in the colony are crossed without deviation. The structure of a shear band is clearly resolved in figure 2.10; the aligned microbands of the rolled iron have been swept into the shear band and local elongation and thinning have contributed to the shear strain. The distinct lattice curvature of figure 2.10 is characteristic of a shear band that has been associated with only light to moderate shearing; a more typical microstructure is shown in figure 2.11. The shear in bands of this type is large. Values as high as 6 have been reported but 2 - 3 is a more usual value. At still higher levels of strain, larger shear bands develop which cross a rolled sheet from one surface to the other and when eventually a uniform population of these bands exists, failure occurs along them.

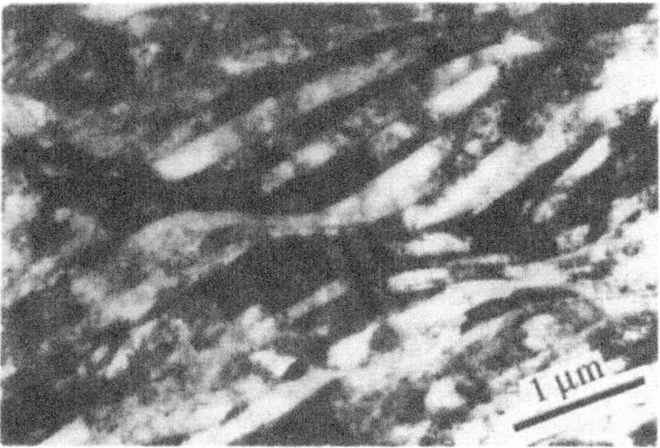

Fig. 2.10. Shear band in 85% cold rolled iron (Willis 1982). TD-plane section; (1μm marker parallel to rolling direction).

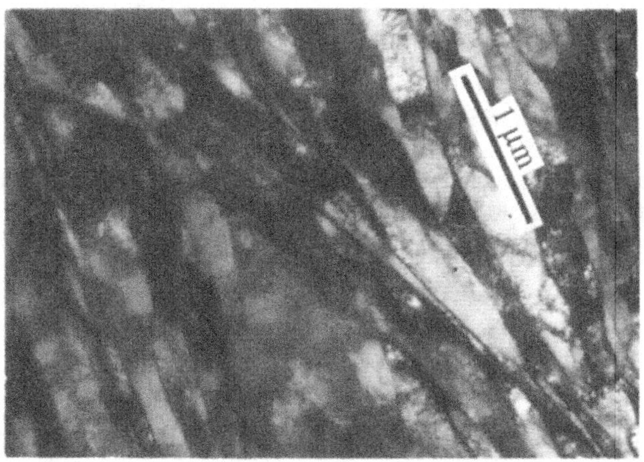

Fig. 2.11. Shear band in 97% cold rolled copper (Malin and Hatherly 1979). TD-plane section; (1μm marker parallel to rolling direction).

2.3.2.4 Aluminium.

The microstructures of deformed aluminium and its alloys warrant special mention. In high purity aluminium, with melting point 660°C, deformation at 20°C corresponds to working at $0.31T_m$, i.e. at the upper limit of what is normally regarded as a cold working operation. In the case of the common aluminium alloys the deformation behaviour is affected also by the presence of additional phases that vary in both morphology and size. We are concerned here mainly with the single phase material and the special case of the alloy, Al-5%Mg.

It has been known for many years that the substructure of even moderately deformed, high purity aluminium is better described in terms of subgrains rather than cells. This is a direct consequence of extensive dynamic recovery effects at room temperature. The most extensive recent work has been that of Hansen and his colleagues at Risø National Laboratory (see, for example, Bay et al. (1992), Hansen (1990), Hansen and Juul Jensen, (1991)). In pure aluminium and alloys with low solute content a cell structure evolves from a tangled dislocation array as described above. At $\sim 10\%$ reduction by rolling, a number of structural inhomogeneities are distinguished which correspond to volume elements in which the morphology and crystallography indicate that the operating slip systems vary from one element to another. The boundaries of these volumes are identified as dense dislocation walls (DDWs) and non-crystallographic features like microbands (MB1) evolve from them. It is to be noted that crystallographically aligned microbands (MB2 in the Risø nomenclature) were not found in this investigation.

With respect to alloys, there has been considerable recent interest in Al-5Mg. It is well known that dynamic recovery is slow in this alloy and a cell structure does not develop (Korbel et al. 1986, Hughes 1993). In the latter work the dislocations were found to be arranged in a geometric pattern along $\{111\}$ slip planes that defined a so-called Taylor lattice (Kuhlmann-Wilsdorf 1989). Crystallographically associated microbands (MB2s) developed from these domain boundaries. Korbel et al. did not identify the Taylor lattice type structure and reported only crystallographically aligned microbands on multiple $\{111\}$ planes. As the strain increased the microbands broadened and crossed grain boundaries to become shear bands.

2.3.3. The effect of twinning on the microstructures.

Twinning is a major deformation mode in fcc metals with $\gamma_{SFE} < 25 \text{mJm}^{-2}$ and in all cph metals. It may also occur in fcc metals with high values of γ_{SFE} and in bcc metals if deformation occurs at low temperatures or high strain rates. In all cases the extent of twinning is orientation dependent.

Manifestations of twinning can be seen on polished surfaces, in etched specimens and in the electron microscope. The well known strain markings (fig 2.6g) have been shown by Duggan et al. (1978b) to consist of very fine deformation twins (fig 2.6f). These features develop freely on the $\{111\}$ planes of fcc materials such as 70:30 brass or 18:8 stainless steel. Strain markings appear first in 70:30 brass at $\epsilon \sim 0.03$; they form initially adjacent to the grain boundaries and in compressed specimens are present on more than one system at strains less than 0.1 (Samuels 1954). They are profuse in most grains at $\epsilon \sim 0.8$. The width of the markings is determined primarily by γ_{SFE} and the deformation temperature (Hatherly 1959) and as these increase, the markings and the constituent twins become wider (cf. figs 2.6g and 2.12a). Because of these factors twinning is a prominent feature of the microstructure of copper after deformation at low temperatures (fig 2.12b).

The significance of γ_{SFE} in determining the deformation mode and therefore the nature of the microstructure is also readily apparent in the electron microscope (Duggan et al. 1978b, Wakefield and Hatherly 1981). In low γ_{SFE} materials, cell structures do not develop and microbands do not form. Instead the dislocations are dissociated into partials and planar arrays of stacking faults occur (fig 2.6e). As the strain increases the dislocation content of the faulted structure rises rapidly and bands of very fine twins appear that are consistent with the strain markings (fig 2.6f). The size of the twins and their frequency depend particularly on γ_{SFE} and in the following discussion the values given apply to rolled 70:30 brass. The first twins appear at $\epsilon \sim 0.05$ and clustering into bands follows almost immediately. The

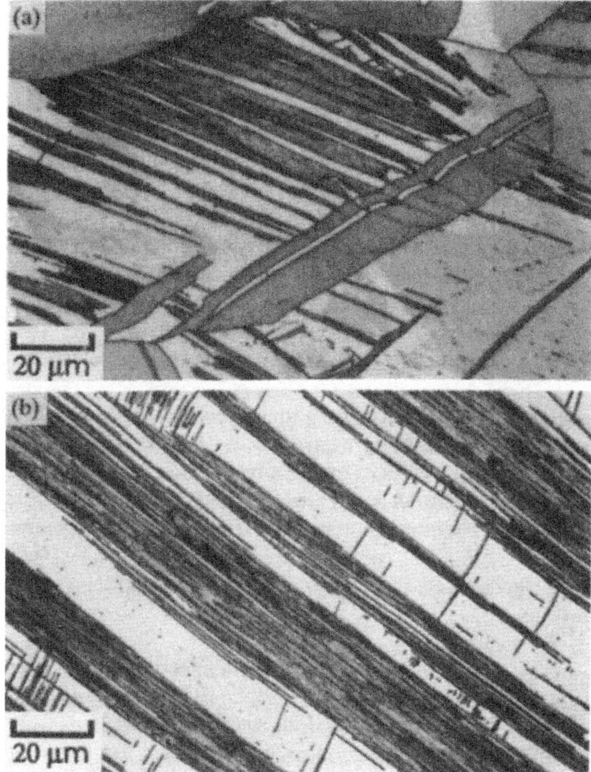

Fig. 2.12. Strain markings in deformed copper and brass showing effects of stacking fault energy and deformation temperature: (a) 90:10 brass compressed 32% at room temperature; (b) (100)[001] single crystal of copper, rolled 60% at -196°C, (Hatherly and Malin 1979).

Fig. 2.13. Aligned deformation twins in 50% cold rolled 70:30 brass and inclined shear band, (Duggan et al. 1978b). TD-plane section; (1μm marker parallel to rolling direction).

twins are extremely fine, with thickness in the range 0.2 - 0.3nm, and the twin-parent repeat distance within the bands ranges from 0.5 to 3nm. These dimensions do not appear to change during deformation but the volume of twinned material increases gradually to about 25%. As rolling proceeds the twins are rotated into alignment with the rolling plane and in some grains and deformation bands this process is complete at $\epsilon \sim 0.8$-1.0 (fig 2.13). The overall twin aligning process continues with further strain and is almost perfect at $\epsilon \sim 2$.

2.3.3.1 Shear bands.

The shear band morphology in materials with low values of γ_{SFE} is quite different to that found in metals which do not develop deformation twins (Hatherly 1982, Hatherly and Malin 1984). In rolled 70:30 brass, isolated shear bands are first seen at $\epsilon \sim 0.8$ and specifically in regions where twin alignment with the rolling plane is already well established (fig 2.13; Duggan et al. 1978b), and soon afterwards ($\epsilon \sim 1$) two families of shear bands are seen in many regions. The bands form initially at approximately $\pm 35°$ to the rolling plane and divide the sheet into rhomboidal prisms with long axes parallel to TD. The twin alignment within the prisms is nearly perfect and most theories of shear band formation are based on the supposition that continued deformation by either slip or twinning is no longer viable in such an aligned structure. As shown in figure 2.14, the bands themselves consist of an array of very small "crystallites" (volumes of almost perfect lattice) which are usually elongated in the direction of shear with aspect ratios in the range of 2:1 - 3:1. Typical shear bands in 70:30 brass range in thickness from 0.1-1μm and the individual crystallites vary in width from 0.02-0.1μm.

The orientations of the crystallites are particularly interesting. Their volume in heavily deformed metals is such that they should make a significant contribution to the rolling texture, and shear bands are a major nucleation site for recrystallized grains. Unfortunately the size is such that orientation determinations are not easily made. Early STEM determinations by Duggan et al. (1978a) reported large ($>20°$) orientation differences between neighbouring crystallites in 70:30 brass that were not cumulative. An overall preference was detected for rolling plane orientations near {110} and a substantial population was found near {110}<001>. This orientation is a major component of the shear texture in fcc metals.

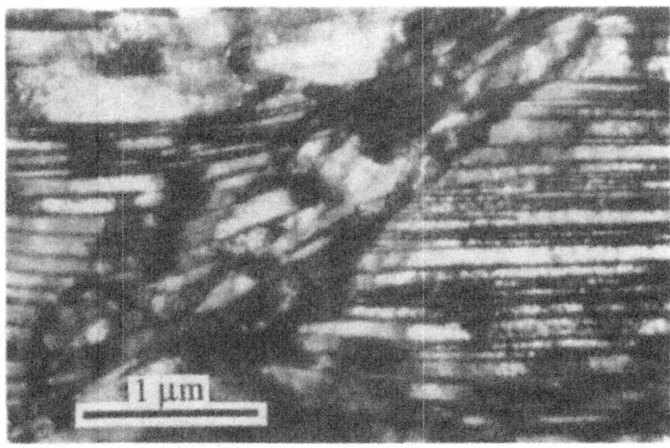

Fig. 2.14. Shear band in (110)[1$\bar{1}$1] copper single crystal, cold rolled 65% at 77°K, showing internal crystallite structure, (Köhlhoff et al. 1988a). (1μm marker parallel to rolling direction).

Fig. 2.15. 70:30 brass specimen initially rolled 50%, then polished and lightly scratched
normal to rolling direction before further rolling of 10% reduction (TD-plane section).
Note magnitude of shear, (Duggan et al. 1978b).

The shear associated with individual bands is high with average values of 3-4 often reported
but shears as high as 10 are sometimes found (Duggan et al. 1978b). The importance of
shear band formation as a deformation process is strikingly illustrated by figure 2.15. A
longitudinal section from a specimen of 50% cold rolled, 70:30 brass, was polished and then
lightly scratched at right angles to the rolling direction. The figure shows the development
of shear bands during a subsequent rolling of 10% reduction and the magnitude of the shear
strain. The number of shear bands increases rapidly with strain and in the range $0.8 < \epsilon < 2.6$
shear band formation appears to be the major deformation mode and at the latter level of
strain only minor amounts of twinned material remain. As in the case of high γ_{SFE} materials
further deformation leads to the formation of large through-thickness shear bands and
eventually fracture occurs along these.

2.3.3.2 The effect of stacking fault energy.

The effect of stacking fault energy on these structures is very marked. In the very low γ_{SFE}
($\sim 3\text{mJm}^{-2}$) alloy Cu-8.8at% Si, Malin et al. (1982a) reported faulting and twinning on all
four $\{111\}$ planes during rolling to $\epsilon = 0.4$ (fig 2.16). In such a structure the gradual
alignment of a set of deformation twins with the rolling plane is clearly impossible and in the
absence of an alternative deformation mode shear band formation begins at $\epsilon = 0.4$-0.5. This
alloy cracks along macroscopic shear bands at quite low strains ($\epsilon \sim 1.2$). In alloys with
intermediate values of γ_{SFE} e.g. 90:10 brass and some stainless steels, the microstructure is
determined additionally by the local orientation (Wakefield and Hatherly 1981). In some

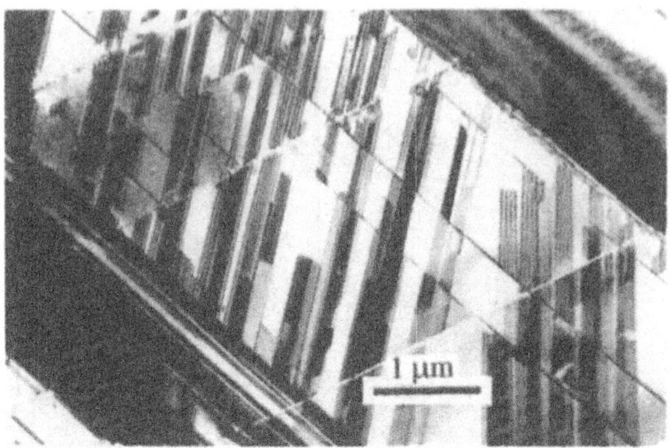

Fig. 2.16. Stacking faults and fine deformation twins in rolled copper-8.8at.%Si (Hatherly and Malin 1979). TD plane section; (1μm marker parallel to rolling direction).

grains (or deformation bands) the orientation favours twinning and in others slip. The resulting microstructure consists of volumes containing cells and microbands adjacent to volumes showing deformation twins. In some grains where the orientation is appropriate, both slip and twinning occur and as the orientation changes so too does the deformation mode. Twinning is the preferred deformation mode during rolling in regions oriented at $\{112\}<111>$ and $\{100\}<001>$ but twinning is not observed in regions oriented at $\{110\}<001>$ or $\{110\}<112>$. For a discussion of the relationship between orientation and twinning in fcc metals see Köhlhoff et al. (1988a).

2.3.4. Hexagonal metals

As with other metals the deformation of polycrystalline cph metals begins with slip, but the symmetry of the lattice and the availability of slip systems are less than with other metals. Under these conditions slip cannot continue as the only deformation mode and twinning becomes significant at quite low strains ($\epsilon<0.2$). The most obvious feature of the microstructure of deformed cph metals is the rapid development of a profuse array of large, broad, lenticular deformation twins (fig 2.17). The twins originate usually as long, thin lamellae at low strain levels but broaden rapidly. Twins are broader in cph metals than in cubic metals because of the smaller value of the twinning shear. Despite their profusion in the microstructure the contribution of twinning to the overall deformation at medium - high strain levels is usually small. The reason for this lies in the limited magnitude of the twinning shear. As twinning ceases, deformation by slip occurs in the newly developed twins and the whole process is repeated. Tables 2.5 and 2.6 give details of the crystallography of the most frequent slip and twinning systems in cph metals.

Unlike slip, twinning occurs in only one direction, and the crystallography is such that elongation is generated in one direction and contraction in others. Inspection of table 2.6 shows that one twinning system is common to all of the cph metals, viz. $\{10\bar{1}2\}<10\bar{1}1>$. If the c/a ratio is >1.633 (the ideal value), as in zinc and cadmium, twinning occurs only on this system but if c/a is <1.633 more than one system may operate. The strain produced by the twinning shear depends on the twinning system and the value of c/a.   For $\{10\bar{1}2\}<10\bar{1}1>$ twinning the strain is greatest for low values of c/a but if other twinning systems are possible these will lead to larger strains.   The sense of the shape change also depends on the perfection of packing.   For the value $c/a = \sqrt{3} = 1.732$ no twinning is possible but for $c/a < 1.733$ the sense of the shape change is opposite to that for $c/a > 1.732$. In practice this means that twinning should occur only if directions that are shortened by twinning lie in the appropriate directions during processing, e.g. radially during wire drawing.

Fig. 2.17. Deformation twins in polycrystalline magnesium deformed 8% in compression at 260°C, (Ion et al. 1982).

Table 2.5
Slip systems in cph metals at room temperature.
(Grewen 1973).

Metal	c/a	Slip system	
		Predominant	Seldom
Cd Zn	1.89 1.88	$\{0001\}<11\bar{2}0>$	$\{11\bar{2}2\}<11\bar{2}3>$
Co,Mg	1.62	$\{0001\}<11\bar{2}0>$	
Zr	1.59	$\{10\bar{1}0\}<11\bar{2}0>$	$\{10\bar{1}1\}<11\bar{2}0>$
Ti	1.59	$\{10\bar{1}0\}<11\bar{2}0>$	$\{10\bar{1}1\}<11\bar{2}0>$

Table 2.6
Twinning systems in cph metals
(Grewen 1973).

Metal	c/a	Twinning shear	Twinning element	
			Plane	Direction
Cd Zn Co	1.89 1.88 1.62	0.17 0.14 0.13	$\{10\bar{1}2\}$	$<10\bar{1}1>$
Mg	1.62	0.13	$\{10\bar{1}2\}$	$<10\bar{1}1>$
Zr,Ti	1.59	0.167	$\{10\bar{1}2\}$	$<10\bar{1}1>$

It will be seen that for metals with c/a greater than ideal, basal slip is favoured while prismatic slip is preferred for those with c/a less than ideal. In general dislocations are confined to the closely packed basal plane and do not move freely from it. There have been only a few detailed examinations of the deformation microstructure in any of these metals and it is difficult to draw general conclusions.

In rolled titanium (c/a < ideal) deformation twinning is found in all grains at a very early stage (Blicharski et al. 1979). The twins are broad and lenticular and usually form on two or more systems. Because of the small twinning shear, further deformation requires the formation of smaller twins between and within those first formed, but eventually this is no longer possible and fresh twins do not form at moderate strain levels ($\epsilon \sim 0.75$). At high strain levels the twins rotate into alignment with the rolling plane so as to produce a microstructure of thin elongated bands. Shear bands begin to form at $\epsilon \sim 2.8$ in this aligned structure but considerable ductility still remains at much higher strain levels.

The microstructure of deformed zinc is interesting. The c/a ratio is high and so too is γ_{SFE} ($\sim 140mJm^{-2}$), but the melting point is low and room temperature deformation occurs at $\sim 0.4T_m$. Large twins are seen in every grain of polycrystalline zinc after very light rolling ($\epsilon = 0.07$) but no further twins develop as rolling continues to $\epsilon = 0.2$ and slip is the preferred deformation mode in this strain range (Malin et al. 1982b). Shear band formation begins in the strain range $0.2 < \epsilon < 0.5$ but the shear bands are detected only by the presence of bands of very small recrystallized grains that outline their positions. It seems clear that recrystallization follows shearing closely in this low melting point metal. The new strain free grains are free to deform by slip and twinning and deformation continues in this way to high strain levels. These changes are easily recognized in the deformation textures and will be referred to later.

2.3.5 Theories of polycrystalline plasticity

The imposed deformation strain and the constraints between neighbouring grains discussed earlier, affect the choice and number of the operating slip systems. With respect to the subject of this book, there are two important consequences of this slip activity. First, the slip processes and their variation both within and between grains, largely determine the **deformation microstructure**, and second, the changes in orientation which are a consequence of the crystal plasticity, determine the **deformation texture**. There have been many attempts to predict the slip activity and the theories fall into two broad groups. For a clear account of this subject which we can mention only briefly, the reader is referred to Reid (1973).

In the earliest of the theories, the lower bound or **Sachs** model (Sachs 1928), it is assumed that each grain deforms independently of its neighbours and on the slip system that has the greatest resolved shear stress, i.e. in precisely the manner of an unconstrained single crystal of the same orientation. In the alternative, upper bound or **full constraints** model of **Taylor** (1938) it is assumed that all grains undergo the same shape change i.e. that of the overall polycrystalline specimen. The strain tensor is always symmetrical and together with the constant volume requirement for plasticity gives 5 independent strain components. To match these exactly requires 5 independent shears, i.e. homogeneous slip on 5 independent slip systems.

Neither theory fully explains the changes taking place during deformation, but there is general agreement that the Taylor model is the better.

There is considerable evidence that, particularly in large-grained specimens, the slip behaviour in the centre of a grain is different from that near the grain boundaries (e.g. Hirth 1972). In the boundary regions where the constraint of neighbouring grains is greatest, more slip systems are active than in the centres of the grains, which are less influenced by the boundaries. Therefore in the boundary regions the plasticity approaches Taylor type behaviour, and in the grain centres it may approach Sachs type behaviour (Kocks and Canova 1981, Leffers 1981). An important consequence of such differences in slip activity across a grain is that the different parts of the grain inevitably rotate to different orientations during the deformation and thus deformation bands develop within a grain (§2.3.6.1).

A more recent development has been the formulation of the so-called **relaxed constraints models** to account for the deformation of non-equiaxed grains (Honeff and Mecking 1978, Kocks and Canova 1981, Hirsch and Lücke 1988b). This term is used in order to distinguish such models, which allow the operation of less than five independent slip systems, from the full constraints model of Taylor and the zero constraints model of Sachs. Like these the relaxed constraints models assume that slip is homogeneous within a grain and that any

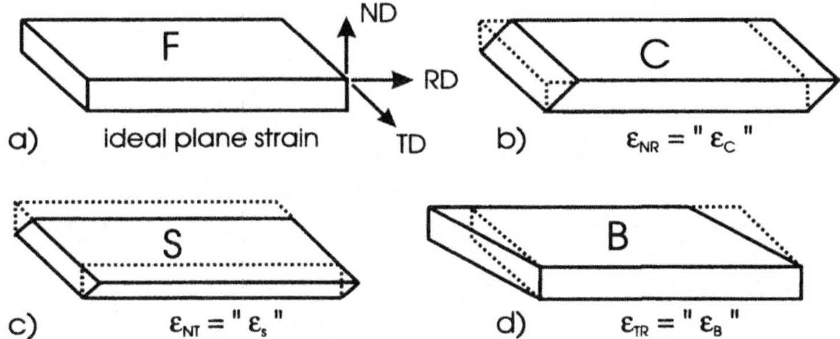

Fig. 2.18. Resulting shape changes from an originally cubic grain deformed under a) full constraints, and b-d) relaxed constraints as defined in the text, (after Hirsch and Lücke 1988b).

effects of individual location within the microstructure can be ignored, i.e. all grains of the same orientation behave similarly. It is then assumed that during rolling some shear is permitted in the planes defined by ND-RD (fig 2.18b), ND-TD (fig 2.18c) and RD-TD (fig 2.18d) either separately or in combination. It will be clear from figure 2.18 that the first two of these are possible at reasonably high strains, because any incompatibility between neighbouring grains must be decreasing as the ratio of grain thickness to length or width decreases. The situation is quite different for RD-TD relaxation where the incompatibility between flat grains increases rapidly with strain. The various relaxations lead to different grain shapes and consequently the models are also referred to as lath (ND-RD relaxation) or pancake (combined ND-RD/ND-TD relaxation) type models.

The predicted influence of crystal plasticity on the development of deformation textures is discussed in §2.4.5.1

2.3.6. The development of microstructural heterogeneity

The earlier discussion of the microstructure of deformed metals was heavily biased towards electron microscopy because it is at this level of observation that the nature of the dislocation configuration is best seen and because the major part of the stored energy is associated with dislocations. There has been only minor reference to the processes by which the various components of the microstructure, and particularly the inhomogeneities, develop. The following discussion gives a brief account of current views about the development of deformation bands, transition bands, microbands, and shear bands.

2.3.6.1 Deformation bands.

The nature of deformation banding has been extensively studied in recent years because of its relevance to the generation of deformation textures (see §2.4.5). Chin (1969) identified two types of deformation bands. One of these originates in the ambiguity associated with the selection of the operative slip systems. In many cases the imposed strain can be accommodated by more than one set of slip systems and the different sets lead to rotations in different senses. In the second type, different regions of a grain may experience different strains if the work done within the bands is less than that required for homogeneous deformation and if the bands can be arranged so that the net strain matches the overall deformation. In their recent model of rolling texture development, Lee et al. (1993) examined the second of these cases and showed theoretically that as few as two independent slip systems may suffice to accommodate the shape change. Their work is consistent with the relaxed constraints models of plasticity discussed above.

2.3.6.2 Transition bands.

A transition band develops when neighbouring volumes of a grain deform on different slip systems and rotate to different end orientations. In its most usual form the transition band consists of a cluster of long narrow cells or subgrains with a cumulative misorientation from one side of the cluster to the other. It is sometimes found, particularly in aluminium alloys (Hjelen et al. 1991), that the width of the transition band is reduced to only one or two units but the large orientation change across the band is retained.

Dillamore and Katoh (1974) considered the case of axisymmetric compression of iron and showed that the rotation paths for various orientations followed those shown in figure 2.19a. For orientation gradients that straddle the line joining [110] and [411] a very sharp transition band develops as part of the gradient rotates towards [100] and part towards [111]. The more general case is shown at A. For the gradient B-B' both extremities rotate towards [111] and the heterogeneity is gradually lost. For a gradient C-C' the extremes follow different paths and misorientation persists in a diminished form. Figure 2.19b shows that the experimentally determined fibre texture after 88% compression is in good agreement with the model used.

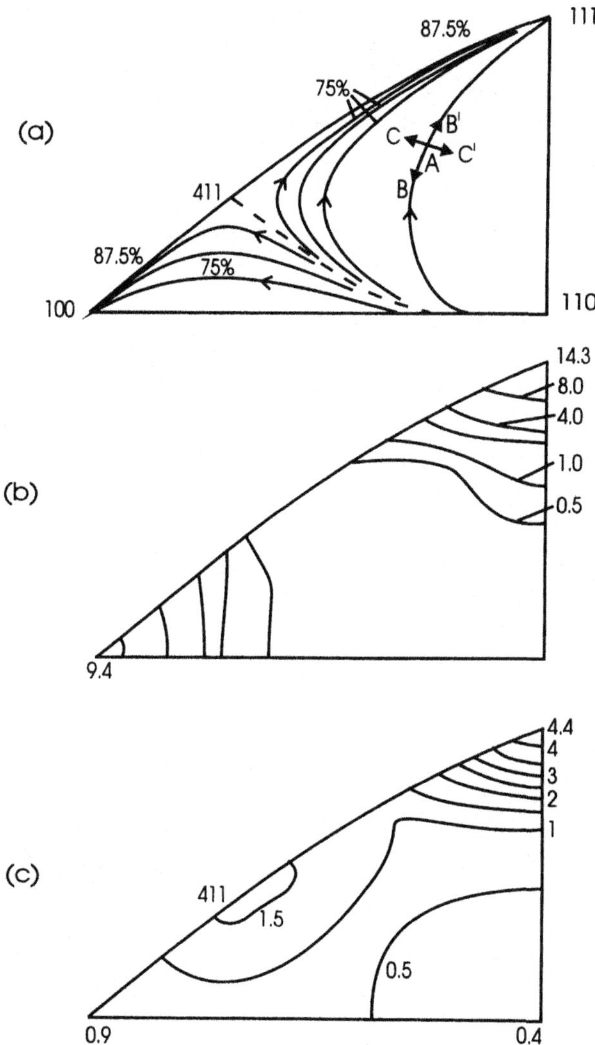

Fig. 2.19. Texture development in axially compressed iron. (a) Predicted rotations from three different initial orientations after 70% and 88% reduction in thickness (b) Deformation texture after 88% compression. (c) Recrystallization texture of (b) material. (after Dillamore et al. 1974).

Dillamore and Katoh also considered the special case of the formation of a transition band in heavily rolled copper having the orientation {100} <001> at its centre. This situation is significant to the generation of the cube texture during recrystallization and is discussed in detail in chapter 10. Some recent reconsiderations of the Dillamore-Katoh model are also discussed in chapter 10.

2.3.6.3 Microbands.

The simple summary of some aspects of microband morphology given in §2.3.2.2 leaves many questions unanswered. These include the relationship between the geometrically aligned bands seen at high strain levels and the visually similar, crystallographically aligned bands formed at lower strains and, more importantly, the mode of deformation within the band. The most generally accepted model of microband formation is that of Jackson (1983). Like other models it recognizes that microbands develop when the cell structure is no longer stable with respect to further strain. This may be due to continuous changes in the strain path within a single grain. It is argued that microband formation is associated with a low work hardening rate and therefore with dynamic recovery and that microbands are a flow instability due to cross-slip that involves mixed dislocations on a series of parallel planes. Formation begins when a dislocation cross-slips to avoid an obstacle. Subsequent dislocations must then cross-slip to avoid the edge dislocations left behind and in so doing leave behind two prismatic mats with a uniform shear across them. A residual tensile stress develops at the centre of the mats which causes slip on the conjugate plane and reaction between dislocations on that plane and those in the walls create Cottrell-Lomer locks that lead to the observed denser walls.

Nes et al. (1986) dispute some aspects of this model and suggest that microband shearing would be confined to the band walls, whereas as stated earlier, shearing occurs uniformly over the band thickness. They point out that Washburn and Murty (1967) have observed similar microstructural features in irradiated metals where clear channels are created by the expansion of glide loops and strain softening develops in the adjacent layers. Nes et al. (1986) propose that the cell structure, like the dislocation loop structure of irradiated metals is unstable during deformation. They suggest that condensation of slip on a set of planes destabilizes the cell walls locally and leads to a clear channel. The walls of the channel are then composed of dislocation debris resulting from reactions between the primary dislocations and those in the cell walls.

The crystallographic habit of newly developed microbands and the associated shear suggests also that the microband is the microstructural manifestation of a surface slip band. In most cases, however, the number of microbands seen at a strain level of 0.2, for example, is considerably less than the number of slip bands. This may be due to experimental problems. Microbands are not usually seen in the thinner parts of a specimen in 100 and 200keV microscopes and even with the latter they are rarely profuse in the thicker regions. Furthermore they are seen more frequently in copper and iron than in the lower melting point metal, aluminium. Finally, they are very much more profuse if HVEM facilities are available (Willis 1982). Taken together these factors suggest that recovery effects and dislocation loss from the thinner parts of the foil may be responsible for the apparent scarcity of microbands. One other factor may be important. It is frequently reported that microbands are not present in grains of some orientations; amongst these are {110} <001> in fcc metals and the complementary orientation {100} <011} in bcc metals. Despite the inability to reconcile the relative frequencies of surface slip bands and microbands it seems likely that the latter is the manifestation of a slip band at the dislocation level.

2.3.6.4 Shear bands.

A shear band is a form of plastic instability and in plane strain deformation (or rolling) the condition for instability is

$$\frac{1}{\sigma} \cdot \frac{d\sigma}{d\epsilon} < 0 \tag{2.10}$$

Dillamore and colleagues (Dillamore 1978a, Dillamore et al. 1979), have written this condition as

$$\frac{1}{\sigma}.\frac{d\sigma}{d\epsilon} = \frac{n}{\epsilon} + \frac{m}{\dot{\epsilon}} + \frac{1+n+m}{M_T}.\frac{dM_T}{d\epsilon} - \frac{m}{\rho}.\frac{d\rho}{d\epsilon} < 0 \tag{2.11}$$

where n and m are the usual strain hardening and strain rate exponents, ρ the dislocation density, and M_T the Taylor factor (defined as $\Sigma s/\epsilon$, where Σs is the total shear strain on all active slip systems, and ϵ is the normal strain). Of the various terms in the equation, that in ϵ is always positive, that in $\dot{\epsilon}$ is usually positive and that in ρ is negative. It follows that if instability is to occur at medium to high strain levels, and in the absence of strain rate effects, it must be because the term in M_T becomes negative through a negative value of $dM_T/d\epsilon$. This latter expression corresponds to geometric softening, in that instability is favoured if it causes a lattice rotation into a geometrically softer condition. $dM_T/d\epsilon$ can be calculated from $dM_T/d\theta$ (the rate of change of hardness with orientation) and $d\theta/d\epsilon$ (the rate of change of orientation with strain). Dillamore et al. have shown that this factor is both negative and a minimum close to $\pm 35°$ for typical rolled metals with well developed textures.

2.4 DEFORMATION TEXTURES.

The orientation changes that take place during deformation are not random. They are a consequence of the fact that deformation occurs on the most favourably oriented slip or twinning systems and it follows that the deformed metal acquires a preferred orientation or texture. If the metal is subsequently recrystallized, nucleation occurs preferentially in association with specific features of the microstructure, i.e. with regions of particular orientation. The ability of the nucleus to grow is determined primarily by the orientations of adjacent regions in the microstructure. Together these features, nucleation and growth, ensure that a texture also develops in the recrystallized metal. Such textures are called **recrystallization textures** to distinguish them from the related but quite different **deformation textures** from which they develop. In this chapter we are concerned only with deformation textures and their relationship to microstructure. Recrystallization textures are considered separately in chapter 10. A detailed account of the voluminous, early literature relating to textures of both types is to be found in the text of Wassermann and Grewen (1962) and the review of Dillamore and Roberts (1965). More recent accounts are to be found in the reports of a regular series of conferences (§1.2.2).

The representation of crystallographic textures is complex. Until recently the standard method of representing textures was by means of **pole figures** (see, for example Barrett and Massalski 1980, Hatherly and Hutchinson 1979). Nowadays there is increasing use of **orientation distribution functions (ODFs)** and **Rodrigues-Frank space**, both of which give a more complete description of the texture. For those unfamiliar with texture analysis, details of the various methods of texture representation are given in the Appendix.

2.4.1. Deformation textures in fcc metals

Most of the texture data available in the literature and almost all of the ODF data refers to rolled materials. For this reason the brief account that follows is concerned mainly with rolling textures. The deformation textures of fcc metals are determined primarily by γ_{SFE}. At one extreme, exemplified by aluminium with $\gamma_{SFE} \sim 170 \text{mJm}^{-2}$ and copper with $\gamma_{SFE} \sim 80 \text{mJm}^{-2}$, slip is the deformation mode and the rolling textures, after large reductions, are very similar to that of figure 2.20a. At the other extreme alloys like 70:30 brass, or metals like silver with $\gamma_{SFE} < 25 \text{mJm}^{-2}$, develop a quite different texture, similar to that of figure 2.20b. The term **texture transition** is widely used to describe the changes that occur over the intermediate range of γ_{SFE}. It is customary to refer to the textures of metals with high values of γ_{SFE} as **pure metal textures** to distinguish them from the **alloy type textures** characteristic of materials with low values of γ_{SFE}.

2.4.1.1 Pure metal texture.

The 100 and 111 pole figures for 95% cold rolled aluminium are shown in figure 2.21 (Grewen and Huber 1978); it is to be noted that the pole figures for rolled copper are almost identical (fig 2.20a). Superimposed on the pole figure of figure 2.21 are the ideal orientations, $\{112\}<111>$, $\{110\}<112>$, $\{123\}<412>$, that are commonly used to describe the texture. Nowadays the last of these has been largely replaced by the slightly different orientation $\{123\}<634>$. A number of the common texture components have acquired the simple names and symbols given in table 2.7 which also gives details of the Euler angles of the main texture components of rolled fcc metals in the first subspace.

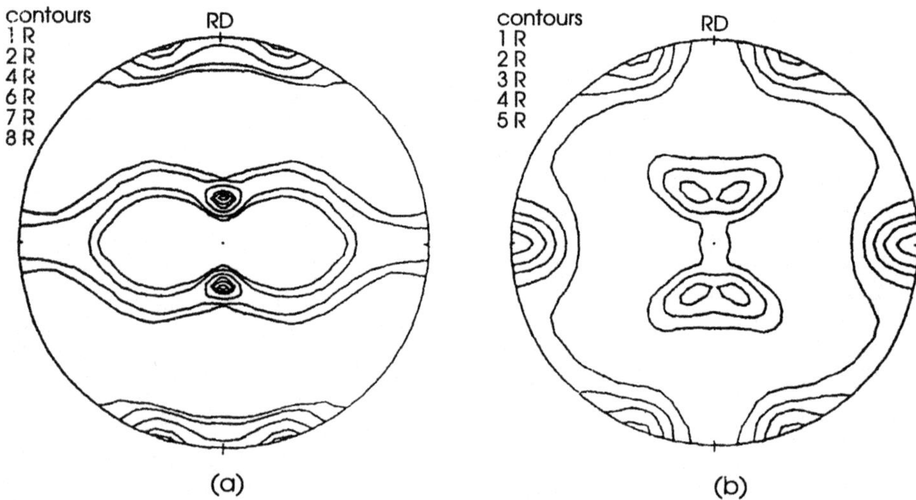

Fig. 2.20. 111 Pole figures of 95% cold rolled fcc metals; (a) copper; (b) 70:30 brass, (Hirsch and Lücke 1988a).

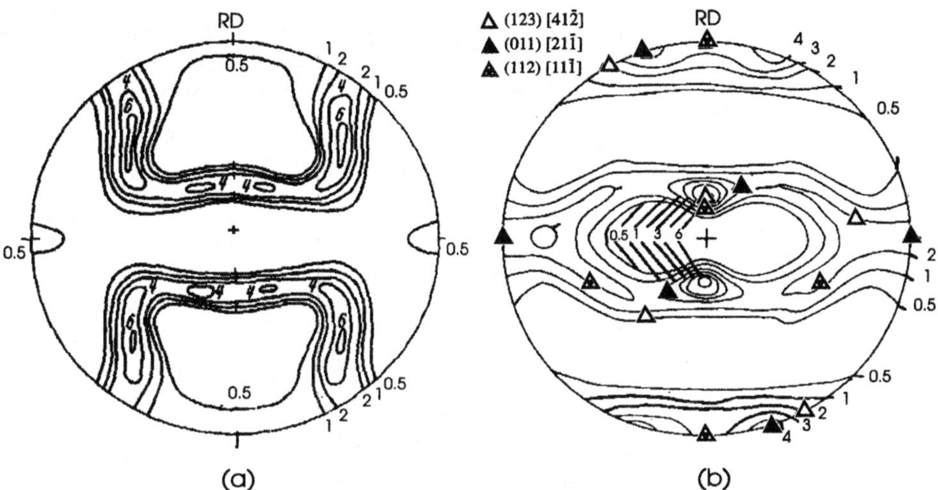

Fig. 2.21. Pole figures of 95% cold rolled aluminium; (a) 100 pole figure; (b) 111 pole figure showing positions of several ideal orientations, (Grewen and Huber 1978).

Table 2.7
Texture components in rolled fcc metals (1st subspace)

Component, Symbol	{hkl}	$<uvw>$	ϕ_1	Φ	ϕ_2
Copper, C	112	111	90	35	45
S	123	634	59	37	63
Goss, G	011	100	0	45	90
Brass, B	011	211	35	45	90
Dillamore, D	4,4,11	11,11,8	90	27	45
Cube	001	100	0	0	0

Examination of figure 2.21 shows that no combination of these orientations corresponds to all of the high intensity regions and therefore a description of the texture in terms of these ideal components is inadequate. More information is given by describing the texture as a spread of orientations from $\{112\}<111>$ through $\{123\}<412>$ to $\{110\}<112>$ but the crystallographic nature of the orientations in the spread remains unclear.

A better description of the texture is provided by the ODF as shown for 95% cold rolled copper in figure 2.22. The first, and most important observation is that the texture is now represented by a continuous **tube** of orientations which runs from $\{110\}<112>$ (B) at $\Phi = 45°$, $\phi_2 = 90°$, $\phi_1 = 35°$ through $\{123\}<634>$ (S) at $\Phi = 37°$, $\phi_2 = 63°$, $\phi_1 = 59°$ to $\{112\}<111>$ (C) at $\Phi = 35°$, $\phi_2 = 45°$, $\phi_1 = 90°$. A schematic representation of the tube is given in figure 2.23 where one of the two branches has been omitted for the sake of clarity. The omitted branch is that observed in the sections $\phi_2 = 0\text{-}45°$. The tube shown in figure 2.23 can also be described in terms of its axis or skeleton line. By convention the axis of the tube shown is called the β fibre and many recent studies of rolled fcc metals report data only in the form of orientation density along this fibre (Hirsch and Lücke 1988a). A second fibre, the α-fibre can be seen in figure 2.23 where it extends from $\{110\}<001>$ (G) at $\Phi = 45°$, $\phi_2 = 90°$, $\phi_1 = 0°$. to $\{110\}<112>$ (B) at $\Phi = 45°$, $\phi_2 = 90°$, $\phi_1 = 35°$.

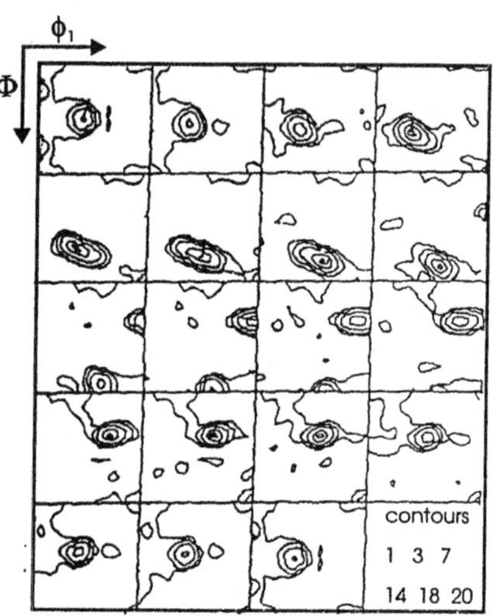

Fig. 2.22. ODF of 95% cold rolled copper, (Hirsch and Lücke 1988a).

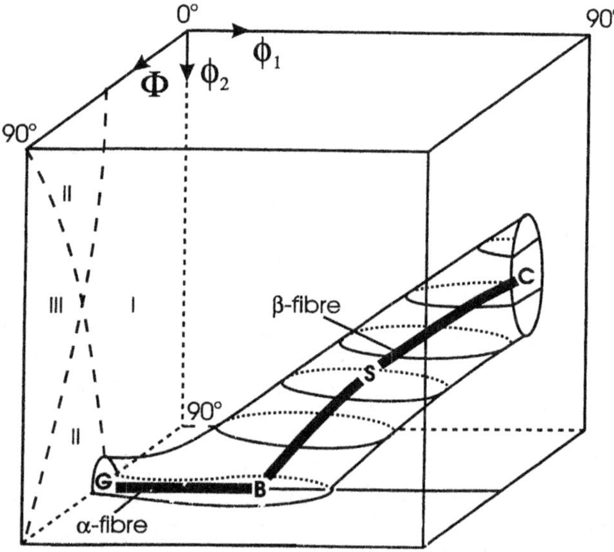

Fig. 2.23. Schematic representation of the fcc rolling texture in the first subspace of three dimensional Euler angle space, (after Hirsch and Lücke 1988a).

Figure 2.24 shows the orientation density values along the α and β fibres for rolled copper. For low reductions the variation of intensity along the fibres is small with values of 2-4 times random (xR). The homogeneity along the fibres deteriorates as rolling continues and this occurs first in the α fibre. By 95% reduction the α tube has almost disappeared from the texture but the β tube is still prominent. If further rolling occurs, deterioration of the β tube begins and pronounced peaks develop. In particular the S component strengthens to become the major component of heavily rolled copper or aluminium. Hirsch and Lücke (1988a) discuss these results in great detail in their analysis of texture development in rolled fcc metals. The essential points are that at low strains the textures are best described by relatively homogeneous orientation tubes and that these deteriorate at high strains into peak type textures.

The ODF can also be used to describe the texture quantitatively in terms of a small number of major components. In this technique the volume fractions, M_i, of the selected components are calculated for the appropriate volumes of ODF space. The derived values for the material described by figure 2.22 are given in table 2.8

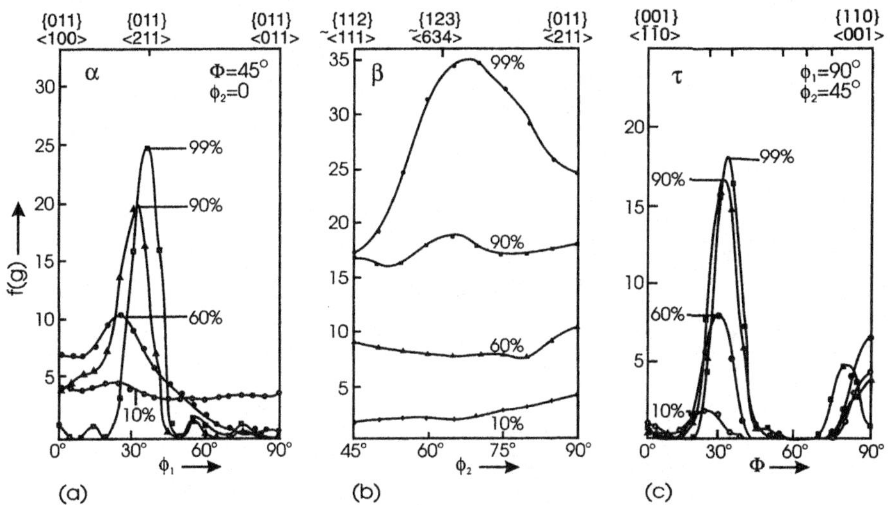

Fig. 2.24. Orientation densities f(g) along the major fibres in 95% cold rolled copper: (a) α fibre; (b) β fibre; (c) τ fibre, (Hirsch and Lücke 1988a).

Table 2.8
Volume percentages of major orientations in 95% cold-rolled copper
(Hirsch and Lücke 1988a).

Component	M_i	Component	M_i, %
Copper, C	27	Brass B	8
S	38	B/S, {168}<211>	18
Goss, G	3	Other	6

2.4.1.2 Alloy texture.

The pole figure of fig 2.20b and the ODF of figure 2.25 show that metals with low values of γ_{SFE} develop sharp $\{110\}<112>$ textures. By comparison with metals with high values of γ_{SFE} the α fibre is much more prominent in the alloy texture (see the $\phi_2 =0°$ section at $\Phi =45°$ and $\phi_1 = 0\text{-}35°$). Also present in the alloy texture are two further fibres referred to as the γ and τ fibres (Hirsch and Lücke 1988a). The τ fibre is a significant feature of the texture of materials with intermediate values of γSFE ($\sim 40mJm^{-2}$). All four fibres associated with rolled fcc metals are shown in figure 2.26. The γ fibre corresponds to volume elements with $\{111\}$ planes parallel to the rolling plane, i,e, to the aligned deformation twins in the microstructure (see fig 2.14) and extends from $\{111\}<112>$ at $\Phi =55°$, $\phi_2 =45°$, $\phi_1 =30/90°$ to $\{111\}<110>$ at $\Phi =55°$, $\phi_2 =45°$, $\phi_1 =0/60°$. The τ fibre corresponds to orientations having a $<110>$ direction parallel to TD and extends along the line $\phi_1 =90°$, in the $\phi_2 =45°$ section from the $\{112\}<111>$, C, orientation to the $\{110<001>$, G, orientation at $\Phi=35°$ and $90°$ respectively.

Figure 2.27 shows the intensity of the fibres for cold rolled 70:30 brass. It may be seen that as deformation proceeds,

    (i) the α tube is retained with the G component remaining strong,
    (ii) the B component strengthens steadily with rolling, and
    (iii) both the C and S components become less significant.

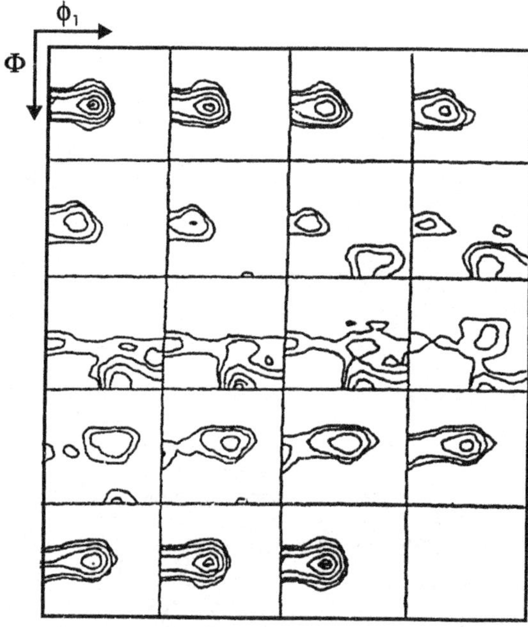

Fig. 2.25. ODF of 95% cold rolled 70;30 brass, (Hirsch and Lücke 1988a).

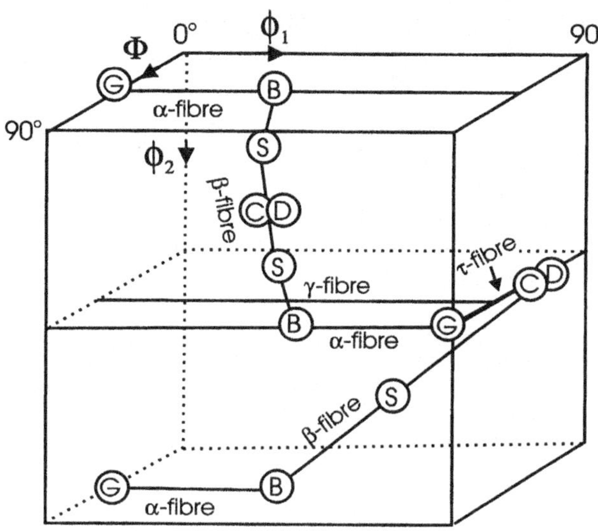

Fig. 2.26. Plots of important fibres in fcc materials, (Hirsch and Lücke 1988b).

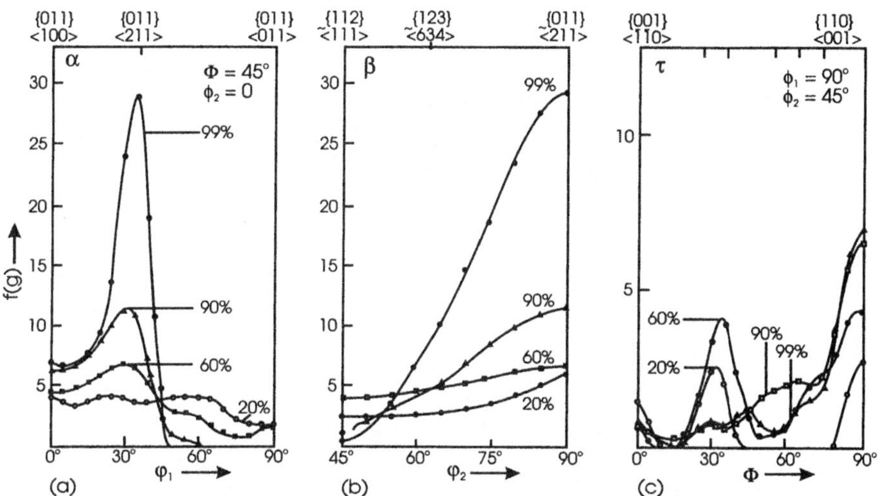

Fig. 2.27. Orientation densities f(g) along the major fibres in 95% cold rolled 70:30 brass: (a) α fibre; (b) β fibre; (c) τ fibre, (Hirsch and Lücke 1988a).

2.4.2 Deformation textures in bcc metals

The deformation textures of bcc metals and alloys are generally more complex than those of fcc metals and despite the overwhelming importance of the steel industry they have been less extensively investigated. The rolling texture of iron and low carbon steel is largely independent of composition and processing variables and even such major microstructural inhomogeneities as shear bands have little effect. A typical 200 pole figure is shown in figure 2.28a. Five orientations have been used to describe the components of this texture: $\{111\}<112>$, $\{111\}<123>$, $\{001\}<110>$, $\{112\}<110>$ and $\{111\}<110>$ (e.g. Hutchinson 1984).

The usual description has concentrated on two prominent orientation spreads which are in turn, described as fibre textures (fig 2.28b). One of these corresponds to a fibre texture with a $<111>$ fibre axis perpendicular to the sheet surface; the above $\{111\}<uvw>$ orientations are prominent in this spread. The other is a partial fibre texture with a $<110>$ fibre axis parallel to the rolling direction; the $\{hkl\}<110>$ orientations are prominent in this spread.

ODF results from a similar material are given in fig 2.29 taken from the work of von Schlippenbach et al. (1986). Before this texture is described it should be noted that the nature of the bcc rolling texture is such that the data are best displayed by sections at constant values of ϕ_1 rather than ϕ_2 as for fcc metals. In many cases, however, the important information can be portrayed by a single section at $\phi_2 = 45°$ (see table 2.9).

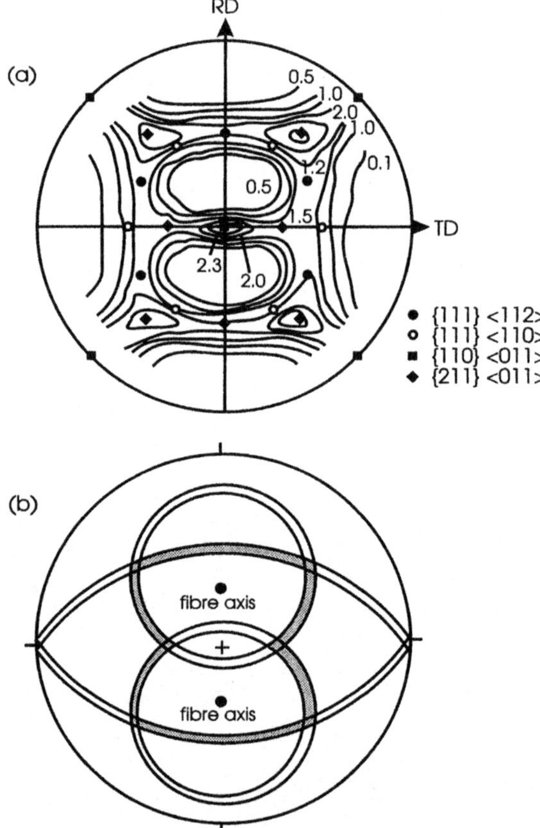

Fig. 2.28. Rolling texture of cold rolled low carbon steel: (a) 200 pole figure after 90% reduction (Hutchinson 1984); (b) partial fibre textures commonly used to describe rolling texture, (Hatherly and Hutchinson 1979).

Table 2.9
Texture components in rolled bcc metals.

{hkl}	<uvw>	ϕ_1	Φ	ϕ_2
001	110	45	0	0
211	011	51	66	63
111	011	60	55	45
111	112	90	55	45
11,11,8	4,4,11	90	63	45
110	110	0	90	45

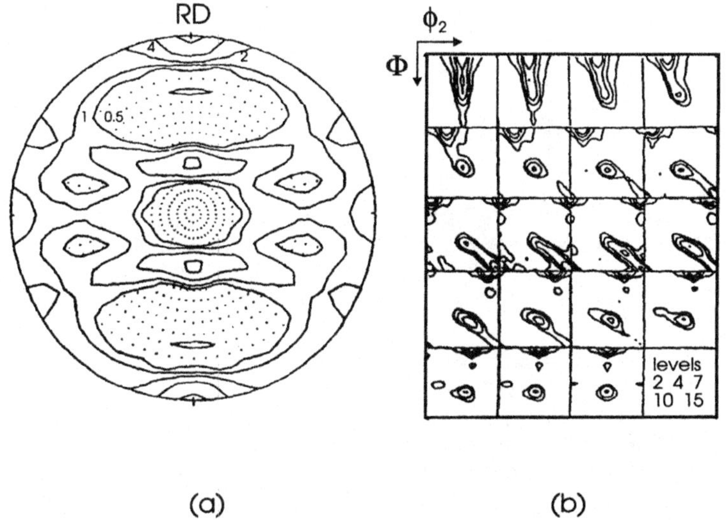

Fig. 2.29. Rolling texture of 90% cold rolled low carbon steel: (a) 110 pole figure; (b) ODF, (Lücke and Hölscher 1991, after von Schlippenbach et al. 1986).

Interpretation of the ODF data has been based mainly on attempts to identify fibres similar to those used earlier for fcc metals. The early work of Schlaffer and Bunge (1974) reported both a β and an α fibre but more recent investigations have disputed this. In particular Lücke and his collaborators (von Schlippenbach et al. 1986 and Emren et al. 1986) have argued that the texture shown in figure 2.29 is best described by use of the α and γ fibres discussed earlier. The orientation density along the α fibre (fig 2.30) increases fairly uniformly with strain up to ~70% reduction but with further rolling {112}<110> and {111}<110> become more prominent. The γ fibre is relatively uniform at reductions up to 80% but thereafter the {111}<110> component strengthens. Table 2.9 gives the Miller indices and Euler angles of the important bcc rolling texture components.

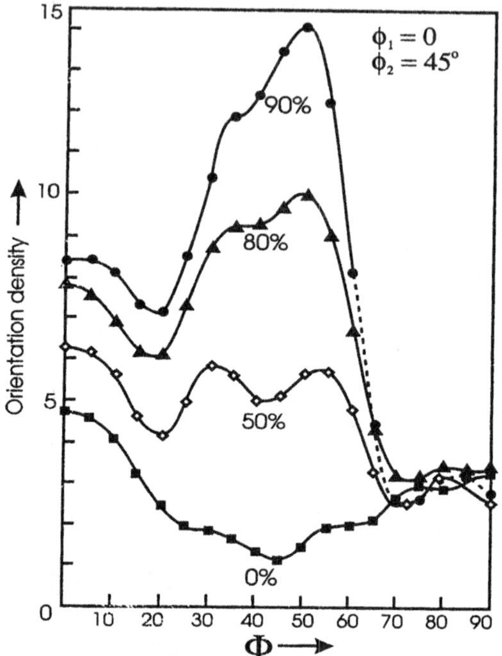

Fig. 2.30. Development of rolling texture in cold rolled, low carbon steel as shown by orientation density along the α fibre, (von Schlippenbach and Lücke 1984).

2.4.3. Deformation textures in hexagonal metals.

The information available for cph metals is limited. However, it is to be expected from the data given in Table 2.5 that the rolling texture will be influenced by the c/a ratio and the particular slip systems that operate (Grewen 1973).

For metals with c/a close to the ideal value of 1.633 the rolling texture has a strong $\{0001\} < 1100 >$ component which is a direct consequence of the favoured basal plane slip (fig 2.31a). In zinc and cadmium which have higher c/a ratios the basal planes are tilted some 20-30° about the transverse direction (fig 2.31b). Such a texture is predictable if basal slip is combined with twinning as shown by the microstructure of deformed cph metals. For the remaining metals with c/a less than ideal the basal planes are again rotated out of the rolling plane but this time the rotation axis is parallel to RD rather than TD and the tilt angle is 30-40° (fig 2.31c); the rotation axis is $< 1010 >$. Several recent investigations using ODF analysis have confirmed the earlier pole figure analyses.

In the case of titanium, for example, the major components of the texture after 90% reduction are $\{1214\} < 1010 >$, $\{1212\} < 1010 >$ and $\{1210\} < 1010 >$ all of which have the $< 1010 >$ RD alignment (Inoue and Inakazu 1988). The first and most important of these can be seen at $\Phi = 37°$, $\phi_1 = 0°$, $\phi_2 = 0°$ in figure 2.32. This component develops only at

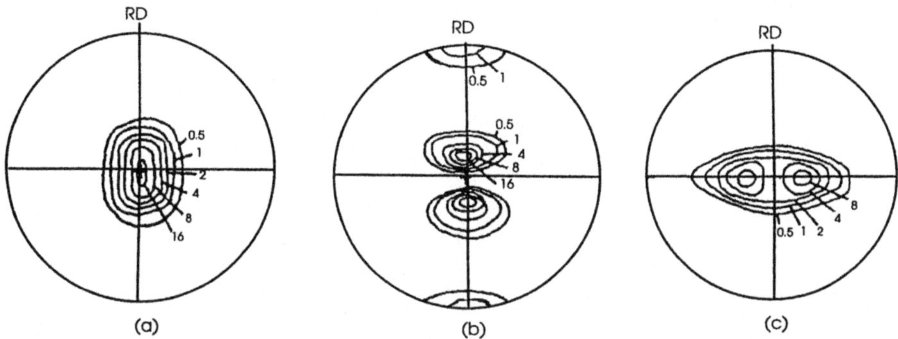

Fig. 2.31. Rolling textures (0002 pole figures) of cold rolled cph metals: (a) magnesium, c/a ideal; (b) zinc, c/a high; (c) titanium c/a low, (Hatherly and Hutchinson 1979).

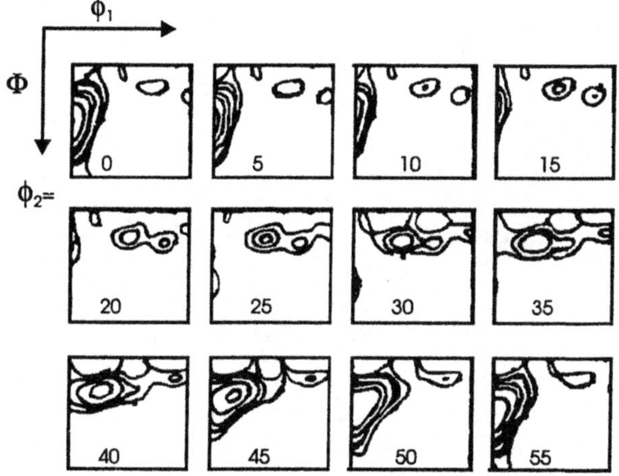

Fig. 2.32. Rolling texture of 90% cold rolled titanium, (Nauer-Gerhardt and Bunge 1988).

reductions greater than ~50%, and at this level of deformation twinning becomes a major deformation mode. It should be noted that the symmetry of cph metals is such that only values of ϕ_2 in the range 0-60° need to be shown in the ODF.

2.4.4. Fibre textures.

The deformation textures of materials deformed by uniaxial processes such as tension, wire drawing, extrusion etc. are invariably fibre textures and the results are most commonly expressed as inverse pole figures. For fcc metals the texture is described simply as a double fibre texture with <111> and <100> parallel to the axis as shown in figure 2.33. The relative proportions of each are determined primarily by γ_{SFE} and vary from none, or almost

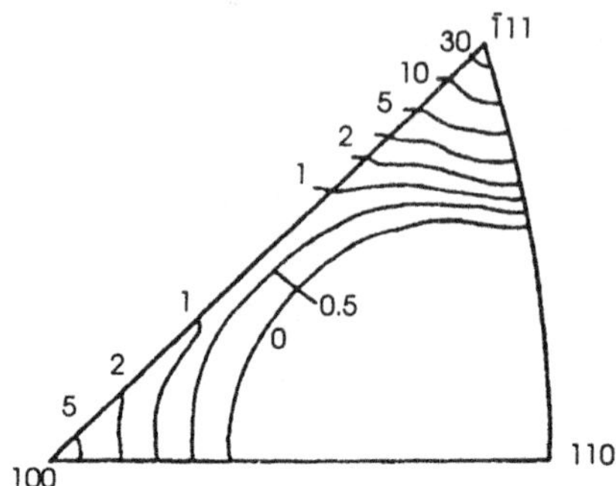

Fig. 2.33. Inverse pole figure for cold drawn aluminium wire,
(Hatherly and Hutchinson 1979).

none of the $<100>$ component in metals like aluminium (γ_{SFE} high) to 100% or nearly
100% in materials with low values of γ_{SFE} like silver (Schmidt and Wassermann 1927). For
bcc metals the fibre axis is always $<110>$ and in cph metals the preferred axis is $<1010>$.
Compression textures are somewhat different and almost opposite to those just described.
For fcc metals a $<110>$ texture is most frequently reported but in some low γ_{SFE} materials
$<111>$ components also form. Most bcc metals develop textures containing mixed $<100>$
and $<111>$ components as shown in figure 2.19b. For a more detailed account of this
subject the reader is referred to Barrett and Massalski (1980).

2.4.5. Theories of deformation texture development.

2.4.5.1 Relationship between texture and plasticity.
Deformation textures are a direct consequence of the lattice rotations that occur during slip
and twinning and there have been many attempts to predict the final texture based on the
theories of polycrystalline plasticity outlined in section 2.3.5.

The Sachs model (§2.3.5) assumes that a grain is unconstrained by its neighbours. In these
circumstances a crystal (grain) deforms on the most highly stressed slip system, and for
example an fcc crystal deformed in uniaxial tension will rotate until a $<112>$ direction is
parallel to the tensile axis, and will rotate towards $<110>$ in compression (see Reid 1973).
If sheet rolling is approximated to a biaxial stress state with compression in the normal
direction, and tension in the rolling direction, then an unconstrained crystal will rotate to a
stable $\{011\}<211>$ orientation (the brass orientation).

On the full constraints Taylor model (§2.3.5), a bcc grain deforming by $\{110\}<111>$ slip
in uniaxial tension is predicted to rotate towards $<110>$ (fig 2.19a), whereas an fcc crystal
will rotate towards either the $<111>$ or $<100>$ orientation depending on the starting
orientation. (Note that the predicted strain path is the same as that shown for bcc in figure
2.19a, but with the arrows reversed). For sheet rolling, the Taylor model predicts a stable
end orientation of $\{4,4,11\}<11,11,8>$, sometimes referred to as the Dillamore orientation.

The considerations above are based on the simplest models of crystal plasticity, and although they give some indication of the deformation textures which develop, for further refinements of the models, the reader should consult a specialist text such as Gil Sevillano et al. (1981).

The agreement between the experimental textures and those predicted by the Taylor model is quite good for fcc metals with medium to high values of γ_{SFE} and for bcc metals, i.e. for metals that deform by slip. However, theory predicts textures that develop more rapidly and are sharper than those observed in practice. In the case of metals with low values of γ_{SFE} the Taylor model is much less satisfactory and a Sachs type model gives better results (Leffers 1981). In both cases the lack of accuracy has its origin mainly in the failure of the simple theories to recognize the heterogeneities of deformation (§2.3.6) which contribute so significantly to the microstructure. The texture predictions of the relaxed constraints theories give improved results for high deformations. The incorporation of grain size inhomogeneities such as deformation bands into models of texture development is relatively straightforward, as these inhomogeneities are directly related to the local crystal plasticity as discussed in §2.3.5.

2.4.5.2 The texture transition.

One of the earliest attempts to incorporate microstructural heterogeneity was that of Wassermann (1963) who suggested that the texture transition in fcc metals could be rationalized if the rolling textures of all fcc metals were considered to be similar at low reductions (up to 50%) and to consist of a spread from $\{112\}<111>$ (C) to $\{011\}<211>$ (B). Wassermann argued that in metals of low stacking fault energy, and at larger strains, twinning occurred preferentially in crystals of the former orientation, which were thereby rotated to $\{552\}<115>$. Normal slip would then result in the development of the observed $\{110\}<001>$ orientation. However, this hypothesis has been negated by subsequent TEM observations which show that the twins produced are extremely fine (fig 2.6f) and bear no relationship to the bulk macroscopic features envisaged by Wassermann.

There have been many subsequent attempts to explain the texture transition. Hutchinson et al. (1979) recognized that the very fine twinned structures seen in metals like 70:30 brass were incompatible with Wassermann's concept of bulk twinning and proposed a five stage pattern of texture development that was more consistent with microstructure:

> (i) 0-50% reduction: rotations due to slip produce a copper type texture in both copper and brass
> (ii) 40-60% reduction: fine twinning occurs in brass in suitably oriented grains,especially near $\{112\}<111>$
> (iii) 50-80% reduction: in twinned volumes slip is restricted to planes parallel to the twin boundaries leading to overshooting and the formation of $\{111\}<uvw>$ components by coupled rotation
> (iv) 60-95% reduction: increasing volume of shear bands destroys the $\{111\}<uvw>$ orientations, inhibits components formed by homogeneous deformation and contributes additional components including $\{110\}<001>$
> (v) above 85% reduction: deformation becomes homogeneous leading to sharpening of the stable $\{110\}<112>$ texture.

Later work has cast some doubt on the similarity of the copper and brass textures in the range 0-50% reduction. While the extensive ODF work of Hirsch and Lücke (1988a) supports this view Leffers and Juul Jensen (1988) have shown that the development of the

$\{112\} < 111 >$ component is different in copper and brass (15%Zn) and detailed work by Gryzliecki et al. (1988) on a copper-germanium alloy (8.8%Ge) supports this result. As shown earlier, twinning is certainly present in many of the grains of 70:30 brass at 50% reduction and some texture difference should be expected. A second problem concerns the role of shear banding. When attention was first focused on shear bands it was often assumed that these non-crystallographic features of the microstructure would explain the lack of precision of the various rolling texture models and contribute significantly to the evolution of the $\{110\} < 112 >$ component in materials in which γ_{SFE} is low. It now seems, however, that shear bands may be less important to the transition problem than previously thought and that their major contribution in metals with low γ_{SFE} may be to produce a structure of very fine crystals which subsequently deform by slip. There is also a difficulty with the orientation changes occurring in the range 50-80% reduction where the orientation $\{110\} < 112 >$ has not been considered significant. The more recent studies just mentioned have shown that this component is already developing at an early stage and Lee et al. (1993) have argued that this can be accounted for by progressive deformation banding. The subsequent rapid development of $\{110\} < 112 >$ at high strains is attributed by Chung et al. (1988) to a homogeneous shear deformation of the fine crystallites formed by shear banding.

Chapter 3

THE STRUCTURE AND ENERGY OF GRAIN BOUNDARIES

3.1 INTRODUCTION

The majority of the book is concerned with the ways in which boundaries separating regions of different crystallographic orientation are formed or are rearranged on annealing either during or after deformation. In this chapter we will introduce some aspects of the structure and properties of these boundaries and in chapter 4 we discuss the migration and the mobility of boundaries. We will concentrate on those aspects of grain boundaries which are most relevant to recovery, recrystallization and grain growth and will not attempt to give a full coverage of the subject. Further information on grain boundaries may be found in the books by Hirth and Lothe (1968), Bollmann (1970), Gleiter and Chalmers (1972), Chadwick and Smith (1976), Balluffi (1980), Wolf and Yip (1992) and Sutton and Balluffi (1995).

If we consider a grain boundary such as that shown in figure 3.1, then the overall geometry of the boundary is defined by the orientation of the boundary plane AB with respect to one of the two crystals (two degrees of freedom) and by the smallest rotation (θ) required to make the two crystals coincident (three degrees of freedom). There are thus **five macroscopic degrees of freedom** which define the geometry of the boundary. In addition to this, the boundary structure is dependent on three **microscopic degrees of freedom**, which are the rigid body translations parallel and perpendicular to the boundary. The structure of the boundary depends also on the local displacements at the atomic level and is influenced by external variables such as temperature and pressure, and internal parameters such as bonding, composition and defect structure. As many of the properties of a grain boundary are dependent on its structure, a knowledge of boundary structure is a necessary prerequisite to understanding its behaviour. Although there has been extensive experimental and theoretical work in this area over the past few decades, there is still a great deal of uncertainty about the structure and properties of boundaries. Most of this work has been carried out for **static boundaries** and there is even more uncertainty over the structure, energy and properties of the **migrating boundaries** which will be important during annealing. In addition, most of the experimental measurements have been made close to the melting temperature. We must therefore recognise that our ability to understand the phenomena of recovery, recrystallization and grain growth may well be limited by our lack of knowledge of the boundaries themselves.

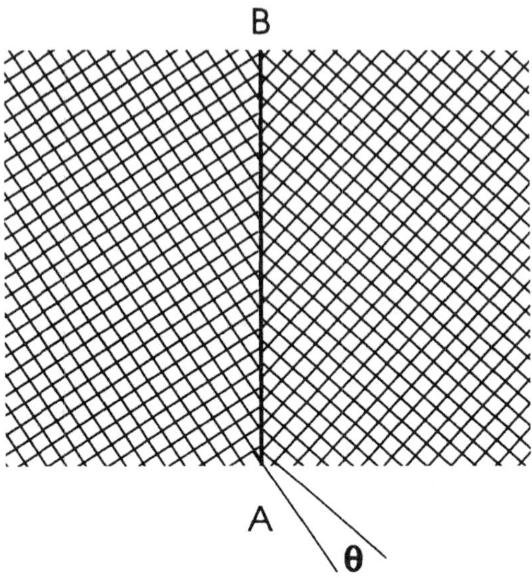

Fig. 3.1. A grain boundary between two crystals misoriented by an angle θ about an axis
normal to the page.

It is convenient to divide grain boundaries into those whose misorientation is greater than a
certain angle - **high angle grain boundaries (HAGB)**, and those whose misorientation is less
than this angle - **low angle grain boundaries (LAGB)**. The angle at which the transition
from low to high angle boundaries occurs is typically taken as **between 10° and 15°** and is
to some extent dependent on what properties of the boundary are of interest. As a very
general guide, low angle boundaries are those which can be considered to be composed of
arrays of dislocations and whose structure and properties vary as a function of misorientation,
whilst high angle boundaries are those whose structure and properties are not generally
dependent on the misorientation. However, as discussed below, there are "special" high
angle boundaries which do have characteristic structures and properties, and therefore it is
clear that a crude division of boundaries into these two broad categories must be used with
caution.

3.2 THE ORIENTATION RELATIONSHIP BETWEEN GRAINS

Although, as discussed above, there are five macroscopic degrees of freedom needed to
define a boundary, in many cases we neglect the orientation of the boundary plane and
consider only the three parameters which define the orientation between the two grains
adjacent to a boundary. **It should however be recognised that the use of such an
incomplete description of a boundary may cause problems in the interpretation of
boundary behaviour.**

The relative orientation of two cubic crystals is formally described by the rotation of one
crystal which brings it into the same orientation as the other crystal. This may be defined
by the rotation matrix

$$R = \begin{bmatrix} a_{11} & a_{12} & a_{13} \\ a_{21} & a_{22} & a_{23} \\ a_{31} & a_{32} & a_{33} \end{bmatrix} \tag{3.1}$$

where a_{ij} are column vectors of direction cosines between the cartesian axes. The sums of the squares of each row and of each column are unity, the dot products between column vectors are zero, so only three independent parameters are involved. The rotation angle (θ) is given by

$$2 \cos\theta + 1 = a_{11} + a_{22} + a_{33} \tag{3.2}$$

and the direction of the rotation axis **[uvw]** is given by

$$[(a_{32} - a_{23}), (a_{13} - a_{31}), (a_{21} - a_{12})] \tag{3.3}$$

In cubic materials, because of the symmetry, the relative orientations of two grains can be described in **24** different ways. In the absence of any special symmetry, it is conventional to describe the rotation by the angle/axis pair associated with the **smallest** misorientation angle. The range of θ which can occur is therefore limited and Mackenzie (1958) has shown that the maximum value of θ is 45° for $<100>$, 60° for $<111>$, 60.72° for $<110>$ and a maximum of 62.8° for $<1,1,\sqrt{2}-1>$. For a polycrystal containing grains of random orientation, the distribution of θ is as shown in figure 3.2, with a mean of 40°.

The **angle/axis** notation described above is commonly used to express misorientations. However, there are several other ways of expressing the orientation relationship, including Euler angles (**misorientation distribution functions**) and **Rodrigues-Frank space**. The use of these methods for describing absolute orientations is outlined in the Appendix and will not be further considered here.

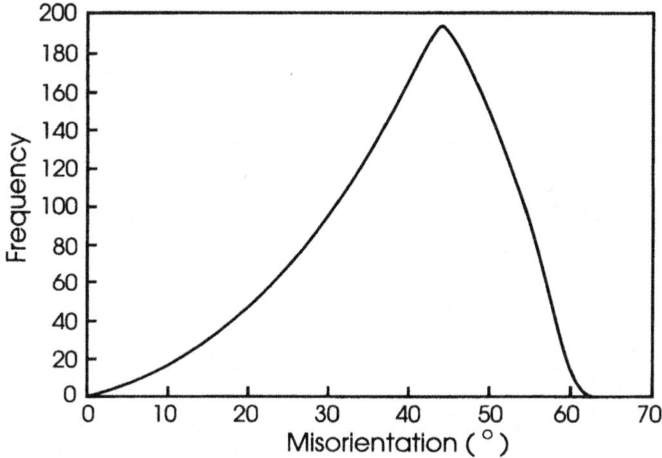

Fig. 3.2. The misorientation distribution for a randomly oriented assembly of grains.

3.3 LOW ANGLE GRAIN BOUNDARIES

A low angle boundary or sub-boundary can be represented by an array of dislocations (Burgers 1940, Read and Shockley 1950). The simplest such boundary is the symmetrical **tilt boundary**, shown schematically in figure 3.3 in which the lattices on either side of the boundary are related by a misorientation about an axis which lies in the plane of the boundary. The boundary consists of a wall of parallel edge dislocation aligned perpendicular to the slip plane. Such boundaries were first revealed as arrays of etch pits on the surface of crystals, but are now more commonly observed by transmission electron microscopy.

3.3.1 Tilt boundaries

If the spacing of the dislocations of Burgers vector **b** in the boundary is **h**, then the crystals on either side of the boundary are misoriented by a small angle θ, where

$$\theta \approx \frac{b}{h} \qquad (3.4)$$

The energy of such a boundary γ_s, is given (Read and Shockley 1950) as:

$$\gamma_s = \gamma_0 \, \theta \, (A - \ln \theta) \qquad (3.5)$$

where $\gamma_0 = Gb/4\pi(1-\nu)$, $A = 1 + \ln(b/2\pi r_0)$ and r_0 is the radius of the dislocation core, usually taken as between b and 5b.

According to this equation, the energy of a tilt boundary increases with increasing misorientation (decreasing h) as shown in figure 3.4. Combining equations 3.4 and 3.5 we note that as θ increases, the energy **per dislocation** decreases as shown in figure 3.4, showing that a material will achieve a lower energy if the same number of dislocations are arranged in fewer, but higher angle boundaries. As shown in figure 3.5, the theory is in good agreement with experimental measurements for small values of θ, although it is unreasonable to use this dislocation model for large misorientations, because when θ exceeds

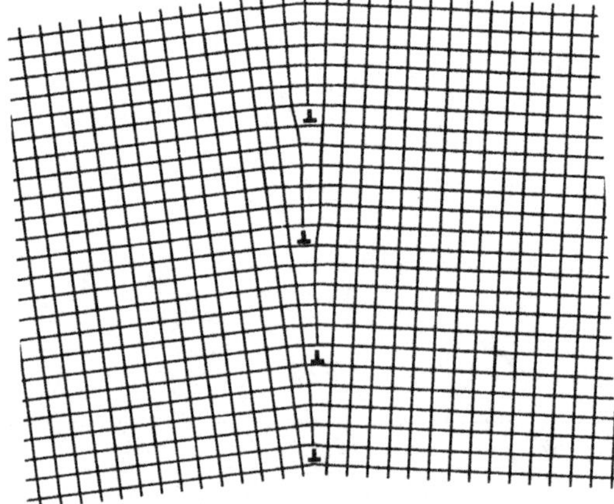

Fig. 3.3. A symmetrical tilt boundary.

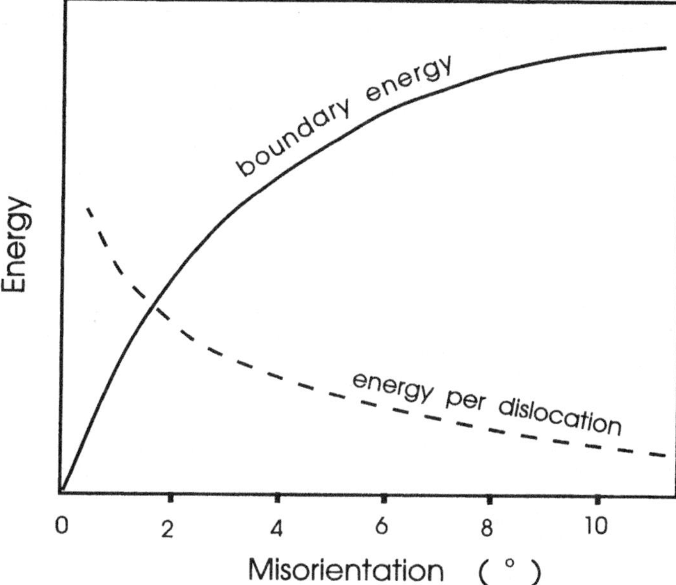

Fig. 3.4. The energy of a tilt boundary and the energy per dislocation as a function of the crystal misorientation.

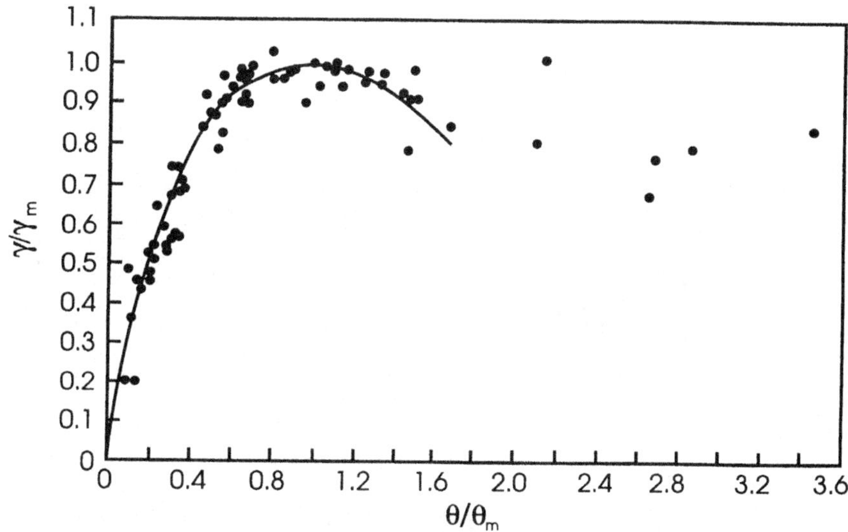

Fig. 3.5. The measured (symbols) and calculated (solid line) energy of low angle tilt boundaries as a function of misorientation, for different metals, (After Read 1953).

~ 15°, the dislocation cores will overlap, the dislocations lose their identity and the simple dislocation theory on which equation 3.5 is based becomes inappropriate.

It is often convenient (Read 1953) to use equation 3.5 in a form where the energy (γ_s) and misorientation (θ) are normalised with respect to the values of those parameters (γ_m and θ_m) when the boundary becomes a high angle boundary (i.e. $\theta \sim 15°$).

$$\gamma = \gamma_m \frac{\theta}{\theta_m}\left(1 - \ln\frac{\theta}{\theta_m}\right) \qquad (3.6)$$

3.3.2 Other low angle boundaries

In the more general case, dislocations of two or more Burgers vectors react to form two-dimensional networks whose character depends on the types of dislocation involved. For example a **twist** boundary, which is a boundary separating crystals related by a misorientation about an axis lying perpendicular to the boundary plane, may be formed by two sets of screw dislocations. If the Burgers vectors of the two sets of dislocations are orthogonal then the dislocations do not react strongly and the boundary consists of a square network of dislocations (fig 3.6a,c). However, if the Burgers vectors are such that the two sets of dislocations react to form dislocations of a third Burgers vector as shown in figure 3.6b,d then a hexagonal network may be formed. If h is the spacing of the dislocations in the network then the misorientation (θ) is given approximately by equation 3.4. The exact shape of the dislocation network will depend on the angle that the boundary plane makes with the crystals. Further details of the reactions involved in forming low angle grain boundaries may be found in textbooks on dislocation theory (e.g. Friedel 1964 , Hirth and Lothe 1982, Hull and Bacon 1984).

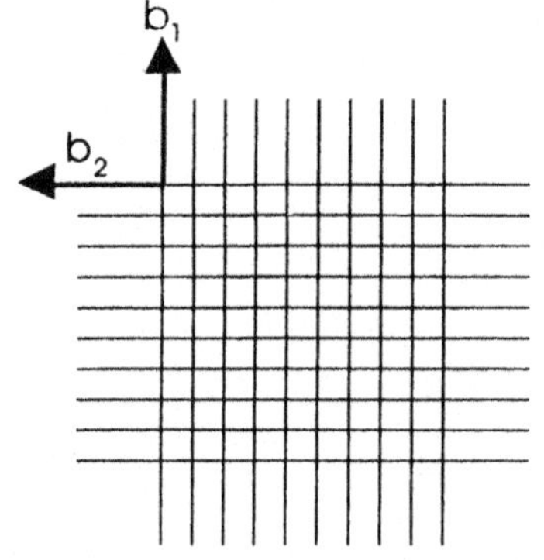

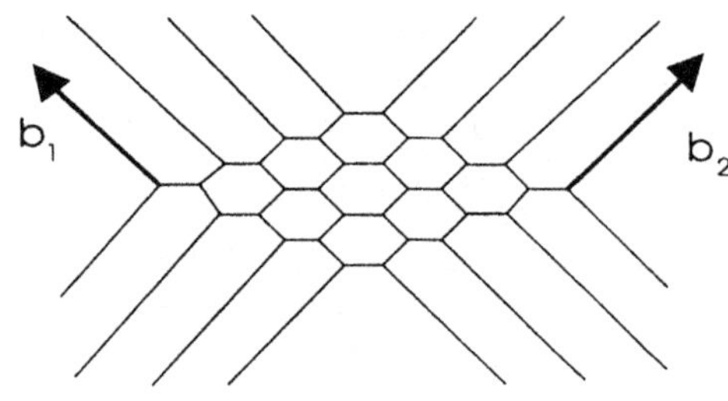

$b_1 = a/2[\bar{1}01] \qquad b_2 = a/2[101]$ $b_2 = a/2[0\bar{1}1] \qquad b_1 = a/2[110]$

(a)   (b)

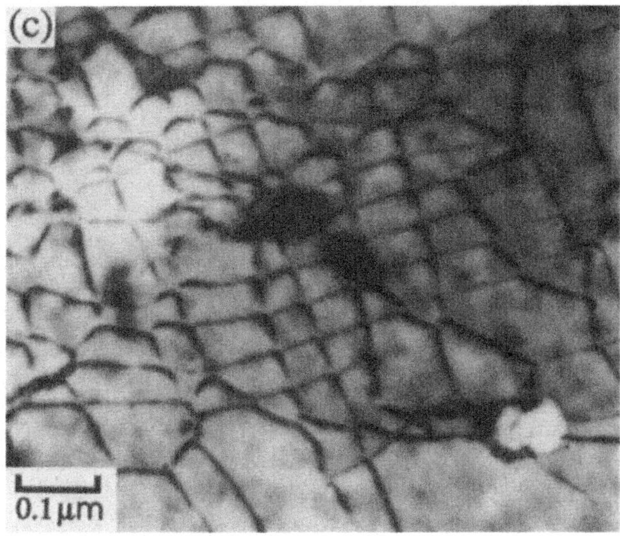

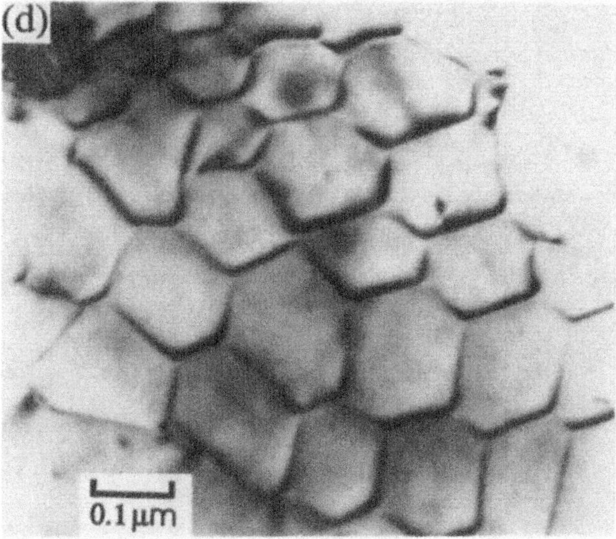

Fig. 3.6. The formation of low angle twist boundaries from dislocation arrays.
 a) A square network formed from screw dislocations of orthogonal Burgers vector.
 b) A hexagonal network formed by screw dislocations with 120° Burgers vectors.
c) TEM micrograph of a square twist boundary in copper, (Humphreys and Martin 1968).
d) TEM micrograph of a hexagonal twist boundary in copper, (Humphreys and Martin 1968).

3.4 HIGH ANGLE GRAIN BOUNDARIES

Although the structure of low angle grain boundaries is reasonably well understood, much less is known about the structure of high angle grain boundaries. Although early theories suggested that the grain boundary consisted of a thin "amorphous layer", it is now known that these boundaries consist of regions of good and bad matching between the two grains. The concept of the **coincidence site lattice (CSL)** (Kronberg and Wilson 1949), and extensive computer modelling, together with atomic resolution microscopy have in recent years considerably advanced the subject.

3.4.1 The coincidence site lattice

Consider two interpenetrating crystal lattices and translate them so as to bring a lattice point of each into coincidence, as in figure 3.7. If other lattice points in the two lattices coincide (the solid circles in fig 3.7), then these points form the coincident site lattice. The reciprocal of the ratio of CSL sites to lattice sites is denoted by Σ. For example in figure 3.7, Σ is seen to be 5. In the general case where there is no simple orientation relationship between the grains, Σ is large and the boundary, which has no special properties is often referred to as a **random boundary**. However, for certain orientation relationships for which there is a good fit between the grains, Σ is small and this may confer some special properties on the boundary. Good examples of this are the coherent twin ($\Sigma = 3$) boundary shown in figure 3.8

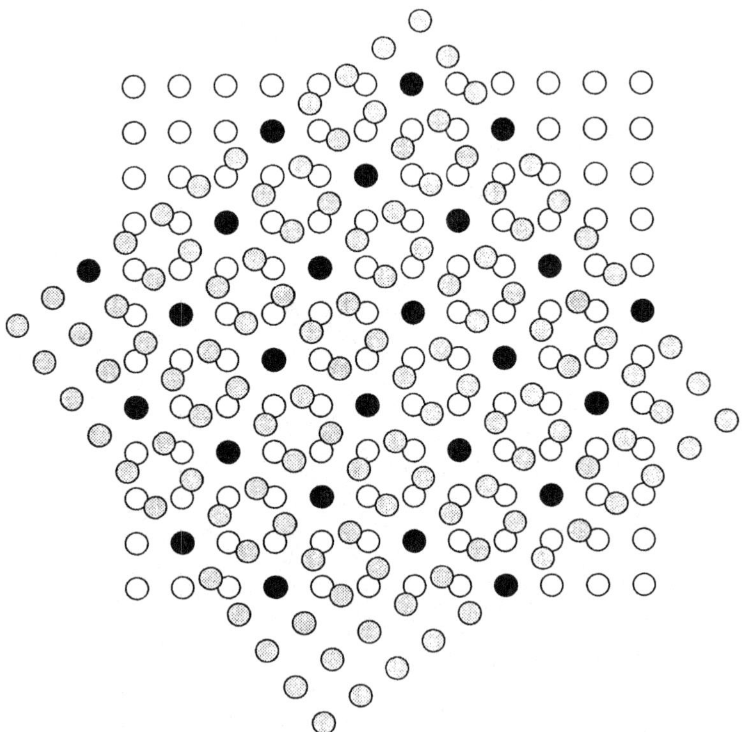

Fig. 3.7. A coincident site lattice ($\Sigma 5$) formed from two simple cubic lattices rotated by 36.9° about an $<001>$ axis. Filled circles denote sites common to both lattices.

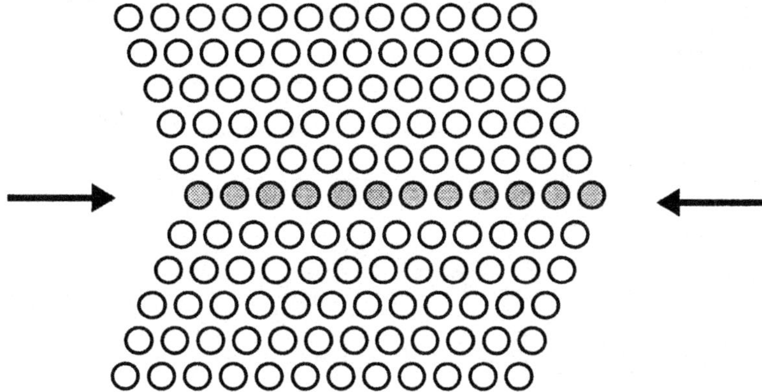

Fig. 3.8. A coherent twin boundary.

and low angle grain boundaries which are close to $\Sigma=1$. Further details of the geometry of CSL boundaries and extensive tables of CSL relationships may be found in Brandon et al. (1964), Grimmer et al. (1974), Warrington (1980) and Mykura (1980). Table 3.1 shows the relationship between Σ and the angle/axis rotation for boundaries up to and including $\Sigma=29$.

3.4.2 The structure of high angle grain boundaries

The atomic structure at the grain boundary is determined by relaxation of the atoms, which is dependent on the nature of the atomic bonding forces, and there has been extensive computer simulation of these structures (e.g. Gleiter 1971, Weins 1972, Vitek et al. 1980, Balluffi 1982, Wolf and Merkle 1992). It is predicted that a high degree of atomic-level coherency is maintained across the boundary and that regular, well defined **structural units** are formed. In the boundary of figure 3.9, the repeating structural units are shaded.

The CSL is a geometric relationship and any deviation from the exact coincidence relationship discussed above will destroy the CSL. However, even in this situation the boundary structure can be maintained by introducing **grain boundary dislocations** which can locally accommodate the mismatch in much the same way as dislocations preserve the lattice in low angle grain boundaries. The Burgers vector of the boundary dislocations can be much smaller than a lattice vector. It is also predicted (King and Smith 1980) that some grain boundary dislocations are associated with **steps** in the boundary. These boundary defects are of importance in the mobility of boundaries and are discussed further in §4.4.1.3.

The structure of grain boundaries has been extensively investigated by high resolution electron microscopy and other techniques (e.g. Pond 1980, Gronski 1980, Sass and Bristowe 1980, Krakow and Smith 1987, Seidman 1992, Wolf and Merkle 1992). The experimental observations have broadly confirmed the computer calculations and show that although the CSL is generally lost during the atomic relaxation at the boundaries, the periodicity of the boundary structure is retained by a network of grain boundary dislocations.

TABLE 3.1
Rotation axes and angles for coincidence site lattices of $\Sigma < 31$
(from Mykura 1980)

Σ	θ_{min} °	Axis
1	0	any
3	60	$<111>$
5	36.87	$<100>$
7	38.21	$<111>$
9	38.94	$<110>$
11	50.48	$<110>$
13a	22.62	$<100>$
13b	27.80	$<111>$
15	48.19	$<210>$
17a	28.07	$<100>$
17b	61.93	$<221>$
19a	26.53	$<110>$
19b	46.83	$<111>$
21a	21.79	$<111>$
21b	44.40	$<211>$
23	40.45	$<311>$
25a	16.25	$<100>$
25b	51.68	$<331>$
27a	31.58	$<110>$
27b	35.42	$<210>$
29a	43.61	$<100>$
29b	46.39	$<221>$

3.4.3 The energy of high angle grain boundaries

On the basis of the structural models outlined above, it might be expected that the energy of the boundary would be a minimum for an exact coincidence relationship and that it would increase as the orientation deviated from this, due to the energy of the network of accommodating boundary dislocations. However, the correlation between the **geometry** and the **energy** of a boundary is more complicated than this (Goodhew 1980) as is illustrated by

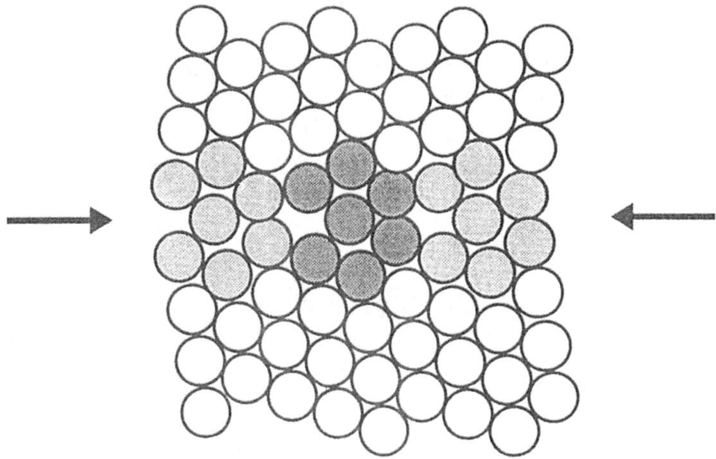

Fig. 3.9. The repeating structural units in a special grain boundary, (after Gleiter 1971).

figure 3.10 which shows a comparison of the measured and calculated energies of symmetrical tilt boundaries in aluminium. It can be seen that low energy cusps are found only for the $\Sigma=3$ (coherent twin) and $\Sigma=11$ boundaries and that the predicted cusps for $\Sigma=5$ and $\Sigma=9$ are not detected.   However, more recent measurements of boundary energies in high purity metals (e.g. Miura et al. 1990, Palumbo and Aust 1990) have found evidence for more low energy special boundaries than were found in earlier work.

The experimental measurements of boundary energy have been reviewed by Palumbo and Aust (1992). It is suggested that the lack of low energy cusps may in some cases be due to an insensitivity in the measurement technique and in other cases be due to small amounts of impurity.   There is evidence that the energy and presumably the structure of high angle boundaries is affected by impurity segregation. Measurements of boundary energy in Ag-Au and Cu-Pb (Sauter et al. 1977, Gleiter 1970a) suggest that with increasing segregation, the energy of special grain boundaries tends towards that of random boundaries as shown schematically in figure 3.11, and more recent experiments (Palumbo and Aust 1992) have confirmed this trend.

Sutton and Balluffi (1987) concluded from a survey of the experimental measurements that there is no simple relationship between the energy of a boundary and the overall **geometry** of the boundary as defined by the macroscopic degrees of freedom, and that parameters such as a low value of Σ were not necessarily indicative of a low energy.   It is likely that the boundary energy is determined primarily by the **microscopic** structure of the boundary and that atomic bonding plays an important role. Computer simulations (Smith et al. 1980, Wolf and Merkle 1992) suggest that the local **volume expansion** or **free volume** associated with a boundary is important, and the latter authors predict a linear relationship between boundary energy and volume expansion.

Some caution is however needed in interpreting the experimental measurements of boundary energy and applying these to the annealing processes considered later in this book, because such measurements are normally made at very high homologous temperatures in order for

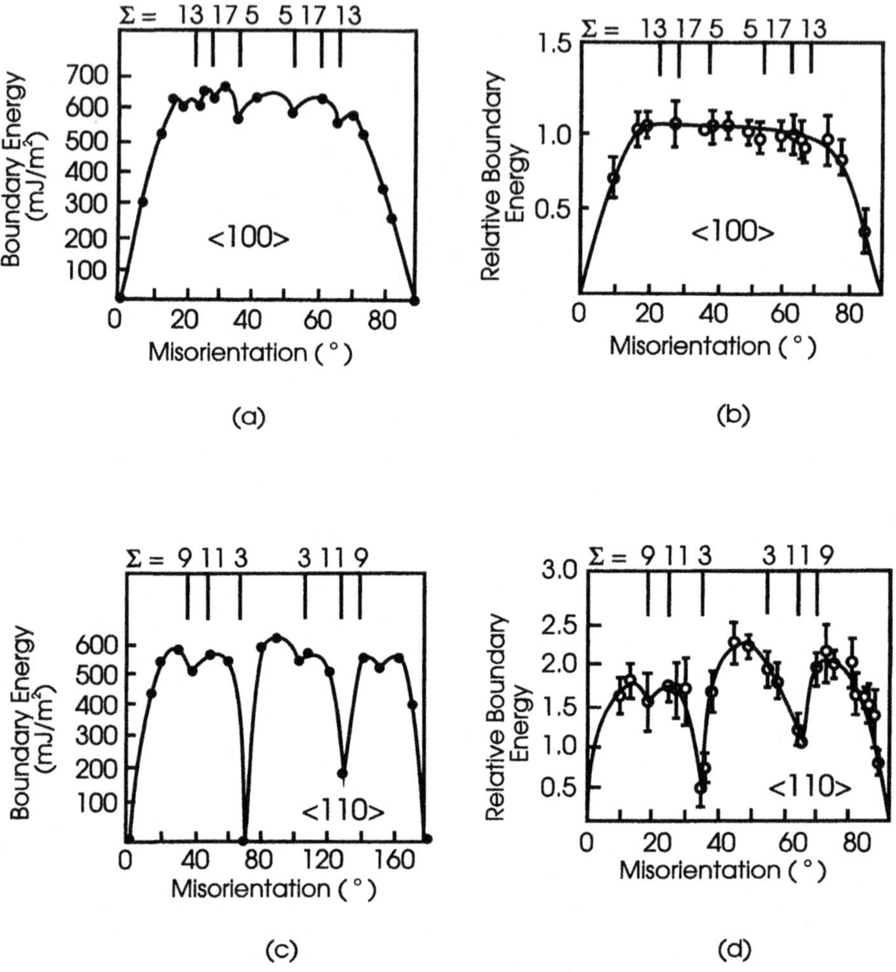

Fig. 3.10. The computed (a and c) and measured (b and d) energies at 650°C for symmetrical < 100 > and < 110 > tilt boundaries in aluminium, (Hasson and Goux 1971).

equilibrium to be achieved. There is some evidence (Gleiter and Chalmers 1972, Goodhew 1980, Shvindlerman and Straumal 1985, Rabkin et al. 1991) that some special boundaries may exhibit a **phase transition** at very high temperatures. This may involve either the loss of their ordered structure, or a transition to a different ordered structure.

It is to be expected that the structure and hence the energy of a special boundary will be dependent on the actual plane of the boundary, as demonstrated by Lojkowski et al (1988). As shown in table 3.2, the energy of the Σ=3 coherent twin boundary is much smaller than that of the non-coherent boundary.

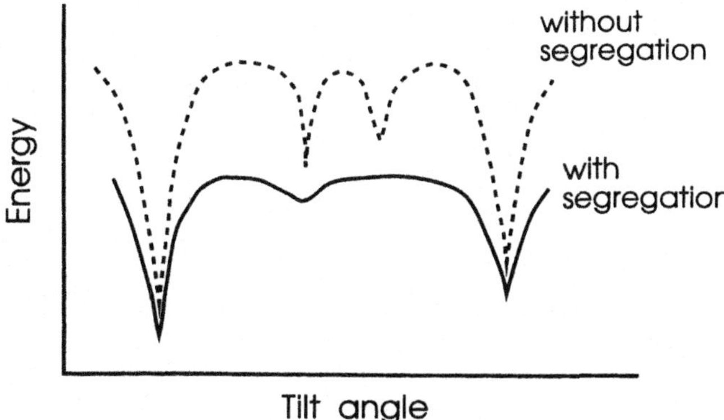

Fig. 3.11. Schematic diagram showing the changes in the energy v misorientation relationship due to solute atoms, with and without segregation at the boundaries, (Sauter et al. 1977).

TABLE 3.2

Measured grain boundary energies (mJm^{-2})
(Data from Murr 1975)

Material	High angle grain boundary energy	Coherent twin boundary energy	Incoherent twin boundary energy
Ag	375	8	126
Al	324	75	-
Au	378	15	-
Cu	625	24	498
Cu-30wt%Zn	595	14	-
Fe (γ)	756	-	-
Fe-3wt%Si	617	-	-
Stainless steel (304)	835	19	209
Ni	866	43	-
Sn	164	-	-
Zn	340	-	-

3.5 THE TOPOLOGY OF BOUNDARIES AND GRAINS

In addition to the structure and properties of individual grain boundaries we need to be aware of the arrangements of boundaries within a material. As boundaries are non-equilibrium defects, a single-phase material is in its most thermodynamically stable state when all boundaries are removed. This is not often achieved and much of this book is devoted to a discussion of the processes by which high and low angle boundaries are eliminated or rearranged into metastable configurations. In this section we consider the nature of these metastable arrangements of boundaries.

C.S. Smith set out the topological requirements of space-filling and the role of boundary tensions in his classic paper of 1952 and these have been discussed recently by Atkinson (1988). In both 2-D and 3-D, the microstructure consists of **vertices** joined by **edges** or **sides** which surround **faces** as shown schematically in figure 3.12. In the 3-D case, the faces surround **cells** or **grains**. The cells (C), faces (F), edges (E) and vertices (V) of any cellular structure obey the conservation law of equation 3.7, known as **Euler's equation**, provided that the face or cell at infinity is not counted.

$$F - E + V = 1 \qquad \text{(2-d plane)} \tag{3.7}$$

$$-C + F - E + V = 1 \qquad \text{(3-d Euclidean space)}$$

where **C** is the number of cells, **E** edges, **F** faces and **V** vertices.

The number of edges joined to a given vertex is its coordination number **z**. For topologically stable structures, $z=3$ in 2-D and $z=4$ in 3-D. Thus in 2-D, a 4-rayed vertex such as that shown at A in figure 3.12 will be unstable and will decompose to two 3-rayed vertices such as B and C.

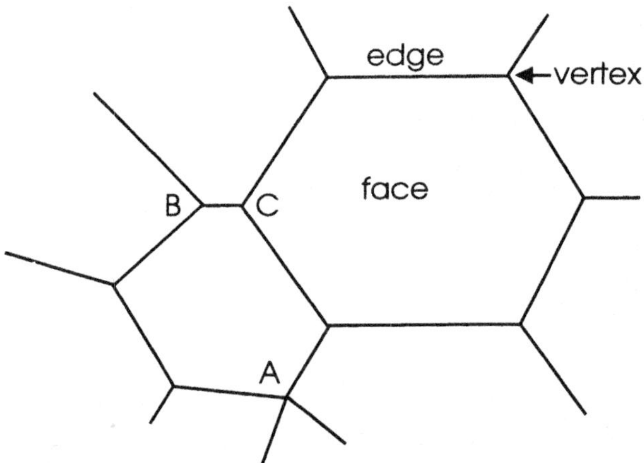

Fig. 3.12. A 2-D section of a grain structure. The 4-rayed vertex at A will tend to decompose into two 3-rayed vertices such as B and C.

3.5.1 Two-dimensional microstructures

In a two-dimensional microstructure, the material will be divided into grains or subgrains separated by boundaries and, if boundaries are mobile, a local mechanical equilibrium will be established at the vertices of grains. Consider the three grains 1, 2 and 3 shown in figure 3.13.   The boundaries have specific energies γ_{12}, γ_{13} and γ_{23} and at equilibrium these energies are equivalent to boundary tensions per unit length.   For the three boundaries of figure 3.13 the stable condition is

$$\frac{\gamma_{12}}{\sin \alpha_3} = \frac{\gamma_{13}}{\sin \alpha_2} = \frac{\gamma_{23}}{\sin \alpha_1} \tag{3.8}$$

If all boundaries have the same energy, then equation 3.8 shows that the three grains will meet at angles of 120°. In this situation an array of equal sized hexagonal grains would be stable. Whatever the actual arrangement of the grains in a two-dimensional microstructure, it follows from equation 3.7 that if z=3, the mean number of sides per grain or cell is 6.

3.5.2 Three-dimensional microstructures

It has been shown (Smith 1952) that there is no three-dimensional plane-faced polyhedron which, when repeated, can simultaneously completely fill space and balance the boundary tensions. Truncated octahedra stacked in a bcc arrangement as shown in figure 3.14 come close, filling the space but not having the correct angles to balance the boundary forces. The **Kelvin tetrakaidecahedron**, figure 3.15, which has doubly curved boundary surfaces satisfies both conditions.

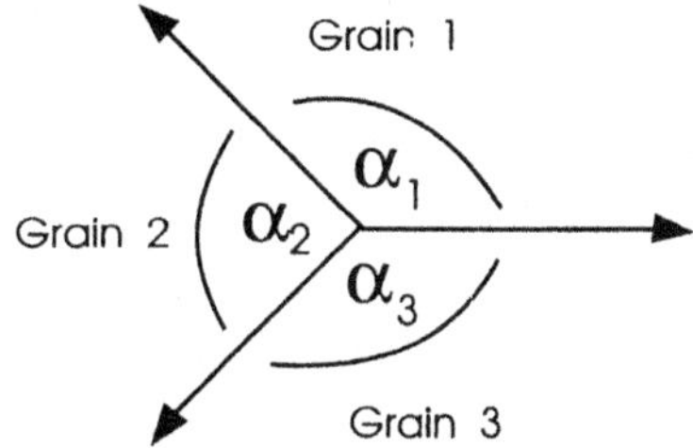

Fig. 3.13. The forces at a boundary triple point.

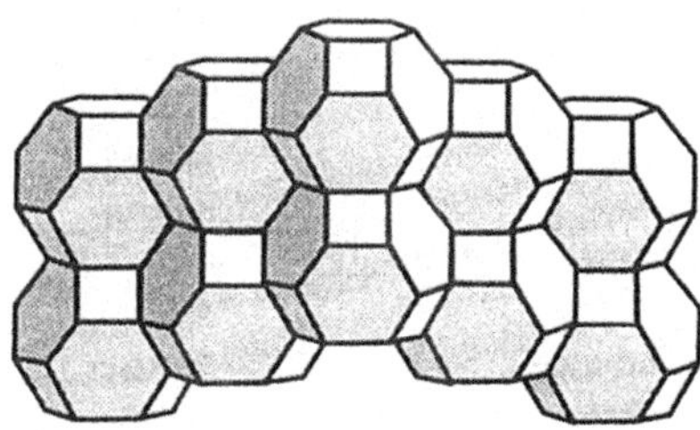

Fig. 3.14. Body centred cubic packing of truncated octahedra, (after Smith 1952).

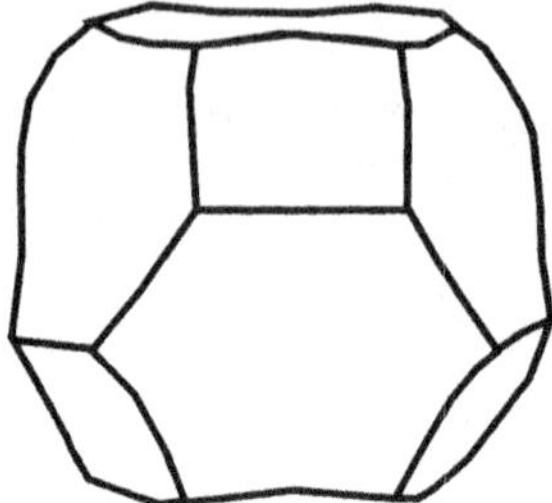

Fig. 3.15.   The Kelvin tetrakaidecahedron.

Polycrystalline materials are usually examined on a random planar section and therefore sectioning effects will mean that the angles measured on the microstructure will not necessarily be the true boundary angles.   However, in microstructures of well annealed single-phase materials, the sectioned grains often appear as hexagons (e.g. fig 9.18), and it has been shown (Smith 1948) that the distribution of measured angles is Gaussian, peaking at the true angle.   Figure 3.16 shows some measurements for high angle grain boundaries in annealed α-brass.   It may be seen that the data peak at 120° which is the equilibrium angle.

If, as is usually the case, the boundary energies are not equal, then the regular geometric structures discussed above will not be stable.   An example of this is the microstructure of recrystallized α-brass, shown in figure 6.33, which contains both normal high angle boundaries and low energy $\Sigma=3$ coherent twins.   The lower energy of the twin boundaries is apparent from the angles, little larger than 90° which they make with the random high angle boundaries at **A**.   This should be compared with the angles at the triple point **B** involving only high energy boundaries.   For similar reasons, low angle grain boundaries,

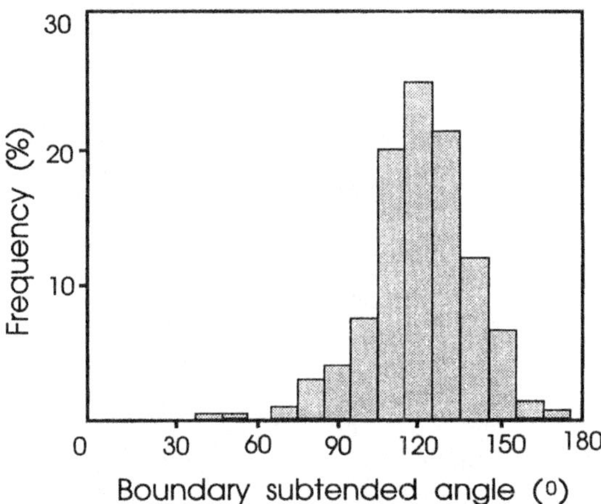

Fig. 3.16.   The frequency of grain boundary angles for high energy boundaries in
α-brass,   (after Smith 1948).

whose energies are strongly dependent on misorientation (fig 3.5) are rarely arranged at 120° to each other as seen in the recovered region of figure 6.29.

The instability resulting from the interaction between the space-filling requirements and the boundary tensions provides the driving pressure for the growth of subgrains and grains which are discussed in later chapters.

3.6 THE INTERACTION OF SECOND-PHASE PARTICLES WITH BOUNDARIES

A dispersion of particles will exert a retarding force on a low angle or high angle grain boundary and this may have a profound effect on the processes of recovery, recrystallization and grain growth. The effect is known as **Zener drag** after the original analysis by Zener which was published by Smith (1948).

3.6.1 The drag force exerted by a single particle

3.6.1.1 General considerations

Let us consider first, the interaction of a boundary of specific energy γ with a spherical particle of radius r which has an incoherent interface.

If the boundary meets the particle at an angle β as shown in figure 3.17 then the restraining force on the boundary is:

$$F = 2\pi r\gamma . \cos\beta . \sin\beta \qquad (3.9)$$

The maximum restraining effect is obtained when $\beta = 45°$, giving

$$F_s = \pi r \gamma \qquad (3.10)$$

As discussed by Nes et al. (1985), there have been many different derivations of this force, but the result is usually similar to the above. It should be noted that when a boundary intersects a particle, the particle effectively removes a region of boundary equal to the intersection area and thus the energy of the system is lowered. Boundaries are therefore attracted to particles.

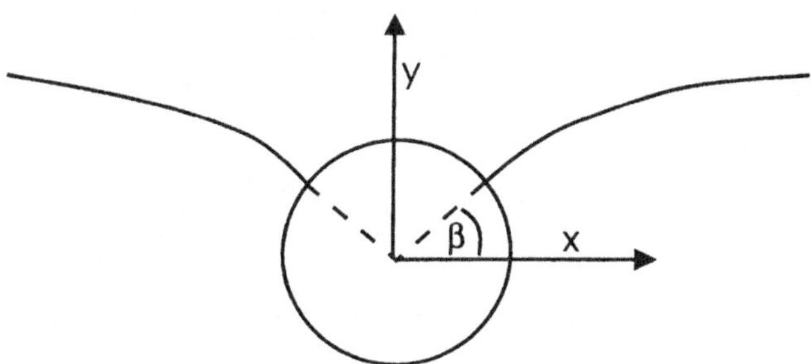

Fig. 3.17. The interaction between a grain boundary and a spherical particle.

3.6.1.2 The effect of particle shape

The particle shape, if not spherical, will have some effect on the pinning force (Ryum et al. 1983, Nes et al. 1985). These authors considered the interaction of a boundary with an ellipsoidal particle as shown in figure 3.18. Their treatment assumes that the grain boundary meets the particle at an angle of 90° and that the particle makes a planar hole in the boundary. They calculated the drag force for the two extreme cases shown in figure 3.18, and showed that the maximum forces were

Case 1

$$F_1 = F_s \left(\frac{2}{(1+e)\, e^{1/3}} \right) \qquad (3.11)$$

Case 2   for $e \geq 1$

$$F_2 = F_s \left(\frac{1 + 2.14\, e}{\pi\, e^{1/2}} \right) \qquad (3.12a)$$

Case 2   for $e \leq 1$

$$F_2 = F_s\, e^{0.47} \qquad (3.12b)$$

where e is the eccentricity of the ellipsoidal particle. When $e = 1$ then the particle is a sphere, and F_s is the drag from a spherical particle of the same volume. Figure 3.19 shows how F_1 and F_2 vary with aspect ratio for a particle of constant volume. It can be seen that the pinning force is only significantly larger than that of a spherical particle for the case of thin plates meeting the boundary flat on and long needles meeting the boundary edge on.

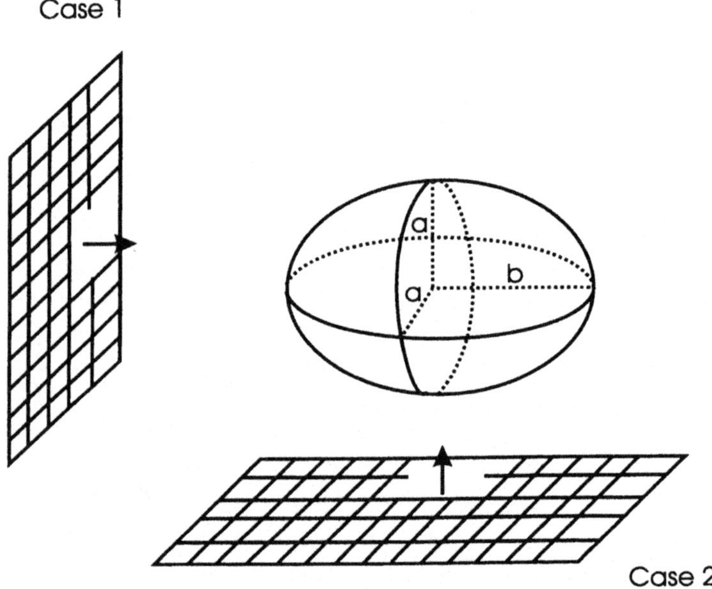

Fig. 3.18. The interaction of boundaries with an ellipsoidal particle, (after Nes et al. 1985).

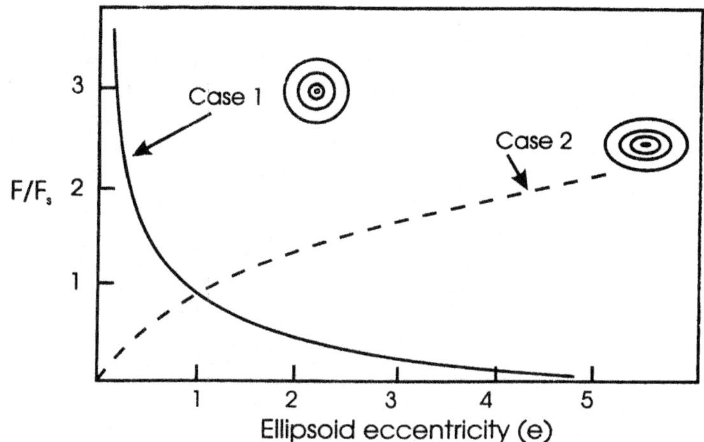

Fig. 3.19. The drag force as a function of particle aspect ratio for the two cases illustrated in figure 3.18, (after Nes et al. 1985).

Ringer et al. (1989) have analyzed the interaction of a boundary with cubic particles. The strength of the interaction depends upon the orientation of the cube relative to the boundary and in the extreme case, when the cube side is parallel to the boundary, the drag force is almost twice that of a sphere of the same volume. However, as this is a special case, this is unlikely to be a very significant factor in practice.

3.6.1.3 Coherent particles

If a high angle grain boundary moves past a coherent particle then the particle will generally lose coherence during the passage of the boundary. As the energy of the incoherent interface is greater than that of the original coherent interface, energy is required to cause this transformation, and this energy must be supplied by the moving boundary. Therefore, as first shown by Ashby et al. (1969), coherent particles will be more effective in pinning boundaries than will incoherent particles.

Following the analysis by Nes et al. (1985), if the grains are denoted 1 and 2 and the particle as 3, as in figure 3.20a, then there are now three different boundary energies γ_{12}, γ_{13} and γ_{23}. These boundaries meet at the particle surface, and if equilibrium is established then

$$\gamma_{23} = \gamma_{13} + \gamma_{12} \cos\alpha \qquad (3.13)$$

and

$$\cos\alpha = \frac{\gamma_{23} - \gamma_{13}}{\gamma_{12}} \qquad (3.14)$$

The drag force is then

$$F_C = 2\pi r \gamma \cos(\alpha - \theta).\cos\theta \qquad (3.15)$$

F_C is a maximum when $\theta = \alpha/2$ and $\alpha = 0$, giving

$$F_C = 2\pi r \gamma \tag{3.16}$$

Thus coherent particles are twice as effective in pinning a grain boundary as incoherent particles of the same size.

The passage of a high angle boundary past a coherent particle, results in the particle losing coherency. As a small incoherent particle will be less stable (Gibbs-Thomson effect), there are a number of alternative interactions. For example the particle may dissolve during

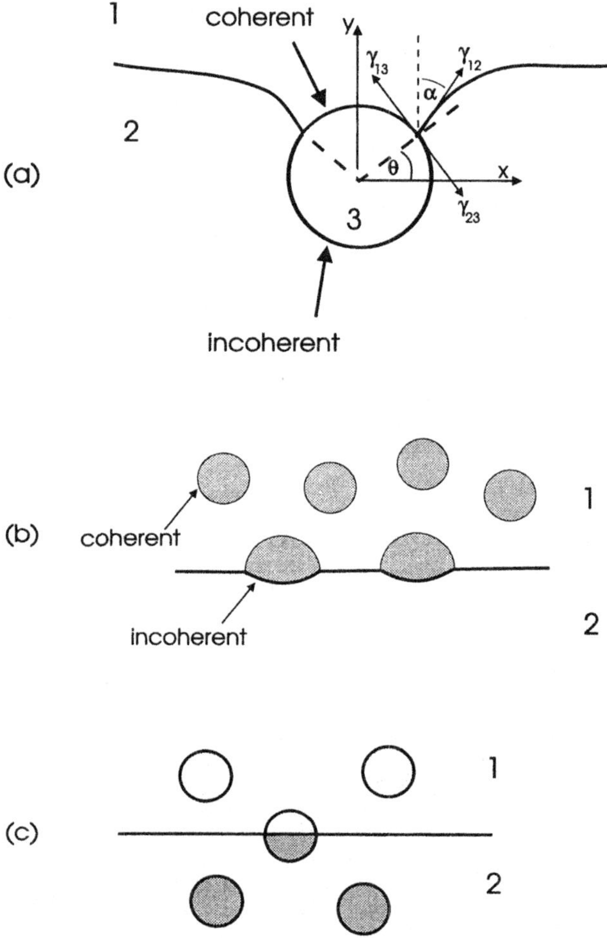

Fig. 3.20. The interaction between a coherent particle and a high angle grain boundary. a) The boundary by-passes the particle. b) The boundary halts at the particles. c) The boundary cuts through the particle.

passage of the boundary and re-precipitate in a coherent orientation, it may reorient itself to a coherent orientation, or the boundary may cut through the particle (Doherty 1982).

Dissolution - There is experimental evidence that small coherent particles can be dissolved by a moving boundary. The problem has been discussed by Doherty (1982) and by Nes et al. (1985). The latter authors show that in these circumstances the pinning force due to a coherent particle is dependent not only on the particle size, but also on the concentration of the alloy. The pinning force is given by

$$F = \frac{2\pi}{3} \cdot \frac{Ak}{V} \cdot Tr^2 \cdot \ln\left(\frac{C_0}{C_{eq}}\right) - 2\pi\gamma' r \tag{3.17}$$

where A = Avogadro's number, k = Boltzmann's constant, V is the molar volume of the precipitate phase, C_0 is solute concentration, C_{eq} is the equilibrium concentration and γ' is the energy of the coherent interface.

Following dissolution, the particles may re-precipitate coherently behind the grain boundary or alternatively discontinuous precipitation may occur at the boundary.

Coherent particles on a stationary boundary - If there is insufficient driving force for a boundary to move past the coherent particles or to dissolve them then the particles will become incoherent along the boundary. As a consequence of this, the equilibrium shape of the particles on the boundary will alter as shown in figure 3.20b. The increase in radius of curvature of the particles on the boundary will then cause them to coarsen at the expense of the smaller spherical coherent particles, thus increasing the pinning of the boundary (Howell and Bee 1980).

Passage of boundary through coherent particles - In some circumstances the precipitate may be cut by the boundary and undergo the same orientation change as the grain surrounding it, as shown in figure 3.20c. This is a rare occurrence, but it has been observed in nickel alloys containing the coherent γ' phase (Porter and Ralph 1981, Randle and Ralph 1986).

3.6.2 The drag pressure due to a distribution of particles

Having considered the pinning force from a single particle, we now need to calculate the restraining pressure on the boundary due to an array of particles. This is a complicated problem which has not yet been completely solved.

3.6.2.1 Drag from a random distribution of particles

For a volume fraction F_V of randomly distributed spherical particles of radius r, the number of particles per unit volume (N_V) is given by

$$N_V = \frac{3F_V}{4\pi r^3} \tag{3.18}$$

If the boundary is planar, then particles within a distance r on either side of the boundary will intersect it. Therefore the number of particles intersecting unit area of the boundary is

$$N_s = 2r\,N_v = \frac{3\,F_v}{2\,\pi\,r^2} \qquad (3.19)$$

The pinning pressure exerted by the particles on unit area of the boundary is given by

$$P_z = N_s\,F_s \qquad (3.20)$$

and hence from equations 3.10 and 3.19

$$P_z = \frac{3\,F_v\,\gamma}{2\,r} \qquad (3.21)$$

This type of relationship was first proposed by Zener (Smith 1948), although in the original paper because N_s was taken as $N_v.r$, the pinning pressure was half that of equation 3.21. P_z as given by equation 3.21 is commonly known as the **Zener pinning pressure**.

It is clear that this calculation is not rigorous, because if the boundary is rigid as is assumed, then as many particles will be pushing the boundary one way as will be pulling it the other as shown in figure 3.21a, and the net pinning pressure will be nil. (This is a similar problem to that encountered in calculating the interaction of a dislocation with an array of solute atoms). Therefore the boundary must relax locally from a planar configuration, as shown in figure 3.21b, if pinning is to occur.

More rigorous calculations of the Zener drag have been attempted by many authors, and the reader is referred to the reviews by Nes et al. (1985), Hillert (1988) and Doherty et al. (1989) for further details. However, it is concluded that the more sophisticated calculations do not lead to relationships which differ significantly from equation 3.21, which remains widely used.

3.6.2.2 Effects of boundary-particle correlation
The assumption of a planar or near planar boundary as required to give the Zener pinning pressure of equation 3.21 will only be reasonable if the grain or subgrain size is very much larger than the interparticle spacing as pointed out by Hellman and Hillert (1975), Anand and

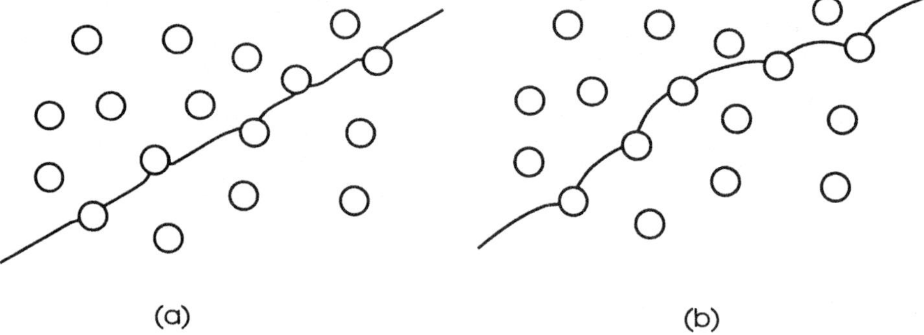

Fig. 3.21. Interaction of particles with a) a rigid planar boundary b) a flexible boundary.

Gurland (1975), Hutchinson and Duggan (1978), Hillert (1988), and Hunderi and Ryum (1992a). If this is not the case, then we need to examine, albeit in a simplified way, the consequences. In figure 3.22 we identify four important cases. In fig 3.22a the grains are much smaller than the particle spacing, in figure 3.22b the particle spacing and grain size are similar, in figure 3.22c the grain size is much larger than the particle spacing and in figure 3.22d, the particles are inhomogeneously distributed so as to lie only on the boundaries. In all these cases there is a strong correlation between the particles and the boundaries, although an accurate assessment of the pinning force is difficult and is dependent on the details of the particle and grain (or subgrain) arrangement.

Consider the particles and boundaries which form the three-dimensional cubic arrays shown in figure 3.22. For case **a**, it is reasonable to assume that all the particles lie not only on boundaries, but at vertices in the grain structure, because in these positions the particles, by removing the maximum boundary area, minimise the energy of the system.

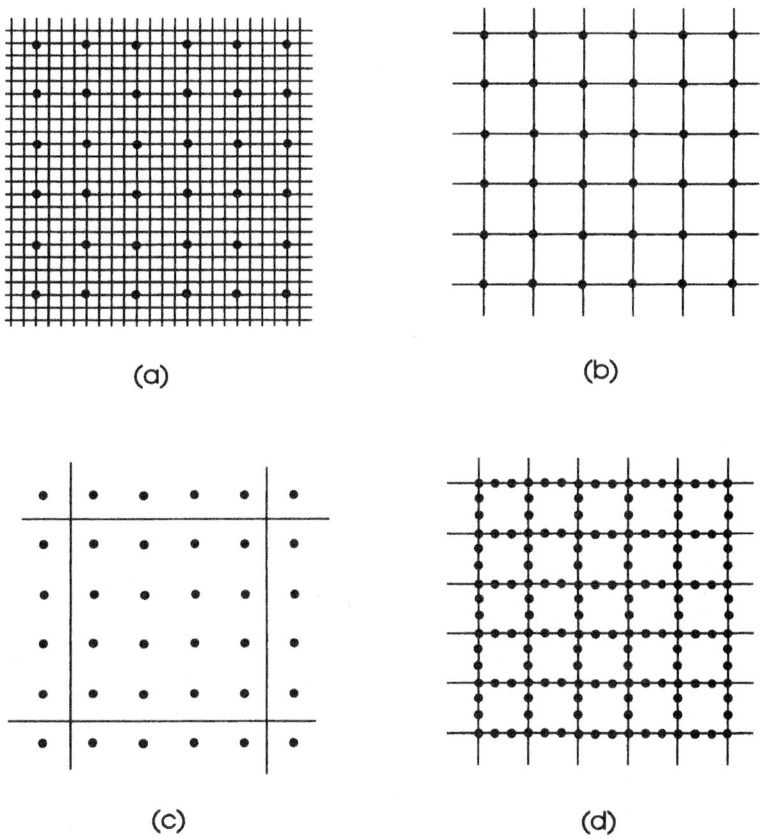

(a)   (b)
(c)   (d)

Fig. 3.22. Schematic diagram of the correlation between particles and boundaries as a function of grain size.

If the grain edge length is **D**, then the grain boundary area per unit volume is 3/D. The number of particles per unit area of boundary (N_A) is given by

$$N_A = \frac{\delta\, N_V\, D}{3} \qquad (3.22)$$

where δ is a factor which depends on the positions of the particles in the boundaries. For particles on boundary faces, $\delta = 1$. For particles at the vertices as shown in figures 3.21 a) and b), $\delta = 3$ and hence, using equation 3.18

$$N_A = \frac{\delta\, N_V\, D}{3} = \frac{3\, D\, F_V}{4\,\pi\, r^3} \qquad (3.23)$$

For incoherent spherical particles the pinning pressure on the boundary (P'_z) will then be given by

$$P'_z = F_S\, N_A = \frac{3\, D\, F_V\, \gamma}{4 r^2} \qquad (3.24)$$

This relationship will be valid for grain sizes up to and including that shown in figure 3.22b, in which the grain size and particle spacing are the same (D_C). In this situation, the spacing of the particles on the cubic lattice (L) is equal to D_C and is given by

$$L = D_C = N_V^{-1/3} = \left(\frac{4\,\pi\, r^3}{3\, F_V} \right)^{1/3} \qquad (3.25)$$

at which point the pinning force reaches a maximum $P'_{z\max}$ given by

$$P'_{z\max} \approx \frac{1.2\,\gamma\, F_V^{2/3}}{r} \qquad (3.26)$$

As the grain size increases beyond D_C, the number of particles per unit area of boundary (N_A) will decrease from that given by equation 3.23 and eventually reach N_S as given by equation 3.19. We can treat this transition from correlated to non-correlated boundary in the following approximate manner.

The number of boundary corners per unit volume in a material of grain size **D** is given approximately by $1/D^3$, and the fraction of particles lying on these potent pinning sites is therefore for the condition $D \geq D_C$.

$$X = \frac{1}{N_V\, D^3} \qquad (3.27)$$

When $D = D_C$ then from equation 3.25 we find $X = 1$. Thus as the grains grow, a diminishing fraction of particles is able to sit at the grain corners.

The number of corner-sited particles per unit area of boundary is then

$$N_C = X\, N_V\, D \qquad (3.28)$$

The remaining particles will sit on grain boundaries or grain edges or lie within the grains. In this simple analysis we assume that the particles which are not sitting at grain corners are intersected at random by boundaries and therefore the number per unit area is

$$N_r = 2 r N_V (1 - X) \qquad (3.29)$$

Thus the total number of particles per unit area of boundary is $N_c + N_r$ and, using equation 3.20, the pinning pressure is given by

$$P''_z = \pi r \gamma [X N_V D + (1 - X) 2 N_V r] \qquad (3.30)$$

or, when $D \geq D_c$

$$P''_z = \pi r \gamma \left(\frac{1}{D_2} + (1 - \frac{1}{N_V D^3}) 2 N_V r \right) \qquad (3.31)$$

In figure 3.23, we show the pinning pressure on the boundary as a function of the grain size. It can be seen that as the grain size increases, the pinning pressure rises according to equation 3.24, peaks when the grain size approaches D_c and eventually falls (equation 3.30) to the Zener pinning pressure as given by equation 3.21.

It should be noted that actual values of the peak pinning pressure discussed here are based on a very simple geometry and are not accurate. However, the principle that the pinning pressure varies in a manner similar to that shown in figure 3.23 is undoubtedly correct and has some important implications for recovery, recrystallization and grain growth of particle-containing materials which will be discussed in later chapters.

3.6.2.3 Drag from non-random particle distributions
In many alloys the particles may not be randomly distributed. Of particular importance is the situation in which the particles are concentrated in planar bands as shown in figure 3.24. This type of structure is often found in rolled products such as Al-killed steels, most

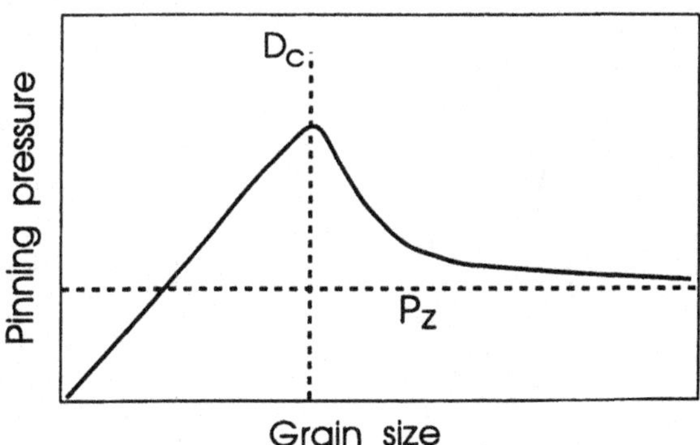

Fig. 3.23. The effect of grain size on the Zener pinning force for a given particle dispersion.

commercial aluminium alloys, and products made by powder metallurgy. In many cases the individual particles are also rod or plate-like and aligned parallel to the rolling plane. Nes et al. (1985) have modelled the case shown in figure 3.24a and have calculated the grain aspect ratio following primary recrystallisation.

Let us assume that spherical particles of radius r are aggregated in bands of thickness t, which are spaced a distance of L apart. A boundary AB lying parallel to the planar bands experiences a drag

$$P_{PZ} = \frac{P_Z L}{t} \tag{3.32}$$

where P_Z is the drag pressure if the particles are uniformly distributed as given by equation 3.20.

The drag on a boundary lying perpendicular to the bands depends strongly on the shape of the boundary. For example the curved boundary segment BC in figure 3.24a does not experience any pinning. However, if the boundary is long, e.g. segment DE, and if the driving pressure is uniformly distributed the average pinning pressure is P_Z. In either case, the pinning force is significantly greater in the direction normal to the bands and this anisotropy will be reflected in the shape of the grains or subgrains. Nes et al. (1985) have shown that the boundary is expected to propagate in a staggered manner as shown in figure 3.24b. Figure 3.25a is a transmission micrograph of a rolled powder metallurgy aluminium sheet product showing how bands of small oxide particles interact with the **low angle boundaries** and affect the shape of the subgrains during recovery, and figure 3.25b is an optical micrograph of the same material after recrystallization, which clearly shows how interaction of the aligned oxide particles with **high angle boundaries** results in an elongated grain structure after recrystallization.

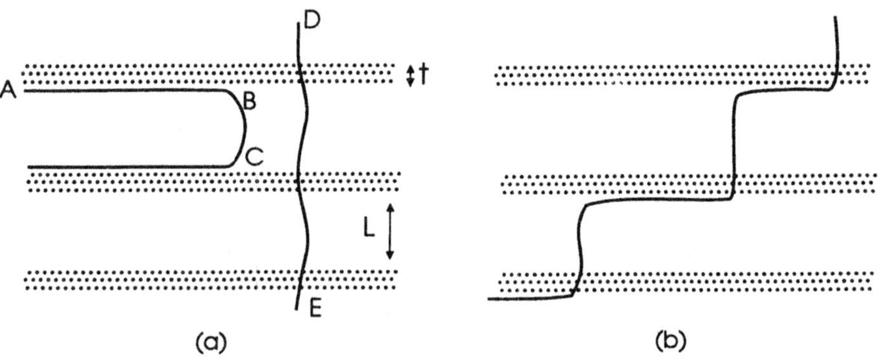

Fig. 3.24. The interaction of boundaries with planar arrays of particles. a) The effect of boundary orientation on pinning. b) Propagation of a boundary through planar arrays of particles, (after Nes et al. 1985).

(a)

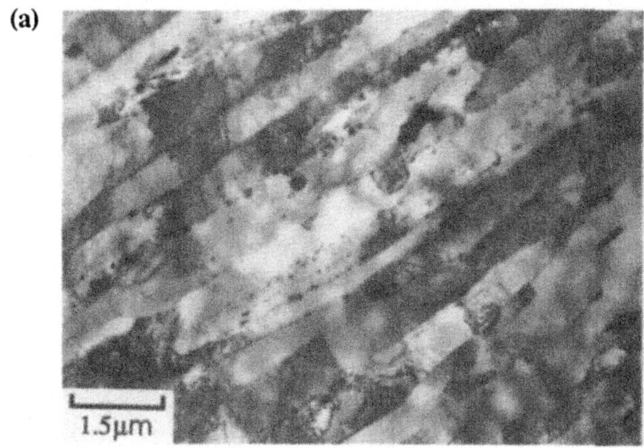

(b)

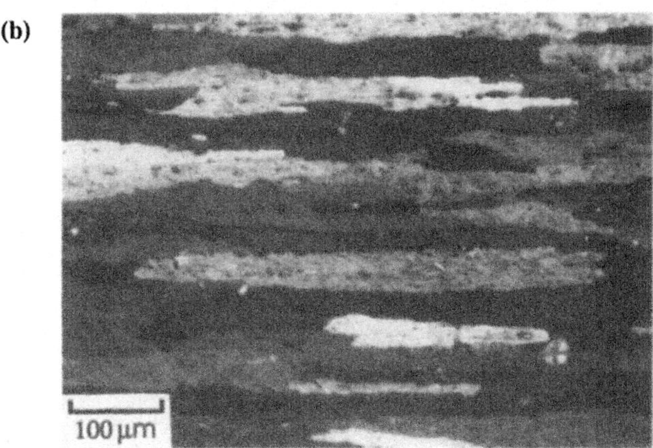

Fig. 3.25. a) Transmission electron micrograph illustrating the pinning of low angle boundaries by planar arrays of small oxide particles in a rolled aluminium billet produced by powder metallurgy, (TD plane sections). b) Optical micrograph of the same material after recrystallization showing an elongated grain structure, (Bowen et al. 1993).

Another important situation in which non-random particle distribution may occur is when the second-phase particles are precipitated onto pre-existing low or high angle boundaries (e.g. Hutchinson and Duggan 1978). In this case, as shown schematically in figure 3.22d, all (or most) of the particles lie on the boundaries. The effective number of particles per unit area of boundary is given approximately by equation 3.24 with δ in equation 3.23 equal to 1, and the pinning force (P_x) by

$$P_X = F_S N_A = \frac{\gamma\, D\, F_V}{4\, r^2} \qquad (3.33)$$

Chapter 4

THE MOBILITY AND MIGRATION OF BOUNDARIES

4.1 INTRODUCTION

The structure and energy of grain boundaries were discussed in chapter 3. In this chapter we examine the models and mechanisms of boundary migration and in particular the mobilities of boundaries. This provides a foundation for the discussions of recovery, recrystallization and grain growth in later chapters.

4.1.1 The role of grain boundary migration during annealing

The migration of low angle and high angle grain boundaries plays a central role in the annealing of cold worked metals. Low angle boundary migration occurs during recovery and during the nucleation of recrystallization, and high angle boundary migration occurs both during and after primary recrystallization.

Despite the importance of grain boundary migration during annealing, the details of the process are not well understood. This is mainly because boundary migration involves atomistic processes occurring rapidly, at high temperatures, and under conditions which are far from equilibrium. It is therefore very difficult to study boundary migration experimentally or to deal with it theoretically. Although there is a wealth of experimental evidence accumulated over many years, in many cases results are contradictory and no clear pattern emerges. However, it is clear that even very small amounts of solute have an extremely large influence on the boundary mobility and it is therefore very difficult to determine the **intrinsic mobility** of a boundary. Although a great deal is now known about the structure of grain boundaries (chapter 3), and models for grain boundary migration based on boundary structure have been formulated, there is as yet little evidence of quantitative agreement between the models and the experimental results.

4.1.2 The micromechanisms of grain boundary migration

Low angle and high angle grain boundaries migrate by means of atomistic processes which occur in the vicinity of the boundary. In this section we briefly outline the various mechanisms which have been proposed, in order to prepare for the more detailed considerations which appear later in this chapter.

The mechanism of boundary migration depends on several parameters including the **boundary structure** which, in a given material, is a function of **misorientation** and **boundary plane**. It also depends on the **experimental conditions**, in particular the **temperature** and the **forces on the boundary**, and it is also strongly influenced by **point defects** in the material such as **solutes** and **vacancies**.

It is well established that low angle grain boundaries migrate by the movement by **climb and glide of the dislocations** which comprise the boundary, and many aspects of LAGB migration may therefore be interpreted in terms of the theory of dislocations.

The basic process during the migration of high angle boundaries is the transfer of atoms to and from the grains which are adjacent to the boundary, and models based on such **thermally activated atomic jumps** constitute the earliest models of boundary migration. As discussed in §3.4, the structure of a grain boundary is related to the crystallography of the boundary, and there is evidence that boundaries may migrate by movement of intrinsic boundary defects such as **ledges or steps** or **grain boundary dislocations**. In some cases boundary migration may involve **diffusionless shuffles**, which are processes involving the cooperative movement of groups of atoms (cf. diffusionless phase transformations). In certain circumstances **combined sliding and migration** of boundaries may also occur.

Because the atomic packing at grain boundaries is less dense than in the perfect crystal, grain boundaries are associated with a **free or excess volume** which depends on the crystallography of the boundary (§3.4.3). This leads to a strong interaction with solute atoms and to the formation of a **solute atmosphere** which, at low boundary velocities, moves with the boundary and impedes its migration. The range of boundary velocities is very large. For example, in the case of **grain growth** the boundary may move relatively slowly through almost perfect crystal. However, during **recrystallization** the boundary moves at a very high velocity (often much higher than during diffusion-controlled phase transformations). In this case, the boundary migrates in highly defective material, leaving perfect crystal behind it. The mechanism by which the boundary removes the defects during recrystallization is not known, but it is reasonable to suppose that the boundary energies and structures may differ in the two cases.

4.1.3 The concept of grain boundary mobility

A grain boundary moves with a velocity (v) in response to the net pressure ($P = \Sigma P_i$) on the boundary (§1.3). It is generally assumed that the velocity is directly proportional to the pressure, the constant of proportionality being the **mobility (M)** of the boundary, and thus

$$v = M P \tag{4.1}$$

This type of relationship is predicted by reaction rate theory (§4.4.1) if the mobility is independent of the driving force and if $P \ll kT$, and should be independent of the details of the mechanism of boundary migration.

There are however proposed boundary migration mechanisms such as step or boundary dislocation controlled migration (Gleiter 1969b, Smith et al. 1980) where the mobility depends on the driving force, and under certain conditions it is then predicted that this relationship may not hold. If the boundary mobility is controlled by solute atoms then, as discussed in §4.4.2, the **retarding pressure due to solute drag (P_{sol})** may be a function of the boundary velocity. Although a complex relationship between **driving pressure (P_d)** and velocity may be found (e.g. fig 4.27), equation 4.1 may still be obeyed if **P** in equation 4.1 is replaced by **(P_d - P_{sol})**.

Surprisingly, there are few experimental measurements of the relationship between velocity and driving pressure. Results consistent with equation 4.1 have been obtained by Viswanathan and Bauer (1973) for copper and Demianczuc and Aust (1975) for aluminium. However, Rath and Hu (1972) and Hu (1974) fitted their data for zone refined aluminium to a power law relationship $v \propto P^n$ with $n \gg 1$. Gottstein and Shvindlerman (1992) argued that the Rath and Hu experiments were strongly influenced by thermal grooving and by re-analyzing their data and taking the effect of thermal grooving into account, showed that they were consistent with equation 4.1. A recent re-examination of the early experimental results by Vandermeer and Hu (1994) has also concluded that they can be reconciled with equation 4.1.

4.1.4 Measuring grain boundary mobilities

Our current knowledge of grain boundary mobility is very much limited by the lack of good experimental data. In addition to the problems arising from the effects of trace amounts of impurities on mobility (§4.3.3), the actual measurements are extremely difficult to make.

Measurements of boundary mobility are made by determining the velocity of a boundary in response to a well defined driving force. Experiments in which the velocity of a boundary is measured during recrystallization are very difficult to interpret. This is because the driving pressure which arises from the stored energy, and is typically 10-100MPa, is not easy to measure accurately, varies throughout the microstructure and does not remain constant with time, decreasing as recovery proceeds. For this reason, most measurements of boundary mobility have been carried out on materials with a better characterised driving force, such as an as-cast substructure or by using bicrystals of carefully controlled geometry (e.g. Masteller and Bauer 1978). In such experiments however the driving pressure is much lower (around 10^{-2}MPa), and it is not clear to what extent measurements of boundary mobility in materials with such low driving forces are directly applicable to migrating boundaries in recrystallizing material. In addition, there is evidence that moving boundaries interact with dislocations, and that defects affect boundary mobility (§4.3.4), indicating that there may be a real physical difference between boundary migration in deformed and undeformed material.

It should be noted that many of the measurements of boundary migration are conducted under conditions where the driving force is constant and the results are expressed as **boundary velocities** rather than mobilities (e.g. fig 4.7). As equation 4.1 is expected to be obeyed, the mobility will be directly proportional to the velocity, and it is therefore reasonable to discuss such results in terms of boundary mobility even though this parameter has not actually been measured.

4.2. THE MOBILITY OF LOW ANGLE BOUNDARIES

4.2.1 Low angle symmetrical tilt boundaries

It has been found that a symmetrical tilt boundary, such as shown in figure 3.3 can move readily under the influence of stress (Washburn and Parker 1952, Li et al. 1953), and it is presumed that the boundary migrates by glide of the edge dislocations which comprise the boundary. The mobility of such boundaries has been shown to depend on both temperature and boundary misorientation, and the effect of misorientation on the migration under constant stress of symmetrical tilt boundaries in zinc is shown in figure 4.1. It may be seen that as the misorientation increases and the dislocations in the boundary become closer together (equation 3.4), the mobility decreases.

If a boundary moves by dislocation glide under the influence of an applied stress, this implies that movement of a boundary by dislocation glide must inevitably lead to a shape change. In unconstrained single crystals such as were used for these experiments, this may have little effect, but in a **grain constrained within a polycrystal** this will lead to the generation of internal stresses which will oppose further boundary migration.

The symmetrical low angle tilt boundary is of course untypical of the more general low angle boundaries found in a deformed and recovered metal (§5.4.2), and therefore **the behaviour of symmetrical tilt boundaries is probably not of great relevance to annealing processes in polycrystals.**

4.2.2 General low angle boundaries

4.2.2.1 Measurements of the mobility of low angle boundaries
It has long been recognised that the mobilities of general low angle boundaries are significantly lower than those of high angle boundaries. However, there are few systematic

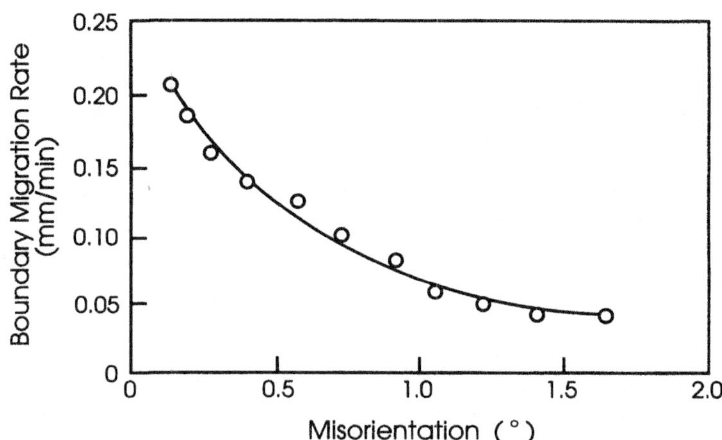

Fig. 4.1. The effect of misorientation on the mobility of symmetrical tilt boundaries in zinc at 350°C under a constant stress, (after Bainbridge et al. 1954).

investigations of the mobility of such boundaries and of the effect of misorientation. This is a major problem, because, as discussed in §5.5.3, a knowledge of the mobility of general low angle boundaries is a critical factor in developing an understanding of the annealing of deformed metals.

Early work which has shown that low and medium angle boundaries have a low mobility includes that of Tiedema et al. (1949) who found that recrystallized grains oriented within a few degrees of the matrix grew very slowly and were left as island grains and Graham and Cahn (1956) on aluminium, Rutter and Aust (1960) on lead, and Walter and Dunn (1960b) on iron. It should be emphasised that the references cited provide only a few largely qualitative observations and do not include a systematic investigation of the effect of misorientation on mobility.

Sun and Bauer (1970) and Viswanathan and Bauer (1973) used a bicrystal technique, in which the boundary energy provided the driving force for migration, to measure the mobility of high and low angle boundaries in NaCl and copper. Their results were reported as a mobility parameter $\mathbf{K'}$ which is the product of the mobility $\mathbf{M}$ and the boundary energy γ.

$$K' = 2\,\gamma\,M \tag{4.2}$$

The results for sodium chloride are shown in figure 4.2 and for copper in figure 4.3. Both figures clearly show a large difference in mobility between low and high angle boundaries. From figure 4.3 it may be seen that K' for the high angle boundaries in copper is some three orders of magnitude greater than that for low angle (2-5°) boundaries. As the boundary energy for 20° tilt boundaries in copper is only around 4 times that of 2° boundaries (Gjostein and Rhines 1959), the difference in mobility between high and low angle boundaries is therefore a factor of ~ 250. It is also seen that the activation energies for migration of the low angle boundaries are in both materials very much higher than that of the high angle boundaries. In the case of copper, the value of ~ 200kJ/mol for the low angle boundaries is close to that for self diffusion (table 4.1) and for NaCl, the value of 205kJ/mol is close to that for the diffusion of Na^+ in NaCl.

Measurements on zinc and aluminium bicrystals by Fridman et al. (1975) have confirmed that the mobility of medium angle boundaries is low and increases with increasing misorientation as shown in figure 4.9. Figure 4.9c also confirms that the activation energy for LAGB migration is higher than for HAGB migration, although the reported activation energies for migration of LAGBs appear to be considerably higher than those for self diffusion.

Although it is difficult to obtain information about the orientation dependence of LAGB mobility from measurements of subgrain coarsening, it is possible to estimate mobilities under recovery conditions from measurements of isothermal subgrain coarsening kinetics. If the subgrain size is $\mathbf{2R}$ then the boundary velocity $\mathbf{v} = \mathbf{dR/dt}$ may be assumed to be related to the driving pressure $(\mathbf{P})$ and the mobility $(\mathbf{M})$ by equation 4.1, with the driving pressure $(\mathbf{P})$ given by equation 5.26. If the average subgrain misorientation is determined at the same time as the subgrain size is measured, for example by using the EBSP method discussed in the Appendix, then γ_S may be calculated from equation 3.5. Measurements of high purity aluminium (99.995%) annealed at temperatures between 250° and 350°C (Humphreys and Humphreys 1994) showed the activation energy for LAGB migration to be ~ 130 kJ/mol, which is similar to that for self diffusion in aluminium. This approach may

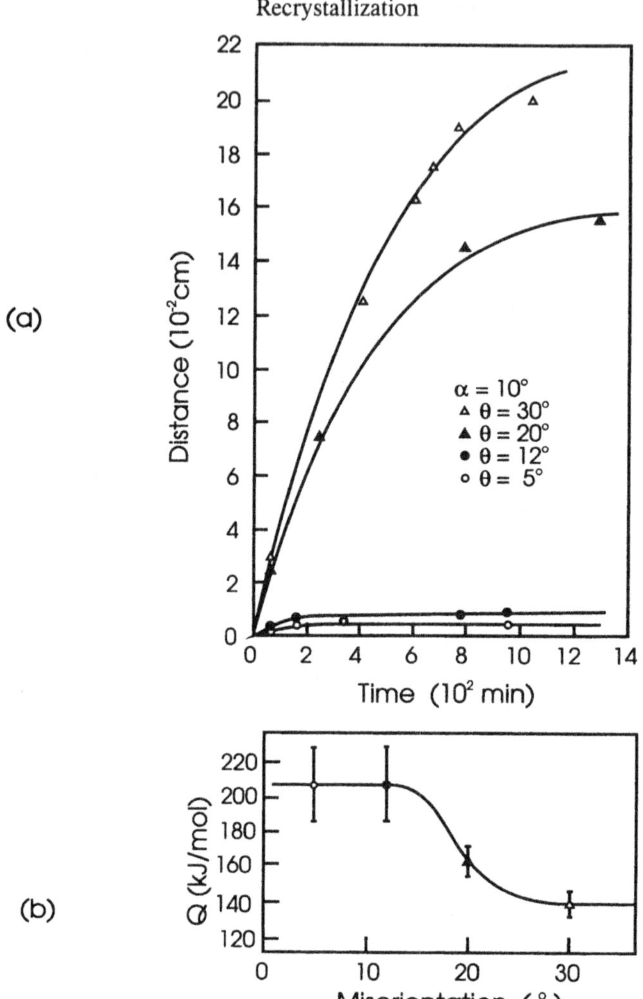

Fig. 4.2. a) Displacement of a curved boundary with time at 750°C in a NaCl bicrystal as a function of misorientation about a <100> axis. b) Activation energy for boundary migration, (after Sun and Bauer 1970).

be extended to compare the mobilities of high and low angle boundaries if the specimen recrystallizes during the anneal. If the subgrain boundary velocity is measured as the specimen begins to recrystallize, then the driving forces for both subgrain coarsening and recrystallization are similar, and the ratio of the HAGB and LAGB velocities therefore gives the relative mobilities. Data for 99.995% aluminium (Humphreys and Humphreys 1994) and commercial purity aluminium (Furu 1992) gave the ratio M_{HAGB}/M_{LAGB} as ~19 and ~25 respectively at 300°C, which is much lower than the ratio of ~250 found for copper (fig 4.3).

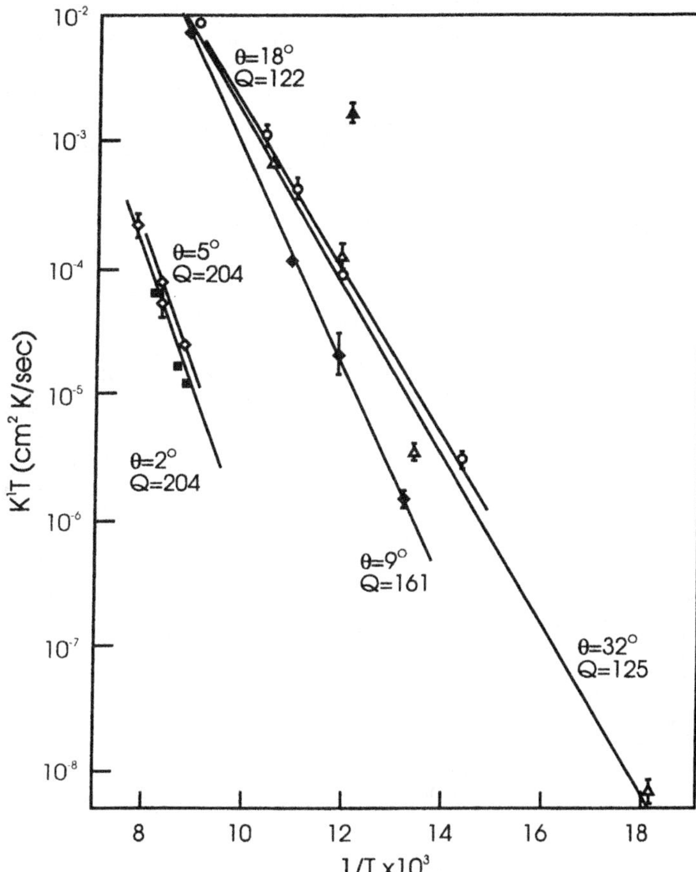

Fig. 4.3. Variation of boundary mobility parameter K′ with temperature and misorientation for 99.999% copper. The activation energies (Q in kJ/mol) for each group of boundaries are also shown, (after Viswanathan & Bauer 1973).

It should be emphasised, as indeed it was in 1963 by Aust and Rutter, that there is an urgent need for more systematic experimental studies of the mobilities of low angle boundaries.

4.2.2.2 Theories of the mobility of low angle boundaries

As the energy and structure of low angle boundaries are sensitive to misorientation, the intrinsic boundary mobility is also expected to depend on these parameters. The migration of low angle boundaries in general requires climb of dislocations in the boundary and this is expected to be the rate determining process. The theory of the mobility of low angle grain boundaries has been discussed by Sandström (1977b) and Ørsund and Nes (1989), and these

TABLE 4.1

Activation energies for self, core and boundary diffusion for a selection of materials

Material	T_m (K)	G (GPa)	b (nm)	Q_S (kJ/mol)	Q_C (kJ/mol)	Q_B (kJ/mol)	γ_b (Jm^{-2})
Al	933	25.4	0.286	142	82	84	0.324a
Au	1336	27.6	0.288	164			0.38a
Cu	1356	42.1	0.256	197	117	104	0.625a
α-Fe	(1810)	69.2	0.248	239	174	174	0.79a
γ-Fe	(1810)	81	0.258	270	159	159	0.76a
NaCl	1070	15	0.399	217 (Na$^+$)	155	155	0.5c
Pb	601	7.3	0.349	109	66	66	0.2b
Sn	505	17.2	0.302	140b		37b	0.16b
Zn	693	49.3	0.267	91.7	60.5	60.5	0.34a

Data are from Frost and Ashby (1982) unless otherwise indicated.
a) Murr (1975)
b) Friedel (1964)
c) Sun and Bauer (1970)

authors suggest that the mobility of low angle boundaries is given by

$$M = \frac{C\,D_s\,b}{k\,T} \tag{4.3}$$

where C is a constant of the order of unity and D_s is the coefficient of self diffusion.

Ørsund and Nes also consider the possibility that at low temperatures boundary migration may be controlled by dislocation pipe or core diffusion.

An important factor discussed by Hirth and Lothe (1968), and not considered by the above investigators, is that there is no net flow of matter as the boundary moves and therefore the vacancy source-to-sink spacing is approximately equal to the dislocation spacing (**h**) in the boundary, which is related to the boundary misorientation (equation 3.4).

In view of the large orientation dependence of the boundary structure on misorientation (θ), the mobility of low angle boundaries is likely to show some dependence on θ, which is not found in equation 4.3, and the following approximate analysis, (which is based on discussions with E. Nes) is suggested.

Consider a boundary bowing to a radius of curvature R. The pressure (P) on the boundary due to the curvature is given by

$$P = \frac{2\gamma_s}{R} \tag{4.4}$$

where γ_s is the boundary energy as given by equation 3.5. For small values of θ we use the approximation $\gamma_s \sim c_1 Gb\theta$, where c_1 is a small constant.

If we assume that equation 4.1 applies then the boundary velocity is given by

$$v_b = \frac{dR}{dt} = \frac{2M\gamma_s}{R} = \frac{2c_1 MGb\theta}{R} \tag{4.5}$$

Case A - Very low angle boundaries ($\theta \to 0$)

We first consider the movement of boundaries in which the dislocations are spaced far apart. In this limiting situation we can assume that the behaviour of the individual dislocations in the boundary is dominant. An individual dislocation in the boundary is bowed to a radius of curvature R and the force on unit length of the dislocation is given approximately by

$$F = \frac{Gb^2}{2R} \tag{4.6}$$

For dislocation climb under conditions of negligible vacancy supersaturation, $Fb^2 \ll kT$ and the climb velocity (v_d) of a dislocation under the influence of a force F is given approximately by (Friedel 1964, Hirth and Lothe 1968)

$$v_d = D_s c_j \cdot \frac{Fb}{kT} \tag{4.7}$$

Where c_j is the concentration of jogs. Substituting for F as given by equation 4.6

$$v_d = \frac{D_s c_j G b^3}{2kT} \cdot \frac{1}{R} \tag{4.8}$$

As the dislocations are assumed to be essentially independent in such a low angle boundary, the velocity of all the dislocations in the boundary is given by equation 4.8 and therefore $v_d = v_b$. Thus from equations 4.5 and 4.8

$$v_b = \frac{2c_1 MGb\theta}{R} = \frac{D_s c_j G b^3}{2kT} \cdot \frac{1}{R} \tag{4.9}$$

and hence

$$M = \frac{D_s c_j b^2}{4c_1 kT\theta} \tag{4.10}$$

i.e. in this limiting case:
- **The dominant mechanism is climb of individual dislocations**
- **$v = M.P$ is expected to hold**
- **M is inversely proportional to θ**

Case B. Boundaries where the dislocation cores overlap ($\theta > 15°\text{-}20°$)

In this situation we have effectively reached the case of a high angle boundary, and for a general boundary we can reasonably assume (§4.4.1.1)

- **The dominant mechanism is atom jumps across the boundary**
- **v = M.P is expected to hold**
- **M is independent of θ**

Case C. The intermediate case

This is clearly the most difficult and probably the most important for the recovery of deformed polycrystals. In this regime the dominant mechanism is now not so clear. There are **two opposing factors**. One is the decrease in mobility with increasing θ due to the factors discussed in **case A**. Secondly, the small dislocation spacing and the fact that the vacancy source-to-sink distance is inversely proportional to θ must mean that climb becomes more rapid as θ increases.

C1 (medium to high angles)

When the boundary angle is large and **h** is small, the movement of the boundary is probably dominated by atom transfer over a distance **h,** and the process is analogous to the diffusion of atoms through a membrane of thickness **h**, which would result in M being inversely proportional to h, i.e M $\propto \theta$.

C2 (low to medium angles)

It is quite possible that in this range the mobility drops with increasing θ, reaching a minimum when the vacancy-source-sink factor begins to dominate. The details of this region are particularly difficult to analyze as it is not known which mechanism is rate controlling. The extent to which a theory based on the behaviour of individual dislocations can be applied is also unknown, because reactions at dislocation nodes may become a dominant factor (Hirth and Lothe 1968). In the absence of a specified mechanism, it is not known if the mobility is independent of driving force as is assumed by equation 4.1. On the basis of the above analysis, it is suggested that the variation of the mobility of a general low angle boundary with misorientation may be as shown in figure 4.4.

Comparison with experiment

Measurements of the activation energy for LAGB migration and recovery generally indicate control by self diffusion as predicted by the model.

The heavy lines in figure 4.4 indicate regions for which there is some experimental evidence as detailed below.

REGION B. There is reasonable evidence that the mobility of random HAGBs is high and independent of θ.

REGION C1. There is some evidence. e.g. figures 4.2, 4.3 and 4.9 that the mobility of boundaries in the range 10-20° increases with θ and saturates at about 20°.

REGION C2. There is presently little evidence of the dependence of velocity or mobility on θ at lower angles. In aluminium at around 300°C the mobility of LAGBs of misorientation

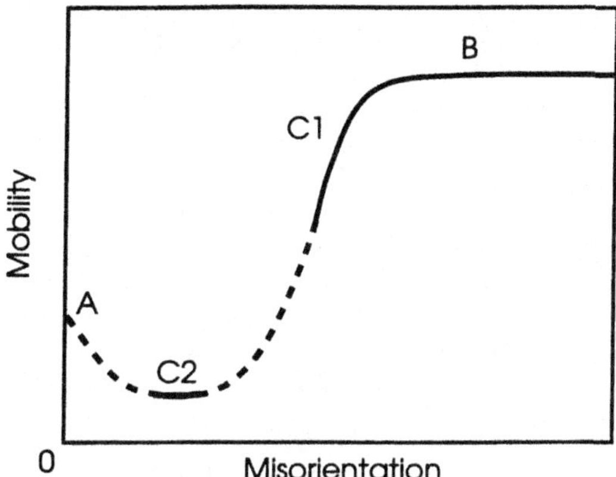

Fig. 4.4. Schematic diagram of the possible variation of the mobility of general low angle boundaries with misorientation. The letters correspond to cases discussed in the text and solid lines are regions for which there is some experimental evidence.

~5° as determined from subgrain growth/recrystallization experiments is a factor of 10-30 lower than that of HAGBs.

REGION A. There are no experimental data for this region.

In summary, it is quite clear that considerable experimental and theoretical work is needed to resolve the question of the mobilities of low angle grain boundaries. As LAGB migration is thought to play an important role in the nucleation of recrystallization (§4.4.3), our current lack of understanding in this area is a significant barrier to the development of detailed models of annealing.

4.3. MEASUREMENTS OF THE MOBILITY OF HIGH ANGLE BOUNDARIES

4.3.1 The effect of temperature on grain boundary mobility in high purity metals

4.3.1.1 Activation energy for boundary migration
The mobility of high angle grain boundaries is temperature dependent and is often found to obey an Arrhenius type relationship of the form

$$M = M_o \exp\left(-\frac{Q}{RT}\right) \tag{4.11}$$

The slope of a plot of **ln(M)** or **ln(v) (for constant P)**, against 1/T therefore yields a value of **Q**, the apparent activation energy which, as is further discussed below, may be related to the atom-scale thermally activated process which controls boundary migration. However, this is not always the case, and great care should be taken in the interpretation of experimentally measured activation energies.

Figure 4.3 is typical of data found for metals of very high purity, and table 4.2 shows some experimental data for the mobility of high angle boundaries, obtained by several techniques in a variety of high purity metals. Comparison of the activation energies for boundary migration in table 4.2 with the activation energies for diffusion given in table 4.1 shows that the activation energy for boundary migration in high purity metals is often close to that for grain boundary diffusion, i.e. about half that for bulk diffusion.

4.3.1.2 Transition temperatures

In impure metals or alloys, as will be discussed in §4.3.3, the variation of mobility with temperature is more complicated than that of figure 4.3, with more than one apparent activation energy being found over a range of temperatures. However, even in high purity metals a change in mobility and activation energy is sometimes found at very high homologous temperatures. There are two likely explanations for this:

(i) Changes in boundary structure, in which coincidence site or ordered boundaries may become more disordered and therefore lose their special properties, may occur at elevated temperatures.

There is some evidence from experiments on high purity lead (Rutter and Aust 1965) and copper (Aust et al. 1963, Ferran et al. 1967) that the higher mobility of special boundaries is lost at high temperatures, and this has been interpreted in terms of a change in grain boundary character at high temperatures (see also §3.4.3). Maksimova et al. (1988) have measured a discontinuous change in mobility with increasing temperature for $\Sigma = 17$, 28° $<001>$ boundaries in 99.9999% tin. Boundaries within $\sim 1°$ of this were found to show

TABLE 4.2

**The activation energy for migration of high angle
grain boundaries in metals of high purity
(Haessner and Hofmann 1978)**

Material	Activation energy for migration (kJ/mol)	Pre-exponential factor (m/s)	Temperature range (K)	Experiment
Aluminium	63	2×10^4	273-350	Recrystallization
Aluminium	67	4.8×10^3	615-705	Capillary
Copper	121	7.5×10^6	405-485	Recrystallization
Copper	123	2.4	700-975	Capillary
Gold	80	5.8×10^{-1}	593-613	Recrystallization
Lead	25	1.1×10^{-2}	473-593	Striated structure
Tin	25	4.5×10^{-1}	425-500	Striated structure
Zinc *	63		525-625	Capillary

Data from Kopetski et al. (1975)

a discontinuous **drop** in mobility with increasing temperature at temperatures between 0.94 and 0.98 T_m as shown in figure 4.5, whereas boundaries misoriented by more than this amount from $\Sigma = 17$ did not show a transition. Boundaries in the higher temperature regime had a **higher** activation energy. The authors claim that this is due to the transition from a special boundary to a general boundary at high temperatures.

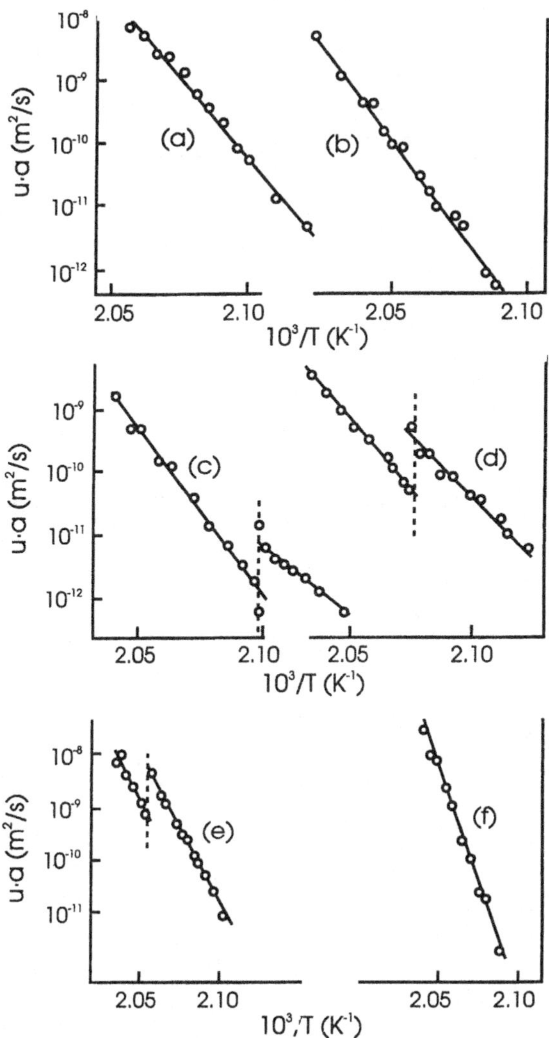

Fig. 4.5. Temperature dependence of the <001> tilt boundary migration rate in tin, for a constant driving force. (The migration rate is expressed as a rate of increase in area). The results show a discontinuous change in mobility for boundaries within ~1° of the 28° $\Sigma = 17$ boundary. a) 26.0°, b) 26.5°, c) 27°, d) 27.7°, e) 28.2°, f) 29.5°.
(after Maksimova et al. 1988).

(ii) The mechanism of boundary migration may change at high temperatures.

Other work has shown a discontinuous transition to a high temperature regime of higher boundary mobility and low activation energy for migration. Gleiter (1970c) has detected a discontinuous change in properties for <100> tilt boundaries in high purity lead at temperatures in the range 0.7 to $0.8T_m$, the actual temperature depending on the crystallography of the boundary. The high temperature regime is characterised by a very low activation energy for migration (26kJ/mol) whereas at the lower temperatures the activation energy is 58kJ/mol which is close to that for boundary diffusion (table 4.1). In zinc, there have been reports of a transition to a very low activation energy of ~17kJ/mol at T>0.91 T_m (Gondi et al. 1992) which is comparable with that for liquid diffusion, and Kopetski et al. (1979) have found a sharp transition to a regime of zero activation energy for boundary mobility in zinc at temperatures above ~0.9Tm as shown in figure 4.6. Although Gleiter (1970c) interpreted his results in terms of a transition to a different boundary structure, there is a similarity between all three sets of measurements, and the last two have been interpreted in terms of a change in the **mechanism of boundary migration** from one controlled by diffusion to one involving cooperative atomic shuffles (§4.4.1.4).

Although all the work discussed above was carried out on high purity materials and the results are qualitatively consistent with changes in intrinsic boundary structure or mechanism of migration, the possibility of small amounts of solute playing a role (§4.4.2) cannot yet be entirely ruled out.

<u>4.3.2 The effect of orientation on grain boundary migration in high purity metals</u>

Because boundary migration involves diffusion processes in and across the boundary, it is to be expected that the structure of the boundary should affect its mobility. In order to define the relationship between two grains it is necessary to specify both the **orientation relationship between the grains** and the **orientation of the boundary** (§3.1). Although,

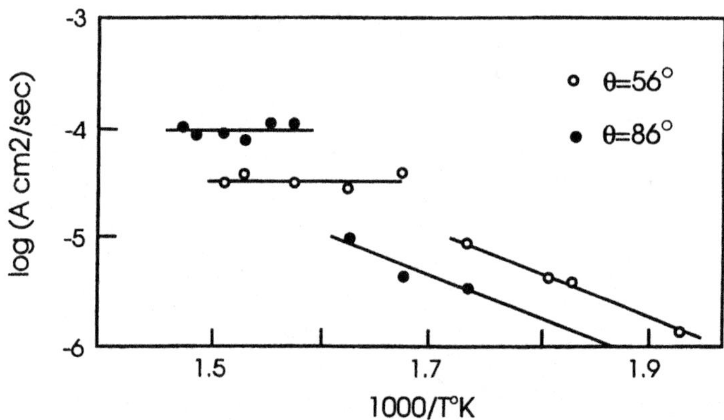

Fig. 4.6. Temperature dependence of the mobility of <11$\bar{2}$0> tilt boundaries in zinc, showing a transition to a high temperature regime of zero activation energy, (after Kopetski et al. 1979).

as discussed below, there are some data on the effect of these parameters on grain boundary migration, the information is far from complete.

4.3.2.1 The orientation dependence of grain boundary mobility
There is extensive evidence that not only do high angle grain boundaries have a greater mobility than low angle boundaries (§4.2.2) but that the mobilities and activation energies for migration of high angle boundaries are dependent on orientation. Early work in this area includes that of Cook and Richards (1940) and Bowles and Boas (1948) who found rapid growth of certain orientations in copper, and Beck et al. (1950) who showed that in lightly rolled high purity aluminium, grains with a misorientation of ~40° about a <111> axis exhibited the largest growth rate.   Kronberg and Wilson (1949) who carried out recrystallization experiments on copper, found that grains related to the deformed matrix by a rotation of 22°- 38° about a common <111> axis, and by 19° about a <100> axis grew most rapidly.   They postulated that there was a relationship between the fast-growing orientations and boundary structure and that **special boundaries**, (later known as **Kronberg-Wilson boundaries**), which had a high density of **coincidence sites** (§3.4.1) often exhibited fast rates of growth.   One of the best known early investigations of the effect of misorientation on mobility is that by Liebmann et al. (1956) who measured the migration rate of <111> tilt boundaries in 99.8% aluminium crystals which had been lightly deformed, and found the mobility to be a maximum for a misorientation of ~40° about a <111> axis as shown in figure 4.7.

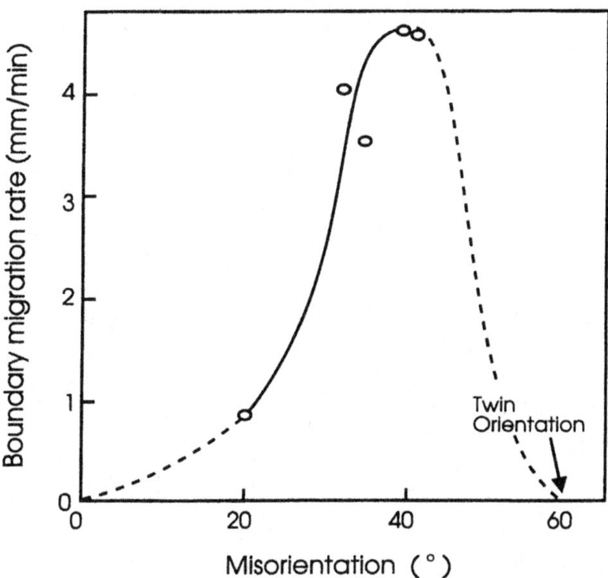

Fig. 4.7. Growth rates of new grains at 615°C into a deformed aluminium crystal, as a function of misorientation about a <111> axis, (after Liebmann et al. 1956).

This result which is of great importance to the theory of recrystallization texture development in fcc metals (§10.7.2), was obtained on lightly strained material in which the impurities were present as dispersed phases. In experiments on similar material in which the impurities were retained in solid solution, a high mobility of 40° $<111>$ boundaries was not observed (Green et al. 1959). However, at larger strains (e.g. 80% by rolling), high mobility of these boundaries was observed irrespective of microstructure, for a wide range of purity levels (from 99.97% - 99.9999%) (Parthasarathi and Beck 1961, Gleiter and Chalmers 1972).

The existence of certain orientation relationships which are associated with a rapid growth rate is well established for a large number of metals (Gleiter and Chalmers 1972), and a summary is given in table 4.3. The fact that a boundary has a large number of coincidence sites is not necessarily an indication that it has a high mobility and it is well known that the boundary with the most coincidence sites, the $\Sigma=3$ twin boundary, has an extremely low mobility (Graham and Cahn 1956). It was shown in chapter 3 that although some low Σ boundaries have a low energy, this is not universally the case, and this is reinforced by the data in table 4.3 which show that high mobility has only been identified for a few values of Σ. As is discussed in §4.3.3, the orientation dependence of boundary migration is closely linked to solute effects, and therefore some care must be taken in interpreting the data of table 4.3.

Extensive experiments on high purity bicrystals of aluminium have been carried out by Shvindlerman and colleagues. Figure 4.8 shows data for the migration of $<111>$ tilt boundaries and figure 4.9 for $<100>$ tilt boundaries. It is seen that for the highest purity material there is a large dependence of mobility on misorientation (fig 4.9), and the minimum activation energy which occurs close to a misorientation of 38° $<111>$ ($\Sigma=7$) is consistent with the mobility maximum in figure 4.7.

The orientation dependence of grain boundary migration has implications for the **texture** developed during recrystallization and grain growth (§10.7.2). In the case of **primary recrystallization**, it is impossible to define accurately, other than on a very local scale, the relationship between the growing grain and the deformed matrix, because any grain of the deformed material will contain a highly misoriented substructure (§2.3). Therefore caution must be exercised in the application of mobility-orientation relationships to recrystallizing material, and in particular in relating a high boundary mobility to a specific boundary structure. The boundary misorientation is normally defined by an **axis/angle** pair (§3.2) and there is abundant evidence (e.g. Ibe and Lücke 1966) that, as indicated by figure 4.7 and table 4.3, there is a significant spread of **angles** for which a high mobility is found. However, information as to the accuracy of the rotation **axis** is lacking (see also §10.7.2).

4.3.2.2 The effect of boundary plane on mobility

It has long been known that for a particular misorientation, the boundary mobility may depend on the actual boundary plane (Gleiter and Chalmers 1972). An extreme example is the $\Sigma=3$ twin orientation in fcc metals where the coherent $\{111\}$ plane boundary is very immobile, whereas the incoherent twin boundary has much greater mobility. It is noticeable that in many cases, grains with an orientation relationship for rapid growth show very anisotropic growth, and an early and very clear example of this is seen in the work of Kohara et al. (1958). For face centred cubic metals the anisotropy is such that the faces parallel to the $<111>$ rotation axis (i.e the twist boundaries) grow much more slowly than the tilt boundaries. Thus an unconstrained grain tends to become plate-shaped as seen in figure 4.10.

Table 4.3
Orientations for which rapid growth is commonly found

Nearest coincidence relationship		Experimental relationship		Metal	Structure	Reference
Sigma	Rotation Axis	Rotation Axis				
$\Sigma = 7$	38.2 <111>	35-45 <111>		Al	fcc	Liebmann et al. (1956)
		38 <111>		Cu	fcc	Kronberg & Wilson (1949)
		36-42 <111>		Pb	fcc	Aust & Rutter (1959a)
$\Sigma = 13a$	22.6 <100>	23 <100>		Al	fcc	May & Erdmann (1959)
		19 <100>		Cu	fcc	Kronberg & Wilson (1949)
$\Sigma = 13b$	27.8 <111>	30 <111>		Cu	fcc	Beck (1954)
		30 <111>		Ag	fcc	Ibe & Lücke (1966)
		20-30 <111>		Nb	bcc	Stiegler et al. (1963)
$\Sigma = 13$	30 <0001>	30 <0001>		Zn	cph	Ibe & Lücke (1966)
		30 <0001>		Cd	cph	Ibe & Lücke (1966)
$\Sigma = 17$	28.1 <100>	26-28 <100>		Pb	fcc	Aust & Rutter (1959,)
		30 <100>		Al	fcc	Fridman et al. (1975)
$\Sigma = 19$	26.5 <110>	27 <110>		Fe-Si	bcc	Ibe & Lücke (1966)

A detailed investigation of growth anisotropy during directional recrystallization of aluminium has been carried out (Gottstein et al. 1978, Gottstein and Schwarzer 1992). The variation of anisotropy, as measured from the shape of the grains, showed a poorly defined dependence on the angle of misorientation about <111> as shown in figure 4.11a, indicating a minimum anisotropy at misorientations close to 40° about <111>. The maximum anisotropy of ~100 was found for the low mobility 60°<111> ($\Sigma = 3$) twin boundary, showing that, as might be expected, the anisotropy is dependent on the absolute

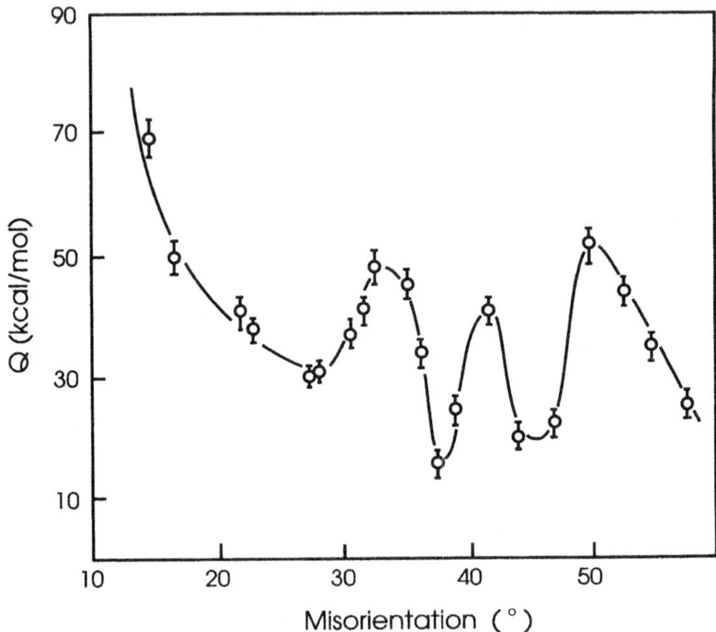

Fig. 4.8. Activation energy for migration of <111> tilt boundaries in aluminium as a
function of misorientation, (after Bokstein et al. 1986).

growth rates. There was a clear dependence of anisotropy on the deviation of the axis from
<111> as shown in figure 4.11b, in which it may be seen that a deviation of 3° from the
<111> axis is sufficient to reduce the anisotropy by ~80%. These experiments indicate
that the anisotropy should be sharply reduced during the recrystallization of heavily deformed
polycrystals where such exact orientation relationships do not occur. For grains with
orientation relationships other than <111> rotations, growth was generally found to be
isotropic.

4.3.3 The influence of solutes on boundary mobility

Solute elements have an enormous effect on boundary migration and very small amounts of
impurity may reduce the mobility by several orders of magnitude (Dimitrov et al. 1978) as
shown in figure 4.12a. Because the effects of solute are so strong, particularly at low
concentrations, it is very difficult to be confident that the mobilities measured in "pure"
metals and discussed in §4.3.1 and §4.3.2 actually represent the intrinsic behaviour of
boundaries.

4.3.3.1 The effect of solute concentration
From figure 4.12 it may be seen that the relationship between mobility and solute content
shows two distinct regimes separated by a transition region:-

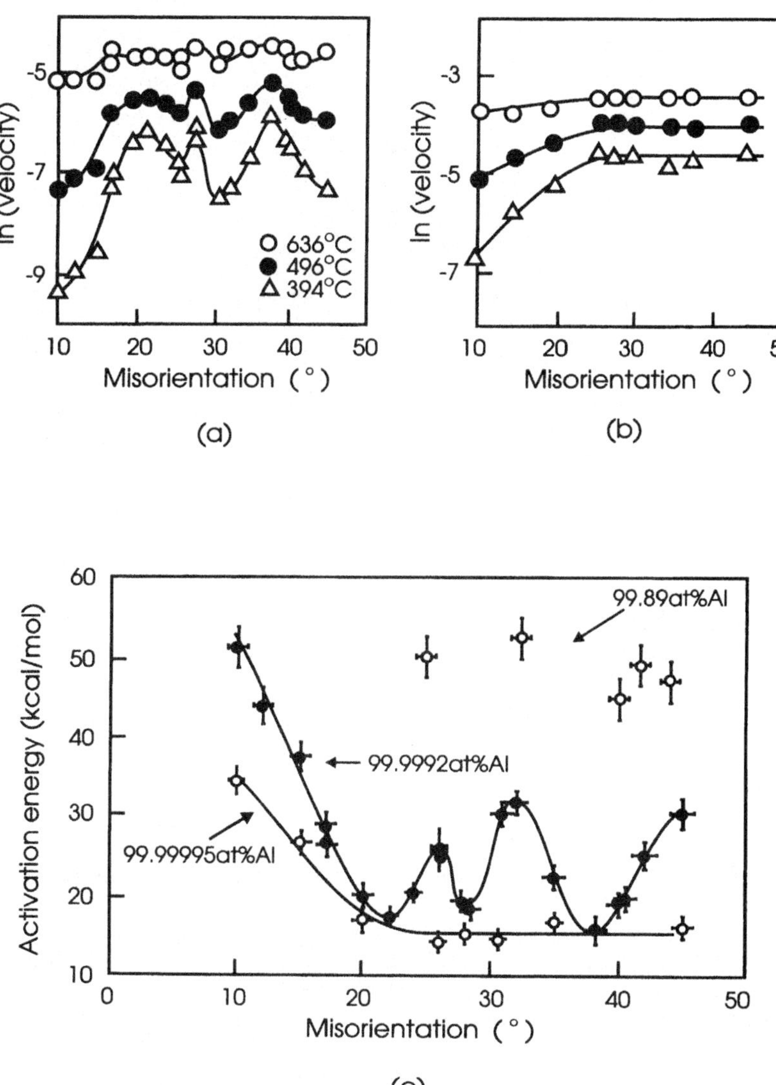

Fig 4.9. The mobility of <100> tilt boundaries in aluminium, (after Fridman et al. 1975).
a) The effect of temperature and misorientation on the mobility of boundaries in 99.9992
at% aluminium. b) The effect of temperature and misorientation on the mobility of
boundaries in 99.99995 at% aluminium. c) The orientation dependence of the activation
energy for migration in samples of different purity.

Fig. 4.10. Optical micrograph of lenticular grains, viewed edge-on in a recrystallizing
single crystal of Al-0.05%Si. The grains are rotated by ~40° to the deformed matrix
about the marked <111> axis, and the broad face of the plate is parallel to the
rotation axis, (Ardakani and Humphreys 1994).

**(i) At high solute concentrations the mobility is low and decreases with increasing solute
concentration.**
In this low mobility regime it is thought that an atmosphere of solute atoms is associated with
the boundary. The boundary velocity is then controlled by the rate of diffusion of the
impurity atoms (§4.4.2). Gordon and Vandermeer (1966) have shown that for aluminium
containing small amounts of copper, in the low mobility regime the boundary velocity at
constant driving force is inversely proportional to the solute concentration (c) as shown in
figure 4.13.

**(ii) At low solute concentrations the mobility is higher and solute has little
influence on the mobility.**
In this high mobility regime, the boundary mobility is little affected by solute and it is
thought that the boundary has broken away from its solute atmosphere (§4.4.2).

For the cases of copper and magnesium in aluminium shown in figure 4.12a, there is a very
sharp transition between the two regimes. However, similar experiments have shown more

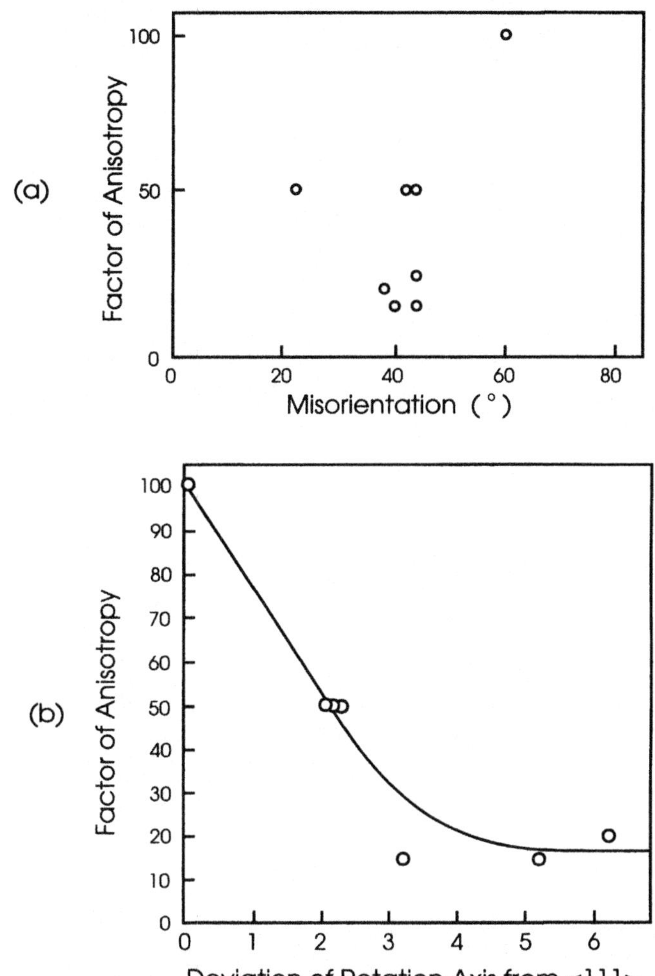

Fig. 4.11. a) The effect of misorientation about a $<111>$ axis on the shape anisotropy of recrystallizing grains a) Dependence on rotation angle. b) Dependence on deviation from the $<111>$ axis, (after Gottstein et al. 1978).

gradual transitions for silver (Figure 4.12c) or iron (Montariol 1963) in aluminium. In the same solvent, different solutes may, as shown in figure 4.12a affect the critical concentration at which the transition occurs and, as is apparent in figure 4.12a may also reduce the mobility in the slow regime, i.e. the parameter dM/dc, by different amounts.

The apparent activation energy for boundary migration is also a function of the solute content, exhibiting a transition at the critical solute content at which boundary breakaway occurs as shown in figures 4.12b and c. The details of the transition are different in both

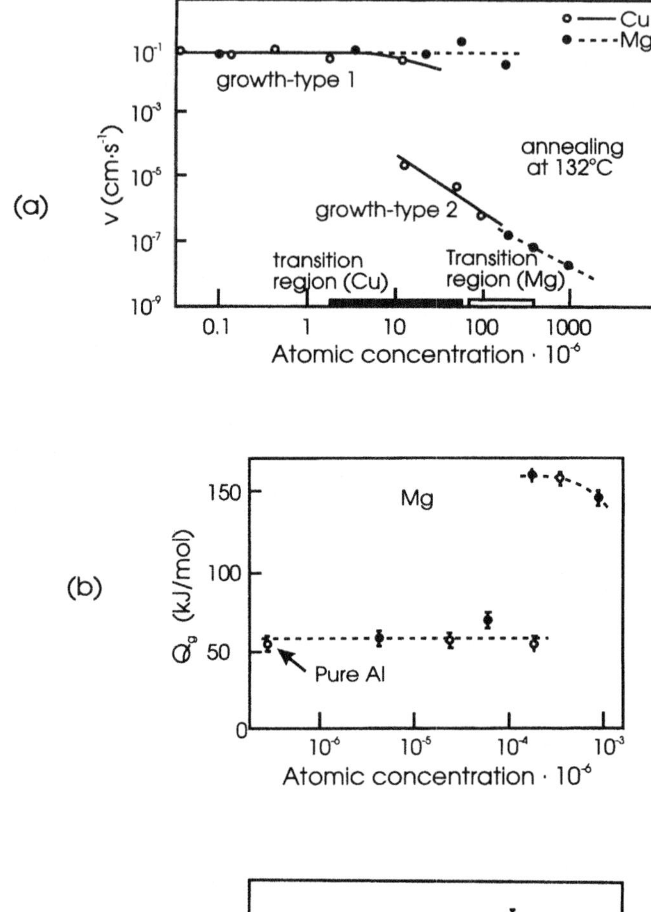

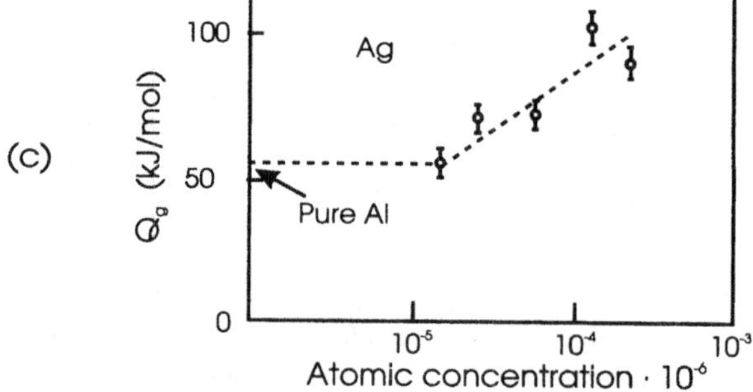

Fig. 4.12. The effect of impurities on the growth rate of new grains in deformed aluminium. a) The effect of copper and magnesium additions on the growth rate. b) The effect of magnesium on the activation energy for boundary migration. c) The effect of silver on the activation energy for boundary migration. (after Frois and Dimitrov 1966).

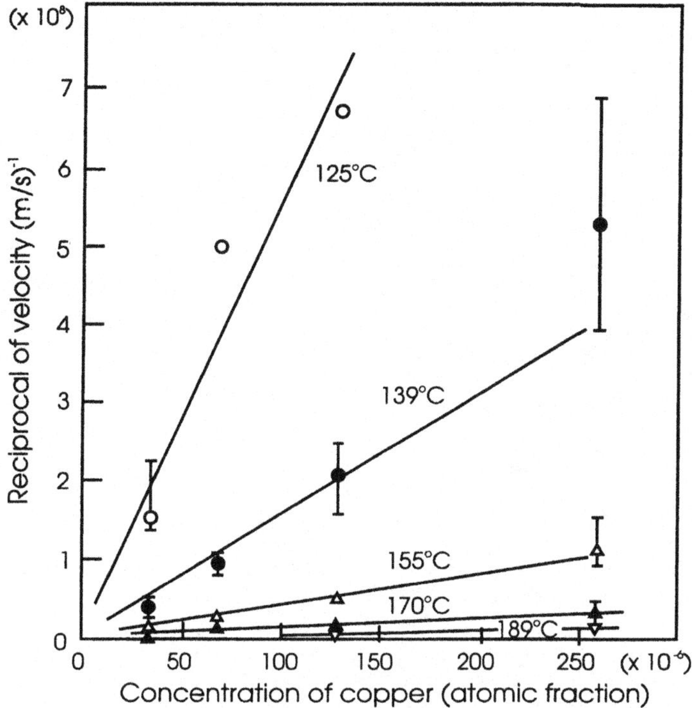

Fig. 4.13. The effect of copper on the migration rate of boundaries in aluminium, (after Gordon and Vandermeer 1966).

cases. For the magnesium solute (fig 4.12b) the transition is abrupt, whereas for the silver solute (fig 4.12c) it is gradual.

In most cases it is found that the activation energies at low solute concentrations are essentially solute independent and correspond to those found for materials of very high purity as shown in table 4.2. The apparent activation energy at higher solute concentrations is typically close to that for self diffusion or for diffusion of the solute in the solvent, and, as seen from figures 4.12b and c, may depend on solute concentration.

4.3.3.2 The effect of temperature

The combined effects of temperature and solutes on mobility are best seen from plots of the logarithm of mobility against $1/T$. As shown in figure 4.14 for boundary mobility in Au with 20ppm Fe (Grunwald and Haessner (1970), such a plot may show both a change to a lower apparent activation energy and a sharp increase in mobility at a higher temperature.

Similar results have been reported for lead (Aust 1969) and copper (Grewe et al. 1973). In most cases it is found that the activation energies at high temperatures correspond to those found for materials of very high purity and the activation energy at low temperature is typically close to that for self diffusion.

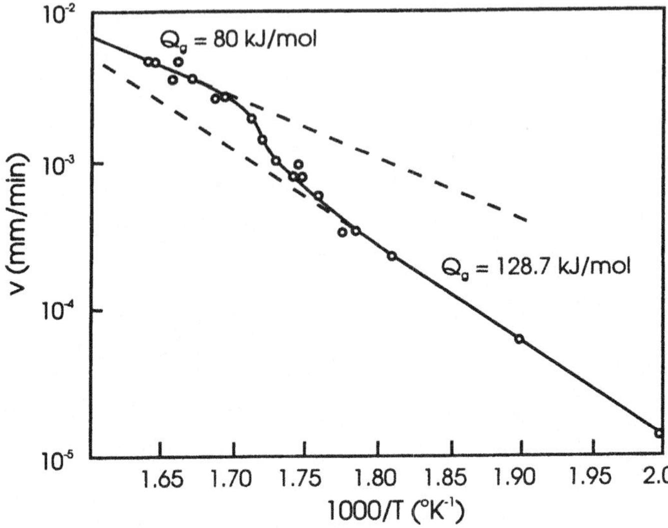

Fig. 4.14. The effect of temperature on the migration velocity of a 30° <111> tilt
boundary in gold, (after Grunwald & Haessner 1970).

In many cases (Gordon & Vandermeer 1962, Fridman et al. 1975) a peak is found in the
activation energy at the transition, as is shown in figure 4.15. The origin of this peak can
be seen in figure 4.14, where during the transition, the slope of the line (i.e. apparent
activation energy) increases. This is clearly an **apparent** activation energy due to a transition
in the effect of solutes on the boundary and does not correspond to the activation energy of
a physical mechanism.

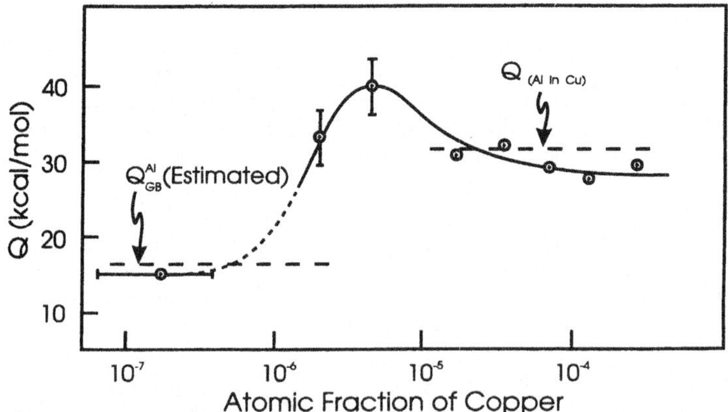

Fig. 4.15. The effect of copper additions on the activation energy for boundary migration
in aluminium, (after Gordon and Vandermeer 1962).

4.3.3.3 The effect of orientation

It has been known for a long time that the effects of solutes on boundary mobility are dependent on the crystallography of the boundary, and that **special boundaries**, i.e. those which are close to a coincidence relationship, are less susceptible to effects of solute than are **random boundaries** (Kronberg and Wilson 1949, Aust and Rutter 1959a,b). The well-known results obtained by Aust and Rutter on tin-doped lead are shown in figure 4.16. At

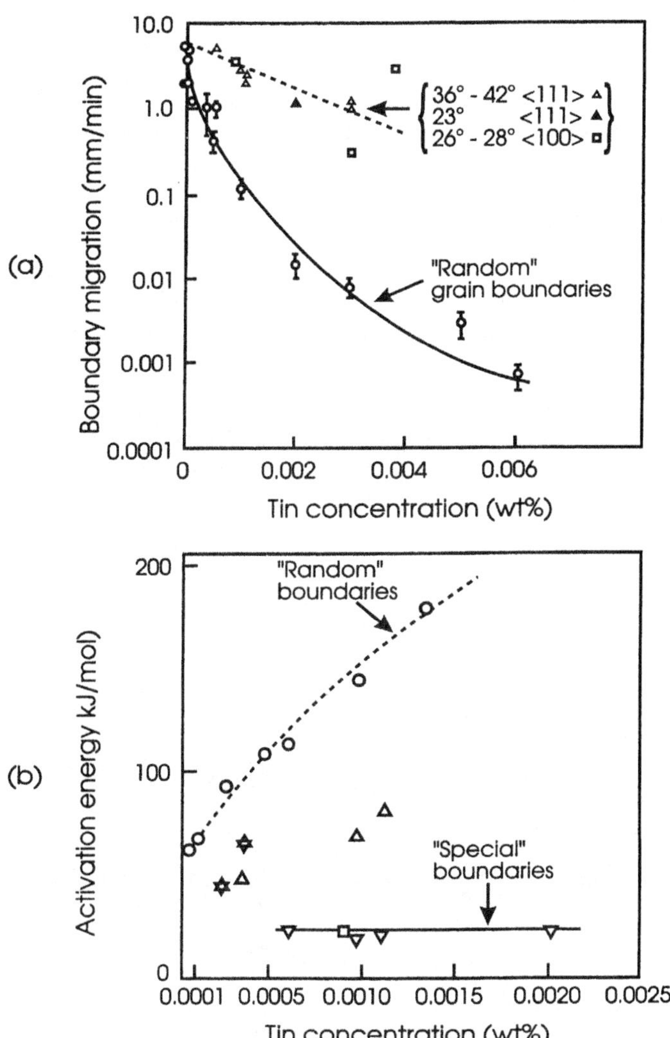

Fig. 4.16. a) The rate of grain boundary migration at 300°C in zone-refined lead crystals doped with tin, (after Aust and Rutter 1959a). b) The effect of tin on the activation energy for the migration of random and special boundaries in lead, (after Aust and Rutter 1959b).

low impurity concentrations the mobilities of the special and general boundaries are similar. As the impurity level is increased the mobility of the random boundaries is affected much more by the impurities. The combined effects of misorientation and solute have also been demonstrated for tilt boundaries in $<111>$ and $<100>$ rotated grains in aluminium, by Fridman et al. (1975) as shown in figures 4.8 and 4.9. For aluminium of 99.9992 at% purity, except at the highest temperature, the mobility fluctuates with misorientation, with maximum growth rates occurring near coincidence rotations (fig 4.9a). However as shown in figure 4.9b, for material of 99.99995at% purity, the mobility is found to be almost independent of orientation for $\theta > 20°$.

These results confirm those of Aust and Rutter (1959a,b) and indicate that:

The orientation dependence of grain boundary mobility (§4.3.2.1) arises primarily from an orientation dependence of solute segregation to the boundary rather than an intrinsic structure dependence of grain boundary mobility.

There is some evidence, such as that shown in figure 4.9c, that the orientation dependence of mobility disappears not only at high purities, but also at lower purity levels. Experiments on zinc (Sursaeva et al. 1976, Gottstein and Shvindlerman 1992), demonstrate a similar effect, which indicates that the **window** within which orientation dependence of mobility is found, may be determined by both the boundary structure and the purity level as shown schematically in figure 4.17.

The evidence for the disappearance of the orientation dependence of boundary mobility at lower purities is not conclusive, and, as discussed by Haessner and Hoffmann (1978), the results of Fridman et al. (1975) for the lowest purity material in figure 4.9c contradict those

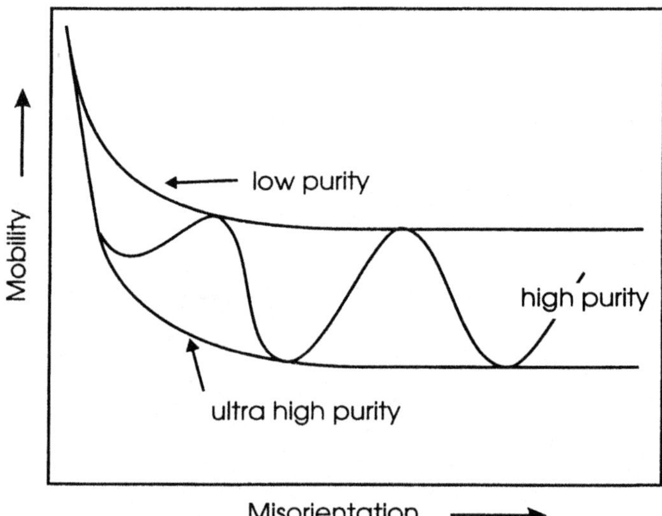

Fig. 4.17. Schematic diagram showing the dependence of the activation energy of grain boundary mobility for different purity levels, (after Gottstein and Shvindlerman 1992).

of Demianczuc and Aust (1975) who found an orientation dependence of mobility in aluminium of similar purity.   In addition, much of the evidence for the orientation dependence of mobility in aluminium (Liebmann et al. 1956, Yoshida et al. 1959) was obtained in material of much lower purity.   The effect of solutes on the orientation dependence of migration warrants further investigation.

Gordon and Vandermeer (1966) and Gottstein and Shvindlerman (1992) speculate that with lower impurity levels than have yet been achieved, it is conceivable that random grain boundaries would move more easily than special boundaries because the lower energy of the special boundaries would provide a greater barrier to the transitory structure modification required when a boundary moves.   However, this argument does not take account of the possibility discussed in §4.4.1 that special boundaries in high purity materials may move by a low energy mechanism involving atomic shuffles, a mechanism which would not be available to random boundaries.

4.3.3.4 The effects of temperature and orientation

There is evidence that in copper (Aust et al. 1963) and lead (Rutter and Aust 1965), the higher mobility of special boundaries in materials of moderate purity may disappear at high temperatures as shown in figure 4.18.   This may be due to the evaporation of the solute atmosphere at boundaries at high temperatures (§4.4.2).

4.3.4. The effect of point defects

The interaction of vacancies and other point defects with **static** high angle grain boundaries has been extensively investigated (e.g. Balluffi 1980), and it has been shown that boundaries may act as sources and sinks for point defects and that point defects interact with grain boundary dislocations.   The effect of point defects on a **migrating** boundary is less clear, although there is evidence (Hillert and Purdy 1978, Smidoda et al. 1978) that the diffusion coefficient of a moving boundary is several orders of magnitude larger than that of a stationary boundary.

4.3.4.1 The effect of vacancies on boundary mobility

The role of defects on boundary mobility has been extensively discussed e.g. Cahn (1983), Gleiter and Chalmers (1972), Haessner and Hofmann (1978).   Although these reviewers are in general agreement that a vacancy flux or supersaturation enhances boundary mobility, the experimental evidence for this is not extensive and in many cases the role of vacancies is only inferred from indirect evidence.

Experiments which show that grain growth during recrystallization is slower in small specimens is typical of such evidence.   The work of In der Schmitten et al. (1960) is one of the most cited references in support of vacancy enhanced boundary migration.   They found that the rate of boundary migration during primary recrystallization in etched wires decreased significantly as the wire diameter was reduced.   They proposed that this effect was due to the swept-up vacancies diffusing along the migrating boundaries and out of the specimen. For similar reasons, the retardation or inhibition of recrystallization found in many in-situ annealing experiments of thin foils in the TEM was initially explained (Gleiter and Chalmers 1972) in terms of vacancy loss.   However, it is now thought that factors such as dislocation loss at free surfaces due to image stresses and other surface effects are a much more likely explanation of these effects.

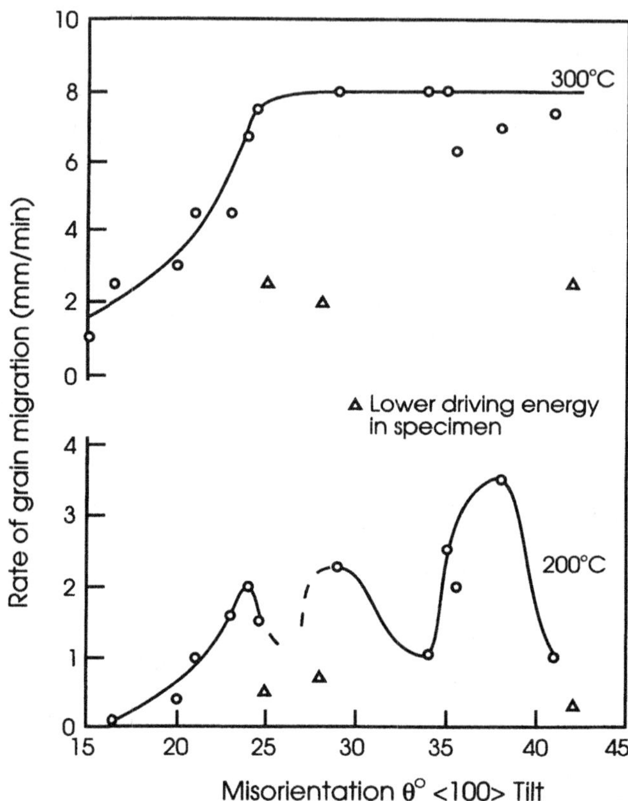

Fig. 4.18. The effect of temperature and orientation on the velocity of tilt boundaries in lead under a constant driving force, (after Rutter and Aust 1965).

More direct evidence which indicates that vacancies enhance migration was obtained by Haessner and Holzer (1974), who showed that neutron irradiation increased the boundary velocity in recrystallizing copper single crystals as shown in figure 4.19. In this material the irradiation produces Frank vacancy loops, and although this increases the driving force for recrystallization, the authors claim that calorimetric measurements showed that this effect was negligible and that the vacancy loops increased the boundary mobility. Support for this is provided by the work of Atwater et al. (1988), who attributed increased grain growth rates in ion-bombarded thin films of gold, germanium and silicon to the effect of the point defects on boundary mobility.

Other indirect evidence indicating that vacancies affect boundary mobility comes from the sintering of copper wires, when the presence of pores on a boundary is found to increase its mobility (Alexander and Balluffi 1957), and from the sintering of alumina, when small pores are found to be consumed by slowly migrating boundaries (Coble and Burke 1963).

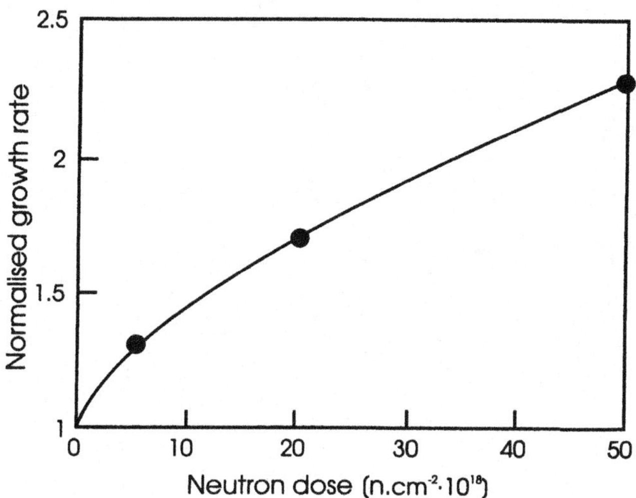

Fig. 4.19. The effect of neutron irradiation on the boundary migration rate in copper,
(after Haessner and Holzer 1974).

The basic concept that the vacancies swept up by a moving boundary will increase the free
volume of the boundary and therefore aid atom movement across the boundary is plausible
(Gordon and Vandermeer 1966), but has not been backed by rigorous theory. Such a model
implies that the structure of static and moving boundaries will be different and also suggests
that boundary structure and hence mobility may be dependent on boundary velocity. This
problem is most likely to be resolved ultimately by dynamic computer simulations of
boundary migration.

4.3.4.2 Generation of defects by moving boundaries

A moving boundary may act as a source of defects, and there is evidence (Gleiter 1980) that
in material of low dislocation density, a higher dislocation density may be left behind a
migrating boundary. This is attributed to the formation of dislocations by growth accidents
rather than to the emission of dislocations by stress relaxation at the boundary. Gleiter
suggests that in a similar manner, growth accidents will lead to vacancies being left behind
a moving boundary, and cites experimental evidence of a vacancy supersaturation behind
moving boundaries in aluminium and nickel alloys.

4.3.5. The scope of experimental measurements

It should be emphasised that for a number of reasons we cannot confidently draw general
conclusions about the mobility of high angle grain boundaries during recrystallization.

(i) It is extremely difficult to carry out reliable experimental investigations of boundary
mobility.

(ii) There are a large number of variables to be investigated including solvent, solute, concentration, misorientation, boundary plane, temperature and driving force, and therefore there have been few systematic investigations.

(iii) Most investigations have been carried out with driving forces which are a factor of $\sim 10^4$ lower than those found in recrystallization.

4.4 THEORIES OF THE MOBILITY OF HIGH ANGLE GRAIN BOUNDARIES

4.4.1. Theories of grain boundary migration in pure metals.

A schematic diagram of a grain boundary is shown in figure 4.20a. Models of grain boundary migration are generally based on the assumption that atoms are continually detached from the grains (e.g. A and B) by thermal activation and move into the more disordered region of the boundary itself (e.g. C), which is shown exaggerated in figure 4.20a. The atoms are then re-attached to one of the grains. If the atom flux in both directions is equal then the boundary is static. However, if there is a driving force for migration then the flux will be greater in one direction. There are several proposed variants on this general model, the main variables being:

Single or group activation. Migration may occur either by the activation of single atoms or by the collective movement of a group of atoms.

The role of atoms in the boundary region. Atoms detached from a parent grain are either immediately reattached to a grain or may remain or move in the boundary region.

Preferential sites. If the crystallography of the boundary is taken into account then there may be preferential sites for detachment and attachment of atoms.

4.4.1.1 Thermally activated boundary migration - early single-process models

The theory of boundary migration based on reaction rate theory, in which boundary movement is controlled by single atom movements, was proposed by Turnbull (1951), and in this section we use this type of approach to illustrate the nature of the problem. Consider the boundary of figure 4.20a, which is considered to have a thickness δ, moving to the left under the influence of a free energy difference of ΔG. In order for an atom to break away from its parent grain it must acquire, by thermal activation, an activation energy of ΔG^a as shown in figure 4.20b. If the frequency of atomic vibration is ν_0, then the number of times per second that the atom acquires this energy is $\nu_0 \exp(-\Delta G^a/kT)$. If there are **n** atoms per unit area of boundary which are suitable sites for a jump, then the number of jumps per second from a grain is $n\nu_0 \exp(-\Delta G^a/kT)$. However, they will not all be in favourable positions to jump, and therefore we include a grain boundary structure dependent factor A_J, which is the fraction of atoms able to jump. As not all atoms may find a suitable site for attachment to the other grain, then we introduce an accommodation factor A_A, which is the fraction of successful attachments. The effective flux of atoms from grain 1 to grain 2 will thus be

$$A_J A_A n \nu_0 \exp\left(-\frac{\Delta G^a}{kT}\right)$$

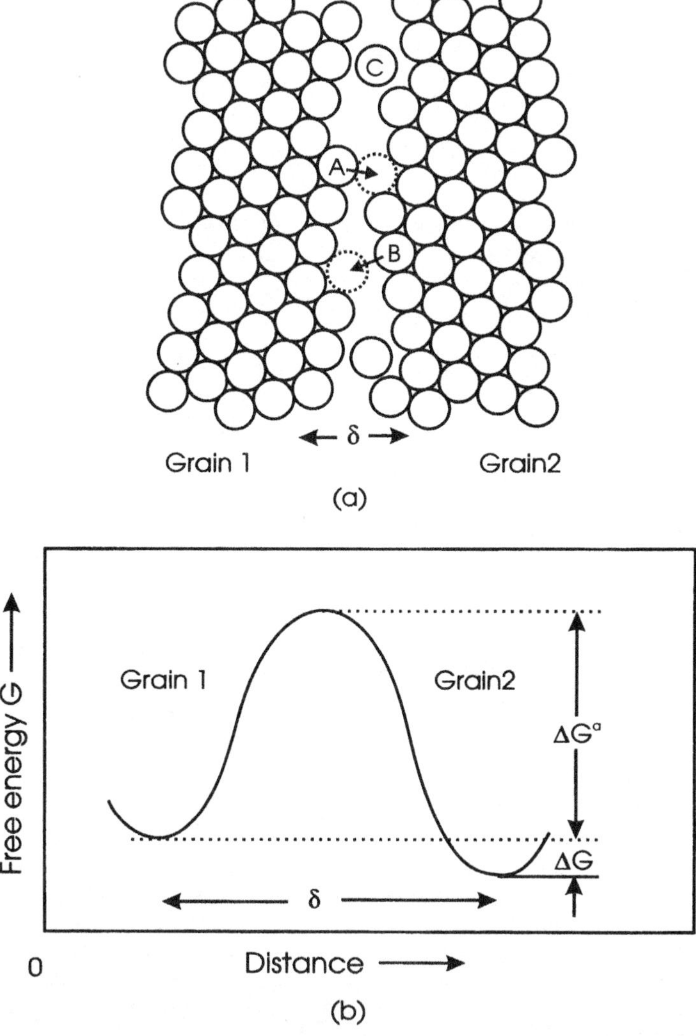

Fig. 4.20. Grain boundary migration by means of atom jumps. a) The mechanism of migration. b) The free energy of an atom during a jump across the boundary.

In the same manner, there will be a flux of atoms from grain 2 to grain 1, given by

$$A_J A_A n \, v_0 \exp \left(- \frac{\Delta G^a + \Delta G}{k T} \right)$$

There will therefore be a net flux from grain 1 to grain 2 of

$$J = A_J A_A n \, v_0 \exp\left(-\frac{\Delta G^a}{kT}\right) . \left[1 - \exp\left(-\frac{\Delta G}{kT}\right)\right] \tag{4.12}$$

If the boundary velocity is $\mathbf{v}$, and the interatomic spacing is $\mathbf{b}$, then,

$$v = J\frac{b}{n} = A_J A_A v_0 b \exp\left(-\frac{\Delta G^a}{kT}\right) . \left[1 - \exp\left(-\frac{\Delta G}{kT}\right)\right] \tag{4.13}$$

As the free energy changes during recrystallization are small, we may assume that $\Delta G \ll kT$ and expand $\exp(-\Delta G/kT)$ giving

$$v = A_J A_A v_0 b \exp\left(-\frac{\Delta G^a}{kT}\right) . \frac{\Delta G}{kT} \tag{4.14}$$

As the driving pressure $P = \Delta G$ then

$$v = A_J A_A v_0 b \exp\left(-\frac{\Delta G^a}{kT}\right) . \frac{P}{kT} \tag{4.15}$$

and substituting $\Delta G = \Delta H - T \Delta S$ then

$$v = A_J A_A v_0 b \exp\left(-\frac{\Delta H^a}{kT}\right) . \exp\left(\frac{\Delta S}{k}\right) . \frac{P}{kT} \tag{4.16}$$

Equation 4.16 is therefore identical in form to equation 4.1 with

$$M = \frac{A_J A_A v_0 b}{kT} . \exp\left(-\frac{\Delta H^a}{kT}\right) . \exp\left(\frac{\Delta S}{k}\right) \tag{4.17}$$

The model discussed above is very general and not specific enough to allow prediction of the parameters such as the activation energy (ΔH^a). For example, although the activated process is often identified with grain boundary diffusion, in this model the atoms move **across** the boundary rather than **within it** and the two processes are not necessarily identical. Also, a better defined basis for the parameters A_J and A_A needs to be developed. These problems, and the question of whether atoms migrate within the boundary region have been addressed more specifically in later theories which take account of the boundary structure and which attempt to relate the mobility to movements in grain boundary defects such as **steps** or **dislocations** as discussed below.

4.4.1.2 Early group-process theories

In the earliest group-process theory by Mott (1948), groups (islands) of atoms move from one grain into the boundary region and similar groups attach themselves to the other grain. Later developments of this type of model are reviewed by Gleiter and Chalmers (1972). The attractiveness of such models was that because groups of atoms were thermally activated, this required a much larger activation energy than that for a single atom (equation 4.17), which was at that time in accord with experimental observations. However, when it was later shown that such large activation energies were due to impurities (§4.3.3), and were not characteristic of boundaries in pure materials (table 4.2), attention switched to single-process theories. As discussed below, there is now some evidence that cooperative atomic movements are important in boundary migration of pure materials and newer versions of group-process models are currently being developed.

4.4.1.3 Step models

An early attempt to incorporate the effects of boundary structure into a model for boundary mobility was made by Gleiter (1969b) who proposed a detailed atomistic model in which boundary migration occurred by the movement of steps or kinks in the boundary as shown in figure 4.21. Evidence for the existence of such steps had been found from transmission electron microscopy (Gleiter 1969a). The steps move by the addition or removal of atoms from the steps, and the atoms are assumed to diffuse for short distances within the grain boundary. This is an analogous process to that occurring during the growth of crystals from a vapour, and interface migration by ledge movement is a well established mechanism for the movement of interphase boundaries during phase transformations (Aaronson et al. 1962).

For driving pressures of the magnitude encountered during annealing, Gleiter calculates the boundary velocity to be

$$v = b\, v_0\, \Psi \exp\left(-\frac{\Delta G^a}{kT}\right).\frac{P}{kT} \tag{4.18}$$

This is similar to equation 4.15, but modified by a factor Ψ which is a function containing details of the step configuration in a boundary of thickness δ, and which is of the form

$$\Psi = \frac{c}{\delta}\left(1 + \frac{b}{\delta}\left(\frac{1}{f_1} - \frac{1}{f_2}\right)\right) \tag{4.19}$$

where c is a constant and f_1 and f_2 are functions of the step density on the crystals either side of the boundary. The mobility of the boundary is predicted to be dependent on both the misorientation and the boundary plane through the terms f_1 and f_2. For example for the case of crystals rotated about a $<111>$ axis in fcc metals, the ledge density and hence the mobility is predicted to be lowest for the $\{111\}$ twist boundaries as is found in practice (§4.3.2.2).

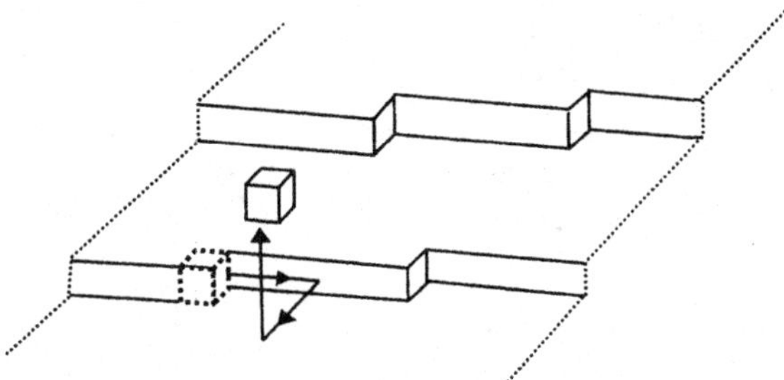

Fig. 4.21. The ledge mechanism of boundary migration. An atom is detached from a kink in the ledge, migrates along the ledge and into the boundary, (after Gleiter 1969b).

4.4.1.4 Boundary Defect Models

An improvement in the understanding of the defect structure of high angle grain boundaries, and recognition that steps and intrinsic boundary dislocations are common features, has led to experimental and theoretical investigations into the role of these defects in the processes of boundary migration. There is a very close relationship between boundary steps and boundary dislocations, and in general, boundary dislocations have steps in their cores (King and Smith 1980). The example shown in figure 4.22 is of a $1/10 < 310 >$ dislocation in a boundary close to $\Sigma = 5$ in an fcc material. The height of these steps depends on the Burgers vector of the dislocation, the boundary plane and the crystallography of the boundary. When such dislocations move, then the steps move and boundary migration inevitably occurs. There is extensive evidence that boundary dislocations move and multiply and that they retain their core structure at elevated temperatures (e.g. Dingley and Pond 1979, Rae 1981, Smith 1992).

In special cases, the dislocations can glide in the boundary plane and therefore athermal boundary migration is possible. A well known example of this is the case of $1/6 < 112 >$ twinning dislocations in fcc materials. These dislocations move by glide on the $\{111\}$ planes and the twin plane advances by one $\{111\}$ planar spacing for the passage of each dislocation. In the more general case, the dislocation cannot glide in the boundary and a combined glide/climb movement of the boundary dislocations is required.

If boundary movement is accomplished by the collective movement of a density ρ of boundary defects each moving at velocity $\mathbf{v}$ and having a step height $\mathbf{h}$, then the boundary velocity (v_b) will be given (Smith 1992) by an expression of the form

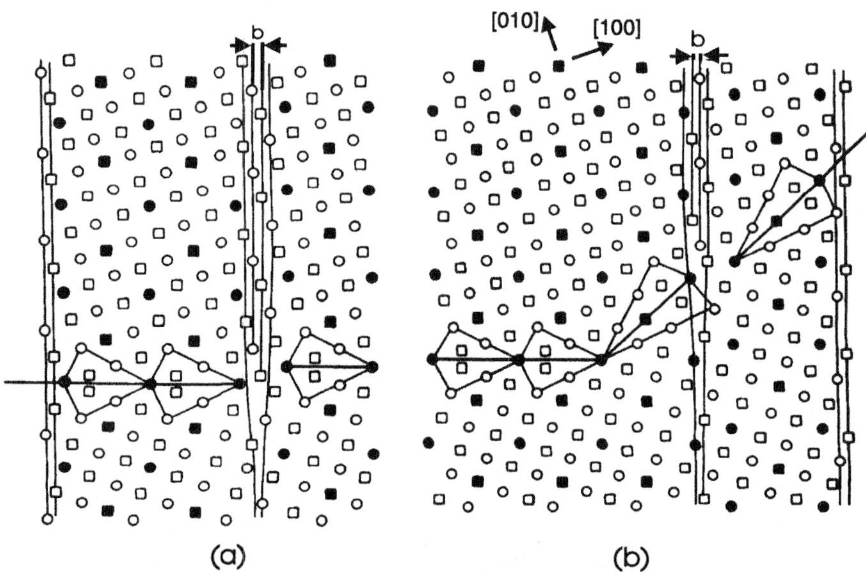

Fig. 4.22. Atomic arrangement of a $\Sigma = 5$ tilt boundary in an fcc lattice which contains dislocations with Burgers vectors a) parallel to the boundary or b) inclined to the boundary. In (a) the boundary dislocations can move by glide, whereas in (b) climb and glide are required. The open symbols represent the ABA... stacking sequence of the $\{001\}$ planes and the filled symbols are coincidence sites, (Smith et al. 1980).

$$v_b \sim h \, \rho \, v \qquad\qquad (4.20)$$

Glide of boundary dislocations is of course associated with shear deformation and the process of boundary dislocation motion results in a combined migration and sliding of the boundary. Such a combined process has been modelled by Bishop et al. (1980) and has been observed in zinc bicrystals by Ando et al. (1990), and on near $\Sigma=5$ boundaries in gold bicrystals by Babcock and Balluffi (1989a), who found good correlation between the movement of boundary dislocations and the migration and sliding of a boundary under stress. It should however be noted that such dislocation motion results in a **shape change**, which if not relaxed can exert a back-stress such as to cancel out the driving pressure for boundary migration (Smith et al. 1980). This suggests that the behaviour of unconstrained bicrystals may be different from that of the constrained polycrystals which are of more general importance (cf. §4.2.1).

Further experiments of Babcock and Balluffi (1989b) in which the boundaries migrated under the influence of capillary forces at high temperature showed quite different behaviour. The boundaries migrated in a jerky fashion and the movement of boundary dislocations was found to account only for a negligible part of the boundary migration, particularly for more general boundaries. Whether or not such jerky boundary migration is genuine or is an artefact of the experiment is not yet clear, although similar behaviour has been found by a number of workers. On the basis of their experiments and from computer simulations (Majid and Bristowe 1987), Babcock and Balluffi (1989b) have interpreted these experiments in terms of **local cooperative shuffling** of atoms in a process similar to that proposed by Bauer and Lanxner (1986). This mechanism is shown schematically for a $\Sigma=5$ <001> twist boundary in figure 4.23.

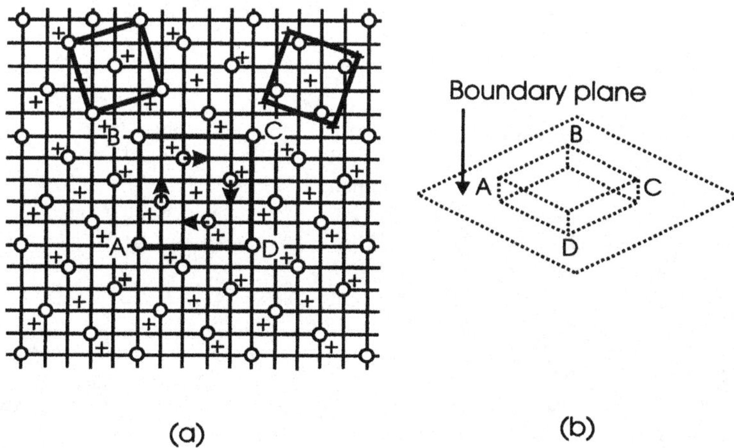

Fig. 4.23. Proposed mechanism of migration of $\Sigma=5$ [001] twist boundary by atomic shuffling. a) View along [001] showing first planes of crystal 1 (O) and crystal 2 (+) facing the boundary. Vectors show the atomic shuffles of atoms in crystal 1 required to displace the boundary by a/2 and produce the square boundary ledge shown obliquely in (b), (after Babcock and Balluffi 1989b).

Babcock and Balluffi (1989b) argue that this is a more important mechanism of boundary migration than is the movement of boundary dislocations. However, the number of atoms which must move in a coordinated manner in such a shuffle is large for high order CSLs, and therefore **this type of mechanism is only likely to be effective in a highly ordered boundary and not in a general boundary.**

4.4.1.5 The status of boundary migration models

Theoretical and experimental work in recent years has largely concentrated on the behaviour of low Σ boundaries, and in particular on symmetrical tilt boundaries in materials of very high purity. As such simple boundaries are not typical of more general boundaries, this has probably led to an over-emphasis on the importance of boundary structure and defects in interpreting the migration of boundaries as discussed by Smith (1992).

Although the movement of steps or boundary dislocations certainly occurs, and can explain some properties of some boundaries, the important process in the more general case appears to be the thermally activated transport of atoms across boundaries and the migration would seem to be controlled by the activation energy for this process. Apart from special cases, there is no good evidence that the sites of attachment and detachment have any significant effect.

It has long been recognised that in general boundaries the **local excess volume** or **porosity** of the boundary is a very important parameter which may determine not only the energy of the boundary (§3.4.2), but also the migration behaviour (Seeger and Haasen 1956, Gordon and Vandermeer 1966), and that except for low Σ boundaries, this is likely to be more important than the atomistic details of the boundary structure. The concept of excess boundary volume can explain a number of important factors which are related to the boundary crystallography, including the higher mobility of tilt boundaries than twist boundaries (§4.3.2.2) and the relative effect of solutes on general and special boundaries (§4.3.3).

4.4.2. Theories of grain boundary migration in solid solutions

Most theories of the effects of solutes on boundary mobility are based on that proposed by Lücke and Detert (1957) for dilute solid solutions. The theory was developed independently by Cahn (1962) and by Lücke and Stüwe (1963). The model was later extended to include higher solute contents by Lücke and Stüwe (1971)

There are several other earlier theories which are reviewed by Lücke and Stüwe (1963), Gordon and Vandermeer (1966) and Gleiter and Chalmers (1972) and there have been several later models, notably those proposed by Bauer (1974) and by Hillert and Sundman (1976) and Hillert (1979) who extended the theory to include high solute contents. However, the **Cahn-Lücke-Stüwe (CLS)** model is still widely accepted as giving a good semi-quantitative account of the effects of solute on boundary migration, and this is the basis for the following discussion.

The CLS theory is based on the concept that atoms in the region of a boundary have a different energy (**U**) to those in the grain interior because of the different local atomic environment. There is therefore a force (**dU/dx**) between the boundary and a solute atom which may be positive or negative, depending on the specific solute and solvent. The total force from the boundary on all solute atoms is $P=\Sigma dU/dx$, and an equal and opposite force is exerted by the solute atoms on the boundary. The result of this interaction is an excess or deficit of solute (an atmosphere) in the vicinity of the boundary, and the solute concentration (c) is given by

$$c = c_0 \exp\left(-\frac{U}{kT}\right) \qquad (4.21)$$

where c_0 is the equilibrium solute concentration.

For a stationary boundary and a solute which is attracted to the boundary, the interaction energy (U), force (F), solute concentration (c) and diffusion coefficient (D) are shown in figure 4.24 as a function of distance (x) from the boundary.

The extent to which an element segregates to a stationary boundary is usually related to its solubility, and as a general rule the tendency for segregation increases as the solubility decreases as is shown in figure 4.25. It is therefore to be expected that the force exerted by solute atoms on a boundary, which is a function of solute boundary concentration and solute diffusivity, will depend strongly on the specific solute and solvent combination.

If moving boundaries have a larger free volume than stationary boundaries (§4.3.4), then it is to be expected that solubility in a moving boundary will be greater than in a static boundary. There is some indirect evidence for this from a study of Pb-Au alloys by Simpson et al. (1970) who found that precipitation occurred at grain boundaries in specimens undergoing grain growth which were quenched and aged, whereas this effect was not found in material with stationary boundaries.

4.4.2.1 Low boundary velocities

For a moving boundary, the solute profile becomes asymmetric as shown in figure 4.24c such that the centre of gravity of the distribution lags behind the boundary.

For the case illustrated in figure 4.24c, where the boundary is moving from left to right, there is therefore a net force due to the solute, dragging the boundary to the left. As the boundary velocity increases, the solute lags further behind the boundary.

The relationship between the driving pressure (P) and boundary velocity (v) at low velocities is found to be

$$P = \frac{v}{M} + \alpha c_0 v \qquad (4.22)$$

where M is the mobility of the boundary in the absence of solute, and α is a constant which depends on the details of the model.

It may be seen from equation 4.22 that the velocity is predicted to be inversely proportional to the solute concentration (c_0). The apparent activation energy for boundary migration will be the mean value for solute diffusion in the boundary region, which will depend both on the value of U and the assumed U-x and D-x profiles (fig 4.24). Hillert (1979) has shown that the choice of these parameters has a large effect on the results. For example, the CLS model

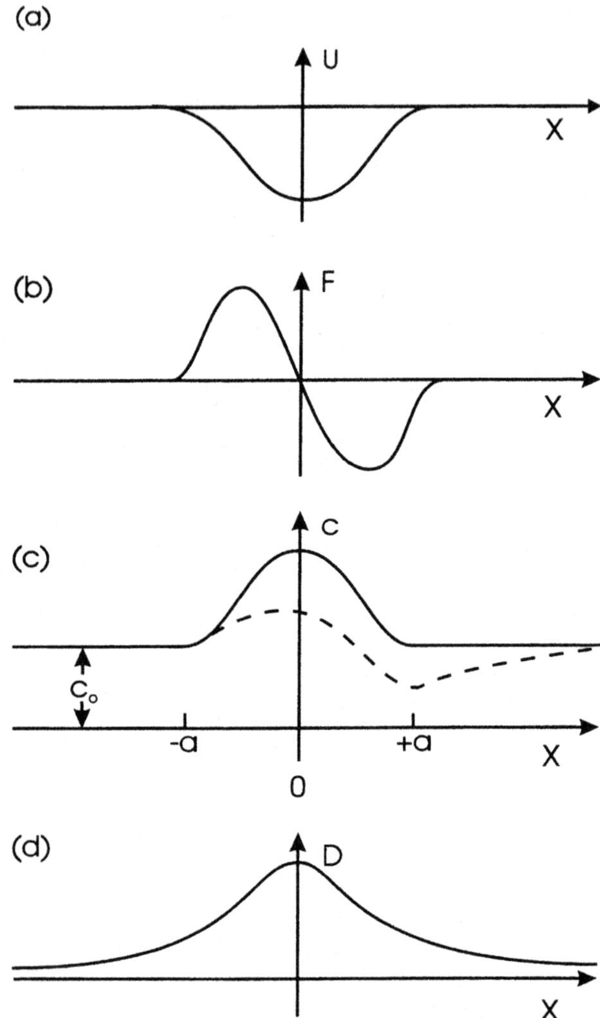

Fig. 4.24. Schematic diagrams of the interaction of solute with a boundary. a) The potential U(x). b) The interaction force F(x) between a solute atom and the boundary. c) The resulting distribution of solute atoms for a stationary boundary (full line) and a boundary moving from left to right (dotted line), d) Diffusivity in the boundary region D(x).

assumed a wedge shaped energy-well rather than that shown in figure 4.24a, and it also assumed that diffusivity is constant throughout the material. It should be emphasised that this theory is limited to dilute solid solutions ($c_0 < 0.1\%$). Hillert (1979) calculated the boundary mobility using a model which allows the boundary diffusivity to differ from that of the bulk and found that the maximum value of the solute drag was much smaller than predicted by equation 4.22 if an increased boundary diffusivity was assumed.

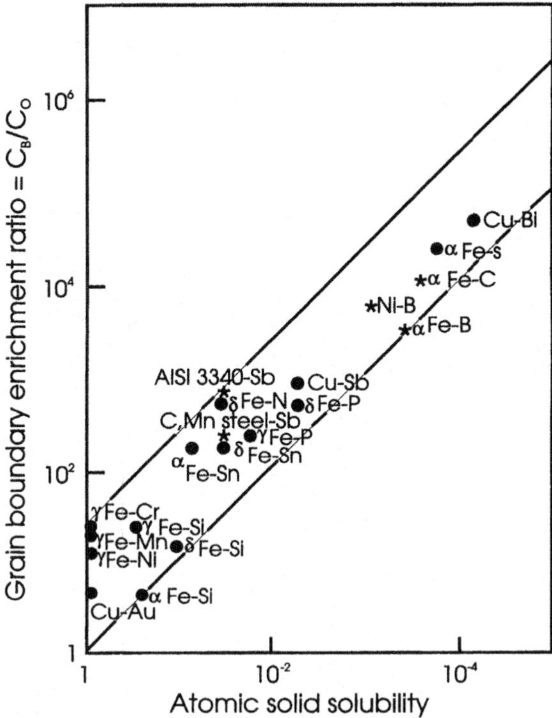

Fig. 4.25. Increasing grain boundary enrichment with decreasing solid solubility in a range of systems, (after Hondros and Seah 1977).

4.4.2.2 High boundary velocities

The relationship between boundary velocity and driving pressure at intermediate velocities is very difficult to calculate as it depends critically on the details and parameters of the model. However a limiting case is reached at high boundary velocities when the solute atoms can no longer keep up with the boundary, which then breaks away from its atmosphere. In this case

$$P = \frac{v}{M} + \frac{c_0}{\alpha' v} \tag{4.23}$$

where α' is a constant.

It is estimated that the first term is dominant and therefore the solute atoms have only a small effect. The mobility and activation energy are then expected to be similar to the intrinsic values for a boundary in a pure metal (§4.3.1).

Equations 4.22 and 4.23 are combined by Cahn (1962) and Lücke and Stüwe (1963) to give

$$P = \frac{v}{M} + \frac{\alpha c_0 v}{1 + \alpha \alpha' v^2} \tag{4.24}$$

4.4.2.3 Predictions of the model

This model predicts that the retarding force due to solute drag varies with boundary velocity, reaching a maximum value as shown in figure 4.26. This figure also shows that the solute drag becomes less effective at high temperatures, and this is because under these conditions the solute concentration near the grain boundary decreases in accord with equation 4.21, so that the atmosphere effectively evaporates.

Although, for the reasons discussed above, it is difficult to fully quantify the CLS model, it is of interest to look at the qualitative predictions regarding the relationships between driving pressure, solute concentration and temperature and boundary velocity, as shown schematically in figure 4.27. In figure 4.27a the effect of three levels of solute content on the relationship between velocity and driving pressure is shown and compared with the behaviour of the pure material (c_0) in which the velocity is proportional to the driving pressure (equation 4.1). It may be seen that for very low solute concentrations the curve is continuous, and only deviates slightly from that predicted for the pure material. However, for high solute concentrations, the curve is S-shaped and predicts a discontinuous change of velocity (dotted lines) from the lower branch (equation 4.22) and the upper branch (equation 4.18). In figure 4.27b, the boundary velocity is plotted as a function of temperature for the cases of low solute **A**, high solute **B** and no solute **C**. The slope of the plot of **ln(v)** vs 1/T may be interpreted in terms of an apparent activation energy. The straight lines **E** and **D** superimposed on **A** and **B,** represent the limiting cases for low velocity as given by equation 4.22.

4.4.2.4 Correlation of experiment and theory

Although the experimental data are very limited and the theories cannot as yet be fully quantified, it appears that many features of the experimental results can be explained in terms of the CLS model.

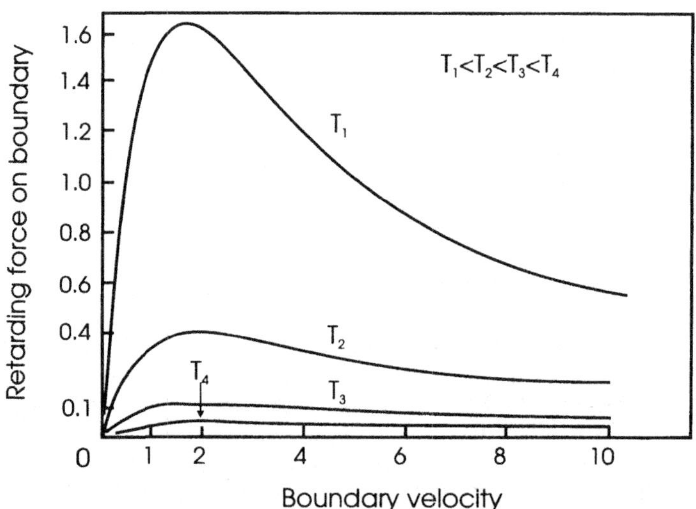

Fig. 4.26. The variation of solute drag force with boundary velocity and temperature, (after after Lücke and Stüwe 1963).

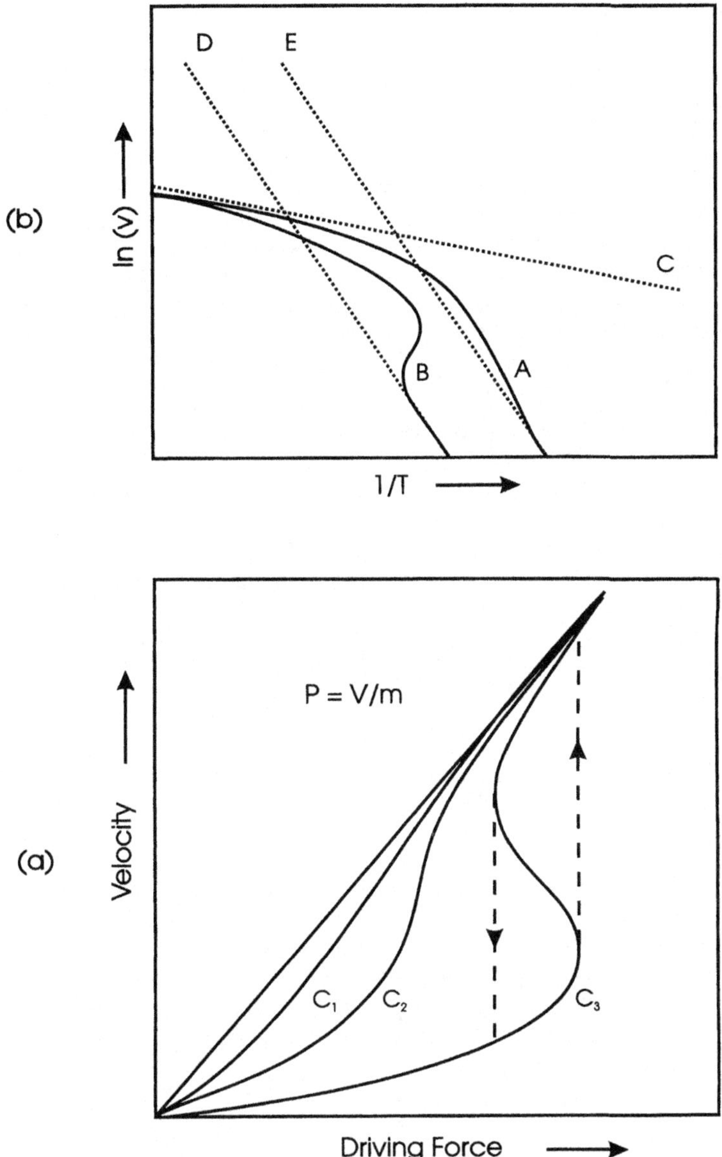

Fig. 4.27. a) Predicted grain boundary velocity as a function of the driving force for different solute concentrations $c_3 > c_2 > c_1$. b) Predicted grain boundary velocity as a function of temperature for low (curve A) and high (curve B) solute concentrations. Curve C represents the behaviour for the pure material and curves D and E the low velocity limiting case described by equation 4.22, (after Lücke and Stüwe 1963).

(i) At low velocities the velocity is inversely proportional to the solute concentration
This result, which is predicted by equation 4.22, is consistent with the experimental results
shown in figures 4.12.

**(ii) At higher driving forces or lower solute concentrations, there is a transition to a high
velocity regime in which the boundary velocity is independent of solute content.**
This is clearly shown in the results of figures 4.12 and 4.14.

(iii) The apparent activation energy will decrease as the temperature increases
This result, which is predicted by the model as shown in figure 4.27b, is in accord with
many of the experimental results discussed in §4.3.3.   The model predicts that at low
boundary velocity, the mobility is controlled by solute diffusion and therefore an activation
energy appropriate for this process is expected.   Figure 4.27b predicts that in moving from
the low to high velocity regimes, the apparent activation energy should exhibit a peak.   The
data shown in figures 4.12, 4.14 and 4.15 are consistent with this.   In some cases, e.g.
figure 4.12, the activation energy in the low velocity regime is a function of solute content.
This may be due to the effect of solute on the boundary free volume and hence on the
activation energy for solute diffusion. The model predicts (figure 4.27b) that in the high
velocity regime, the activation energy for boundary migration should be similar to that for
high purity material.   The experimental results show that this is often the case.

(iv) The nature of the transition depends on the solute concentration and driving force.
Figure 4.27b shows that the nature of the velocity transition with increasing temperature is
predicted to be gradual for low solute concentrations and sharp for higher concentrations.
There is evidence that both gradual (figs 4.14 and 4.15) and discontinuous transitions (fig
4.12) can occur, although a change in the nature of the transition as a function of solute
content has not yet been demonstrated in a single investigation.

(v) The effect of solute is less at higher temperatures
Equation 4.21 shows that at higher temperatures the solute atmosphere will be much weaker.
The results shown in figure 4.18 are in agreement with this.

Although the model appears to be in good qualitative agreement with many of the results,
there are a number of factors which will need to be incorporated into quantitative theories
of grain boundary migration in solute-containing alloys.   These include **the effect of solutes
on boundary structure, energy and diffusivity** and the **interaction of solutes and
boundary crystallography**.

Chapter 5

RECOVERY AFTER DEFORMATION

5.1 INTRODUCTION

5.1.1 The occurrence of recovery

The term **recovery** refers to changes in the properties of a deformed material which occur prior to recrystallization; these changes are such as to partially restore the properties to their values before deformation. It is now known that recovery is primarily due to changes in the dislocation structure of the material, and in discussing recovery it is most convenient to concentrate on the microstructural aspects.

Recovery is not confined to plastically deformed materials, and may occur in any crystal into which a non-equilibrium, high concentration of point or line defects has been introduced. Well known examples of this are materials which have been irradiated or which have been quenched from high temperatures. In these situations recovery will occur on subsequent annealing and this may restore the properties and microstructure completely to the original condition (e.g. Koehler et al. 1957, Baluffi et al. 1963). In this chapter we are concerned only with recovery which occurs on annealing after prior deformation.

Even in the case of deformed metals, point defects as well as dislocations will be formed during deformation. However, as most of the excess point defects will anneal out at low temperatures, this does not usually constitute a separately identifiable stage of the recovery of a cold worked metal and will not be considered further.

Dislocation recovery is not a single microstructural process but a series of micromechanisms which are schematically shown in figure 5.1, and a detailed discussion of these recovery processes is given in §5.3, §5.4 and §5.5. Whether any or all of these occur during the annealing of a particular specimen will depend on a number of parameters, including the material, purity, strain, deformation temperature and annealing temperature, In many cases some of these stages will have occurred during the deformation as **dynamic recovery**. Although the recovery stages tend to occur in the order shown, there may be significant overlap between them.

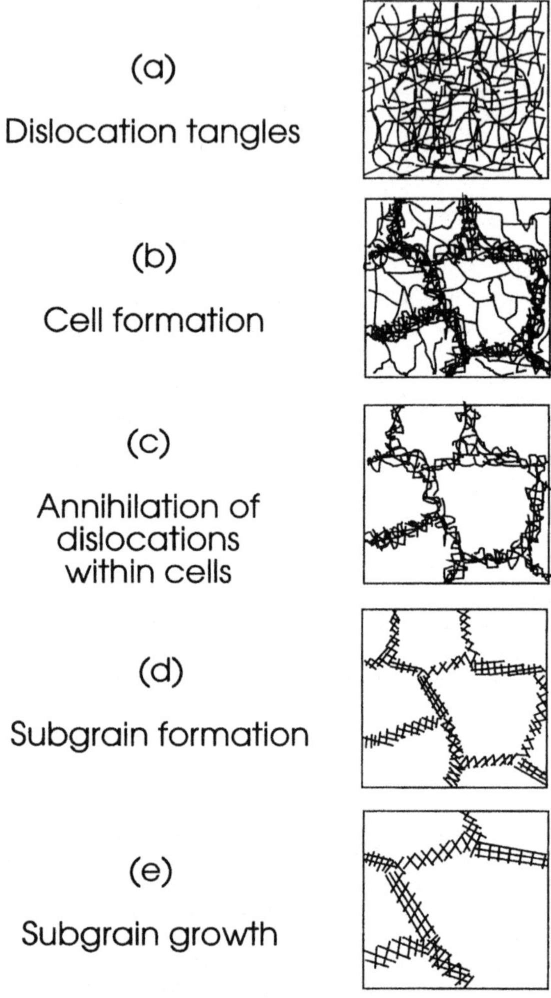

Fig. 5.1. Various stages in the recovery of a plastically deformed material.

Before considering the details of recovery, we should note that recovery and recrystallization
are competing processes as both are driven by the stored energy of the deformed state (§2.2).
Once recrystallization has occurred and the deformation substructure has been consumed,
then clearly no further recovery can occur. The extent of recovery will therefore depend on
the ease with which recrystallization occurs. Conversely, because recovery lowers the
driving force for recrystallization, a significant amount of prior recovery may in turn
influence the nature and the kinetics of recrystallization. The division between recovery and
recrystallization is sometimes difficult to define, because recovery plays an important role
in nucleating recrystallization. Additionally, as will be discussed at the end of this chapter,
there are circumstances in which there is no clear distinction between the two phenomena.

The extensive early work on the recovery of deformed metals has been reviewed by Beck (1954) and Bever (1957), but despite its obvious importance, recovery has attracted little interest during the past 30 years. However, the current industrially driven move to produce quantitative physically-based models for annealing processes has resulted in renewed interest in recovery (e.g. Nes 1995).

5.1.2 Properties affected by recovery

During recovery, the microstructural changes in a material are subtle and occur on a small scale. The microstructures as observed by optical microscopy do not usually reveal much change and for this reason, recovery is often measured indirectly by some bulk technique, for example by following the change in some physical or mechanical property.

Measurement of the changes in stored energy by calorimetry is the most direct method of following recovery because the stored energy is directly related to the number and configuration of the dislocations in the material (§2.2). In a classic series of experiments Clareborough and colleagues (Clareborough et al. 1955, 1956, 1963) studied the stored energy release in copper, nickel and aluminium, and the results of the early work in this area are reviewed by Titchener and Bever (1958). The development of high-sensitivity differential scanning calorimeters has, in recent years, led to their being increasingly used to measure annealing phenomena (Haessner 1990). However, because of the small amount of stored energy in a deformed material, calorimetric measurements of recovery and recrystallization can only be made on materials in which no phase transformations occur over the temperature range of the experiment. For example, the stored energy of deformed aluminium is $\sim 20 J/mol$ (§2.2), and the latent heat released by precipitating a volume fraction of only 0.01 of $CuAl_2$ in aluminium is also $\sim 20 J/mol$. It is now thought that some early measurements of stored energy are suspect because of the occurrence of small amounts of phase transformation during the experiments.

Figure 5.2 from the work of Schmidt and Haessner (1990), shows the annealing of high purity aluminium which was deformed at -196°C and maintained at that temperature until the calorimetry was performed. The peak at around -70°C is due to the recovery of point defects and that at ~-20°C is due to recrystallization. A separate peak due to dislocation recovery is not detected. The 99.999% high purity aluminium recrystallizes at such a low temperature that either no significant dislocation recovery occurs prior to recrystallization, or else the peak is too close to that due to recrystallization to be detected.

Figure 5.3 shows the heat evolved on annealing 99.998% aluminium after deformation at room temperature. In this case two stages are detected, the broad peak at 100-250°C corresponds to the heat evolved during dislocation recovery and the peak at 300°C to that evolved during recrystallization.

Several other physical properties which are altered during plastic deformation are subsequently modified by recovery, and these include density and electrical resistivity (fig 5.3). It is however difficult to relate these quantitatively to the microstructural changes which occur during recovery, and as is the case for stored energy, these parameters are sensitive to any small amounts of phase transformation which may occur on annealing.

The microstructural changes which occur during recovery affect the mechanical properties,

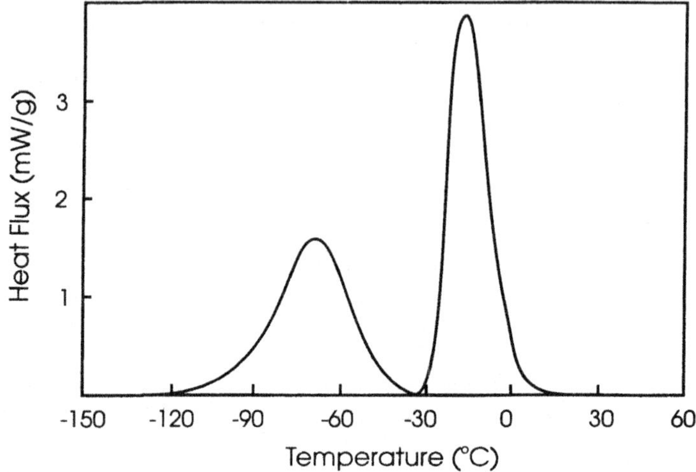

Fig. 5.2. The recovery of 99.999% Al deformed to a strain of 6.91 at 77K as measured by differential scanning calorimetry, (Schmidt & Haessner 1990).

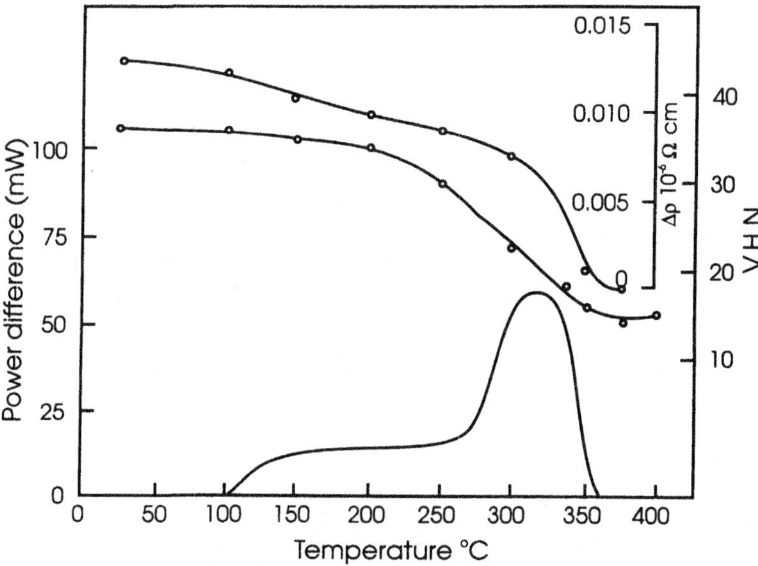

Fig. 5.3. The recovery of 99.998% Aluminium deformed 75% in compression, measured by calorimetry, electrical resistivity and hardness. Heating rate 6°C/min, (Clareborough et al. 1963).

and therefore recovery is often measured by changes in the yield stress or hardness of the material (fig 5.3), although these changes are often small. The mechanical properties of a recovered material are of some practical importance and their relationship to the microstructural changes has been extensively studied (§5.5.2.3).

5.2 EXPERIMENTAL MEASUREMENTS OF RECOVERY

5.2.1 The extent of recovery

5.2.1.1 The effect of strain

For polycrystalline metals, complete recovery can only occur if the material has been lightly deformed (Masing and Raffelsieper 1950, Michalak and Paxton 1961). However, single crystals of hexagonal close-packed metals such as zinc, which can be deformed to large strains on only one slip system, are able to completely recover their original microstructure and properties on annealing (Haase and Schmid 1925, Drouard et al. 1953). Single crystals of cubic metals if oriented for single slip and deformed in stage 1 of work hardening may also recover almost completely on annealing (Michalak and Paxton 1961). However, if the crystals are deformed into stages II or III of work hardening, then recrystallization may intervene before any significant amount of recovery has occurred (Humphreys and Martin (1968). This behaviour is explainable in terms of the annealing behaviour of the various types of dislocation structure produced during deformation (§5.3).

The fraction of the property change which may be recovered on annealing at a constant temperature is usually found to increase with strain (Leslie et al. 1963). However, this trend may be reversed in more highly strained materials, because of the earlier onset of recrystallization.

5.2.1.2 The effect of annealing temperature

The annealing curves of lightly deformed iron (fig 5.4) and zinc (fig 5.5) indicate that more complete recovery occurs at higher annealing temperatures.

5.2.1.3 The nature of the material

The nature of the material itself also determines the extent of recovery. One of the most important parameters is the **stacking fault energy** γ_{SFE}, which, by affecting the extent to which dislocations dissociate, determines the rate of dislocation climb and cross slip. These are the mechanisms which usually control the rate of recovery (§5.3). In metals of low stacking fault energy (table 2.4) such as copper, α-brass and austenitic stainless steel, climb is difficult, and little recovery of the dislocation structure normally occurs prior to recrystallization. However, in metals of high stacking fault energy such as aluminium and α-iron, climb is rapid, and significant recovery may occur as seen in figures 5.3 and 5.4. Solutes may influence recovery by their effect on the stacking fault energy (as discussed above), by pinning dislocations, or by affecting the concentration and mobility of vacancies. For example, magnesium is known to pin dislocations in aluminium, and to retard dynamic recovery (§2.3.2.4). Because of the resultant high stored energy, recovery in Al-Mg alloys is rapid (Barioz et al. 1992), and may be even faster than in pure aluminium (Perryman 1955). This effect is presumably due to the pure aluminium having already extensively recovered after deformation. In a similar manner, small amounts of manganese pin the dislocations in iron (Leslie et al. 1963), and by inhibiting dynamic recovery and retaining a high dislocation content, promote subgrain formation on subsequent annealing.

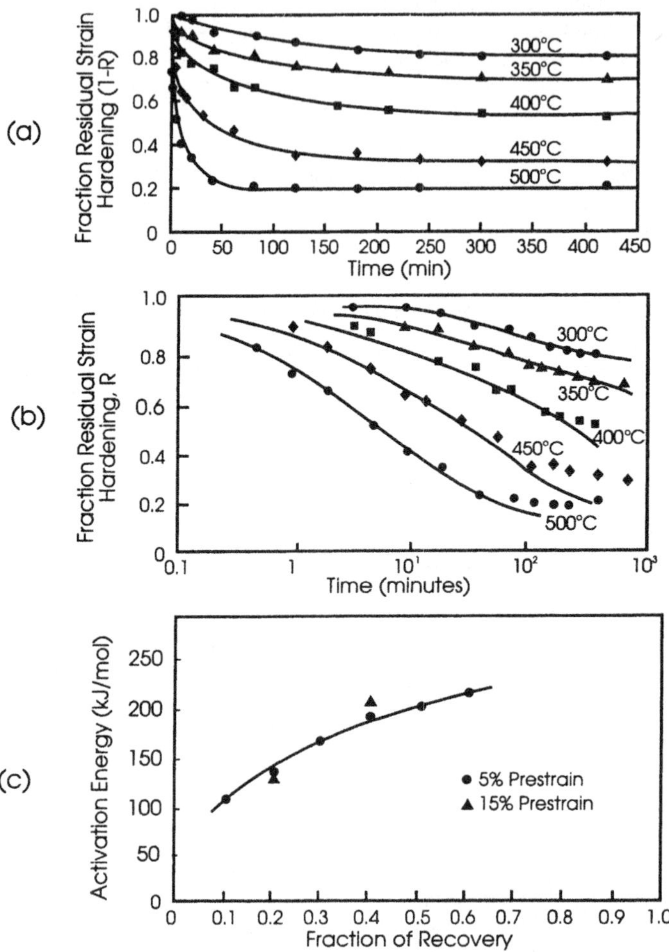

Fig. 5.4. Isothermal recovery of iron deformed 5% at 0°C. a) Recovery of hardness as a function of time. b) Logarithmic plot c) Change of activation energy during recovery, (Michalak and Paxton 1961).

5.2.2 Measurements of recovery kinetics

Experimentally, recovery is often measured by the changes in a single parameter, such as hardness, yield stress, resistivity or heat evolution. If the change of such a parameter from the annealed condition is X_R, then the kinetics of recovery, dX_R/dt may be determined from experiment. It is often difficult to obtain a fundamental insight into recovery from such analyses, firstly because the relationship of the parameter X_R to the **microstructure** is usually very complex, and secondly because recovery may involve several concurrent or consecutive atomistic mechanisms (fig 5.1) each with its own kinetics.

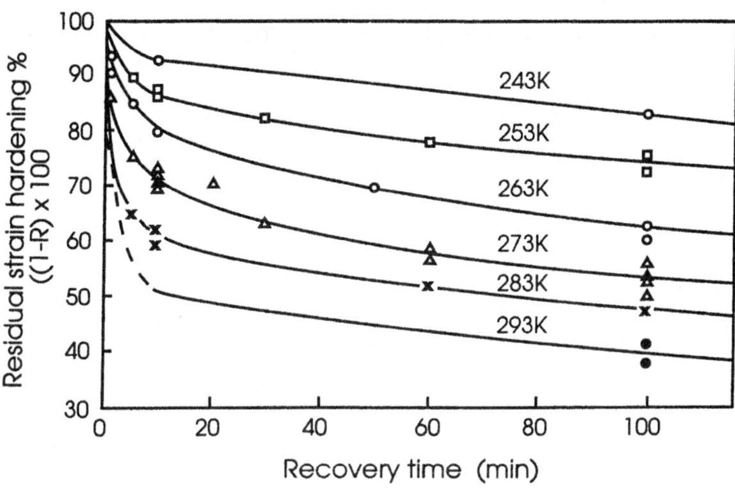

Fig. 5.5. Recovery in zinc crystals deformed to a shear strain of 0.08,
(Drouard et al. 1953).

5.2.2.1 Empirical kinetic relationships

Experimental results have been analyzed in terms of several different empirical relationships between X_R and t. The two most commonly reported isothermal relationships, which we will refer to as **types 1 and 2** are as follows:

Type 1 kinetics:

$$\frac{dX_R}{dt} = -\frac{c_1}{t} \tag{5.1}$$

which integrates to

$$X_R = c_2 - c_1 \ln t \tag{5.2}$$

where c_1 and c_2 are constants.

It is clear that this form of relationship cannot be valid during the early stages of recovery $(t \to 0)$ when $X_R \to X_0$ or at the end of recovery $(t \to \infty)$ when $X_R \to 0$.

Type 2 kinetics:

$$\frac{dX_R}{dt} = -c_1 X_R^m \tag{5.3}$$

which integrates to

$$X_R^{-(m-1)} - X_0^{-(m-1)} = (m - 1) c_1 t \tag{5.4}$$

for $m > 1$, and

$$\ln(X_R) - \ln(X_0) = c_1 t \tag{5.5}$$

for $m = 1$.

5.2.2.2 Recovery of single crystals deformed in single slip.

The kinetics of recovery of single crystals of zinc deformed at -50°C in simple shear so that deformation occurred by shear on the basal plane were measured by Drouard et al. (1953) and are shown in figure 5.5. The extent of the recovery (R) was defined in terms of the yield stress of the recovered crystals (σ), the yield stress of the as-deformed crystal (σ_m) and the yield stress of the undeformed crystal (σ_0) as

$$R = \frac{(\sigma_m - \sigma)}{(\sigma_m - \sigma_0)} \qquad (5.6)$$

The relationship between the amount of recovery, time and temperature was found, over a wide range of conditions to be

$$R = c_1 \ln t - \frac{Q}{kT} \qquad (5.7)$$

These are **type 1** kinetics, and the activation energy (Q) was found to be 83.7kJ/mol, which is close to the activation energy for self diffusion of zinc (table 4.1).

Other measurements of recovery in zinc crystals (Cottrell and Aytekin 1950) and of aluminium crystals (Kuhlmann et al. 1949, Masing and Raffelsieper 1950) have also found a relation between X_R and time, of a form consistent with **type 1** kinetics.

An unusual type of recovery has been found in deformed dispersion-hardened single crystals of copper (Hirsch and Humphreys 1969, Gould et al. 1974). If these crystals are deformed to strains of up to ~0.15 at room temperature or below, then on annealing at room temperature, a large fraction of the work hardening recovers, as shown in figure 5.6. The activation energy for the process is found to be ~90kJ/mol, which is less than half th^r activation energy for self diffusion in copper. This recovery effect has been attributed to the annealing out of the unstable high energy **Orowan loops** which form at the particles during deformation (§8.2.3), by a process of dislocation climb controlled by **pipe or core diffusion**. Although similar effects have been reported in particle-containing aluminium single crystals (Stewart and Martin 1975), this type of recovery has not been detected in particle-containing polycrystals.

5.2.2.3 Recovery kinetics of polycrystals

The recovery kinetics of iron in figure 5.4 are typical of metals in which extensive recovery occurs, and show that the rate of recovery is initially rapid and then decreases monotonically.

Type 1 kinetic relationships have frequently been found to describe recovery in polycrystals, and plots of fractional recovery against log(time) for iron are shown in figure 5.4b. However, it may be seen from this figure that a straight line corresponding to equation 5.2 is usually only found over part of the recovery range as would be expected from this type of relationship.

An activation energy for recovery may be obtained from the temperature dependence of the recovery rate at a particular stage of the recovery, using the usual Arrhenius relationship. In some cases the activation energy is found to rise as recovery proceeds. For example, Michalak and Paxton (1961) measuring the recovery kinetics of zone-refined iron deformed

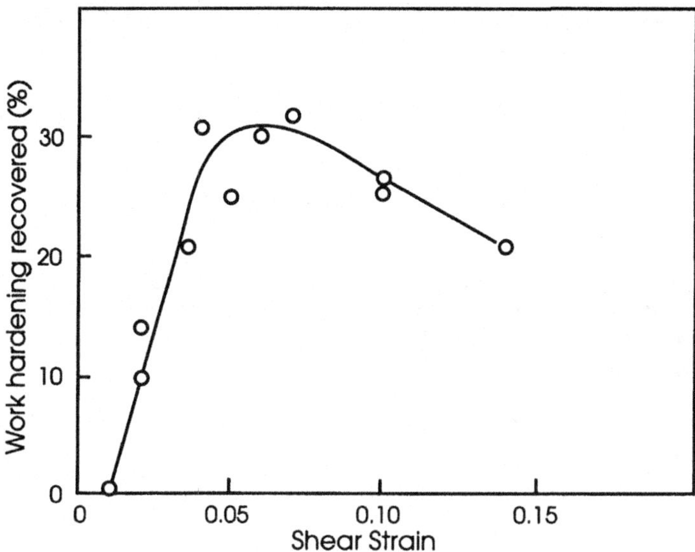

Fig. 5.6. The amount of work hardening which is recovered in 3hrs at room temperature in
Cu-Al$_2$O$_3$ crystals oriented for single slip, and deformed at 77K,
(Humphreys and Hirsch 1976).

5%, found that the activation energy rose from 91kJ/mol at the start of recovery to
~220kJ/mol (close to that for self diffusion in iron), when recovery was essentially complete
as shown in figure 5.4c. Van Drunen and Saimoto (1971) measured the recovery kinetics
of deformed <100> copper crystals at various temperatures. They found kinetics of type
2 with m=2, and an activation equal to that for self diffusion in copper.

An interpretation of the types of kinetics discussed above can only be obtained if the physical
processes occurring during recovery are understood. In the following sections we will
discuss three levels of recovery - **dislocation annihilation, dislocation rearrangement** and
subgrain growth, and where possible relate these mechanisms to the recovery kinetics.

5.3 DISLOCATION MIGRATION AND ANNIHILATION DURING RECOVERY

During recovery the stored energy of the material is lowered by dislocation movement.
There are two primary processes, these being the **annihilation of dislocations** and the
rearrangement of dislocations into lower energy configurations. Both processes are
achieved by glide, climb and cross-slip of dislocations. We do not intend to discuss the
physics of dislocations in detail, and for further information the reader should consult a
suitable reference such as Friedel (1964), Hirth and Lothe (1968) or Hull and Bacon (1984).

5.3.1 General considerations

A schematic drawing of a crystal containing an array of edge dislocations is shown in figure 5.7. The elastic stress fields of the dislocations interact and the resultant forces will depend on the Burgers vectors and relative positions of the dislocations. For example, dislocations of opposite sign on the same glide plane e.g. A and B, may annihilate by gliding towards each other. Such processes can occur even at low temperatures, lowering the dislocation density during deformation and leading to **dynamic recovery**. Dislocations of opposite Burgers vector on different glide planes, e.g. C and D, can annihilate by a combination of glide and climb. As climb requires thermal activation, this can only occur at high homologous temperatures. A similar configuration of screw dislocations would recover by annihilation of dislocations by cross-slip. This would occur at low temperatures in a material such as aluminium with a high stacking fault energy, but at high temperatures for a material of lower stacking fault energy.

The crystal of figure 5.7 contains dislocations of only one type of Burgers vector and it contains equal numbers of dislocations of the two signs. Thus complete recovery by dislocation annihilation is possible. This corresponds to the case of crystals which have deformed by slip on only one system as discussed in §5.2.1.1

5.3.2 The kinetics of dipole annihilation

Two parallel dislocations of opposite Burgers vector, such as C and D in figure 5.7 are known as a dislocation dipole. In a well known paper, Li (1966) discussed the kinetics of annihilation of dislocation dipoles of edge, screw or mixed character.

The attractive force (F) between two parallel screw dislocations of opposite sign, separated by a distance x is

$$F = \frac{G b^2}{2 \pi x} \tag{5.8}$$

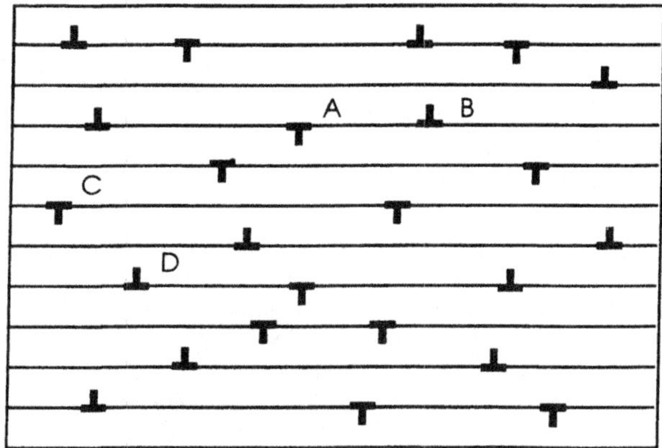

Fig. 5.7. Schematic diagram of a crystal containing edge dislocations.

If the dislocation climb velocity is proportional to F (see equation 4.7), then the dipole separation will decrease as

$$\frac{dx}{dt} = -c_1 F \qquad (5.9)$$

Combining equations 5.8 and 5.9 we obtain

$$\frac{dx}{dt} = -\frac{c_2}{x} \qquad (5.10)$$

Measurements of the life of screw dipoles in LiF (Li 1966) shown in figure 5.8 are in agreement with this relationship. Li also showed that the kinetics of annihilation of edge and mixed dipoles would be similar to those of equation 5.10.

In order to use this theory to calculate the recovery of a material containing a distribution of dipoles, it is necessary first to assume that the dipoles do not interact with each other, and second to know the initial distribution of dipole heights. Li (1966), making assumptions about the dipole size distribution based on their mechanism of formation during deformation showed that

$$\frac{d\rho}{dt} = -c_1 \rho^2 \qquad (5.11)$$

or

$$\frac{1}{\rho} - \frac{1}{\rho_0} = c_1 t \qquad (5.12)$$

where ρ_0 is the initial dislocation density. These kinetics are equivalent to those of equations 5.3 and 5.4 (**type 2 kinetics**) with **m=2**

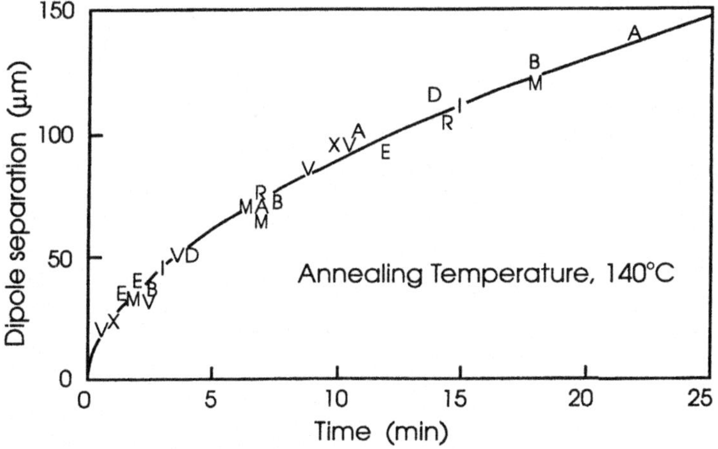

Fig. 5.8. Life of screw dislocation dipoles in LiF crystals, (Li 1966).

Rates of change in dislocation density during recovery consistent with **type 2** kinetics have been found by Li (1966) for LiF (m=2) and by Prinz et al. (1982) in copper and nickel (m=3).

It is of interest to compare the kinetics of the annihilation of non-interacting dislocation dipoles with the kinetics predicted for a random array of parallel edge dislocations such as is shown schematically in figure 5.7 when both dipole and longer range interactions are considered. The elastic forces acting on a single edge dislocation due to a parallel dislocation of the same or opposite Burgers vector are known (see e.g. Hirth and Lothe 1968, Hull and Bacon 1984), and for an array such as shown in figure 5.7, the forces acting on any dislocation may be taken to be the sum of the forces due to the other dislocations. If the velocity of a dislocation is assumed to be proportional to the force acting on it (equation 4.7), an assumption which was shown to be valid for the dipoles discussed above, then the rearrangement of the dislocation array due to its internal forces may be computed. Starting with a computer-simulated array of 500 dislocations of each of the two opposite Burgers vectors, the microstructure evolves as shown in figure 5.9, and the kinetics are shown in figure 5.10. Although the rate of recovery is initially consistent with equation 5.11, it subsequently falls more rapidly with time and, as shown in figure 5.10b, is closer to a relationship of the form $\rho^{-1.5} \propto t$. The reason for the slower recovery rate at longer times may be seen from examination of figure 5.9b. As annealing proceeds, the formation and annihilation of dislocation dipoles occurs, e.g. at **D**. However, in addition to this, the dislocations begin to form low energy tilt boundaries (**B**), and as these are reasonably stable structures, the rate of dislocation annihilation is now much slower. Therefore there are **two** recovery mechanisms operating simultaneously, **dislocation annihilation** and **rearrangement into stable configurations**; this illustrates well the difficulties in analysing recovery data in terms of a single recovery mechanism.

The simulation described above neglects the behaviour of screw dislocations and is therefore unlikely to accurately predict the annealing of a deformed material containing dislocations of one type of Burgers vector. However, it does indicate that in addition to dipole interactions, other longer range interactions may be important in determining the annealing kinetics, and this simulation is in general agreement with the in-situ HVEM annealing experiments of Prinz et al. (1982) which showed low angle boundary formation and dipole annihilation to occur concurrently.

5.3.3 Recovery kinetics of more complex dislocation structures

In the above section we considered only the recovery of simple dislocation arrays consisting of parallel dislocations of similar Burgers vector. In deformed polycrystals and single crystals deformed on more than one slip system, the dislocation structures are much more complex. Friedel (1964) has considered the recovery kinetics of general dislocation structures. It is not obvious which mechanism controls dislocation movement during recovery in various circumstances, as in addition to **dislocation climb** by either **bulk** or **core** diffusion, there is also the possibility that the processes of **thermally activated glide or cross-slip** may be rate controlling.

Following Friedel, we assume that the dislocation structure is in the form of a three-dimensional network (Frank network) of mesh size **R**. From the geometry of the network we expect that the dislocation velocity (**v**) is related to the change in scale of the network by

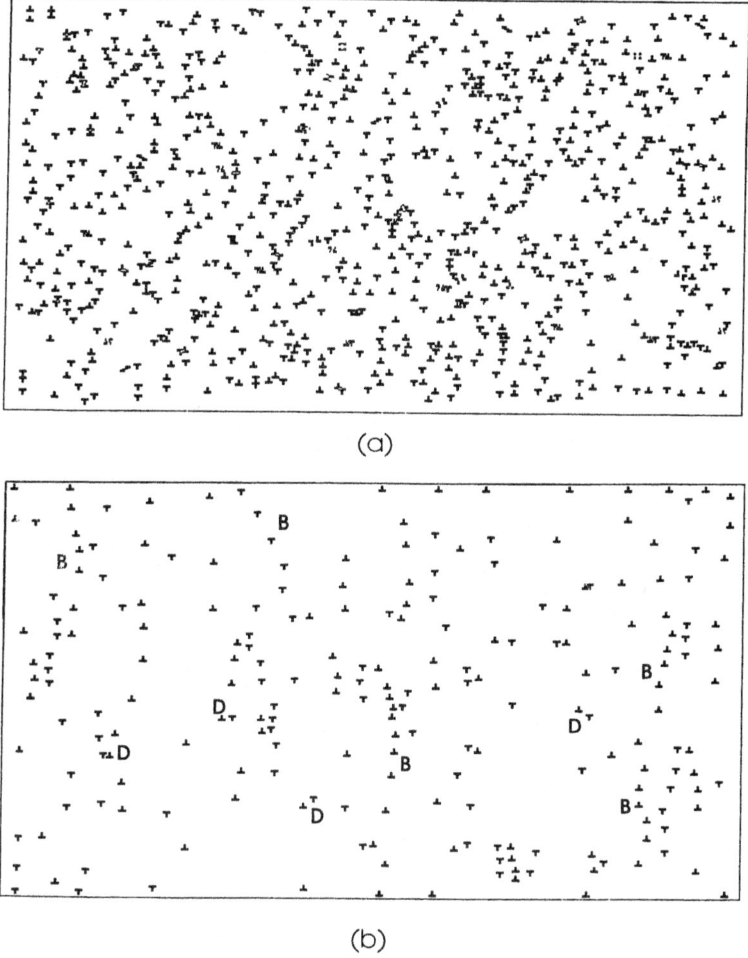

Fig 5.9. Computer simulation of the annealing of edge dipoles a) initial microstructure b) later stage of recovery showing the incipient formation of low angle boundaries.

dR/dt ~ v, and considerations of the energy of the system show that the driving force for coarsening of the network by dislocation migration is then

$$F \approx \frac{G b^2}{R} \qquad (5.13)$$

5.3.3.1 Control by dislocation climb

Under conditions of small driving force and a negligible vacancy supersaturation then the dislocation velocity (**v**) is given by equation 4.7. Thus, from equations 4.7 and 5.13

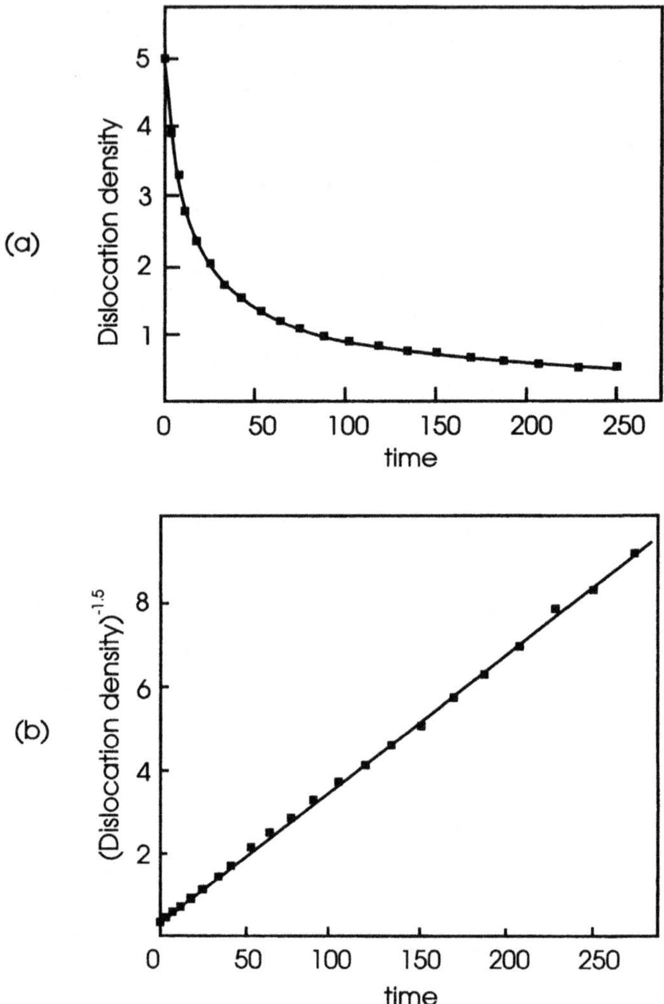

Fig. 5.10. Kinetics of the computer simulation of figure 5.9, a) The variation of dislocation density (ρ) with time, b) The variation of $\rho^{-1.5}$ with time.

$$v = \frac{dR}{dt} = \frac{D_s\,G\,b^3\,c_j}{k\,T}\cdot\frac{1}{R} \qquad (5.14)$$

and hence

$$R^2 = R_0^2 + c_1 t \qquad (5.15)$$

where $c_1 = 2D_s\,Gb^3c_j/kT$

The dislocation density ρ is related to the network spacing by $\rho \sim R^{-2}$, and hence equation 5.15 may be written as

$$\frac{1}{\rho} - \frac{1}{\rho_0} = t\,c_1 \qquad\qquad (5.16)$$

which is of the same form as the **type 2** empirical kinetics of equation 5.4 with **m=2**.

By assuming the relationship between flow stress (σ) and ρ given by equation 2.2, this can also be written as

$$\frac{1}{\sigma_0^2} - \frac{1}{\sigma^2} = c_3\,t \qquad\qquad (5.17)$$

where $c_3 = (2\,D_S\,c_j)/(\alpha^2\,GbkT)$

At high temperatures climb is expected to be controlled by the formation (Q_{VF}) and movement (Q_{VM}) of vacancies, as is self diffusion (Q_S). However at lower temperatures, jog formation (Q_j) may also be a rate determining step. Therefore the activation energy for climb (Q_{CL}) is expected to be

$$Q_{CL} = Q_{VF} + Q_{VM} = Q_S \ \text{(at high temperatures)} \qquad\qquad (5.18)$$

$$Q_{CL} = Q_{VF} + Q_{VM} + Q_j \ \text{(at low temperatures)}$$

There are however, circumstances in which the climb may be controlled by diffusion along the dislocation lines (core or pipe diffusion), rather than through the lattice. In these circumstances the activation energy (Q_c) will be substantially lower than Q_s, and typically $Q_c/Q_s \sim 0.5$ (table 4.1). Prinz et al. (1982) have suggested that for extended dislocations, the climb during recovery is likely to be controlled by core diffusion.

5.3.3.2 Control by thermally activated glide of dislocations

Several authors (Kuhlmann 1948, Cottrell and Aytekin 1950, Friedel 1964) have considered the possibility that recovery may be controlled by thermally activated glide or cross-slip. It is proposed by these authors that the activation energy (Q) is a decreasing function of the internal stress (σ) and that the rate of recovery is

$$\frac{d\sigma}{dt} = -c_1 \exp\left(-\frac{Q(\sigma)}{kT}\right) \qquad\qquad (5.19)$$

It is suggested that the stress dependence of the activation energy at low stresses may be of the form

$$Q(\sigma) = Q_0 - c_2\,\sigma \qquad\qquad (5.20)$$

and hence

$$\frac{d\sigma}{dt} = -c_1 \exp\left(-\frac{(Q_0 - c_2\,\sigma)}{kT}\right) \qquad\qquad (5.21)$$

On integration this gives (Cottrell and Aytekin 1950)

$$\sigma = \sigma_0 - \frac{kT}{c_2} \ln\left(1 + \frac{t}{t_0}\right) \tag{5.22}$$

where σ_0 is the flow stress at $t=0$ and t_0 is given by

$$\frac{kT}{c_1 c_2} \exp\left(\frac{Q_0 - c_2 \sigma}{kT}\right)$$

This relationship is similar in form to the **type 1** kinetics of equation 5.2.

Therefore if recovery is controlled by thermally activated glide, we expect an activation energy which **decreases with increasing dislocation density** (or cold work), and which **increases during the anneal**. This relationship is consistent with the experimental results of Kuhlmann et al. (1949), Cottrell and Aytekin (1950) and Michalak and Paxton (1961) (figure 5.4c). It has been claimed that recent measurements of recovery in Al-Mg alloys, shown in figure 5.11 (Barioz et al. 1992), are consistent with this model. Q_0 was found to be independent of strain, but to depend on the magnesium content of the alloy, and ranged from 277kJ/mol for Al-1%Mg to 231kJ/mol for Al-5% Mg. In all cases Q_0 was found to much larger than that for self diffusion for aluminium (~ 142kJ/mol).

Hirth and Lothe (1968) have discussed the velocity of dislocations under various other circumstances, and in particular have considered the thermally activated glide of jogged dislocations. The particular relationship between the dislocation velocity and the driving force is found to depend not only on the type of diffusion (core or bulk), but also on the way in which the jogs are assumed to move and on the effect of solutes.

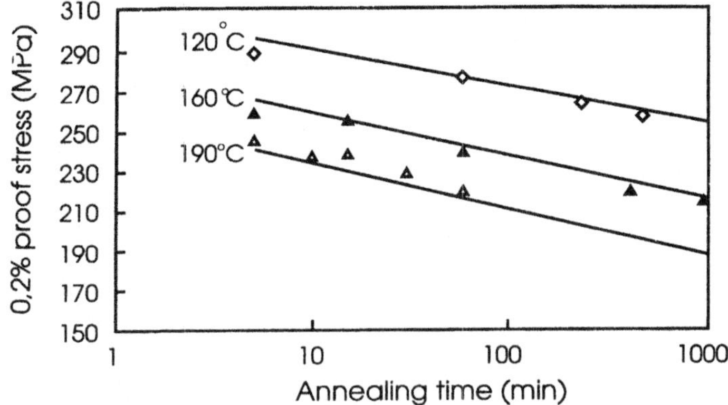

Fig. 5.11. Recovery of Al-3%Mg ($\epsilon=3$) plotted according to equation 5.22, (Barioz et al. 1992).

Stacking fault energy will also be an important parameter in determining the rate of climb or cross-slip.   Argon and Moffat (1981) proposed that the climb of extended edge dislocations in fcc metals is controlled by vacancy evaporation at extended jogs, and under these conditions the rate of climb is given by

$$ v = \frac{c' D_x}{kT} \cdot F \gamma_{SFE}^2 \tag{5.23} $$

where D_x may be either core or bulk diffusivity.

Reference to the discussion of climb-controlled recovery in §5.3.3.1 shows that if the climb velocity is given by equation 5.23 instead of equation 4.7, then the recovery rate will be proportional to γ_{SFE}^2.

In conclusion we see that there are several possible kinetic equations to describe uniform dislocation recovery, some of which predict kinetics similar to those which have been observed experimentally. In principle, it should be possible to identify the applicable model from the form of the experimental annealing response. However, much more work is needed before this can be done with confidence, and we highlight four areas of difficulty.

(i) The relationships between mechanical properties and microstructure are in reality much more complex than those used in current recovery models (§5.2.2).
Although recovery is frequently measured by changes in mechanical properties, reliable interpretation of such data in terms of the dislocation content and arrangement is not possible in most cases.

(ii) The mechanisms of dislocation annihilation and dislocation rearrangement may not be consecutive, but may occur concurrently (fig 5.9).
Because of this, a model of recovery based solely on dislocation annihilation or growth of a Frank dislocation network may not be appropriate, and the effect of the recovery mechanisms to be discussed in §5.4 and §5.5 will also need to be incorporated as appropriate into the overall recovery model.

(iii) The effects of solutes and stacking fault energy on dislocation mobility need more accurate formulation than has been possible to date.

(iv) The heterogeneity of recovery throughout the microstructure cannot be ignored.
As is evident in considering the **deformed state** (§2.3) and **primary recrystallization** (chapter 6), inhomogeneities in the deformed state lead to inhomogeneous annealing. Therefore changes in the rate of recovery, as measured by changes of a bulk property, during an anneal may well be due to different recovery rates in different parts of the microstructure. This factor, which was first suggested by Kuhlmann (1948) in order to explain the change in activation energy during recovery, needs to be incorporated into any comprehensive model of recovery.

5.4 REARRANGEMENT OF DISLOCATIONS INTO STABLE ARRAYS

5.4.1 Polygonization

If, as shown in figure 5.12a, unequal numbers of dislocations of two signs are produced during deformation, then, the excess cannot be removed by annihilation (fig 5.12b). On annealing, these excess dislocations will arrange into lower energy configurations in the form of regular arrays or **low angle grain boundaries** (LAGBs). The simplest case is that shown in figure 5.12c in which dislocations of only one Burgers vector are involved. Such a structure may be produced by bending a single crystal which is deforming on a single slip system, as was first demonstrated by Cahn (1949), and this mechanism is often known as **polygonization**.

The dislocation walls shown in figure 5.12c are known as **tilt boundaries**. These are a particularly simple type of **low angle grain boundary**, and such boundaries are discussed more fully in chapter 3. According to the Read-Shockley equation (equation 3.5), the energy of a tilt boundary **increases** with increasing misorientation and the energy per dislocation **decreases** with increasing misorientation (fig 3.4). Therefore there is a driving force to form fewer, more highly misoriented boundaries as recovery proceeds.

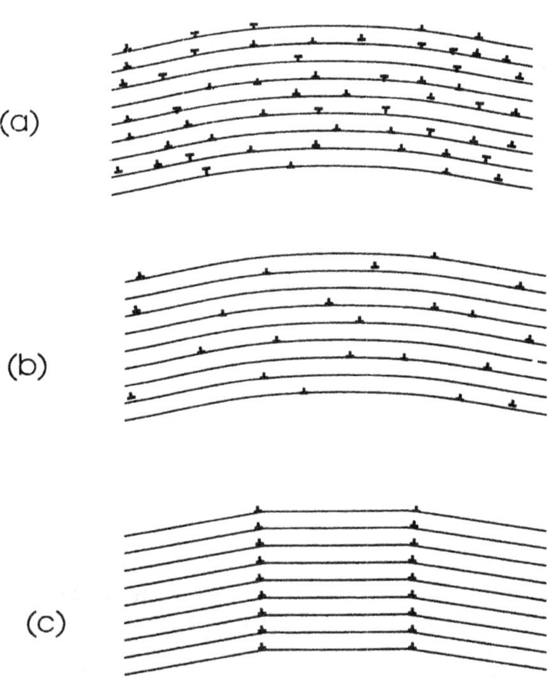

Fig 5.12. Polygonisation of a bent crystal containing edge dislocations. a) As deformed, b) After dislocation annihilation, c) Formation of tilt boundaries.

5.4.2. Subgrain formation

In the case of a polycrystalline material subjected to large strains, the dislocation structures produced on deformation and on subsequent annealing are much more complex than the simple case shown in figure 5.12c because dislocations of many Burgers vectors are involved. Dislocations of two or more Burgers vectors react to form two-dimensional networks whose character depends on the types of dislocation involved (§3.2).

In an alloy of medium or high stacking fault energy, the dislocations are typically arranged after deformation in the form of a three dimensional **cell structure**, the cell walls being complex dislocation tangles (§2.3.2.1 and fig 2.6a) and the size of the cells depending on the material and on the strain.

The transmission electron micrograph of figure 5.13a shows equiaxed cells of diameter $\sim 1\mu$m in deformed aluminium. By annealing this specimen **in situ** in an HVEM, the recovery processes can be followed directly. Figure 5.13b shows the same area after annealing. The tangled cell walls, e.g. at **A** and **B** have become more regular dislocation networks or low angle grain boundaries and the number of dislocations in the cell interiors has diminished. The cells have now become **subgrains** (§2.3.2.1), and at this stage there is little change in the scale of the structure.

The transition from loose tangles of dislocations in cell walls to subgrain boundary formation as shown in figure 5.13, can be considered to be a distinct stage in the recovery process.

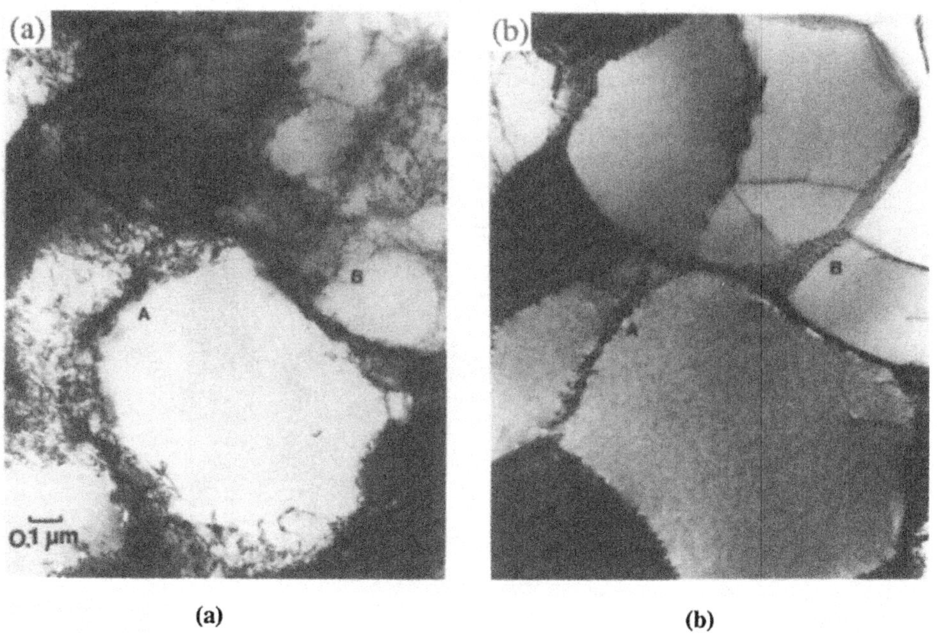

(a)   (b)

Fig. 5.13. Transmission high voltage electron micrographs (HVEM) of aluminium deformed 10% and annealed in-situ. a) Deformed structure. b) Same area after 2 minute anneal at 250°C.

It involves some annihilation of redundant dislocations and the rearrangement of the others into low angle grain boundaries. There have been many TEM studies of this stage of recovery, including those of Bailey and Hirsch (1960) on nickel, Carrington et al. (1960) and Hu (1962) on iron, and Lytton et al. (1965) and Hasegawa and Kocks (1979) on aluminium. As this process occurs, the mechanical properties are altered, the yield stress being lowered, but the work hardening rate rising, as shown by Lytton et al. (1965) and Hasegawa and Kocks (1979).

In some cases **dynamic recovery** during deformation occurs to such an extent that the dislocations are already in the form of a well developed subgrain structure after deformation, and post-deformation recovery then involves mainly a coarsening of the subgrain structure as discussed in the next section. The factors which tend to promote the formation of a subgrain structure during deformation are :- high stacking fault energy, low solute content, large strain and high temperature of deformation (§2.3.2.1).

Well developed subgrain structures are not usually observed in metals of lower stacking fault energy such as stainless steel, because, as discussed above, recrystallization occurs before significant recovery can occur. However, if recrystallization is inhibited, for example by the presence of finely dispersed second-phase particles, then at a sufficiently high temperature, recovery will occur leading to the formation of well defined subgrains (Humphreys & Martin 1967) as shown in figure 5.14. If the deformation is relatively uniform so that the heterogeneities required for the nucleation of recrystallization are not formed, then pronounced recovery may occur even in pure copper, and van Drunen and Saimoto (1971) found that subgrain formation rather than recrystallization occurred during the annealing of $<100>$ copper crystals which had been deformed in tension.

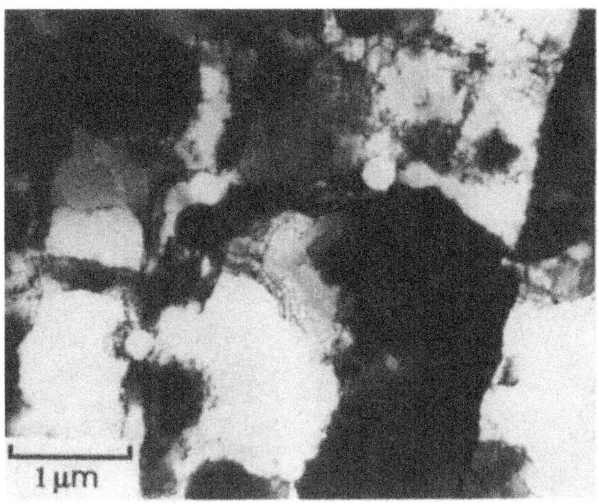

Fig. 5.14. Subgrain formation in copper containing a dispersion of SiO_2 particles, deformed 50% by cold rolling and annealed at 700°C, (Humphreys and Martin 1968).

5.5 SUBGRAIN COARSENING

The stored energy of a recovered substructure such as that of figure 5.13b is still large compared with that of the fully recrystallized material, and can be further lowered by coarsening of the substructure, which leads to a reduction in the total area of low angle boundary in the material.

5.5.1 The driving force for subgrain growth

The driving force for subgrain growth arises from the energy stored in the subgrain structure. If the microstructure can be approximated to an array of subgrains of radius R and boundary energy γ_s, then the stored energy per unit volume (E_d) is $\alpha\gamma_s/R$, where α is a shape factor of the order of 1.5 (equation 2.7). The driving force F for subgrain growth arises from the reduction in stored energy as the subgrain grows and thus, following Ørsund and Nes (1989), we find

$$F = -\frac{dE_D}{dR} = -\alpha \cdot \frac{d}{dR} \cdot \left(\frac{\gamma_s}{R}\right) \tag{5.24}$$

If it is assumed that the stored energy is uniform and that the force is distributed evenly then the driving pressure P is given by

$$P = \frac{F}{A} = -\alpha R \frac{d}{dR}\left(\frac{\gamma_s}{R}\right) \tag{5.25}$$

If γ_s is constant during subgrain growth then

$$P = \frac{\alpha \gamma_s}{R} \tag{5.26}$$

However, as γ_s is a function of the misorientation of the adjacent subgrains (§3.3), it may not remain constant during subgrain growth.

5.5.2 Experimental measurements of subgrain coarsening

5.5.2.1 The kinetics of subgrain growth

Many measurements of the change of subgrain size (D) with time (t) at constant temperature reveal kinetics of the form

$$D^n - D_0^n = c_1 t \tag{5.27}$$

where n is a constant, c_1 is a temperature dependent rate constant and D_0 is the subgrain size at $t=0$.

In many cases the value of the exponent n in equation 5.27 is found to be ~ 2, which is a similar form to the kinetics of grain growth following recrystallization (§9.2). Work in which an exponent close to 2 has been reported includes Smith and Dillamore (1970) on high purity iron, Sandström et al. (1978) on Al-1%Mn, Varma and Willits (1984) and Varma and Wesstrom (1988) on 99.99% aluminium, and Varma (1986) on Al-0.2%Mg. Figure 5.15 shows some typical experimental results.

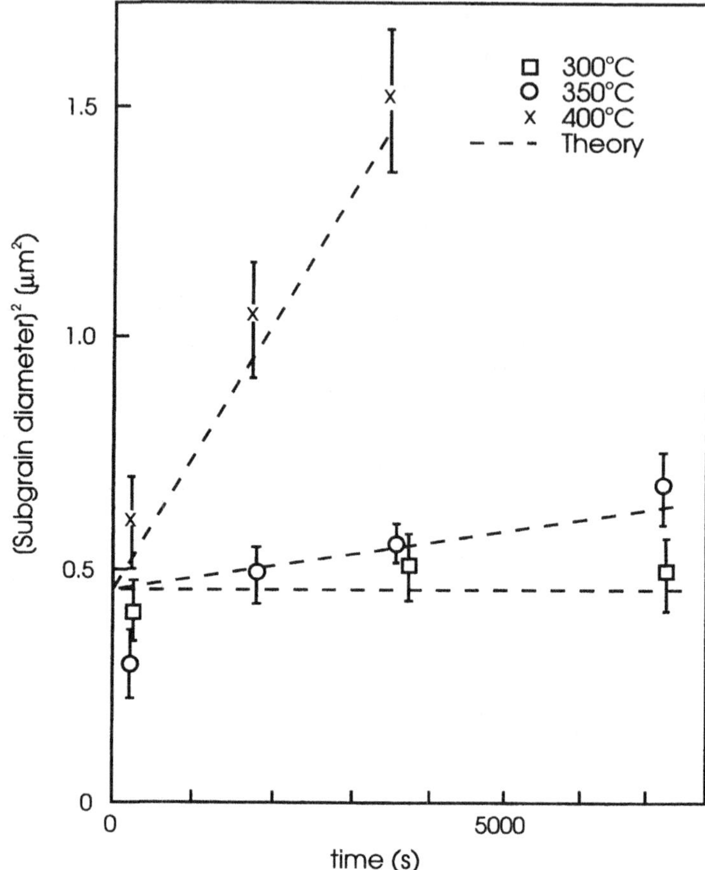

Fig. 5.15. Subgrain growth in an Al-1%Mn alloy plotted according to equation 5.27,
(Sandström et al. 1978).

There are several other investigations of subgrain growth in aluminium in which the subgrain
growth rate decreases much more rapidly than implied by equation 5.27. For example Furu
and Nes (1992), whose results are shown in figure 5.16a, reported an exponent of ~4 in
commercial purity (99.5%) aluminium. Other investigations of high purity (>99.995%)
aluminium (Beck et al. 1959, Sandström et al. 1978, Humphreys and Humphreys 1994) have
found growth kinetics which are consistent with either a large value of **n** in equation 5.27,
or else a logarithmic relationship of the form

$$\log D = c_2 t \qquad\qquad (5.28)$$

The number of experimental investigations of subgrain growth kinetics is rather small. In
addition, there is often considerable scatter of the data, and the error bars shown in figure
5.15 are typical. We should also note that the amount of subgrain growth which is actually
measured is usually very small, because of the onset of recrystallization. In the

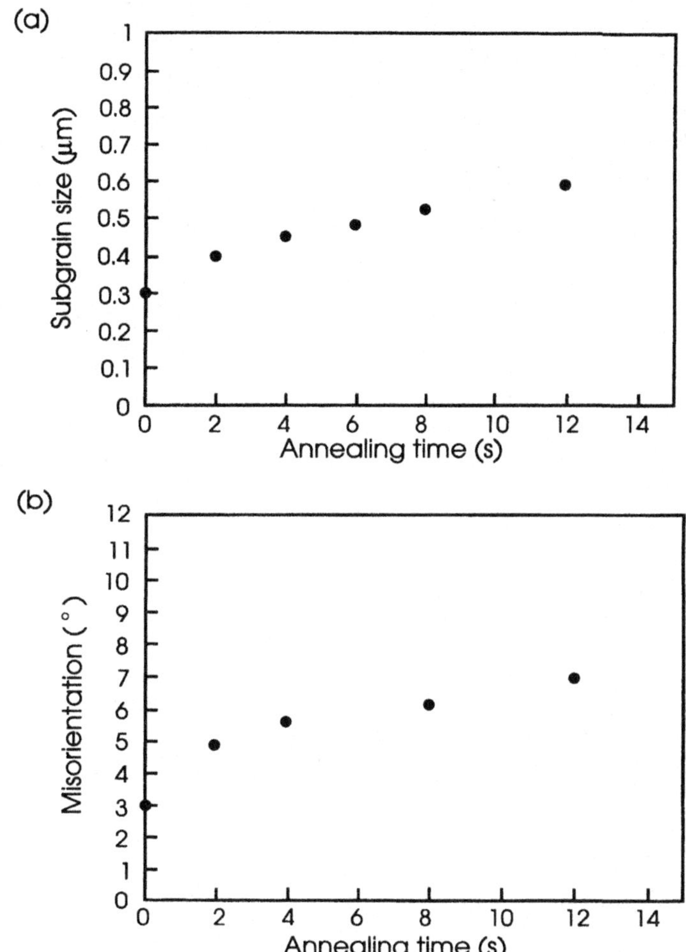

Fig. 5.16. Subgrain growth in commercial purity aluminium at 325°C. a) Subgrain growth kinetics. b) Variation of mean misorientation during annealing, (Furu and Nes 1992).

investigations quoted above, the ratio D/D_0 is rarely much greater than 2. The combination of scatter and limited data range makes it very difficult to place much confidence on the quoted values of the exponent **n**, or indeed to determine whether the data fit equation 5.27 with a very high exponent **n** or whether they fit equation 5.28.

5.5.2.2 Correlation of orientation and subgrain growth

From equation 5.25, it may be seen that the driving force for subgrain growth is expected to be proportional to the energy of the low angle grain boundaries (γ_S). Therefore the growth rate will be a function of the subgrain misorientation (θ). This is a parameter which has not often been measured during subgrain growth experiments, but which is expected to be a function of prior strain and grain orientation (§2.3.2.1). Smith and Dillamore (1970) showed that the rate of subgrain growth varied significantly between different texture components and this was assumed to be a result of different subgrain orientation distributions. Failure to measure this parameter during measurements of subgrain growth must inevitably lead to decreased reliability of the experimental data and undoubtedly accounts for some of the scatter in the available results as discussed above.

Furu and Nes (1992) have measured the changes of both subgrain size and misorientation in commercially pure aluminium as shown in figure 5.16. It may be seen that as subgrain growth proceeds, the average subgrain misorientation increases. This result is consistent with there being **orientation gradients** in the specimen, a point which is further discussed in §5.5.3.4.

Humphreys and Humphreys (1994) have recently measured the misorientation changes during subgrain growth in 99.998% Al as shown in figure 5.17. In order to avoid the effects of grain orientation, all measurements at 300°C were made on the same large grain. There was no overall orientation gradient, and the mean subgrain misorientation (and the orientation spread) decreased slightly during the anneal as shown in figure 5.17b.

5.5.2.3 Relationship between subgrain size and mechanical properties

In view of the use of mechanical properties to measure recovery and of the intrinsic importance of substructures in contributing to the strength of structural materials, we should briefly consider the relationship between the flow stress of a material and the subgrain structure. In the following discussion we assume that the dislocation density within the subgrains is negligible. Substructure strengthening has been reviewed by McElroy and Szkopiak (1972), Thompson (1977) and Gil Sevillano et al. 1980).

We will consider three simple approaches to the problem.

(i) **It is assumed that subgrains behave like grains** and that the yield stress will be given by a Hall-Petch relationship

$$\sigma = \sigma_0 + c_1 D^{-1/2} \tag{5.29}$$

(ii) **It is assumed that the flow stress is determined by the operation of dislocation sources** (Kuhlmann-Wilsdorf (1970), whose length will be closely related to the subgrain diameter, in which case

$$\sigma = \sigma_0 + c_2 D^{-1} \tag{5.30}$$

Equations 5.29 and 5.30 may both be written in the form

$$\sigma = \sigma_0 + c_3 D^{-m} \tag{5.31}$$

(iii) **It is assumed that the dislocations which comprise subgrain boundaries can be averaged over the microstructure**. In this case the area of subgrain per unit volume (A) is $\sim 3/D$. For small misorientations $\theta = b/h$ (equation 3.4), and the length of dislocation per unit area of boundary (L) is $1/h$. Therefore

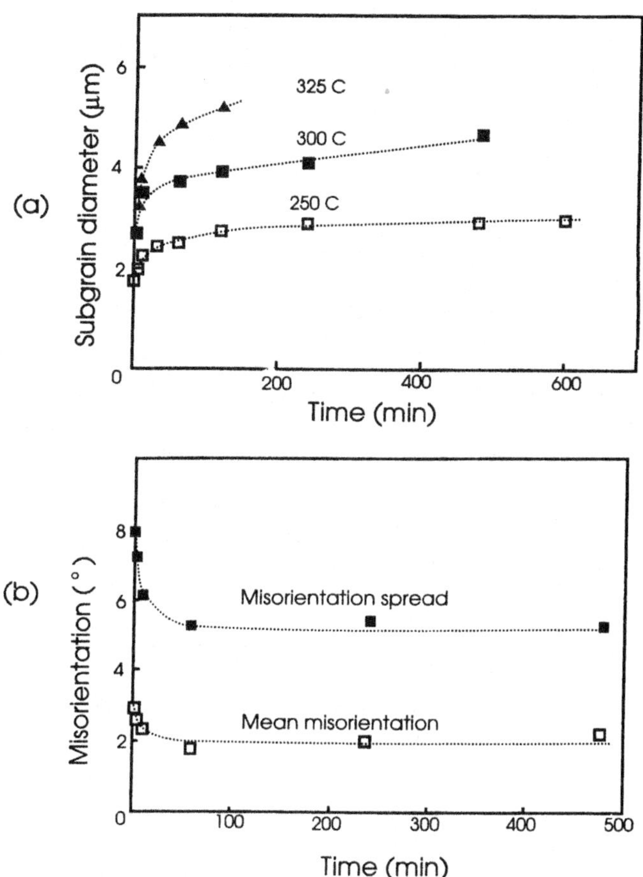

Fig. 5.17. Subgrain growth within single grains in 99.995% Al deformed 20% by cold rolling. a) Subgrain growth kinetics at various temperatures. b) Variation of misorientation at 300°C, (Humphreys and Humphreys 1994).

$$\rho = A L = \frac{3}{Dh} = \frac{3\,b\,\theta}{D} \qquad (5.32)$$

Assuming the relationship between flow stress and dislocation density of equation 2.2, then

$$\sigma = \sigma_0 + c_4 \left(\frac{\theta}{D}\right)^{1/2} \qquad (5.33)$$

which is of the same form as equation 5.31 with $m = 1/2$ and $c_3 = c_4\,\theta^{1/2}$.

All these equations have been used to analyze substructure strengthening and there is little agreement as to their application. Thompson's review indicates that for well developed subgrain boundaries m ~ 1, whilst for cell structures m ~ 0.5. However, he also cites experimental evidence in support of equation 5.33, and there is clearly scope for further work in this area.

5.5.3 Subgrain growth by boundary migration.

5.5.3.1 General considerations
Two quite different mechanisms by which coarsening of the substructure can occur have been proposed - **subgrain boundary migration**, which will be considered in this section, and **subgrain rotation and coalescence** which is considered in §5.5.4.

In a deformed and recovered polycrystal, the subgrain structure may be topologically rather similar to the grain structure of a recrystallized material, and coarsening by the migration of low angle grain boundaries is sometimes considered to be a similar process to that of **grain growth following recrystallization** which is discussed in chapter 9.

The local driving forces for the migration of low angle grain boundaries (LAGBs) arise from the energy and orientation of adjacent boundaries. In the two-dimensional case shown schematically in figure 5.18, the triple point **A** is subjected to forces F_1, F_2 and F_3 from three boundaries. If the specific energies of the boundaries are equal, then the triple point will be in equilibrium when the boundaries make angles of 120° with each other (§3.5.1). The boundaries are therefore forced to become curved (dotted lines), and will tend to migrate in the direction of the arrows so as to minimise their length. Thus the triple point will migrate, and will become stable if it reaches a position (**A′**) in which the boundaries are straight (dashed lines) and the angles are at the equilibrium value of 120°.

The considerations of the topology of growth of a grain assembly in chapter 9, show that large subgrains will grow at the expense of small ones which will shrink and disappear. The subgrain structure adjusts to the forces arising from the boundary energies by the movement of triple point **Y** junctions, and the detailed mechanism of this will depend on the dislocations which comprise the boundaries. For example, in the simple case shown in figure 5.19, vertical movement of the junction **A** may be accomplished by dissociation of the higher angle

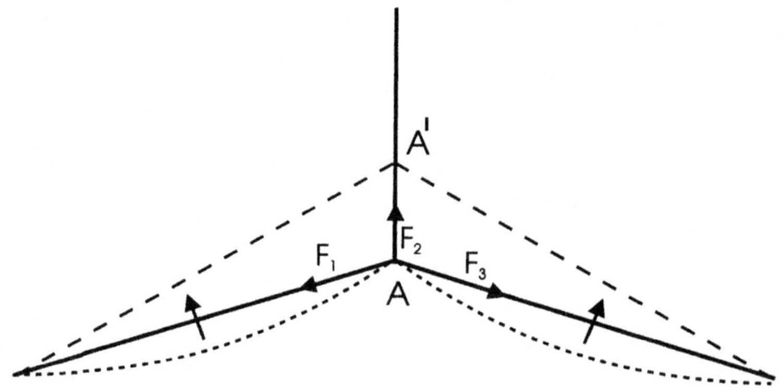

Fig. 5.18. The migration of low angle boundaries

boundary **AB** to provide dislocations for the boundaries **AC** and **AD**. In the general case, more complex dislocation interactions are required as the **Y** junctions move and it is possible that under certain conditions these could be rate determining. However, no analysis of this process is yet available.

5.5.3.2 Growth of polygonized structures

Symmetrical tilt boundaries, such as those shown in figure 5.12c can move by the glide of the edge dislocations which comprise the boundary (§4.2.1). The mobility is high and migration may occur at low temperatures (Parker & Washburn 1952). Subgrain growth of polygonized substructures (the arrays of tilt boundaries formed on the annealing of bent single crystals), has been extensively investigated, and growth of these subgrains has been shown to occur by the migration of **Y** junctions as shown in figure 5.19. The first quantitative study of this process was by Gilman (1955) on zinc crystals. Other investigations have been made on crystals of zinc (Sinha and Beck 1961), NaCl (Amelinckx and Strumane 1960), LiF (Li 1966) and silicon iron (Hibbard and Dunn 1956). In general these measurements show that both the boundary misorientation and spacing increase with the logarithm of time and that the activation energy increases as growth proceeds. The process has been analyzed theoretically by Li (1960) and Feltner and Loughhunn (1962) and reviewed by Friedel (1964) and Li (1966). It should be emphasised that polygonization of bent single crystals is a very special case of subgrain formation and growth, and is expected to have little relevance to the growth of the subgrain structures which occur in polycrystals or in single crystals which have undergone more complex deformation.

5.5.3.3 Subgrain growth in the absence of an orientation gradient

There is a significant difference between **grain growth** which is discussed in chapter 9, and **subgrain growth** which is considered here. In the former case the boundaries are mainly high angle and have similar energies, the equilibrium angle between the boundaries is typically 120°, and the grain structure (in two dimensions) tends towards an assembly of equiaxed hexagonal grains (§3.5.1). In contrast, the energies of low angle grain boundaries

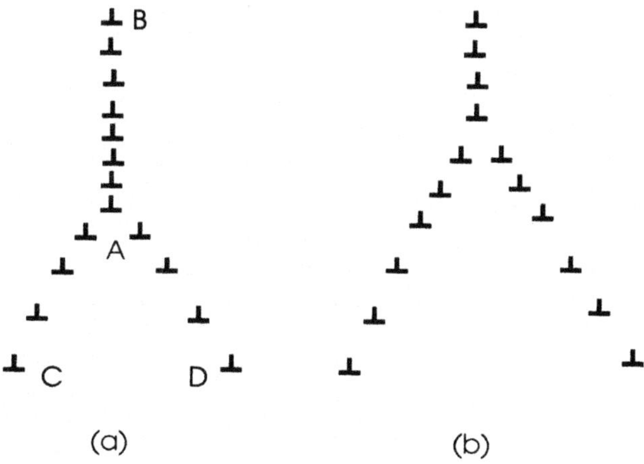

Fig. 5.19. Movement of a Y-junction during subgrain growth.

are strongly dependent on both the misorientation and the boundary plane (§3.3.1), and the equilibrium angles between the boundaries are in general unlikely to be 120°. In lightly deformed materials the subgrains sometimes appear almost rectangular (e.g. figure 6.29). However, in heavily deformed materials the subgrains often appear topologically similar to a grain structure, although caution should be used in interpreting subgrain growth kinetics in terms of models developed for high angle boundaries. Nevertheless, in modelling the kinetics of subgrain growth by boundary migration, it is usually assumed that if the appropriate values of boundary mobility (**M**) and the driving pressure for growth (**P**) are taken, then the growth of the subgrain structure may be treated in a similar manner to **grain growth** involving high angle boundaries, which is discussed in chapter 9 (Smith and Dillamore 1970, Sandström 1977b).

If there is no orientation gradient and θ remains constant during growth then using equations 5.26 and 4.1 and putting **dR/dt=v** we find

$$\frac{dR}{dt} = \frac{\alpha \, M \, \gamma_s}{R} \qquad (5.34)$$

which on integration gives

$$R^2 - R_0^2 = c \, t \qquad (5.35)$$

where $c = 2\alpha M \gamma_s$ and R_0 is the subgrain radius at $t=0$.

If c remains constant during growth, then equation 5.35 is of the same form as the usual grain growth relationship (equation 9.5) and is similar to equation 5.27, with **n=2**, which was shown to be consistent with some of the experimental measurements of subgrain growth kinetics.

5.5.3.4 Subgrain growth in an orientation gradient

In many deformed or recovered substructures there is an **orientation gradient** ($\beta = d\theta/dR$) (§2.3.2.1). If this is the case, then the average subgrain misorientation (θ) will increase during subgrain coarsening (fig 5.16). Recovery in an orientation gradient is thought to play a particularly important role in the **nucleation of recrystallization** (§6.6).

Starting with equation 5.26, we now note that the LAGB energy varies with θ according to the Read-Shockley equation (3.6). For small values of θ, $\gamma_s = \theta \gamma_m / \theta_m$ and hence

$$P = \frac{\alpha \, \gamma_s}{R} = \frac{\alpha \, \theta \, \gamma_m}{\theta_m \, R} \qquad (5.36)$$

(γ_m is the boundary energy predicted by the Read-Shockley equation when the misorientation reaches θ_m, the transition to HAGB behaviour).

Because of the orientation gradient, θ increases as the subgrain grows and

$$\theta = \theta_0 + \Omega \, (R - R_0) \qquad (5.37)$$

where θ_0 and R_0 are the initial values of θ and R, and Ω is the orientation gradient.

Hence, from equation 4.1

$$\frac{dR}{dt} = \frac{\alpha\,M\,\theta\,\gamma_m}{\theta_m\,R} = \frac{\alpha\,M\,\gamma_m}{\theta_m\,R}\,[\theta_0 + \Omega(R - R_0)] \tag{5.38}$$

Initially, and for small orientation gradients, equation 5.38 reduces to equation 5.34. However, for the large orientation gradients found at deformation heterogeneities ($\Omega > 20°/\mu m$), a rapid subgrain growth rate, which decreases more slowly with increasing R than does the rate in the absence of an orientation gradient, is predicted. The exact form of equation 5.38 depends on the orientation dependence of **M** and hence the model assumed for boundary mobility. However, as may be seen from figure 4.4 and §4.2.2.2, the dependence of mobility on misorientation is complex and poorly understood, which is a major obstacle to the development of models for the growth of subgrains in an orientation gradient.

5.5.3.5 Decreasing misorientation during subgrain growth
Even in the absence of an overall orientation gradient, the average misorientation θ may not remain constant. Using a two-dimensional network model (§13.2.3) to simulate subgrain growth in the absence of an orientation gradient, Humphreys (1992b) found that the average misorientation decreased as the subgrains grew (fig 5.20), and that the exponent **n** in equation 5.27 was >3.

It is not unreasonable to expect that low angle boundary migration will occur in such a way as to lower the local energy by removing or shortening boundaries of higher angle (higher energy). Consider the annealing of the boundary configuration shown in figure 5.21a, in

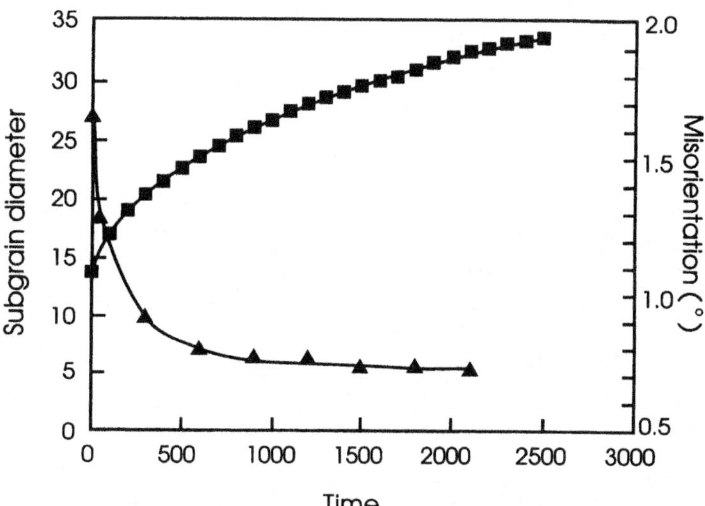

Fig. 5.20. Computer simulation of subgrain growth and misorientation decrease in the absence of a misorientation gradient, (Humphreys 1992b)

which higher energy boundaries are shown as bold lines. The boundary angles are initially 120°, but line tension forces will tend to increase the angles DEC and GFH and decrease the angles CAK and HBL, resulting in the microstructure of figure 5.21b. In responding to the line tensions, the total length of higher angle boundary is thus decreased. There is as yet little evidence that this occurs in practice, although the results of figure 5.17 showed both a very high value of n and a slight decrease in θ on annealing. The role of boundary character in grain growth is discussed further in §9.3.2.

5.5.4 Subgrain growth by rotation and coalescence.

An alternative mechanism for subgrain growth was proposed by Hu (1962), based on in-situ TEM annealing experiments of Fe-Si alloys. He suggested that subgrains might rotate by boundary diffusional processes until adjacent subgrains were of similar orientation. The two subgrains involved would then coalesce into one larger subgrain with little boundary migration, the driving force arising from a reduction in boundary energies, as discussed by Li (1962). The process is shown schematically in figure 5.22. Consider two adjacent low angle boundaries, and let rotation occur in such a direction as to increase the misorientation of the higher angle boundary (AB) and reduce the misorientation of the lower angle boundary (BC). It can be seen from equation 3.4 and figure 3.5 that the change in boundary energy, $d\gamma/d\theta$ is greater the lower the angle of the boundary. Therefore it is energetically favourable for rotation to occur in such a way as to decrease the misorientation of boundary BC and increase the misorientation of boundary AB, leading eventually to coalescence.

Experimental evidence for the occurrence of subgrain rotation and coalescence during recovery has not been accepted as unambiguous. The evidence has been obtained from two types of experiment:

(i) Direct observation of the annealing of thin foils in the TEM.

(ii) Post-mortem observation in the TEM of material annealed in the bulk and then thinned.

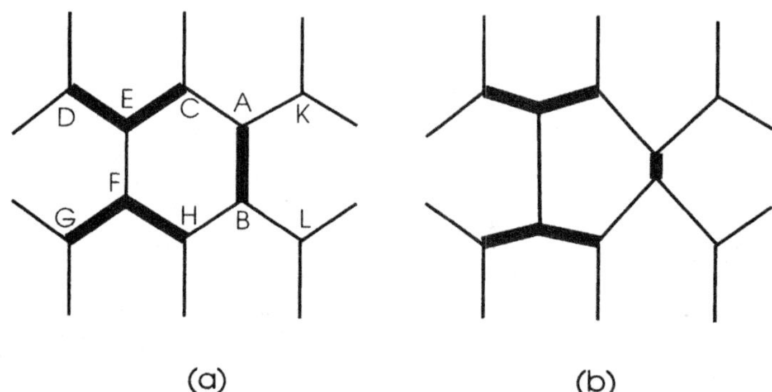

Fig. 5.21. The boundary movements which may lead to a decrease in the average misorientation if there is no orientation gradient.

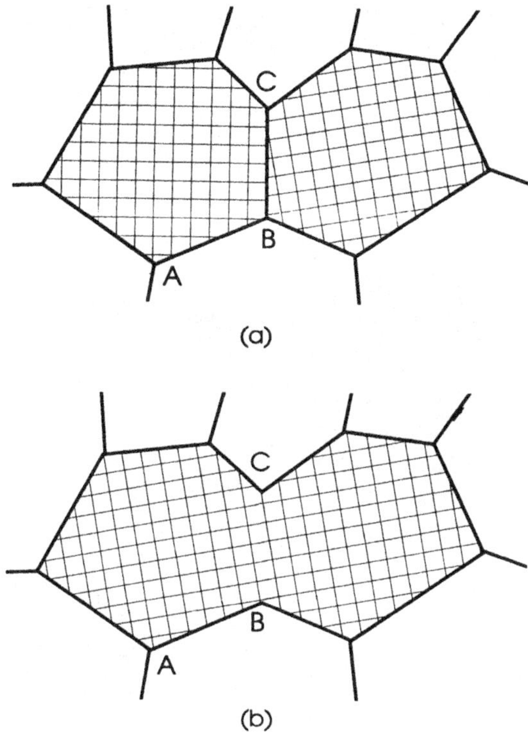

Fig. 5.22. The rotation and coalescence of subgrains.

5.5.4.1 Evidence from in-situ TEM observations

This technique is suspect because events occurring in thin foils are not generally representative of bulk behaviour. This is because dislocations can easily escape to the surface of a thin foil (which even in an HVEM is seldom thicker than ~2μm) and alter the dislocation content and hence the misorientation at low angle grain boundaries. As an example, consider the dislocations in the boundaries in figure 5.12c. The forces between the edge dislocations in the tilt boundaries are such as to push them apart vertically (Hirth and Lothe 1968). It is easy to see that if the specimen were heated to a temperature at which climb would occur, then dislocations would be lost and the boundary would eventually disappear. In addition, a thin specimen allows the stresses arising from the changes in dislocation configuration to be accommodated by bending. Several in-situ annealing experiments have observed dislocation extraction from very low angle boundaries (e.g. Sandström et al. 1978).

Evidence of subgrain rotation in deformed Al-6wt%Ni at very high temperatures was obtained by Chan and Humphreys (1984a) and Humphreys and Chan (1996) during in-situ annealing in the HVEM. This alloy contained a dispersion of second phase particles which coarsened but which did not dissolve at high temperatures. On annealing, well defined subgrains were formed (fig 5.23a), and at lower temperatures of annealing these were seen to grow by **the migration of low angle boundaries** at a rate controlled by the coarsening of the particles (see §9.4.3). However, at temperatures in excess of 550°C, **subgrain reorientation** was found to occur. Clear microscopic evidence of a change in the dislocation content of the boundaries is seen in figure 5.23b at boundary X. In addition, measurements of the orientations of boundaries before and after reorientation clearly established that the contrast changes in the subgrains were due to changes in relative orientations and not to artifacts such as bending of the thin foil.

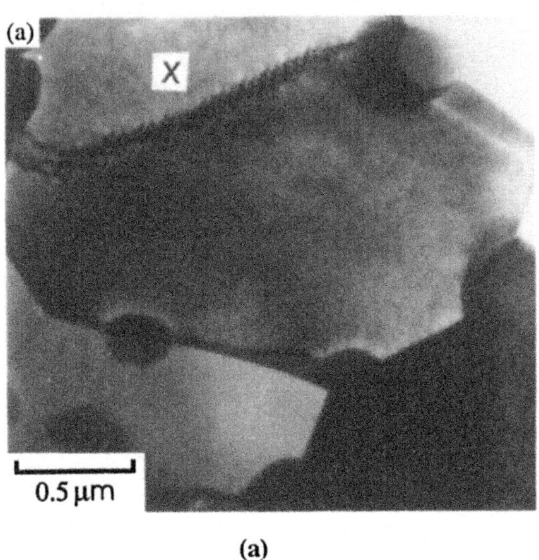

(a)

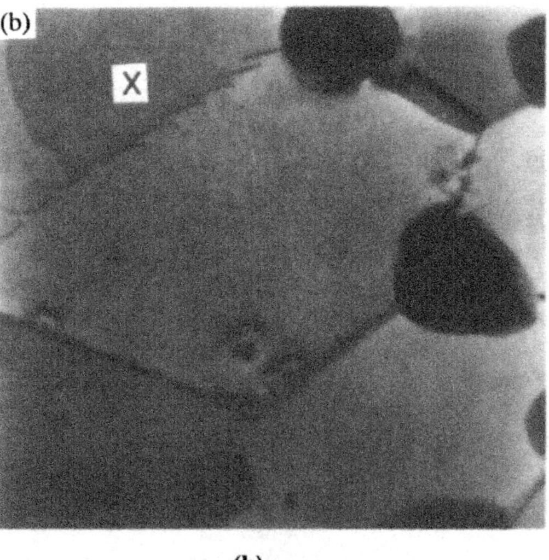

(b)

Fig. 5.23. Rotation of subgrains during the in-situ annealing of a thin foil of Al-6%Ni in the HVEM at 550°C. Note the dislocation loss at boundary X, (Humphreys and Chan 1996)

Although the experiments described above showed that subgrain rotation occurred, it should be emphasised firstly that this only took place at very high temperatures ($0.9T_m$), far above those where LAGB migration occurs readily, and thus the relevance to normal recovery at relatively low temperatures is questionable. Secondly, for the reasons discussed above, the dislocation rearrangement in the boundaries is almost certainly aided by the free surfaces of the specimen.

Most in-situ annealing experiments, particularly those in which the mechanisms of recrystallization were being investigated, have found clear evidence of LAGB **migration**, e.g. Kivilahti et al. (1974), Ray et al. (1975), Bay and Hansen (1979), Humphreys (1977), Berger et al. (1988), but have found no evidence of subgrain **rotation**.

5.5.4.2 Evidence from bulk annealed specimens.

The difficulty with determining the annealing mechanism from examination of bulk-annealed samples is that only the structure before **or** after the event can be seen and the evidence is therefore indirect. Evidence cited in favour of coalescence is usually a micrograph of a large subgrain containing a very low angle boundary. This latter boundary is then stated to be in the process of disappearing by coalescence. Several clear instances of such structures have been obtained for example by Hu (1962), Doherty (1978), Faivre and Doherty (1979) and Jones et al. (1979). The problem with such observations is that during deformation, subgrains or cells are continually being formed and reformed, so that new subgrain boundaries will often form within old subgrains (§2.3.2.1). Therefore micrographs of poorly developed low angle boundaries within well defined subgrains are as likely to be showing a boundary **forming** as a boundary **disappearing** and therefore do not provide convincing evidence of subgrain coalescence.

There have been several observations of recovered microstructures which show that the subgrain size in the vicinity of a grain boundary is larger than in the interior of the grain (Goodenow 1966, Ryum 1969, Faivre and Doherty 1979 and Jones et al. 1979). Whether this effect is a result of the deformation microstructure adjacent to the HAGB being different to that in the grain interior (§2.3.5) or whether it is an indication of a difference in annealing behaviour near the HAGB is not yet clear. Doherty and Cahn (1972) have suggested that subgrain coalescence is more likely to occur near high angle boundaries than in areas away from such a boundary. They argued that the energy change resulting from reorientation ($d\gamma/d\theta$) is generally very small for a HAGB, whereas it is large for a LAGB. Therefore if the elimination of a low angle boundary adjacent to the HAGB were to increase the angle of the HAGB, this would not incur an energy penalty.

Jones et al. (1979) have presented clear evidence that the dislocation content of very low angle LAGBs connected to HAGBs is not uniform, and that the dislocation content of the boundary may be lower adjacent to the HAGB, as shown in figure 5.24. It is envisaged that the HAGB is able to absorb dislocations from the LAGB, and the authors suggest that this is the early stages of subgrain rotation.

5.5.4.3 Modelling reorientation at a single boundary

The kinetics of subgrain coalescence are best considered in two stages. In this section, we examine the reorientation at a single boundary, ignoring the constraints of adjacent subgrains, and in the next, we consider the total microstructure.

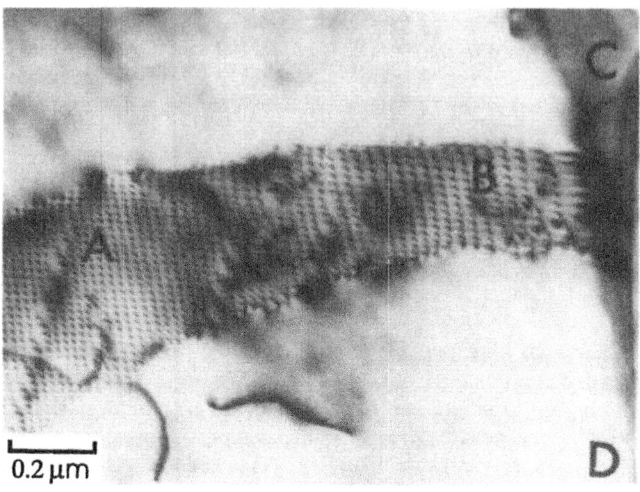

Fig. 5.24. Transmission electron micrograph of annealed aluminium showing that the spacing of dislocations in the low angle boundary AB becomes larger as the high angle boundary CD is approached, (Jones et al. 1979).

Li (1962) has discussed the kinetics of reorientation at a single boundary in terms of two possible mechanisms - **cooperative climb of edge dislocations** and **cooperative vacancy diffusion**. For small boundary misorientations, such as those typically found in recovered structures, dislocation climb is likely to be rate controlling, and we will only consider this mechanism.

Li examined the case of a single symmetrical tilt boundary of misorientation θ and height L. If the dislocation spacing remains uniform during climb, then the work done by the dislocations in time **dt** during which the orientation changes by **dθ** is given by

$$dW = \frac{2\,L^3\,d\theta}{3\,b\,\theta\,B}\,\frac{d\theta}{dt} \tag{5.39}$$

where **B**, the mobility of dislocations by climb, is given by $B = D_x b^3 c_j / kT$. D_x is the diffusivity and c_j the concentration of jogs on the dislocations.

This is equal to the energy change (**dE**) when the boundary misorientation changes by dθ

$$dE = 2\,L\,\gamma_m\,\ln\!\left(\frac{\theta_m}{\theta}\right)\,d\theta \tag{5.40}$$

where γ_m and θ_m are constants from the Read-Shockley equation (3.6).

Equating dE and dW we obtain

$$\frac{d\theta}{dt} = \frac{3\gamma_m\,\theta\,Bb}{L^2}\,\ln\!\left(\frac{\theta}{\theta_m}\right) \tag{5.41}$$

which on integration gives

$$\ln \ln \left(\frac{\theta_m}{\theta} \right) = \left(\frac{3Bb\gamma_m}{L^2} \right) t + K \qquad (5.42)$$

Li (1962) also analyzed the reorientation of twist and other low angle boundaries and found that this differed from equation 5.41 by only a small factor.

Doherty and Szpunar (1984) have re-examined the kinetics of reorientation by dislocation climb, and suggest that it is more appropriate to consider the behaviour of dislocation loops encircling a subgrain than to consider isolated boundaries as was done by Li. For climb controlled by bulk diffusion, they obtained kinetics very similar to those of Li, i.e. equation 5.41. However, they showed that core diffusion, which modifies the term **B** in equation 5.41, would lead to much faster rates of reorientation at low temperatures ($T < 350°C$) in aluminium. Their computer simulations also indicated that if the condition of uniform dislocation spacing in the boundary was relaxed, as suggested by the observations of Jones et al. (1979), then a further increase in kinetics would be obtained.

Sandström (1977a) has discussed the possibility that boundary reorientation might occur by other mechanisms. He has analyzed the case in which dislocations are extracted from a boundary into the interior of a subgrain, thus lowering the orientation of the boundary. In this mechanism, the dislocation density in the interior of the subgrain controls the rates of extraction of the dislocations and their absorbtion on other boundaries. A dynamic equilibrium is achieved, and subgrain growth due to this mechanism is predicted to cease after a time.

5.5.4.4 Modelling the kinetics of subgrain coalescence

During a coalescence event in a recovered subgrain structure, reorientation of one subgrain affects **all** the boundaries around the subgrain, and therefore we would expect some 14 LAGBs to be involved. Li (1962) showed that the driving force for rotation of a subgrain, calculated from the energy changes of all the boundaries involved is always finite about all axes except one, and that therefore the process is thermodynamically feasible. The total driving force from the α boundaries surrounding a subgrain is

$$\sum_{\alpha} L_{\alpha}^2 \frac{d\gamma}{d\theta} = \sum_{\alpha} L_{\alpha}^2 \, \gamma_m \ln\left(\frac{\theta_m}{\theta_{\alpha}} \right) \qquad (5.43)$$

It may be seen from equation 5.43 that the boundary which provides the most driving force will be that with the smallest θ and largest area.

In order to properly calculate the kinetics of rotation of an enclosed subgrain, the driving force from equation 5.43 must be used in conjunction with the appropriate climb rate of the dislocations in all the participating boundaries, which is a difficult problem. However, reference to Li's climb kinetics of equation 5.41, show that the boundary which has the lowest rate of rotation also has the largest driving force. Li therefore argued that this boundary could be regarded as controlling the rate of rotation, and thus equation 5.41 could, to a first approximation be used to predict the rate of rotation of a subgrain. This approach has been widely adopted by later workers. However, it should be borne in mind that the values of L and θ to be used in the equation refer not to the **average subgrain**, but to that with the largest value of $L^2 \ln(\theta_m/\theta)$. Another problem in using this analysis is that rotation of one subgrain affects the environment and hence the subsequent rotation kinetics of all adjacent subgrains. Therefore, in order to determine the kinetics of subgrain rotation we need to know not only the initial distribution of subgrain sizes and orientations, but all subsequent values of these parameters.

There have been several attempts to compare the experimentally measured kinetics of subgrain growth (§5.5.2) with those predicted from subgrain coalescence. The usual method of predicting growth kinetics from equation 5.41 has been to equate θ with the measured mean misorientation and to assume that at time **t**, all subgrains of diameter less than **L** will have coalesced, so that $\mathbf{L^2}$ is proportional to **t**. These kinetics are of similar form to those often found in practice (equation 5.27 with n=2), and are of course also similar to those which have been predicted for subgrain coarsening by LAGB migration (e.g. equation 5.35).

Smith and Dillamore (1970) found that the kinetics of subgrain growth in iron were faster than predicted on Li's model by several orders of magnitude, and cited this as an argument against subgrain coalescence, and Faivre and Doherty (1979) found a similar result in aluminium. However, the later theoretical analysis by Doherty and Szpunar (1984) discussed above, has shown that the coalescence model can predict more realistic kinetics. In the light of the uncertainties over the theoretically predicted growth rates for both subgrain rotation and subgrain growth by migration, it would be unwise to use kinetics as a basis for distinguishing between these mechanisms.

5.5.4.5 Simulations of subgrain coalescence

Computer simulation of subgrain rotation and coalescence within a two-dimensional array of subgrains has been carried out. The driving force for the process is given by equation 5.43, and the rate of rotation is as calculated by Li (1962) on the assumption of control by dislocation climb. Saetre and Ryum (1992) used, as a starting structure, an array of square subgrains. We have used an extension of the network model (§13.2.3.2), which allows a more realistic subgrain structure. Figure 5.25 shows the results of our simulation for identical microstructures but with various misorientation spreads. As shown in figure 5.25a, there is an initial sharp increase in subgrain size, due to a large number of coalescence events involving the elimination of very low angle boundaries. Thereafter there is a linear increase in subgrain size in agreement with the results of Saetre and Ryum (1992). Note that after the initial period the rate of growth is little affected by the misorientation. The mean subgrain misorientation is seen in figure 5.25b to increase almost linearly with time. The simulation was also run with a high angle boundary in the microstructure, but with the same spread of subgrain misorientations. It was found that coalescence took place preferentially at the HAGB as first predicted by Doherty and Cahn (1972). An analysis of the data showed that throughout the simulation, the chance of a coalescence event occurring at the high angle boundary was **2.5 times greater** than within the grain interior.

5.5.4.6 Recovery mechanisms and the nucleation of recrystallization

There is no doubt that subgrain rotation and coalescence are thermodynamically feasible, and, as discussed above, there is evidence that in thin foils heated in the electron microscope, dislocations may be emitted from very low angle boundaries and that this may lead to coalescence under these special conditions. There is also some evidence that dislocation

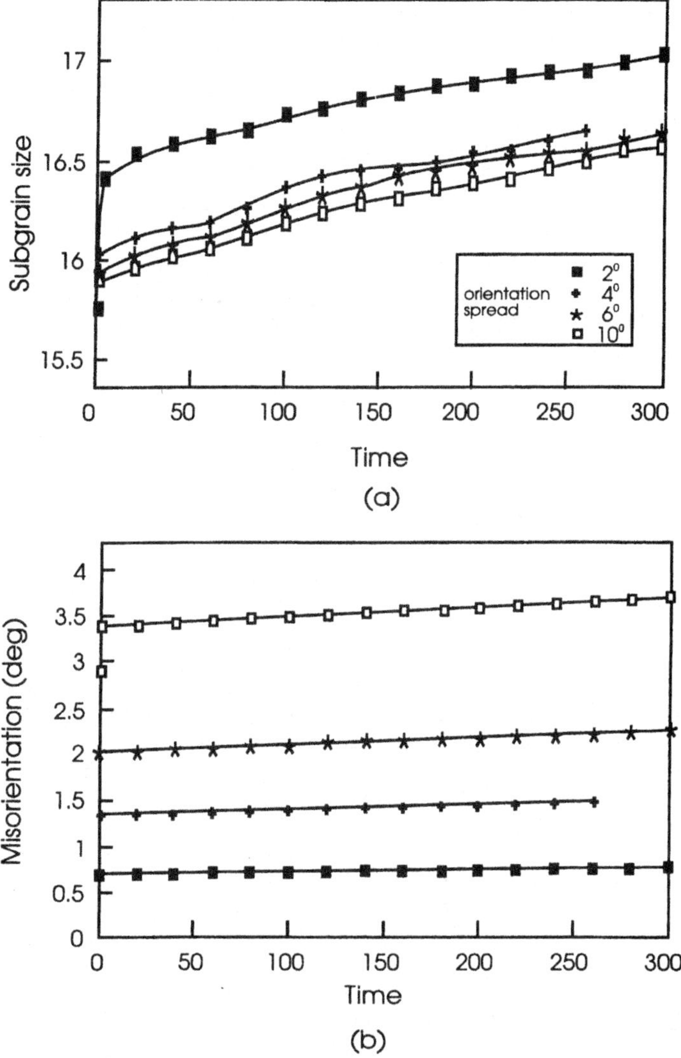

Fig. 5.25. Computer simulation of the kinetics of subgrain coalescence in a two dimensional microstructure. Orientation spreads of from 2 to 10 degrees have been introduced into the microstructure. a) Subgrain growth. b) Mean misorientation.

rearrangement may occur in low angle boundaries adjacent to high angle boundaries. However, there is no convincing evidence that this mechanism plays any significant role in subgrain coarsening either during uniform recovery, or during recovery in regions of large orientation gradient, which, as is discussed in §6.6, is thought to be the way in which recrystallization originates.

It is therefore reasonable at the current time, on the available theoretical and experimental evidence to assume that both subgrain coarsening and the nucleation of recrystallization from pre-existing subgrains are controlled by the migration of low angle grain boundaries.

5.6 THE EFFECT OF SECOND-PHASE PARTICLES ON RECOVERY

Second-phase particles may either have been present during the deformation of the material, in which case they are often distributed reasonably uniformly in the microstructure, or they may precipitate during the anneal, in which case their distribution is often related to the dislocation structure. In this section we will be primarily concerned with a stable distribution of particles which does not alter during the anneal. The effects of concurrent precipitation and annealing are considered in §8.9.

Second-phase particles may affect recovery mechanisms in several ways. During the annihilation and rearrangement of dislocations to form low angle boundaries (§5.3), the particles may pin the individual dislocations and thus inhibit this stage of the recovery. The driving force (F) for the annealing of a three-dimensional dislocation network is given by equation 5.13. This will be opposed by the force due to the pinning effect of the particles (F_p). To a first approximation, this is given by

$$F_p = \frac{c_1 \, G \, b^2}{\lambda} \tag{5.44}$$

where c_1 is a constant whose value depends on the magnitude of the interaction between particles and dislocations, and λ is the spacing of particles along the dislocation lines (dislocation-particle interaction are discussed in more detail in §8.2). The resultant driving force then becomes $F-F_p$. This approach is only valid if the particle spacing is small compared to the scale of the dislocation network, and as this rarely occurs in practice, there has been little investigation of the effect of particles on this stage of recovery.

There has however been much more interest in the effect of particles on subgrain growth during recovery, as this is relevant to the origin of recrystallization in particle-containing alloys as discussed in chapter 8.

5.6.1 The effect of particles on the rate of subgrain growth

There is substantial evidence that a fine particle dispersion may exert a strong pinning effect on subgrains (Humphreys and Martin 1968, Ahlborn et al. 1969, Jones & Hansen 1981), and an example is shown in figure 5.14. The use of particles to pin and stabilize the recovered substructure is a well established method for improving the strength and creep resistance of alloys at high temperatures. The particles used for this purpose need to be stable at high temperatures, and oxide particles, often introduced by **powder metallurgy** or **mechanical alloying** are incorporated into **dispersion hardened alloys**, such as nickel alloys (e.g. Benjamin 1970) and aluminium alloys (e.g. Kim and Griffith 1988).

The electron microscope observations of Jones and Hansen (1981) have revealed the pinning of individual boundary dislocations by particles as shown in figure 5.26. As yet, no theories of boundary mobility based on such interactions have been formulated, and the models used to account for subgrain growth in the presence of particles are essentially similar to those developed to account for the effect of particles on grain growth (§9.4) (Sandström 1977b). There are as yet, few quantitative measurements of the effect of particles on the rate of growth of subgrains.

The interaction of high and low angle grain boundaries with particles is discussed in chapter 3. The pinning force (F_S) due to a single non-coherent spherical particle is given by equation 3.10, and the drag pressure (P_Z) due to the interaction of a random distribution of particles with a planar boundary, **the Zener drag**, is given by equation 3.21. If the interparticle spacing is much smaller than the subgrain diameter (2R), then from equations 5.26 and 3.21, the driving pressure for subgrain growth is given by

$$P = P_D - P_z = \frac{\alpha \, \gamma_s}{R} - \frac{3 \, F_v \, \gamma_s}{2 \, r} \qquad (5.45)$$

and hence, using equation 4.1, and as before, putting $v = dR/dt$

$$\frac{dR}{dt} = M \gamma_s \left(\frac{\alpha}{R} - \frac{3 \, F_v}{2 \, r} \right) \qquad (5.46)$$

This equation predicts a parabolic growth law for short times, and at longer times, as P_D approaches P_Z it predicts a **limiting subgrain size** of $4\alpha r/3F_v$.

In practice there are several difficulties in applying equation 5.46 to the rate of subgrain growth. In particular, as discussed in §8.2, the initial cell or subgrain size formed on deformation is closely related to the interparticle spacing, and therefore the assumption of a planar boundary interacting with randomly distributed particles on which equation 3.21 is based is not justified. In these circumstances, the pinning pressure varies in a much more complicated manner with subgrain size as discussed in §3.5 and shown by figure 3.23.

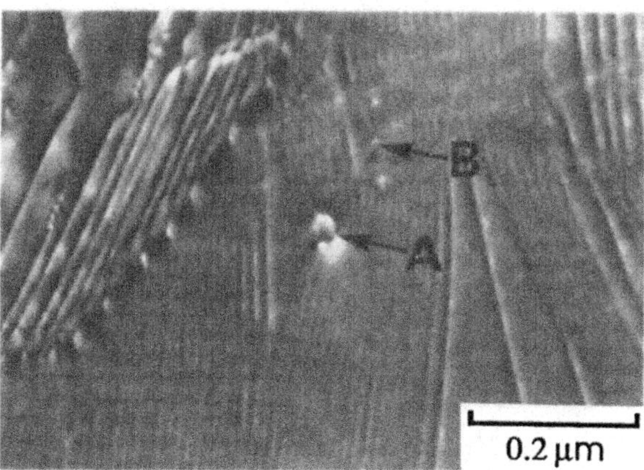

Fig. 5.26. Transmission electron micrograph showing pinning of grain boundary dislocations by Al_2O_3 particles A and B in a low angle boundary in aluminium, (Jones and Hansen 1981).

5.6.2 The particle-limited subgrain size

Because of the industrial importance of particle-stabilised subgrain structures, there is considerable interest in calculating the particle-limited subgrain size. Because of the similarity of the pinning of high and low angle boundaries by particles, we have chosen to discuss the concept of the **limiting grain size** in some detail in §9.4.2, and in this section will comment only on those aspects of particular relevance to subgrains.

5.6.2.1 Stable particle dispersions

From §9.4.2, we expect that for small particle volume fractions the limiting subgrain size will be proportional to F_v^{-1} (equation 9.33), and at large volume fractions it will be proportional to $F_v^{-1/3}$ (equation 9.37). The critical volume fraction at which there is a transition is not known precisely, but may be ~ 0.06 (§9.4.2.3). Anand and Gurland (1975) measured the limiting subgrain size in steels with F_v in the range 0.1-0.2, and showed that their results (fig 5.27) were consistent with the $F_v^{-1/2}$ relationship of equation 9.36. However, as discussed in §9.4.2.3, the subgrain size predicted by equation 9.36 is unlikely to be stable, and these experimental results agree equally well with the preferred $F_v^{-1/3}$ relationship of equation 9.37.

Stable subgrain structures are found in many alloys containing dispersions of small particles, although there have been few systematic investigations of the effect of particle parameters on the limiting subgrain size. Unless the interparticle spacing is small (less than $\sim 1\mu m$), or the ratio F_v/r is greater than $\sim 0.2\mu m^{-1}$, recrystallization often intervenes before the particle-limited subgrain size is reached (§8.3).

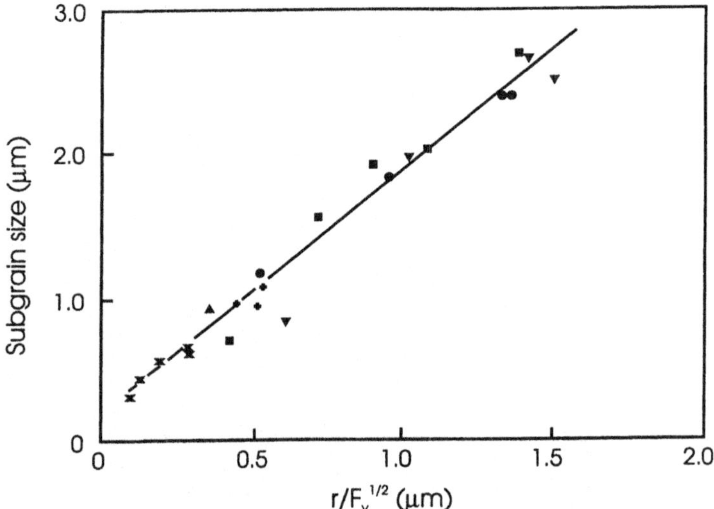

Fig. 5.27. The particle-limited subgrain size in carbon steels plotted according to equation 9.36, (after Anand and Gurland 1975).

5.6.2.2 The inhibition of recrystallization

The particle-limited subgrain size may have a significant effect on the recrystallization behaviour of an alloy. If the particles prevent a subgrain which is in an orientation gradient from growing to a size such that it acquires sufficient misorientation to become a viable recrystallization nucleus, then recrystallization will be inhibited. This topic is considered in §8.5.

5.6.2.3 Precipitation after subgrain formation

Precipitation on the recovered substructure may occur during the annealing of a supersaturated alloy. Examples of this behaviour have been found in the commercially important Al-Mn alloys (Morris and Duggan 1978), Fe-Cu, (Hutchinson and Duggan 1978), and Fe-Ni-Cr (Hornbogen 1977). In this situation, the precipitation generally occurs on the low angle boundaries. Such non-uniformly distributed precipitates (fig 3.22d) exert a strong pinning effect on the boundaries as discussed in §9.4.3.1, and the limiting subgrain size is then given by equation 9.38.

5.7 EXTENDED RECOVERY AND CONTINUOUS RECRYSTALLIZATION

5.7.1 Particle-containing alloys

There is now considerable evidence that in certain circumstances the differences between the processes of recovery and recrystallization, and the terminology used to describe them are not clear, and this is particularly the case for particle-containing alloys in which normal discontinuous recrystallization is inhibited. There is no evidence that any micromechanisms other than the migration of low and high angle boundaries occur, and it would appear that the scale and homogeneity of the annealing phenomena are the important factors.

If, after a particle-stabilised subgrain structure has formed, the particles subsequently coarsen during the recovery anneal, then homogeneous subgrain growth, whose kinetics are controlled by the coarsening of the particles, may occur as shown schematically in figure 5.28. An example of such a microstructure is seen in figure 5.23a. The kinetics of subgrain growth in these circumstances and the relationship between the particle parameters and the subgrain size are discussed in §9.4.3.1.

This phenomenon was first studied in Al-Cu alloys by Hornbogen and colleagues (e.g. Köster and Hornbogen 1968, Ahlborn et al. 1969, Hornbogen 1970). As this process continues, the scale of the substructure may become comparable with that of a normal grain structure. Although the phenomenon has been referred to as "continuous recrystallization" or "recrystallization in situ", there is no evidence that the resulting boundaries are predominantly high angle, or that new HAGBs are created during the process, suggesting that the process is one of recovery, uninterrupted by the onset of recrystallization, and the term **extended recovery** is preferable.

Figure 5.29a shows an idealised one-dimensional deformed microstructure. The orientation is represented by the vertical axis and the distribution of second-phase particles and boundaries is shown in the horizontal direction. The microstructure therefore represents two grains, each containing small subgrains.

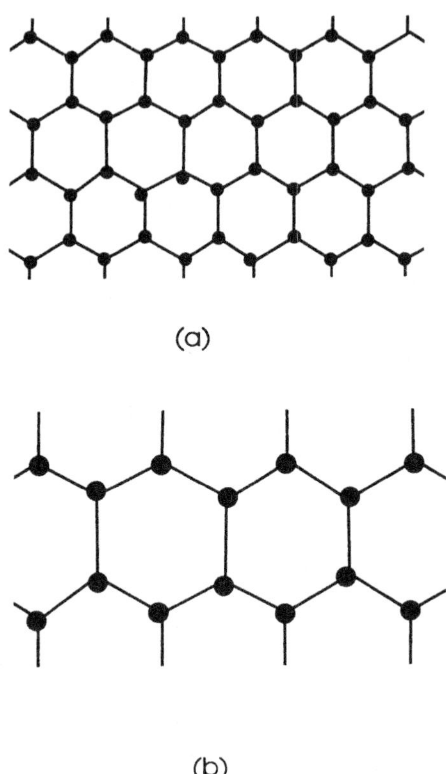

(a)

(b)

Fig. 5.28. Extended recovery controlled by the coarsening of second-phase particles.

On annealing at a low temperature, recovery will occur, and the initial subgrain growth will be largely unaffected by the particles until the subgrains have grown to a size comparable with the interparticle spacing (fig 5.29b), at which point the recovered structure is stabilised as discussed in §5.6.2.1. The orientation spread within a grain is not likely to change significantly during subgrain growth, and as there is no overall orientation gradient within the grains, the subgrain misorientations remain small. Further annealing at higher temperatures may coarsen the particles, thus allowing the subgrains to grow as shown in figure 5.29c, and this process is **extended recovery**. If the particle coarsening continues, then, eventually (fig 5.29d), the microstructure will comprise mainly high angle boundaries. This is therefore similar to a recrystallized microstructure and it is not unreasonable to describe it by the term **continuous recrystallization** (cf. continuous and discontinuous phase transformations). However, it should be noted that this term does not imply any particular micromechanism and may be applied to any process in which a high angle boundary structure evolves without any obvious nucleation and growth sequence. The difference between extended recovery and continuous recrystallization is therefore defined in terms of the **relative amounts of low and high angle boundary** in the microstructure, and there will be cases which are intermediate between both. An example of extended recovery is the microstructural evolution of deformed Al-6%Ni alloys (Morris 1976, Humphreys and Chan

1996). After hot extrusion, this alloy contains a volume fraction of 0.1 of $\sim 0.3\mu$m NiAl$_3$ particles in a deformed matrix. On annealing, discontinuous recrystallization does not occur, but the particles and hence the substructure coarsen at high temperatures (see figs 9.12 and 9.13). Although the subgrains grow from $\sim 0.8\mu$m at 100°C to $\sim 2\mu$m at 500°C, the distribution of grain/subgrain misorientations is largely unchanged as shown in figure 5.30.

The transition between extended recovery and continuous recrystallization is also affected by the strain and the initial grain size. If the strain increases, the amount of high angle boundary in the microstructure increases both for geometric reasons (§2.2.1), and because more heterogeneities such as deformation bands will be formed. With reference to figure 5.29a we therefore see that the ratio of the separation of high angle boundaries to the interparticle spacing (A/B) decreases with increasing strain, and the transition from recovery to continuous recrystallization therefore requires less particle coarsening. For very large deformations (fig 5.29e) resulting in the spacing of the high angle boundaries approaching the interparticle spacing (A/B $\sim$ 1), a structure which contains mainly high angle boundaries may be formed at low annealing temperatures without any particle coarsening (fig 5.29f).

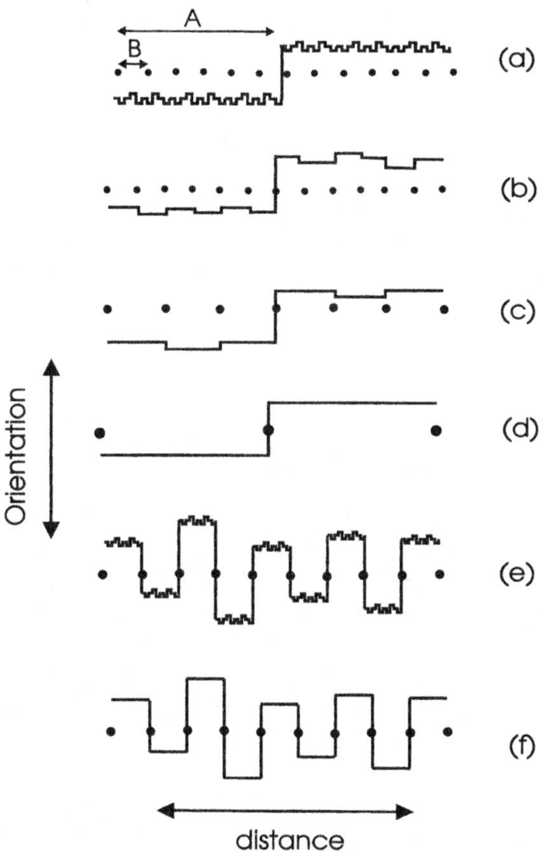

Fig. 5.29. Schematic diagram of extended recovery and continuous recrystallization.

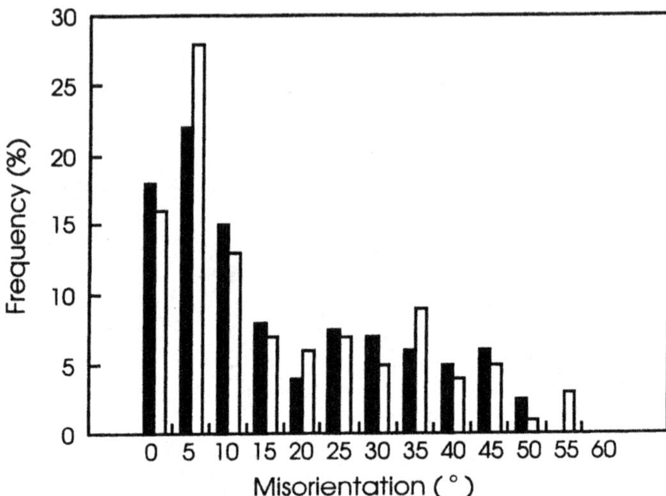

Fig. 5.30. The distribution of grain/subgrain misorientations in deformed Al-6%Ni
annealed at 100°C (filled) and 500°C (open), (Humphreys and Chan 1996).

Such structures will be stable, because the initial spacing of high angle grain boundaries is
less than the particle-limited grain size (§9.4.2). A good example of this phenomenon in
two-phase aluminium-iron alloys has been given by Oscarsson et al. (1992), who found that
after very large rolling reductions ($>95\%$), a stable fine-grained microstructure was formed
on annealing, whereas at lower strains, normal discontinuous recrystallization occurred.

5.7.2 Single-phase alloys

The continuous annealing phenomena discussed above may also occur in single-phase alloys.
If the nuclei for recrystallization are not available then recovery will proceed and very large
subgrains may form. This type of **extended recovery** has been noted in grains of specific
orientations in aluminium by Hjelen et al. (1991).

Recent research has clearly shown that severe cold working of a metal can result in a
substructure which consists almost entirely of high angle grain boundaries. Restoration
processes at ambient or low annealing temperatures after deformation then result in a fine
grained "recrystallized" microstructure, a process similar to that shown in figures 5.29 e and
f, which falls into the category of **continuous recrystallization**. This type of behaviour has
been reported for a variety of metals including copper and nickel (Smirnova et al. 1986),
magnesium and titanium alloys (Kaibyshev et al. 1992)) and Al-Mg (Wang et al. 1993).
These materials were all subjected to true strains in the range 3-7, and grain sizes in the
range 0.1 to 0.5μm have been reported. As these microstructures have a large stored energy
and lack the second-phase particles which pin the boundaries, they are unlikely to be stable
at very high temperatures. Nevertheless, initial results suggest that some of these materials
may have interesting mechanical properties such as low temperature superplasticity (Wang
et al. 1993).

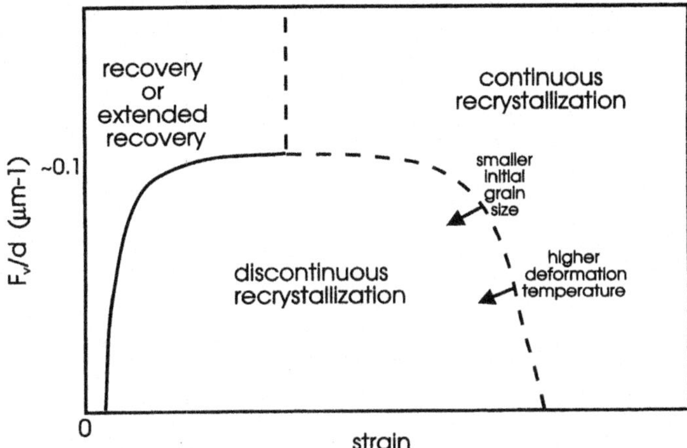

Fig. 5.31. Suggested relationship between continuous and discontinuous annealing phenomena.

We should note that the phenomenon of **geometric dynamic recrystallization** (§11.3.1) which may occur during high temperature deformation is essentially a high temperature version of the phenomena discussed above.

The discussions of this section, taken together with those in §11.3.1, show that conventional definitions of recovery and recrystallization are sometimes inadequate to describe the range of annealing behaviour in alloys, particularly those containing a dispersed second phase. In figure 5.31 we indicate a possible relationship between some of these phenomena in terms of the parameters of the particle dispersion and the deformation strain. Although the relationship between F_v/d, strain and the onset of discontinuous recrystallization (indicated by the solid line) is reasonably well known (§8.3), the right-hand side of the diagram is speculative. In addition, other parameters are important, and in particular, the deformation temperature and the original grain size will, as indicated, affect the size of the discontinuous recrystallization field.

We conclude that the fine-grained microstructures produced by the deformation of conventional alloys to very large strains appear to be broadly explainable in terms of existing models and do not involve any new mechanisms. However, the production and processing of materials in this way is an exciting new development of great scientific interest and potential commercial importance.

Chapter 6

RECRYSTALLIZATION OF SINGLE-PHASE ALLOYS

6.1 INTRODUCTION

Recovery, which was discussed in the last chapter, is a relatively homogeneous process in terms of both space and time. When viewed on a scale which is larger than the cell or subgrain size, most areas of a sample are changing in a similar way. Recovery progresses gradually with time and there is no readily identifiable beginning or end of the process. In contrast, recrystallization involves the formation of new strain-free grains in certain parts of the specimen and the subsequent growth of these to consume the deformed or recovered microstructure (fig 1.1c,d). The microstructure at any time is divided into recrystallized or non-recrystallized regions as shown in figure 6.1, and the fraction recrystallized increases from 0 to 1 as the transformation proceeds.

Recrystallization of the deformed microstructure is often called **primary recrystallization** in order to distinguish it from processes of exaggerated grain growth which may occur in fully recrystallized material and which are sometimes called **secondary recrystallization** or **abnormal grain growth** (§9.5). The unqualified term **recrystallization** is always taken to mean primary recrystallization.

It is convenient to divide primary recrystallization into two regimes, **nucleation** which corresponds to the first appearance of new grains in the microstructure and **growth** during which the new grains replace deformed material. Although these two events occur consecutively for any particular grain, both nucleation and growth may be occurring at any time throughout the specimen. The kinetics of recrystallization are therefore superficially similar to those of a phase transformation which occurs by nucleation and growth. In discussing recrystallization, we tend to use the term "nucleation" in a rather loose sense, and as will be further discussed in §6.6.1, it is not thought that nucleation in the classic thermodynamic sense occurs.

The progress of recrystallization with time during isothermal annealing is commonly represented by a plot of the **volume fraction of material recrystallized (X_v)** as a function of **log(time)**. This plot usually has the characteristic sigmoidal form of figure 6.2 and typically shows an apparent incubation time before recrystallization is detected. This is

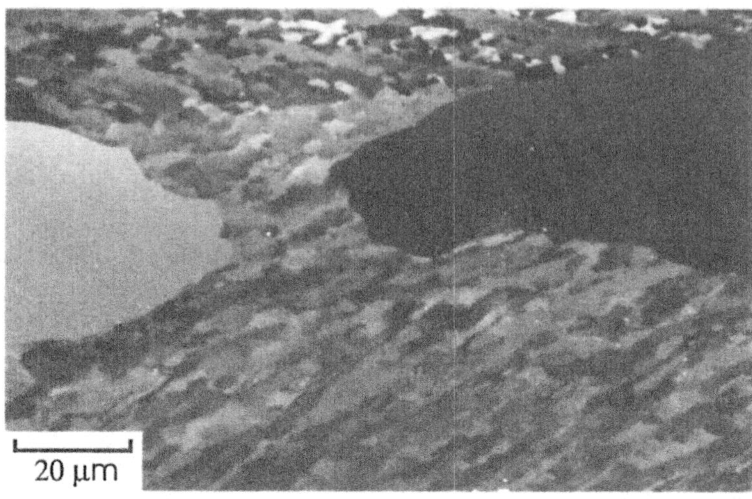

Fig. 6.1. An SEM channelling contrast micrograph of a partly recrystallized specimen of aluminium.

followed by an increasing rate of recrystallization, a linear region, and finally a decreasing rate of recrystallization.

In this chapter we will consider the origin of recrystallization, and general aspects such as the kinetics and other factors which affect both the process of recrystallization and the final microstructure. We will primarily be concerned with the behaviour of single-phase alloys, as the recrystallization of two-phase alloys is considered separately in chapter 8. The mobility and migration of the grain boundaries during recrystallization is considered in more detail in chapter 4, and the orientation of the recrystallized grains (the recrystallization texture), is discussed in chapter 10.

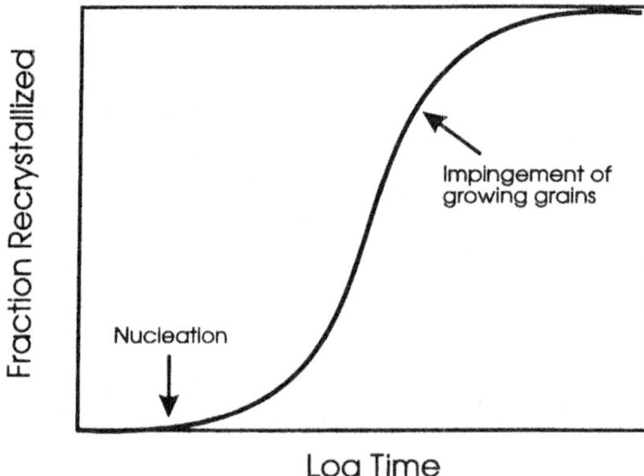

Fig. 6.2. Typical recrystallization kinetics during isothermal annealing.

6.1.1 Quantifying recrystallization

Recrystallization is a microstructural transformation, which is most directly measured by quantitative metallography. However, it is possible to follow the progress of recrystallization by calorimetry or by measurement of other physical or mechanical properties (see §5.1.2).

6.1.1.1 The overall transformation

The extent of recrystallization is often described by X_V, and for isothermal experiments it is convenient to use the time at which recrystallization is 50% complete ($t_{0.5}$), as a measure of the rate of recrystallization. For recrystallization at different temperatures, where annealing is carried out for a constant time, e.g. 1hr, then the **recrystallization temperature**, is usually defined as the temperature at which the material is 50% recrystallized.

Although such measurements of the recrystallization process have practical value, more fundamental measurements of recrystallization are obtained by separate reference to the constituent **nucleation and growth processes**.

6.1.1.2 Nucleation

Our first difficulty lies in defining what we mean by a **recrystallization nucleus.** A working definition might be **a crystallite of low internal energy growing into deformed material from which it is separated by a high angle grain boundary.**

If the number of nuclei per unit volume (N) remains constant during recrystallization, then this is the key parameter. However, if this is not the case, then the **nucleation rate $\dot{N}$ = dN/dt** also needs to be considered. As nucleation of recrystallization is a very complex process, the nucleation rate is not expected to be a simple parameter. Early work indicated that the rate was not always constant during recrystallization as shown in figure 6.3.

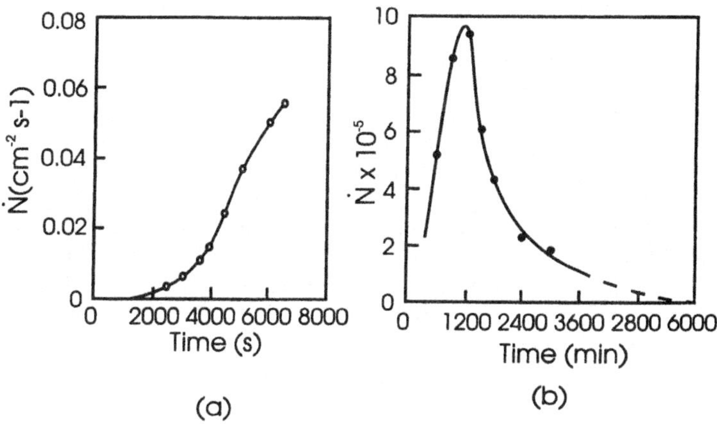

Fig. 6.3. The variation of nucleation rate ($\dot{N}$) with time during the annealing of 5% deformed aluminium, annealed at 350°C. a) Initial grain size 45μm, b) Initial grain size 130μm, (Anderson and Mehl 1945).

However, such measurements need to be regarded with caution, particularly at the early stages of recrystallization, because the number of nuclei observed depends strongly on the technique used to detect them. For example in figure 6.3a are there really no nuclei present at times shorter than 1000 seconds or were they just too small to be detected by optical microscopy? Is the initial rising nucleation rate shown in figure 6.3b a real effect or does it reflect the difficulty of detecting small nuclei, in which case perhaps this figure really represents a monotonically decreasing nucleation rate?

We must recognise that although formal models of the recrystallization process (§6.3) usually refer to nucleation rates, these are often very difficult to measure experimentally. When we come to consider the mechanisms and sites of nucleation (§6.6), it will be clear that in most cases, it is not yet possible to obtain a realistic quantitative model for the number of nuclei and its variation with time and other parameters.

6.1.1.3 Growth

The growth of the new grains during recrystallization is more readily analyzed than is their nucleation. It is generally accepted (§4.1.3) that the velocity (**v**) of a high angle grain boundary is given by

$$v = \dot{G} = M P \qquad (6.1)$$

where M is the boundary mobility and P is the net pressure on the boundary (§4.1.3).

Driving pressure

The driving pressure (P_d) for recrystallization is provided by the dislocation density (ρ), which gives rise to a stored energy given by equation 2.6. The resultant driving force for recrystallization is then given approximately as:

$$P_d = E_D = \alpha \, \rho \, G \, b^2 \qquad (6.2)$$

where α is a constant of the order of 0.5

For a dislocation density of $10^{15} m^{-2}$, typical of a metal, and taking Gb^2 as $10^{-9} N$, the driving pressure for primary recrystallization is of the order of 1MPa. If the dislocations are in the form of low angle boundaries, then, using equation 2.8, the driving pressure may be expressed in terms of subgrain size and misorientation.

Boundary curvature

If we consider a small, spherical, new grain of radius R, growing into the deformed structure, then there is an opposing force which comes from the curvature of a high angle grain boundary of specific energy γ_b. The grain boundary area will be reduced and the energy lowered if the grain were to shrink, and there is thus a retarding pressure on the boundary given by the Gibbs-Thomson relationship:

$$P_c = \frac{2 \gamma_b}{R} \qquad (6.3)$$

This pressure is however only significant in the early stages of recrystallization when the new grains are small. For a grain boundary energy of 0.5 Jm^{-2} and the driving force discussed above, we find that P_c is equal to P_d when R is $\sim 1\mu m$. Below this grain size there would therefore be no net driving force for recrystallization. During the recrystallization of alloys, other retarding forces arising from solutes or second-phase particles may be important, and these will be discussed elsewhere.

The pressure on a grain boundary and therefore the boundary velocity, may not remain constant during recrystallization. In particular, as will be discussed later, the driving force may be lowered by concurrent recovery and both the driving force and the boundary mobility may vary through the specimen. Therefore the growth rate $(\dot{G})$, which is an important parameter in any model of recrystallization is also a complex function of the material and the deformation and annealing conditions.

6.1.2 The Laws of Recrystallization

In this chapter we are primarily concerned with the principles of recrystallization, and although we will refer to selected work from the literature, no attempt will be made to present a detailed survey of the numerous experimental investigations which have been carried out over the past 50 years. Cotterill and Mould (1976) provide an extensive bibliography of the earlier work on a wide variety of materials.

In order to introduce the subject, it is of interest to recall one of the earliest attempts to rationalise the recrystallization behaviour of materials, viz. the formulation of the so-called **laws of recrystallization** (Mehl 1948, Burke & Turnbull 1952). This series of qualitative statements, based on the results of a large body of experimental work, expresses the effect of the initial microstructure (grain size), and processing parameters (deformation strain and annealing temperature), on the time for recrystallization and on the grain size after recrystallization. These rules are obeyed in most cases and are easily rationalised if recrystallization is considered to be a nucleation and growth phenomenon, controlled by thermally activated processes, whose driving force is provided by the stored energy of deformation.

(i). A minimum deformation is needed to initiate recrystallization. The deformation must be sufficient to produce a nucleus for the recrystallization and to provide the necessary driving force to sustain its growth.

(ii). The temperature at which recrystallization occurs decreases as the time of anneal increases. This follows from the fact that the microscopic mechanisms controlling recrystallization are thermally activated and the relationship between the recrystallization rate and the temperature is given by the Arrhenius equation.

(iii). The temperature at which recrystallization occurs decreases as strain increases. The stored energy, which provides the driving force for recrystallization, increases with strain. Both nucleation and growth are therefore more rapid or occur at a lower temperature in a more highly deformed material.

(iv). The recrystallized grain size depends primarily on the amount of deformation, being smaller for large amounts of deformation. The nucleation rate is more affected by strain than is the growth rate. Therefore a higher strain will provide more nuclei per unit volume and hence a smaller final grain size.

(v). For a given amount of deformation the recrystallization temperature will be increased by:

A larger starting grain size. Grain boundaries are favoured sites for nucleation, therefore a large initial grain size provides fewer nucleation sites, the nucleation rate is lowered, and recrystallization is slower or occurs at higher temperatures.

A higher deformation temperature. At higher temperatures of deformation, more recovery occurs during the deformation (dynamic recovery), and the stored energy is thus lower than for a similar strain at a lower temperature.

6.2 FACTORS AFFECTING THE RATE OF RECRYSTALLIZATION.

When Burke and Turnbull (1952) reviewed the subject, they were able to rationalise a large body of experimental data into the comparatively simple concepts embodied in the above "laws of recrystallization". However, later research has shown that recrystallization is a much more complex process than was envisaged at that time, and in addition, we now know that it is difficult to extrapolate some of the earlier work, much of which was carried out on materials of high purity deformed to low strains by uniaxial deformation, to industrial alloys deformed to large strains. Although the laws of recrystallization still provide a very useful guide to the overall behaviour, we now recognise there are so many other important material and processing parameters that need to be taken into account that any attempt to understand recrystallization needs to address the phenomena at a much finer microstructural level than was possible 40 years ago.

In this section, we will consider some of the important factors which are known to affect the rate at which recrystallization occurs in materials.

6.2.1 The deformed structure

6.2.1.1 The amount of strain
The amount, and to some extent the type of deformation, affect the rate of recrystallization, because the deformation alters the amount of stored energy and the number of effective nuclei. As will be discussed in §6.6, the type of nucleation site may also be a function of strain. There is a minimum amount of strain, generally 1-5%, below which recrystallization will not occur. Above this strain the rate of recrystallization increases, levelling out to a maximum value at true strains of $\sim$2-4. The effect of tensile strain on the recrystallization kinetics of aluminium is shown in figure 6.4.

6.2.1.2 Mode of deformation
The mode of deformation also affects the recrystallization rate. Single crystals deformed in single glide recover on annealing, but may not recrystallize, because the dislocation structure (similar to that shown in figure 5.7) does not contain the heterogeneities and orientation gradients needed to provide a nucleation site. The best examples of this effect are found in the hexagonal metals which deform by glide on the basal plane. For example Haase and Schmid (1925) showed that even after extensive plastic strain, crystals of zinc would recover their initial properties completely without recrystallization. At the other extreme, a metal which has undergone a deformation resulting in little or no overall shape change (i.e. most

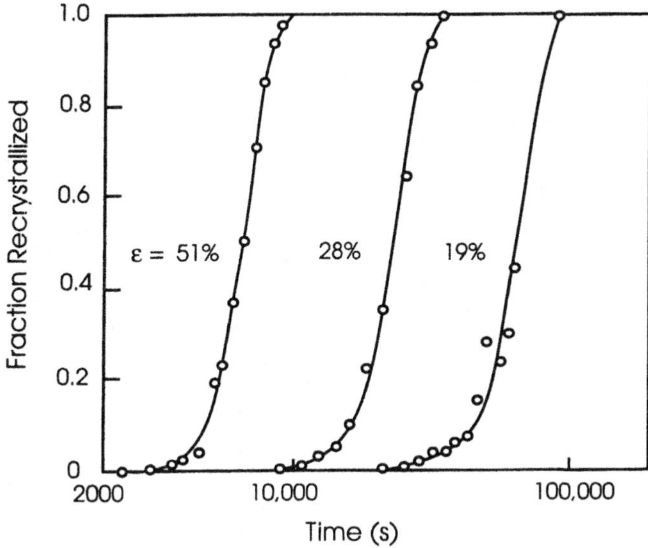

Fig. 6.4. The effect of tensile strain on the recrystallization kinetics of aluminium annealed at 350°C, (Anderson and Mehl 1945).

of the deformation is **redundant**) may nevertheless recrystallize readily. The effect of the deformation mode on recrystallization is complex. For example in both iron (Michalak and Hibbard 1957) and copper (Michalak and Hibbard 1961) it was found that straight rolling induces more rapid recrystallization than does cross rolling to the same reduction in thickness. Barto and Ebert (1971) deformed molybdenum to a true strain of 0.3 by tension, wire drawing, rolling and compression and found that the subsequent recrystallization rate was highest for the tensile deformed material and decreased in the above order for the other samples.

Experiments to determine the role of strain path on the recrystallization of high purity copper have recently been reported by Lindh et al. (1993). They carried out tension and combined tension and compression experiments and measured the recrystallization temperature. For specimens deformed solely by tension, the recrystallization temperature decreased with increasing tensile strain (fig 6.5) as would be expected. However, specimens deformed to the same **total permanent strain** (ϵ_{perm}) by tension and compression are seen to recrystallize at higher temperatures. By reading across from the data points for the tension/compression samples to the curve for the tensile specimens of figure 6.5, an equivalent **recrystallization strain** (ϵ^R_{equ}) could be determined for the tensile/compression samples. It was found that this was given by

$$\epsilon^R_{equ} = \epsilon_{perm} + \eta\,\epsilon_{red} \qquad (6.4)$$

where ϵ_{red} is the redundant strain and η a constant found to be 0.65. This shows that the redundant strain is only 0.65 times as effective in promoting recrystallization as is the permanent strain.

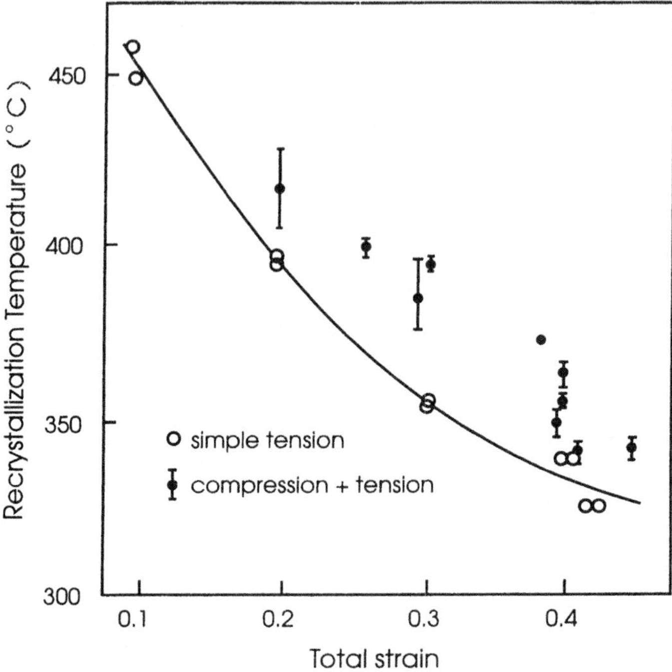

Fig. 6.5. The recrystallization temperature of copper as a function of the total applied strain in specimens deformed by tension or by a combination of tension and compression, (Lindh et al. 1993).

Embury et al. (1992) have compared the recrystallization behaviour of cubes of pure aluminium deformed in uniaxial compression with similar specimens which were deformed incrementally in different directions so as to produce no overall shape change. Their results (fig 6.6a) show that for the same equivalent strain the specimens deformed by multi-axis compression recrystallized more slowly than the specimens deformed by monotonic compression, in agreement with the results of figure 6.5. However, if the specimens were deformed to the same level of **stress**, as shown in figure 6.6b, the recrystallization kinetics of the specimens were similar. A simple interpretation of these results is that the flow stress, which is closely related via the dislocation density to the stored energy (equation 2.2) is the most important factor in determining the recrystallization kinetics. However, a detailed explanation of the effect of deformation mode on recrystallization cannot yet be given, as the effects of redundant deformation on the dislocation density and the microstructure are not adequately understood.

A remarkable example of how a highly deformed microstructure can be "undeformed" is found in the work of Farag et al. (1968). These authors found that the elongated grain structure produced in pure aluminium by a strain in torsion of 2.3 at 400°C could be returned to an equiaxed grain structure if an equal and opposite strain was subsequently given.

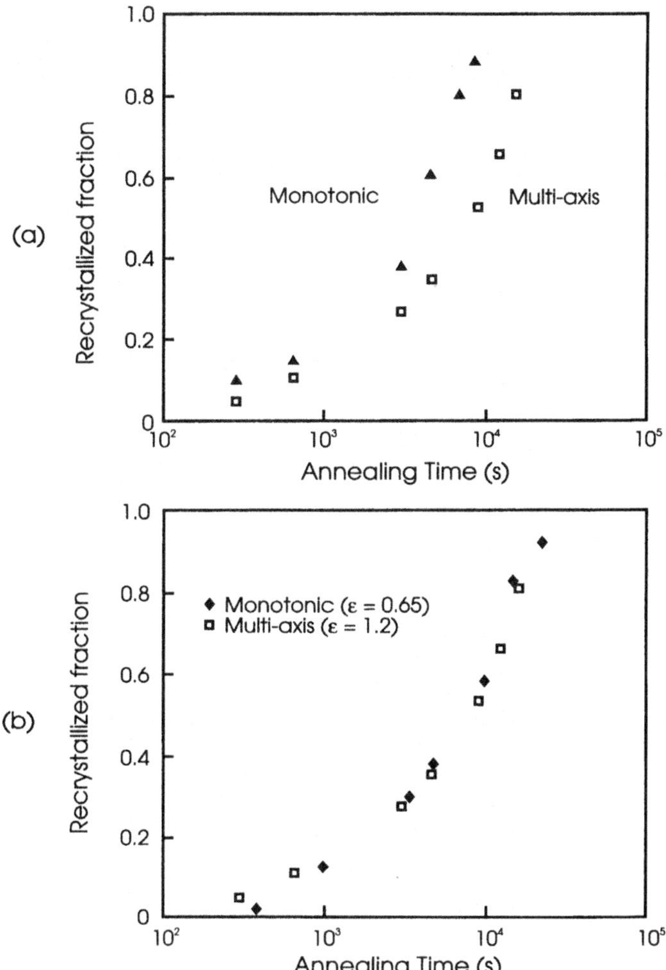

Fig. 6.6. The effect of redundant deformation on the recrystallization of pure aluminium. The recrystallization kinetics for specimens deformed monotonically are compared with those of specimens deformed by multi-axis compression. a) Specimens deformed to the same equivalent strain (0.7). b) Specimens deformed to the same equivalent stress, (Embury et al. 1992).

6.2.2 The grain orientations

6.2.2.1 Single crystals

As discussed in chapter 2, the microstructure and stored energy of a deformed grain depends strongly on the active slip systems and hence on its initial orientation. Therefore the initial orientation will affect both the nucleation sites and the driving force for recrystallization.

Hibbard and Tully (1961) measured the recrystallization kinetics of both copper and silicon-iron single crystals of various orientations, which had been cold rolled 80%, and found significant differences in rates of recrystallization at 600/700°C, which are summarized in table 6.1.

Brown and Hatherly (1970) investigated the effect of orientation on the recrystallization kinetics of copper single crystals which had been rolled to a reduction of 98.6%. The recrystallization times on annealing at 300°C varied between 5 and 1000 minutes. These compared with a recrystallization time of 1 minute for a polycrystalline specimen deformed to the same strain. It was found, in agreement with the first two samples in table 6.1, that the recrystallization kinetics could not be clearly associated with the texture resulting from deformation. For example, both (110)[1$\bar{1}$2] and (110)[001] crystals developed similar {110}<1$\bar{1}$2> rolling textures, but due to differences in the nature of the deformed structure, the former recrystallized 50 times more slowly.

Differences in the recrystallization kinetics and textures of the single crystals and the reference polycrystal in the above investigation serve to highlight the importance of grain boundaries during recrystallization and show that care must be taken in using data from single crystal experiments to predict the behaviour of polycrystals.

6.2.2.2 Polycrystals

In polycrystalline materials there is evidence that the overall recrystallization rate may be dependent on both the starting texture and the deformation texture. In iron, the orientation dependence of the stored energy after deformation has been determined (§2.2.3), and it has been predicted on this basis that the various texture components will recrystallize in the sequence shown in figure 6.7. Recent work (Hutchinson and Ryde 1995) has confirmed the trends shown in this figure. In aluminium alloys, Blade and Morris (1975) have shown that differences in the texture formed after hot rolling and annealing lead to differences in recrystallization kinetics when the material is subsequently cold rolled and annealed.

The observations of inhomogeneous recrystallization discussed in §6.4 provide further evidence of the variation of recrystallization rate for different components of the deformation texture.

Table 6.1

Recrystallization of silicon-iron single crystals at 600°C
(Hibbard and Tulley 1961)

Initial orientation	Final orientation	Time for 50% Recrystallization (s)	Orientation after Recrystallization
(111)[$\bar{1}\bar{1}$2]	(111)[$\bar{1}\bar{1}$2]	200	(110)[001]
(110)[001]	(111)[11$\bar{2}$]	1000	(110)[001]
(100)[001]	(001)[210]	7000	(001)[210]
(100)[011]	(100)[011]	No recrystallization	(100)[011]

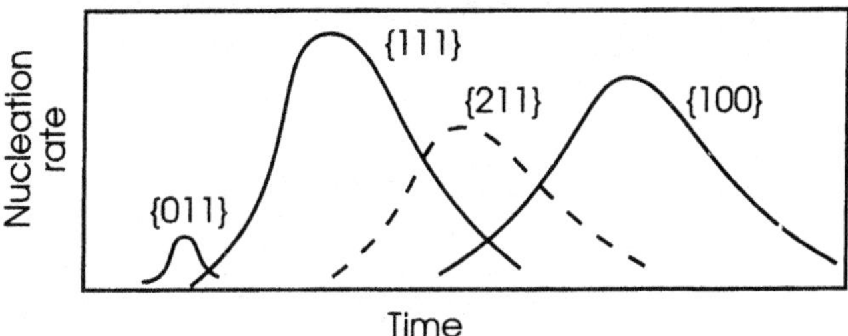

Fig. 6.7. Schematic representation of the proposed orientation dependence of the nucleation rate in deformed iron, (after Hutchinson 1974).

There are two important general conclusions to be drawn from this section:

(i) Selective recrystallization of different texture components will inevitably lead to inhomogeneous recrystallization (§6.4).

(ii) The recrystallization behaviour will be affected not only by the orientations present **after** deformation, but also by the orientations present **before** deformation. The **strain path history** affects both the stored energy and the heterogeneities of the microstructure, and thus apparently identical texture components may recrystallize quite differently. Without knowledge of the starting texture it may not be possible to predict the recrystallization behaviour of a particular component of the deformation texture.

6.2.3 The original grain size

It is generally found that a fine-grained material will recrystallize more rapidly than a coarse-grained material, as seen in figure 6.8.

There are several ways in which the initial grain size may be expected to affect the rate of recrystallization.

(i) The stored energy of a metal deformed to low strains ($\epsilon < 0.5$) tends to increase with a decrease in grain size (§2.2.2).

(ii) Inhomogeneities such as deformation and shear bands are more readily formed in coarse grained material (e.g. Hatherly 1982) and thus the number of these features, which are also sites for nucleation, decreases with decreasing grain size.

(iii) Grain boundaries are favoured nucleation sites (§6.6.4), and therefore the number of available nucleation sites is greater for a fine grained material.

(iv) The deformation texture and hence the recrystallization texture is influenced by the initial grain size, and as discussed in §6.2.2, the recrystallization kinetics will then be affected.

Evidence for all of these effects may be found in the literature, although it should be noted that in the investigation of Hutchinson et al. (1989a) shown in figure 6.8, where the growth

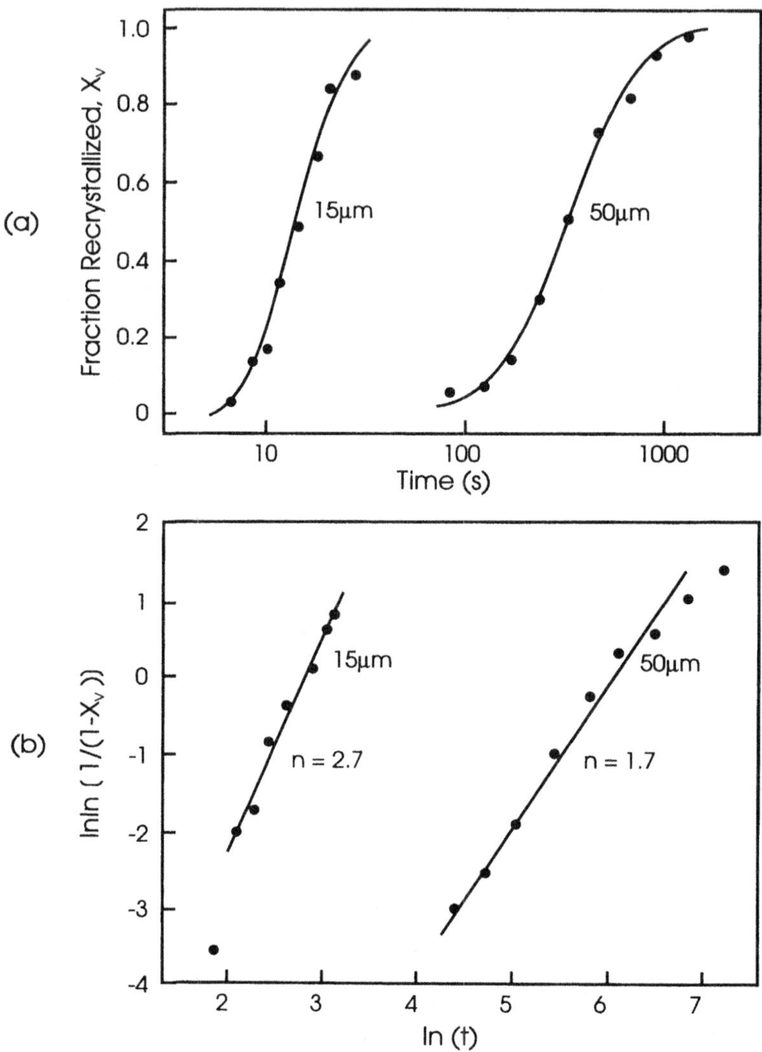

Fig. 6.8. The recrystallization kinetics at 225°C of copper of different initial grain sizes cold rolled 93% a) Fraction recrystallized b) JMAK plot, (Hutchinson et al. 1989a).

rate of the grains in the fine grained material was some 20 times faster than that of the coarse grained material at the same temperature, the difference in kinetics was ascribed primarily to texture effects.

6.2.4. Solutes

The usual effect of solutes is to hinder recrystallization (Dimitrov et al. 1978). Table 6.2 shows the effect of zone refining on the recrystallization temperatures of various metals after large rolling reductions. Although the amount and type of impurities in the commercial materials is not specified, and in some cases they may contain second-phase particles, the marked effect of purifying the metals is clearly seen.

The quantitative effect of a solute on the recrystallization behaviour is dependent on the specific solvent/solute pair. A good example of a very potent solute which has an important influence on commercial alloys is iron in aluminium. Figure 6.9 shows how very small concentrations of iron in solid solution in high purity aluminium may increase the recrystallization temperature by $\sim 100°C$.

An extensive survey of the effect of transition element solutes on the recrystallization of iron has been carried out by Abrahamson and colleagues (e.g. Abrahamson & Blakeny 1960). Although this work has indicated a correlation with the electron configuration of the solute, the effect remains unexplained.

Impurities may affect both the nucleation and growth of recrystallizing grains. It is difficult to measure or even define a **nucleation rate** for recrystallization, and the effect of impurities has not been quantified. Solutes may in some cases have a strong influence on the recovery processes which are an integral part of nucleation and which may affect the driving force for recrystallization. However, the majority of experimental work suggests that the main influence of solutes is on the grain boundary mobility (§4.3.3) and hence on the growth rate of recrystallizing grains.

6.2.5 The deformation temperature and strain rate

At temperatures where thermally activated restoration processes such as dislocation climb occur, then the microstructure will be dependent on the **deformation temperature** and **strain rate** in addition to the strain. After deformation at high temperatures and low strain rates, the stored energy will be reduced and recrystallization will occur less readily than after deformation to a similar strain at low temperature. These effects are considered in §11.6.

Table 6.2.

**The effect of purity on the temperature of recrystallization
(Dimitrov et al. 1978)**

Metal	Recrystallization Temperature (°C)	
	Commercial Purity	Zone Refined
Aluminium	200	-50
Copper	180	80
Iron	480	300
Nickel	600	300
Zirconium	450	170

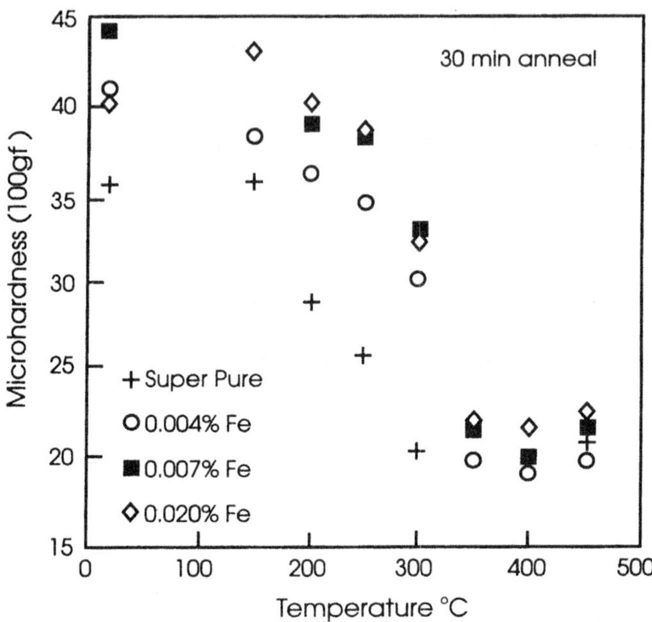

Fig. 6.9. The effect of trace amounts of iron in solid solution on the annealing response of aluminium deformed 80% by cold rolling, (Marshall and Ricks 1992).

6.2.6 The annealing conditions

6.2.6.1 The annealing temperature

As may be seen from figure 6.10a, the annealing temperature has a profound effect on the recrystallization kinetics. If we consider the transformation as a whole and take the time for 50% recrystallization ($t_{0.5}$) to be a measure of the rate of recrystallization then we might expect a relationship of the type

$$\text{Rate} = \frac{1}{t_{0.5}} = C \exp\left(-\frac{Q}{kT}\right) \tag{6.5}$$

In figure 6.10b, we see that a plot of $\ln(t_{0.5})$ v $1/T$ gives a good straight line in accordance with equation 6.5, with a slope corresponding to an activation energy of 290kJ/mol. Although such an analysis may be of practical use, an activation energy obtained from it is not easy to interpret, as it refers to the transformation as a whole, and is unlikely to be constant. For example such activation energies have been found to depend on strain (Gordon 1955) and to vary significantly with small changes in material purity. Vandermeer and Gordon (1963) found that an addition of only 0.0068 atomic% of copper to pure aluminium raised the activation energy for recrystallization from 60 to 120 kJ/mol.

An activation energy is only interpretable if it can be related to thermally activated processes at the atomic level. Consider the separate nucleation and growth processes which constitute recrystallization.

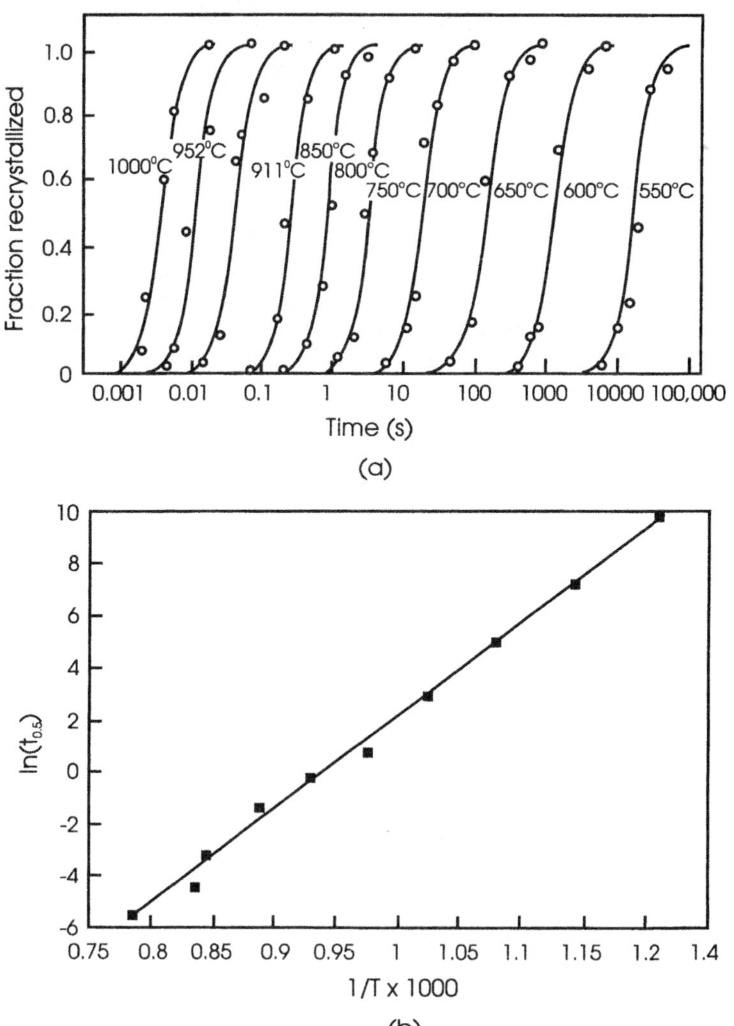

Fig. 6.10. a) The effect of annealing temperature on the annealing of Fe-3.5%Si deformed 60%. b) Arrhenius plot of the time for 50% recrystallization as a function of temperature. (data from Speich and Fisher 1966).

$$\dot{N} = C_1 \exp\left(-\frac{Q_N}{kT}\right) \tag{6.6}$$

$$\dot{G} = C_2 \exp\left(-\frac{Q_G}{kT}\right) \tag{6.7}$$

Although equation 6.7 has some validity because grain boundary migration rates are known to follow this type of relationship (§4.3.1), equation 6.6 is of little use because we have already noted that a constant nucleation rate is rarely found in practice. If, as discussed in chapter 5, we may assume that the rate controlling process in recrystallization nucleation is subgrain growth, then the activation energy for this process is the most relevant to nucleation.

6.2.6.2 The heating rate

The rate of heating of the specimen to the annealing temperature may also be important because a rapid heating rate may reduce the amount of recovery occurring before recrystallization and hence by preserving a large driving force, promote recrystallization. Such an effect would be predicted if the temperature dependence of the controlling mechanisms in recovery and recrystallization were different, e.g. there was a higher activation energy for recrystallization than recovery. However, there is little firm evidence for this effect in the single-phase alloys with which we are primarily concerned in this chapter.

There is however extensive published work on the effects of heating rate on the recrystallization of complex alloys. It has been found that a rapid heating rate accelerates recrystallization and results in a smaller grain size in Al-Li alloys (Bowen 1990) and in Al-Zn-Mg-Cu-Si alloys (AA7075), (Wert et al. 1981), and these effects have been attributed to the role of solutes and second-phase particles in retarding recovery. Conversely, it is found that in steels, rapid heating may slow down recrystallization (Ushioda et al. 1989). In this case the heating rate determines the amount of carbon in solid solution, and this in turn affects the recrystallization (§12.3).

6.3 THE FORMAL KINETICS OF PRIMARY RECRYSTALLIZATION

6.3.1 The Johnson-Mehl-Avrami-Kolmogorov (JMAK) model

The type of curve shown in figure 6.2 is typical of many transformation reactions, and may be described phenomenologically in terms of the constituent nucleation and growth processes. The early work in this area is due to Kolmogorov (1937), Johnson and Mehl (1939) and Avrami (1939) and is commonly known as the JMAK model. A more general discussion of the theory of the kinetics of transformations may be found for example in Christian (1965).

6.3.1.2 Theory

It is assumed that nuclei are formed at a rate $\dot{N}$ and that grains grow into the deformed material at a linear rate $\dot{G}$. If the grains are spherical, then their volume varies as the cube of their diameter, and the fraction of recrystallized material (X_V) rises rapidly with time. However, the new grains will eventually impinge on each other and the rate of recrystallization will then decrease, tending to zero as X_V approaches 1.

The number of nuclei (dN) actually appearing in a time interval (dt) is less than $\dot{N}dt$ because nuclei cannot form in those parts of the specimen which have already recrystallized. The number of nuclei which would have appeared in the recrystallized volume is $\dot{N}.X_V.dt$ and therefore the total number of nuclei (dN') which would have formed, including the "phantom" nuclei is given by:

$$dN' = \dot{N}\,dt = dN + \dot{N}\,X_V\,dt \qquad (6.8)$$

If the volume of a recrystallizing grain is V at time t, then the fraction of material which would have recrystallized if the phantom nuclei were real (X_{VEX}) is known as the **extended volume** and is given by:

$$X_{VEX} = \int_0^t V \, dN'$$

(6.9)

If the incubation time is much less than t, then

$$V = f \dot{G}^3 t^3$$

(6.10)

where f is a shape factor ($4\pi/3$ for spheres). Thus

$$X_{VEX} = f G^3 \int_0^t t^3 \dot{N} \, dt$$

(6.11)

If $\dot{N}$ is constant then

$$X_{VEX} = \frac{f \dot{N} \dot{G}^3 t^4}{4}$$

(6.12)

During a time interval dt, the extended volume increases by an amount dX_{VEX}. As the fraction of unrecrystallized material is $1-X_V$, it follows that $dX_V = (1-X_V).dX_{VEX}$ or

$$dX_{VEX} = \frac{dX_V}{1 - X_V}$$

(6.13)

$$X_{VEX} = \int_0^{X_V} dX_{VEX} = \int_0^{X_V} \frac{dX_V}{1 - X_V} = \ln \frac{1}{1 - X_V}$$

(6.14)

$$X_V = 1 - \exp(-X_{VEX})$$

(6.15)

Combining equations 6.12 and 6.15

$$X_V = 1 - \exp\left(\frac{-f \dot{N} \dot{G}^3 t^4}{4}\right)$$

(6.16)

This may be written more generally in the form

$$X_V = 1 - \exp(-B t^n)$$

(6.17)

which is often called the **Avrami, Johnson-Mehl or JMAK equation**

In the case considered above, in which the growing grains were assumed to grow in three dimensions, then the exponent **n** in equation 6.17, which we will refer to as the **JMAK or Avrami exponent** is seen from equation 6.16 to be **4**. The above treatment assumed that the rates of nucleation and growth remained constant during recrystallization. Avrami (1939) also considered the case in which the nucleation rate was not constant, but a decreasing function of time, $\dot{N}$ having a simple power law dependence on time. In this situation **n** lies between 3 and 4, depending on the exact form of the function.

Two limiting cases are of importance, that already discussed in which $\dot{N}$ is constant and **n = 4**, and that where the nucleation rate decreases so rapidly that all nucleation events effectively occur at the start of recrystallization. This is termed **site saturated nucleation.** In this case, X_{VEX} in equation 6.12 is given by f N $(Gt)^3$, where N is the number of nuclei. This may be seen to give a JMAK exponent of **3** in equation 6.17.

The above analyses assume that until impingement, the grains grow isotropically in three-dimensions. If the grains are constrained either by the sample geometry or by some internal microstructural constraint to grow only in one or two-dimensions, then the JMAK exponent is lower as shown in table 6.3.

Cahn (1956) has extended the theory to include nucleation at random sites on grain boundaries and found that **n** fell from 4 at the start of the transformation to 1 at the end. However, no general analytical treatment of non-randomly distributed nucleation sites is available.

It should be noted that the essential feature of the JMAK approach is that the nucleation sites are assumed to be randomly distributed.

6.3.1.2 Comparison with experiment

Experimental measurements of recrystallization kinetics are usually compared with the JMAK model by plotting $\mathbf{log_e[log_e\{1/(1-X_V)\}]}$ against $\mathbf{log_e}$ **t**. According to equation 6.17, this should yield a straight line of slope equal to the exponent **n**. This method of data representation is termed a **JMAK plot**. It should incidentally be noted that equation 6.14 shows that this is equivalent to a plot of $\mathbf{log_e}$ $\mathbf{X_{VEX}}$ against $\mathbf{log_e}$ **t**.

The growth rate is not easily determined from measurements of X_V. However, if the interfacial area between recrystallized and unrecrystallized material per unit volume (S_V) is measured as a function of time, then Cahn and Hagel (1960) showed that the global growth rate $\dot{G}$ is given by

$$\dot{G} = \frac{1}{S_V} \cdot \frac{dX_V}{dt} \qquad\qquad (6.18)$$

Table 6.3
Ideal JMAK exponents

Growth dimensionality	Site saturation	Constant nucleation rate
3-d	3	4
2-d	2	3
1-d	1	2

There are a number of early experimental investigations of recrystallization in which JMAK kinetics are found with n~4. These include the work of Anderson and Mehl (1945) on aluminium, Reiter (1952) on low carbon steel and Gordon (1955) on copper. All these investigations were for fine grained material deformed small amounts in tension. As pointed out by Doherty et al. (1986), in such conditions an increase in grain size on recrystallization is to be expected, with less than one successful nucleus arising from each old grain. Therefore the overall spatial distribution of these nuclei is likely to approach random (fig 6.14a), thus fulfilling the conditions for JMAK kinetics. Figure 6.8b, from the work of Hutchinson et al. (1989a) on copper, demonstrates that the fine grained material exhibits JMAK kinetics, whilst the logarithmic plot for the coarse grained material shows a strong negative deviation at long times.

It is in fact very unusual to find experimental data which, on detailed analysis, show good agreement with JMAK kinetics. Either the JMAK plot is non-linear, or the slope of the JMAK plot is less than 3, or both. A very clear example is the work of Vandermeer and Gordon (1962) on 40% cold rolled aluminium containing small amounts of copper, shown in figure 6.11. The JMAK slopes were found to be 1.7, and the data at the two lower annealing temperatures are seen to fall below the straight line at long annealing times. There are numerous investigation of aluminium and of other materials in which values of **n** of the order of 1 have been found. These include copper (Hansen et al. 1981) and iron (Michalak and Hibbard 1961, Rosen et al. 1964).

It is therefore concluded that the JMAK analysis discussed above is too simple to quantitatively model a process as complex as recrystallization. In particular, the change in microstructure during recrystallization needs to be described by more parameters than the fraction of material transformed (X_V or X_{VEX}).

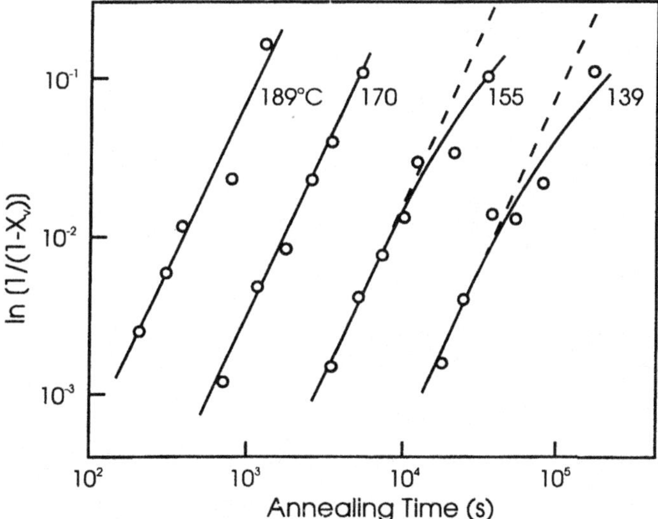

Fig. 6.11. JMAK plot of the recrystallization kinetics of aluminium containing 0.0068 at% Cu, deformed 40% by rolling, (Vandermeer and Gordon 1962).

6.3.2 Microstructural path methodology

A significant attempt to improve the JMAK approach has recently been made by Vandermeer and Rath (1989a) using what they have termed **Microstructural path methodology (MPM)**. In this approach, more realistic and more complex geometric models are employed by using additional microstructural parameters in the analysis and, if required, relaxing the uniform grain impingement constraint.

As for the JMAK model, it is convenient to use the concept of **extended volume** (X_{VEX}) whose relation to the recrystallized fraction (X_V) is given by equation 6.15. In addition, the microstructure is characterised by the **interfacial area** per unit volume between recrystallized and unrecrystallized material (S_V) and this is related to the extended interfacial area (S_{VEX}) by the relationship (Gokhale and DeHoff 1985).

$$S_{VEX} = \frac{S_V}{1 - X_V} \qquad (6.19)$$

As before, this relationship is valid only for randomly distributed recrystallized grains.

The growth of individual recrystallizing grains may be written in integral form

$$X_{VEX} = \int_0^t V_{(t-\tau)} \cdot \dot{N}_{(\tau)} \, dt \qquad (6.20)$$

$$S_{VEX} = \int_0^t S_{(t-\tau)} \cdot \dot{N}_{(\tau)} \, dt \qquad (6.21)$$

where $V_{(t-\tau)}$ and $S_{(t-\tau)}$ are the volume and interfacial area respectively at time **t**, of a grain which was nucleated at time τ.

If it is assumed that the grains are spheroids which do not change shape, then the volume and interfacial area of a grain are

$$V_{(t-\tau)} = K_V \cdot a_{(t-\tau)}^3 \qquad (6.22)$$

$$S_{(t-\tau)} = K_S \cdot a_{(t-\tau)}^2 \qquad (6.23)$$

where **a** is the major semi-axis of the spheroid and K_V and K_S are shape factors. The radius function $a_{(t-\tau)}$ is related to the interface migration rate $G(t)$ by

$$a_{(t-\tau)} = \int_\tau^t G(t) \cdot dt \qquad (6.24)$$

Whereas X_{VEX} and S_{VEX} are global microstructural parameters which may be measured metallographically, $a_{(t-\tau)}$ is a local property, and may be estimated by measuring the diameter of the largest unimpinged grain intercept (D_L) on a plane polished surface. If it can be assumed that this is the diameter of the earliest nucleated grain, then $a_{(t-\tau)} = D_L/2$.

Vandermeer and Rath (1989a), using the method of Gokhale and Dehoff (1985) have shown that the nucleation and growth characteristics of the recrystallizing material are contained in the time dependence of equations 6.20 and 6.21, and have demonstrated how these equations may be evaluated to extract the **nucleation rate**, **growth rate**, and **size** of the recrystallizing grains. The method involves inverting the equations, based on the mathematics of Laplace transforms.

For isothermal annealing, it is assumed that the time dependence of X_{VEX}, S_{VEX} and D_L may be represented by power law functions of the form

$$X_{VEX} = B\,t^{\,n} \qquad\qquad (6.25)$$

$$S_{VEX} = K\,t^{\,m} \qquad\qquad (6.26)$$

$$D_L = S\,t^{\,s} \qquad\qquad (6.27)$$

It is also postulated that the derived functions can be represented by power law functions, thus

$$\dot{N}_{(\tau)} = N_1\,\tau^{\,\delta-1} \qquad\qquad (6.28)$$

$$a_{(t-\tau)} = G_a\,(t-\tau)^{\,r} \qquad\qquad (6.29)$$

where N_1, G_a δ and r are constants.

If the shapes of the grains remain spheroidal during growth then $a_{(t)} = D_L/2$ and hence $r = s$ and $G_a = S/2$

Vandermeer and Rath show that for spheroidal grains which retain their shape during growth

$$\delta = 3\,m - 2\,n \qquad\qquad (6.30)$$

$$r = s = n - m \qquad\qquad (6.31)$$

Reference to equations 6.25 to 6.29 shows that:
 $\delta = 1$ corresponds to a constant nucleation rate
 $\delta = 0$ to site saturated nucleation
 $r = s = 1$ corresponds to a constant growth rate.

Therefore, by measuring X_V, S_V and D_L experimentally as a function of annealing time, n, m and s may be determined, δ and r calculated and the form of the nucleation kinetics identified.

If the simple JMAK model discussed above applies, then, for constant nucleation and growth rates we would obtain n = 4, m = 3, s = 1 and hence $\delta = 1$ and $r = 1$.

Figure 6.12, from the work of Vandermeer and Rath shows data for the recrystallization of a deformed iron crystal plotted to give n, m and s from the slopes of the lines in figures 6.12 a,b,c respectively. The data have been normalised using an Arrhenius relationship so as to include results obtained at several temperatures. The values obtained are n = 1.90, m = 1.28, s = 0.60, which give δ = 0.04 and r = 0.62. As δ is almost zero, it is concluded that nucleation is effectively site saturated, and as r $\sim$ s it is concluded that the grains grow as spheroids. The low value of r indicates a growth rate decreasing with time.

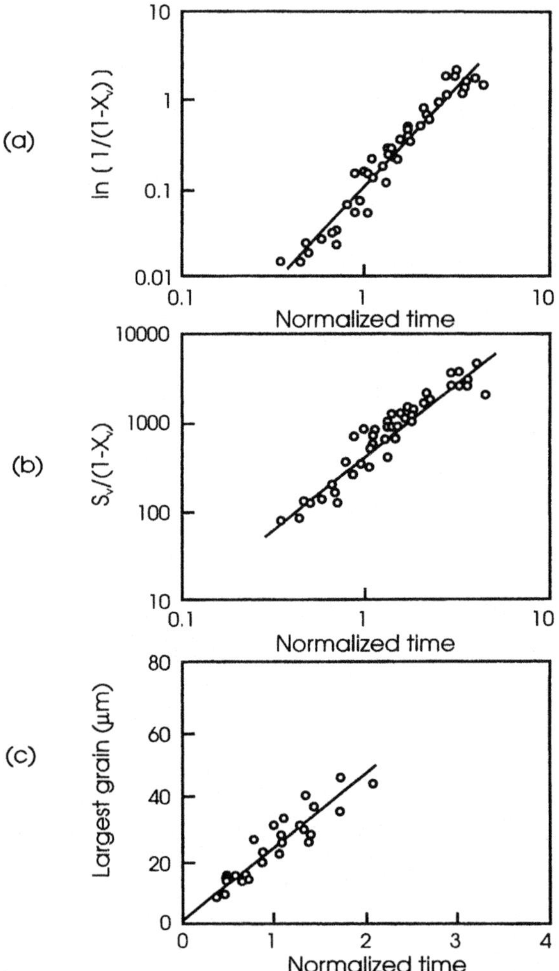

Fig. 6.12. Annealing kinetics of an iron crystal deformed 70% and annealed at various temperatures. By normalising the time axis by means of an Arrhenius relationship, data from a range of annealing temperatures are found to fall on a common curve. a) Variation of X_V, b) Variation of S_V. c) Variation of largest recrystallized grain diameter, (Vandermeer and Rath 1989a).

If $s \neq r$ then the grain shape is changing during recrystallization, and further information about the shapes of the grains may be extracted (Vandermeer and Rath 1990. The model has also been extended to include the effects of recovery during the recrystallization anneal (Vandermeer and Rath 1989b).

Microstructural path methodology allows more detailed information about nucleation and growth rates to be extracted from experimental measurements than does the original JMAK analysis. However, it should be noted that whilst the MPM approach is more flexible than the JMAK model, it is still based on the assumption that the spatial distribution of nuclei is random and that growth rates are global rather than local parameters.

From the above discussion of the JMAK and MPM models it is clear that the two factors most likely to cause problems in realistic modelling of recrystallization kinetics are **concurrent recovery** and **inhomogeneous recrystallization,** and these are considered in the next section.

6.4 RECRYSTALLIZATION KINETICS IN REAL MATERIALS

6.4.1 Non-random distribution of nuclei

It is widely recognised that the nucleation sites in recrystallization are non-randomly distributed. However, the analytical treatments of kinetics discussed above are unable to satisfactorily take this into account.

Evidence for non-random site distribution is found for all materials, particularly those with a large initial grain size, .nd examples include **iron** (Rosen et al. 1964), **copper** (Hutchinson et al. 1989a), **brass** (Carmichael et al. 1982) and **aluminium** (Hjelen et al. 1991). Figure 6.13 shows the recrystallized microstructure of high purity aluminium. The inhomogeneity of the transformation is very marked, and the positions of the original grains can still be seen.

On the scale of a single grain, nucleation of recrystallization is inhomogeneous as will be discussed in more detail in §6.6. Nucleation occurs at preferred sites such as prior grain boundaries, at transition bands, and shear bands. At lower strains, fewer of these deformation heterogeneities are formed and fewer of those formed act as nucleation sites. Whether or not these small-scale inhomogeneities lead to an overall inhomogeneous distribution of nuclei will depend on the number of nuclei relative to the number of potential sites, as shown schematically in figure 6.14. If nucleation only occurs at grain boundaries, then a given number of nuclei in a fine grained material (fig 6.14a) will lead to more homogeneous recrystallization than the same number in a coarse grained material (fig 6.14b). The limiting case is that of single crystals in which recrystallization nucleation is rarely randomly distributed.

On a larger scale, it is also found, that not all grains in a material recrystallize at a similar rate, and it is this larger-scale heterogeneity that is frequently the main cause of inhomogeneous recrystallization. This effect arises from the orientation differences between grains. As discussed in chapter 2, the slip systems and strain path which operate during deformation are dependent on crystallographic orientation. Thus the distribution and density

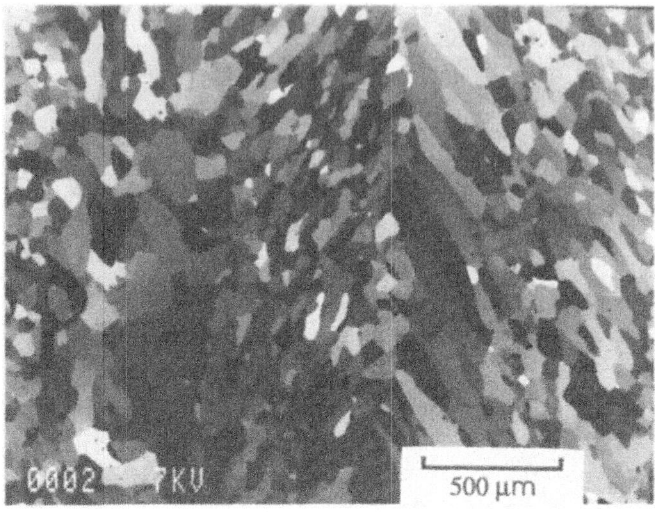

Fig. 6.13. SEM channelling contrast image of recrystallized high purity aluminium,
(Hjelen et al. 1991).

of dislocations and of larger scale microstructural inhomogeneities is orientation dependent.
Therefore the availability and viability of nucleation sites and also the growth rate of the
recrystallizing grains depends strongly on orientation. The inhomogeneous recrystallization
in figure 6.13 is due to this effect. An extreme example of this is in low zinc α-brass
(Carmichael et al. 1982) in which, depending on their initial orientation, grains may deform
either by reasonably uniform slip or by deformation twinning and shear banding. On
annealing, the latter grains recrystallize first and are often completely recrystallized before
the former have started to recrystallize as shown in figure 6.15b.

In alloys containing large second-phase particles which stimulate the nucleation of
recrystallization (§8.4.4), clustering of the particles may, for similar reasons, lead to
macroscopically inhomogeneous recrystallization.

6.4.2 The variation of growth rate during recrystallization

6.4.2.1 Experimental observations
There is considerable evidence that the growth rate ($\dot{G}$) is not a constant, and therefore a
realistic treatment of recrystallization kinetics must allow for the variation of $\dot{G}$ in both space
and time. Although there may be some variation of the boundary mobility (M) during
recrystallization, for example through an orientation dependence of mobility (§4.3.2), the
main reason for such changes in growth rate is thought to be variations in the **driving
pressure**. The driving pressure may vary through the material due to microstructural
inhomogeneity and may also decrease with time due to recovery occurring concurrently with

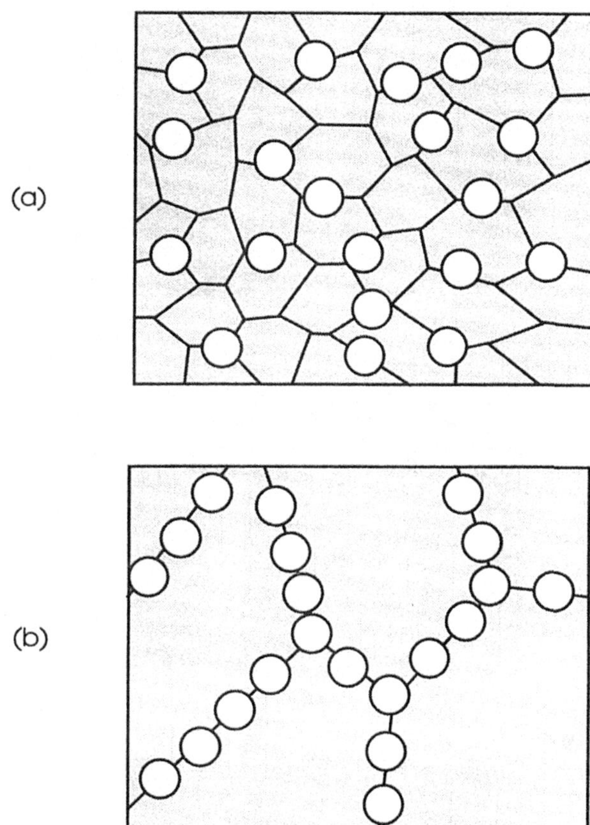

Fig. 6.14. Schematic representation of the effect of the initial grain size on the
heterogeneity of nucleation.

recrystallization. It is now thought that in many cases **both** these effects occur, making the
interpretation or prediction of growth rates very difficult.

There have been several detailed measurements of grain boundary velocities during
recrystallization, including those on **aluminium** (Vandermeer and Gordon 1959, Furu and
Nes 1992), **iron** (Leslie et al. 1963, English and Backofen 1964, Speich and Fisher 1966,
Vandermeer and Rath 1989a) and **titanium** (Rath et al. 1979). The growth rates were
usually determined either by measurement of the size of the largest growing grain, or by the
Cahn-Hagel analysis (equation 6.18). In all these investigations the growth rates were found
to decrease significantly with annealing time.

Various equations have been used to express the variation of $\dot{G}$ with time such as

$$\dot{G} = \frac{A}{1 + Bt^r} \qquad (6.32)$$

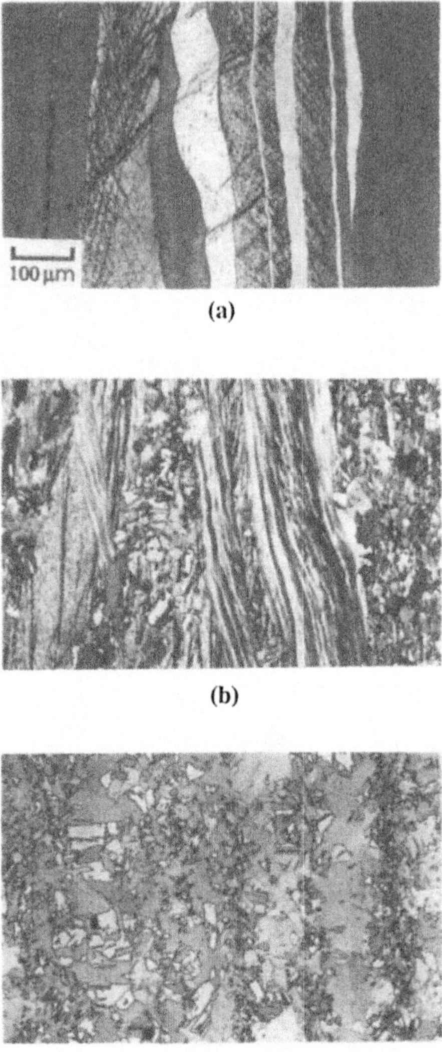

(a)

(b)

(c)

Fig. 6.15. Inhomogeneous recrystallization in 95:5 brass cold rolled 75%. and annealed at 350°C, (TD section with rolling direction vertical). a) As deformed. b) 6 min at 250°C. c) 24 hrs at 250°C, (Carmichael et al. 1982).

which, at long times reduces to

$$\dot{G} = C\, t^{-r} \tag{6.33}$$

In several cases **r** has been found to be close to 1, although in their recent investigation of the recrystallization kinetics of iron, Vandermeer and Rath (1989a) found r=0.38.

The basic JMAK model of §6.3.1 assumes a constant growth rate, and the following analysis illustrates the effect of recovery on the growth rate and on the kinetics of recrystallization. More detailed treatments are discussed by Furu et al (1990).

If we allow $\dot{G}$ to vary, then for site-saturated nucleation with N nuclei, equation 6.16 becomes

$$X_v = 1 - \exp\left[-fN\left(\int_0^t \dot{G}\,dt\right)^3\right] \qquad (6.34)$$

If the variation of growth rate with time is as given by equation 6.33, then combining equations 6.33 and 6.34 we find

$$X_v = 1 - \exp\left[-fN\left(C\,\frac{t^{(1-r)}}{(1-r)}\right)^3\right] \qquad (6.35)$$

The JMAK plots for the recrystallization kinetics given by equation 6.35 are shown in figure 6.16 for various values of r.

Note that as the growth rate is increasingly slowed (increasing r), the JMAK plot remains a straight line. However the slope decreases and, from equation 6.35 is equal to 3(1-r).

6.4.2.2 The role of recovery

Until recently, a decreasing growth rate during recrystallization or a non-linear JMAK plot was thought to be due entirely to the effects of recovery on the driving force for recrystallization. The classic early work in this field is that of Vandermeer and Gordon (1962) who carried out an extensive metallographic and calorimetric study of the

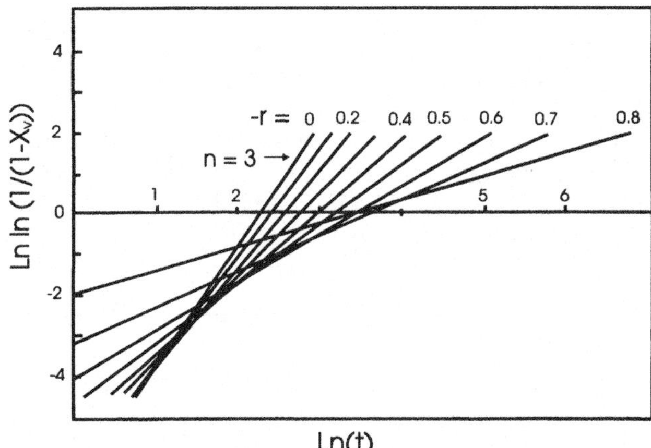

Fig. 6.16. The predicted effect of the recovery parameter **r** in equation 6.35 on the JMAK plot for recrystallization under conditions of site saturated nucleation.

recrystallization of aluminium alloys containing small amounts of copper. In this work the low JMAK slopes and the deviation of these plots from linearity at long times (fig 6.11) were attributed to recovery. However, it is now thought that in many cases, factors other than recovery may be more important.

Consider the measurements of the softening during annealing of cold worked copper and aluminium shown in figure 6.17. For the case of copper (fig 6.17a) there is no softening prior to that due to recrystallization, but for aluminium (fig 6.17b), there is extensive prior softening, and these observations are consistent with our knowledge of recovery in these materials. Although we may expect some recovery during the recrystallization of aluminium

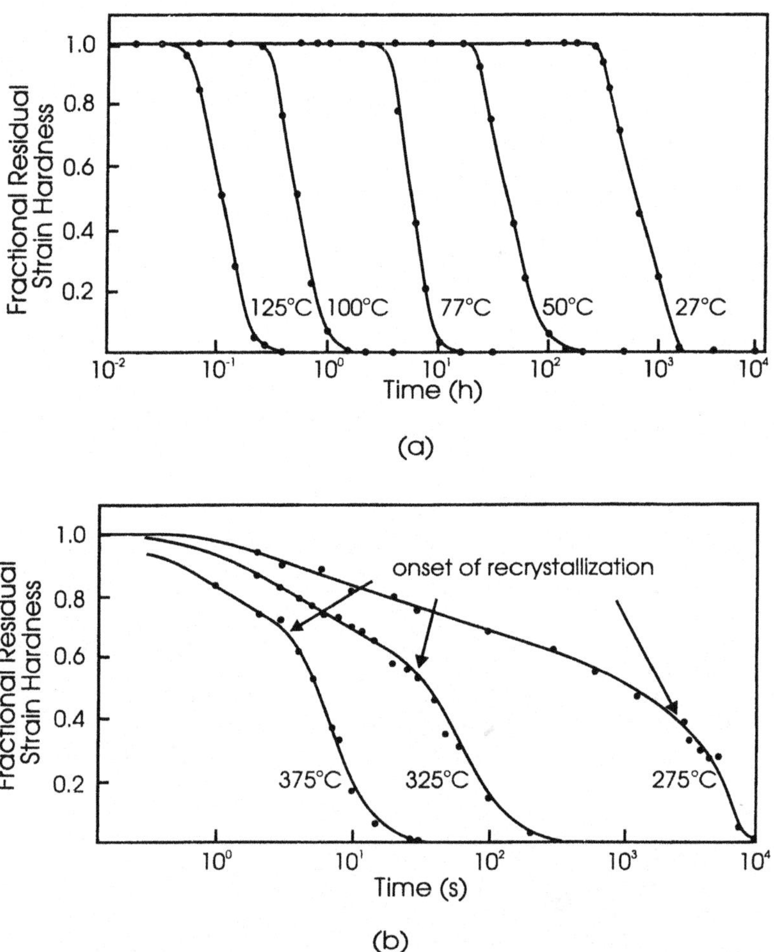

Fig. 6.17. The variation of the fractional residual strain hardening during annealing for a) cold worked copper (Cook and Richards 1946) and b) commercial purity aluminium, (Furu et al. 1990).

which will lower the driving force, it is unlikely that there will be significant recovery in copper, or other metals of medium to low stacking fault energy. In such materials, recovery is not a likely explanation for the low JMAK exponents, nor the non-linearity of the JMAK plots at longer times, such as shown in figure 6.8b.

We must now question the role of recovery in affecting recrystallization even in those materials for which recovery is known to occur readily.

Would uniform recovery explain the observed JMAK plots?
As discussed in chapter 5, recovery lowers the driving pressure (P) for recrystallization and is therefore expected to reduce $\dot{G}$ according to equation 6.1. Although our knowledge of recovery kinetics is far from complete, the available evidence suggests that recovery processes are broadly consistent with the kinetics of growth rate reduction as given by equations 6.32 and 6.33. For example a JMAK slope of 2, which is commonly observed experimentally, implies a not unreasonable value of $r=0.33$ in equation 6.33. However, recovery on the model discussed above or on the model of Furu et al. (1990), leads to linear JMAK plots and would therefore not explain the frequent observation of non-linearity such as shown in figure 6.11.

Is the temperature dependence of $\dot{G}$ consistent with recovery?
In their investigation of the recrystallization kinetics of iron, Vandermeer and Rath (1989a) found that data over a wide range of annealing temperatures could be fitted to equation 6.33 if the data were normalised to take account of the temperature dependence of the growth rate (fig 6.12c). If this decrease in growth rate were due to recovery, then it would fit equation 6.33 over the whole temperature range only if the activation energies for recovery and high angle grain boundary migration were equal, which is not necessarily so.

Is the amount of recovery sufficient to explain the recrystallization kinetics?
Several investigations (e.g Perryman (1955) on aluminium and Rosen et al. (1964) on iron), have indicated that prior recovery treatment has little effect on recrystallization. In figure 6.18, the JMAK plots for iron which has been recovered at several temperatures before recrystallization, show that prior recovery has had little effect on the recrystallization kinetics.

More recent experimental measurements of both the extent of recovery and the growth rate of recrystallization in aluminium (Furu and Nes 1992) have shown that the amount of uniform recovery is relatively small (20% at $X_V=0.1$) and is in itself not sufficient to account for the observed decrease of the growth rate.

It is therefore likely that factors other than recovery are contributing to the decrease in growth rate at long times.

6.4.2.3 The role of microstructural inhomogeneity
The nature of the deformation microstructure (§2.3) gives rise not only to the inhomogeneous distribution of recrystallization nuclei discussed in §6.4.1, but also to the variations in stored energy which are responsible for inhomogeneous growth rates.

Hutchinson et al. (1989a) have recently found inhomogeneous recrystallization in cold rolled OHFC copper. They measured the release of stored energy during the recrystallization, and found that it was not proportional to the fraction recrystallized, and that the regions of high

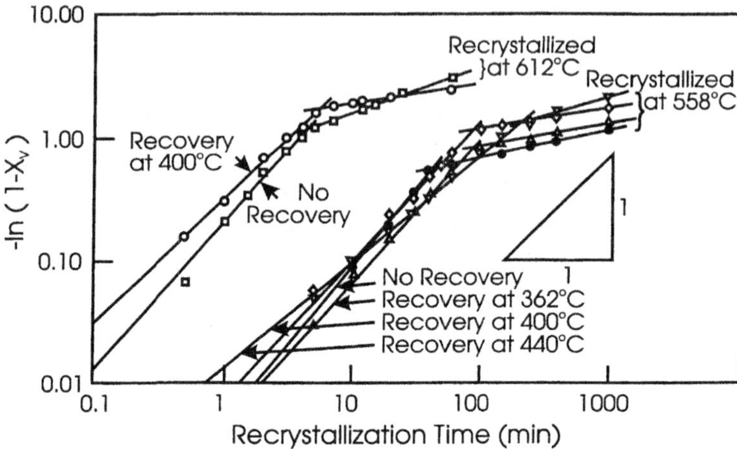

Fig. 6.18. The effect of recovery on the recrystallization of iron, (Rosen et al. 1964).

stored energy recrystallized first (fig 6.19a). The measured growth rate of the recrystallizing grains decreased with time, and they were able to show that this was due entirely to the inhomogeneity of the stored energy distribution. As shown in figure 6.19b, the mean growth rate was found to be proportional to the true driving force, which was calculated from the calorimetric measurements. The recrystallization kinetics from this work, shown in figure 6.8b, indicate that although the fine grained material gave a linear JMAK plot with a slope of 2.7, the coarse grained material, in which the recrystallization was more inhomogeneous had a significantly lower slope of 1.7 and the data fell well below the straight line plot during the later stages of recrystallization.

Earlier work on aluminium (Vandermeer and Gordon 1962), also showed that the amount of heat evolved per unit volume recrystallized, decreased as recrystallization proceeded. Although this was interpreted as being due to recovery, it is quite possible that these results could also be explained at least in part, by an inhomogeneous distribution of stored energy.

As deformation microstructures are very varied and not well understood, it is difficult to give a general treatment of recrystallization in inhomogeneous microstructures, although specific cases have been modelled analytically by Furu et al. (1990) and Vandermeer and Rath (1989b).

In figure 6.20 we show schematically one possible case. In figure 6.20a, spherical regions of high stored energy (dark shaded) are present in the deformed microstructure. These might for example correspond to the highly deformed regions associated with large second-phase particles (§8.2.4). If recrystallization nucleates predominantly in or close to these regions then growth will be initially rapid, but will decrease when these regions of high stored energy are consumed (fig 6.20b). In this case the heterogeneity is **local** and if the stored energy within these regions can be calculated, then the growth rate and hence the recrystallization kinetics may be modelled analytically (Furu et al. 1990) on the basis of equation 6.34.

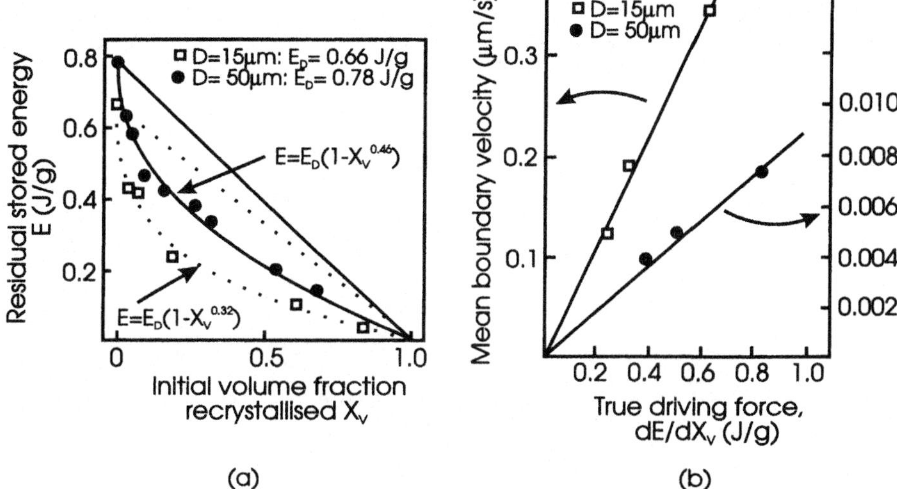

Fig. 6.19. The effect of the inhomogeneous distribution of stored energy on the recrystallization of the copper specimens whose kinetics are shown in figure 6.8. a) Correlation of the residual stored energy and the volume fraction recrystallized shows that a large part of the stored energy is released during the early stages of recrystallization. b) Correlation of the growth rate and the rate of decrease of stored energy shows the boundary velocity to be approximately proportional to the driving force, (Ryde et al. 1990).

However, if few nuclei are formed, for example after low strains (fig 6.20c), then growing grains will experience a driving force which is an average of the high and low energy regions and which will not vary during the recrystallization.

In figure 6.21, we illustrate a situation in which the deformed grains have different microstructures and stored energies. In this case the heterogeneity is on a much **coarser scale** than the example of figure 6.20. Even if nucleation is relatively uniform, growth will occur most rapidly in the grains of highest stored energy, leading to the type of microstructures shown in figure 6.21b. In this case not only will the growth rate decrease during recrystallization as the higher stored energy regions are consumed, but growth of the grains will be restricted by impingement with their neighbours within the original grains.

Although analytical modelling cannot easily deal with the kinetics of inhomogeneous recrystallization, computer simulations of the type discussed in chapter 13 are now being used to address the problem (Rollett et al. 1989a, Furu et al. 1990). As may be seen from the simulations of figure 13.14, inhomogeneous distribution of nuclei leads not only to low values of the JMAK slope, in agreement with many experimental measurements, but also to a decrease of the JMAK slope as recrystallization proceeds.

Thus we conclude that the deviation of recrystallization kinetics from the ideal linear JMAK plots and the low values of the exponent which are found in many experimental investigations, are in most cases directly attributable to the inhomogeneity of the microstructure. This leads to non-random distribution of nucleation sites and stored energy and to a growth rate which decreases with time.

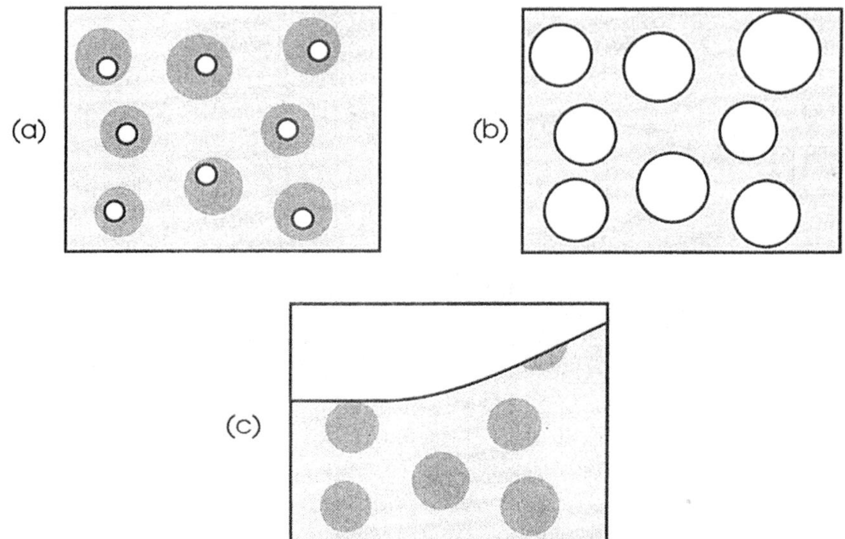

Fig. 6.20. The effect of local variations of stored energy on recrystallization. a) If nucleation (white areas) occurs in regions of high stored energy (dark areas) then the growth rate diminishes as recrystallization proceeds (b) and the regions of high stored energy are consumed. c) If recrystallization nucleation is on a coarser scale than the distribution of stored energy then the growing grains sample an average stored energy which remains approximately constant as recrystallization proceeds.

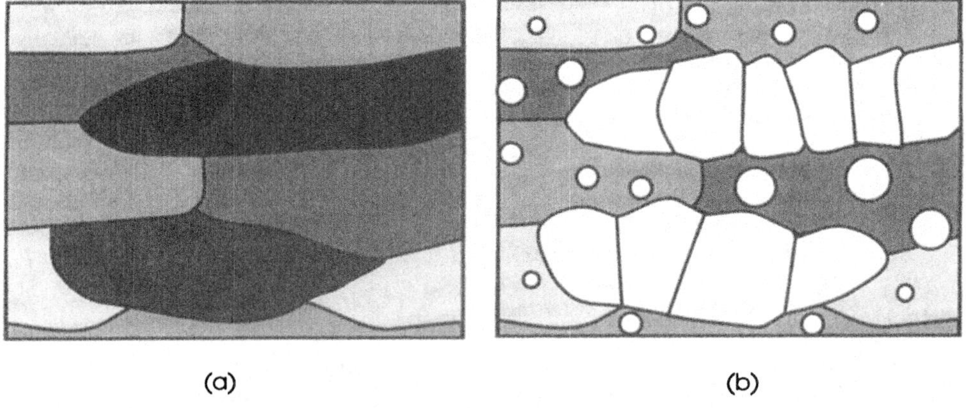

(a)   (b)

Fig. 6.21. Variation of stored energy from grain to grain as shown in (a), where regions of higher stored energy are shaded darker, results in inhomogeneous grain growth during recrystallization as shown in (b).

6.5 THE RECRYSTALLIZED MICROSTRUCTURE

6.5.1 The grain size

The magnitude of the final grain size can be rationalised in terms of the effects of the various parameters on the nucleation and growth processes. Any factor such as a high strain or a small initial grain size, which favours a large number of nuclei or a rapid nucleation rate, will lead to a small final grain size as illustrated for α-brass in figure 6.22, where it may be seen that the annealing temperature has little effect on the final grain size. However if a change of annealing temperature or heating rate alters the balance between nucleation and growth (§6.2.6), then the final grain size will be affected accordingly.

The grain size may not be constant throughout a specimen. Just as the different texture components within a specimen recrystallize at different rates (§6.4), so the final grain size and shape within each texture component may be different as seen in figure 6.13.

The spectrum of grain sizes as measured on a plane section is usually found to be close to a log normal distribution as shown in figure 6.23, and a similar grain size distribution has been reported for partly recrystallized material (Marthinsen et al. 1989).

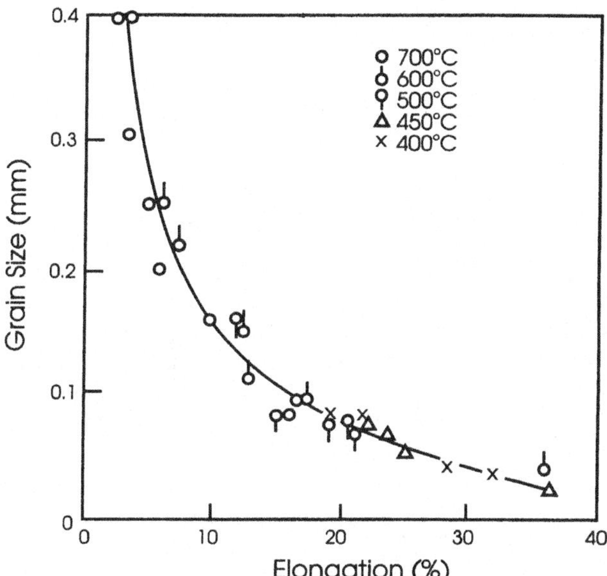

Fig. 6.22. The effect of tensile strain on the final grain size in α-brass recrystallized at various temperatures, (Eastwood et al. 1935).

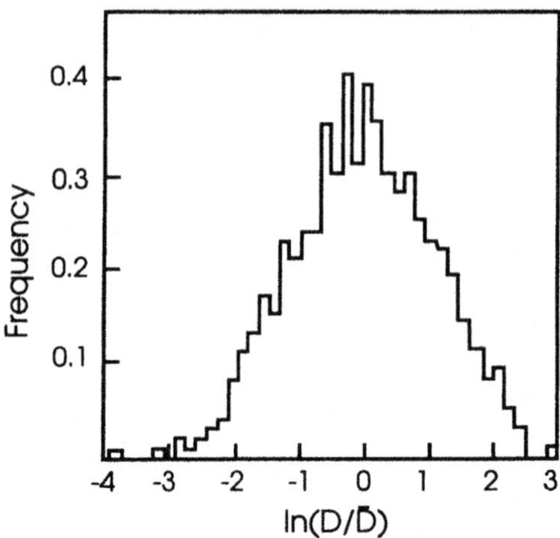

Fig. 6.23. The experimentally measured grain size distribution following the recrystallization of an Al-0.3%Fe alloy, (Saetre et al. 1986a).

6.5.2 The grain shape

If grains are uniformly distributed and growth is isotropic then the recrystallized structure consists of equiaxed polyhedra (§3.5). Although in many cases the recrystallized grains are approximately equiaxed, there are some instances where anisotropic growth leads to plate-shaped grains as may be seen in figure 6.13. This type of growth anisotropy is crystallographic, and a particularly important case is for grains of fcc metals oriented by ~40° about a <111> axis to the deformed matrix. In this situation, which is discussed in more detail in §4.3.2 and §10.7.2, the sides of a growing grain which form **tilt boundaries** to the deformed grain grow much faster than the other faces, resulting in plate-shaped grains such as that shown in figure 4.10.

In industrial, particle-containing alloys, anisotropic distribution of the particles, often along planes parallel to the rolling plane may reduce the growth rate perpendicular to the rolling plane as discussed in §3.6.2.3, resulting in pancake grains as shown in figure 3.25.

6.5.3 The grain orientations

In general the recrystallized grains are not randomly oriented, but have a **preferred orientation** or **texture**. The origin of recrystallization textures is discussed in detail in chapter 10, and some examples of texture control in industrial practice are given in chapter 12.

6.6 THE NUCLEATION OF RECRYSTALLIZATION

There has been a considerable research effort aimed at trying to understand the processes by which recrystallization originates. The importance of recrystallization nucleation is that it is a critical factor in determining both the size and orientation of the resulting grains. In order to control recrystallization effectively, it is necessary to understand the mechanisms of nucleation and the parameters which control it.

6.6.1 Classical nucleation

The possibility that the classical nucleation theory developed for phase transformations might be applicable to recrystallization was considered by Burke and Turnbull (1952). In this situation, nucleation would be accomplished by random atomic fluctuations leading to the formation of a small crystallite with a high angle grain boundary. Such a nucleus would be stable if the difference in energy between the local deformed state and the recrystallized state were larger than the energy of the high energy interface produced in forming the nucleus. Although this theory appears to account for some aspects of recrystallization, such as the incubation period and preferential nucleation at regions of high local strain, it is considered to be highly unlikely because:

> **The driving force is low.** In comparison with phase transformations, the energy which drives the recrystallization process is very small (§2.2).

> **The interfacial energy is large.** The energy of a high angle grain boundary, which is an essential factor in the recrystallization process, is large (§3.4).

Calculations based on nucleation theory (see Christian 1965) indicate that the radius of the critical nucleus at $>0.1\mu m$, is so large that the rate of nucleation will be negligible, and that this is therefore not a viable mechanism for the origin of recrystallization.

It is now accepted that the "nuclei" from which recrystallization originates are therefore not nuclei in the strict thermodynamic sense, but small volumes which pre-exist in the deformed microstructure.

6.6.2 Strain induced grain boundary migration (SIBM)

This mechanism, an example of which is seen in figure 6.24 was first reported by Beck and Sperry (1950) and has been observed in a wide variety of metals. SIBM involves the bulging of part of a pre-existing grain boundary, leaving a dislocation-free region behind the migrating boundary as shown schematically in figure 6.25. The characteristic features are that the new grains have similar orientations to the old grains from which they have grown. This mechanism is particularly important after low strains, and Beck and Sperry found that at reductions $>40\%$ in pure aluminium the mechanism was apparently replaced by one in which grains of orientations different from either of the matrix grains were formed (see §6.6.4). In the early work, the similarity or otherwise of the new and old grain orientations was inferred from the contrast produced by etching, which may not be reliable. However, Bellier and Doherty (1977) were able to determine the grain orientations and, in rolled aluminium, confirmed that SIBM was the dominant recrystallization mechanism for reductions of less than 20%. Because of the orientation relationship between the new and old grains, operation of this mechanism is expected to result in a recrystallization texture which is closely related to the deformation texture.

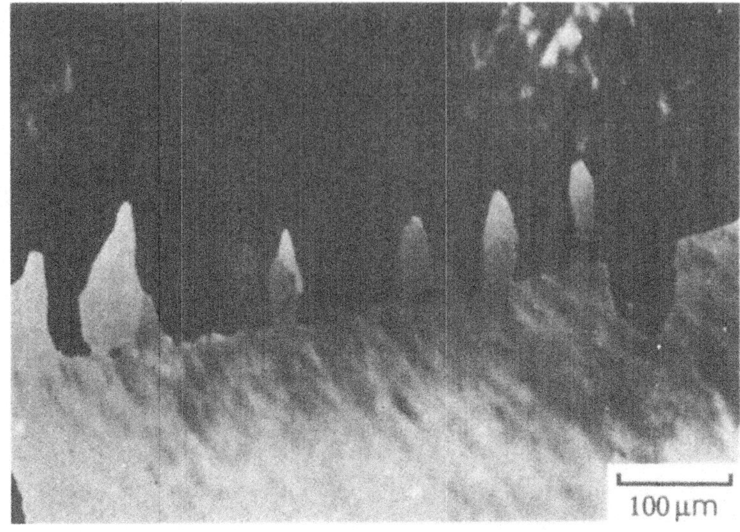

Fig. 6.24. An optical micrograph showing strain induced grain boundary migration in aluminium, (Bellier and Doherty 1977).

The driving force for SIBM is presumed to arise from a difference in dislocation density on opposite sides of the grain boundary, and the kinetics of the process have been analyzed by Bailey and Hirsch (1962). In figure 6.25a, if the deformed grains have stored energies of E_1 and E_2 and $E_1 > E_2$, then the driving force is provided by the energy difference $E_V = E_1 - E_2$. If the bulging boundary is a spherical cap of radius R, with a specific boundary energy γ_b, then the interfacial energy of the boundary is given by

$$E_B = 4\pi R^2 \gamma_b \tag{6.36}$$

and

$$\frac{dE_B}{dR} = 8\pi R \gamma_b \tag{6.37}$$

In the early stages of bulging, the energy difference per unit volume across the boundary is E_V and

$$\frac{dE}{dR} = 4\pi R^2 E_V \tag{6.38}$$

For the bulge to grow $dE/dR > dE_B/dR$ and hence

$$R > \frac{2\gamma_b}{E_V} \tag{6.39}$$

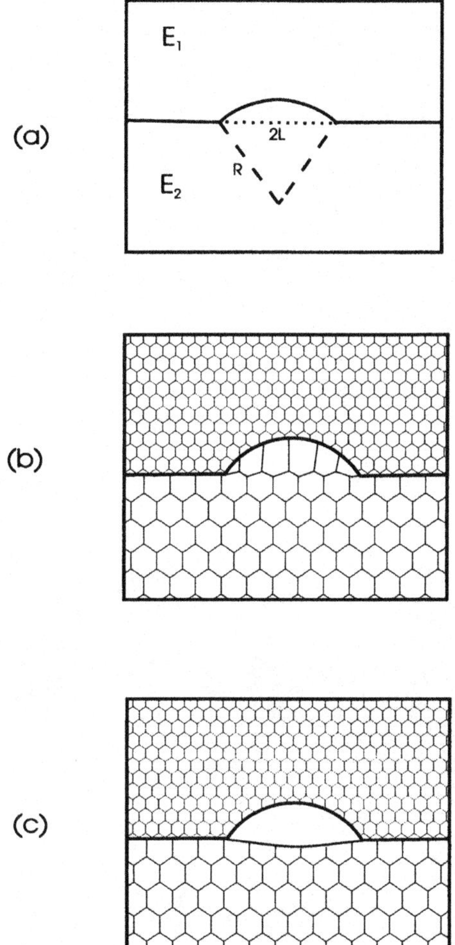

Fig. 6.25. a) SIBM of a boundary separating a grain of low stored energy (E_2) from one of higher energy (E_1). b) Dragging of the dislocation structure behind the migrating boundary. c) The migrating boundary is free from the dislocation structure.

This reaches a critical value when the boundary bulge becomes hemispherical, when $R = L$ and

$$L > \frac{2\gamma_b}{E_V} \tag{6.40}$$

There remain however a number of unanswered questions concerning this mechanism.

(i) How and when does the migrating boundary separate from the dislocation structure of the parent grain?

It is not at all clear whether at the critical stage (equation 6.40), the boundary can still be considered as separating regions of E_1 and E_2 as shown in figure 6.25b, or whether it separates E_1 and perfect crystal 6.25c, in which case E_V in equation 6.40 is now given by E_1. If this is the case, then the critical condition is likely to arise before the hemispherical configuration is reached. There is surprisingly little experimental evidence to resolve this point. The transmission electron micrographs of Bailey and Hirsch (1962) one of which is shown in figure 6.26 show that the dislocation density of the bulge region, although lower than that of the parent grain, is significant, in agreement with figure 6.25b. Simulations of SIBM using the network model described in §13.2.3 show some dragging of dislocations behind the migrating boundary, which is consistent with the Bailey and Hirsch results.

(ii) How does SIBM begin?

The origin of the differential dislocation density required for the operation of this mechanism is not entirely clear. It could result directly from the deformation process, because it is known that the dislocation storage rate may be dependent on grain orientation (§2.2) and be different in the boundary regions (§2.3.5). However, it could also arise from preferential recovery in the vicinity of a grain boundary which might increase the subgrain size and hence

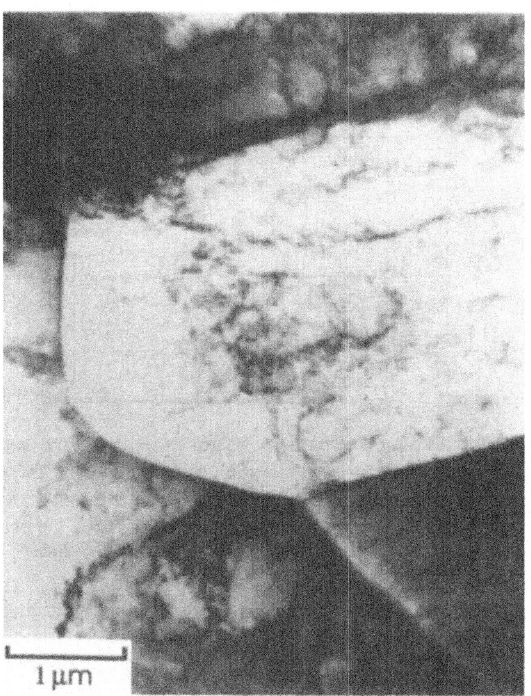

Fig. 6.26. TEM micrograph of SIBM in copper deformed 14% in tension and annealed 5 mins at 234°C, (Bailey and Hirsch 1962).

lower the dislocation density on one side of the boundary. Doherty and Cahn (1972) and Jones et al. (1979) have discussed how recovery by subgrain coalescence in the vicinity of a high angle boundary (§5.5.4) could initiate SIBM as shown schematically in figure 6.27.

Although there is good evidence that SIBM occurs, there have been surprisingly few investigations which have used modern microstructural characterisation techniques to provide the experimental data which are required to clarify the details of this mechanism.

6.6.3 The preformed nucleus model

In the process of strain induced grain boundary migration discussed above, the high angle grain boundary which is a prerequisite for recrystallization, was already in existence. However, in many cases, recrystallization originates in other regions of the material, and we need to consider how a nucleus may be formed in such circumstances.

The possibility that recrystallization might originate at crystallites present in the deformed material was first postulated by Burgers (1941). In his **block hypothesis**, the nuclei could be either crystallites which were highly strained, or ones which were relatively strain free. It is now established beyond reasonable doubt that recrystallization originates from **dislocation cells or subgrains which are present after deformation**. Although there are still uncertainties about how these pre-existing subgrains become nuclei, several points are now clear.

(i) The **orientation of the nucleus is present in the deformed structure**. There is no evidence that new orientations are formed during or after nucleation, except by twinning.

(ii) Nucleation occurs by the **growth of subgrains** by the mechanisms discussed in §5.5.3. All direct in-situ TEM annealing observations of recrystallization nucleation (Ray et al. 1975,

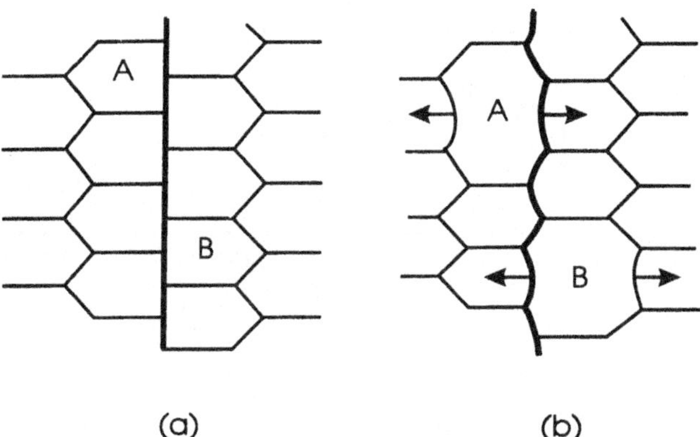

Fig. 6.27. Schematic diagram showing how subgrain coalescence in the vicinity of a high angle boundary might result in suitable conditions for initiating SIBM. a) original substructure b) Enlarged subgrains starting to grow, (after Doherty and Cahn 1972).

Humphreys 1977, Bay and Hansen 1979) have shown the mechanism to be one of sub-boundary **migration**. There is also the possibility that subgrain **coalescence** (§5.5.4) may occur in the early stages, although there is much less evidence for this. In-situ HVEM experiments have shown that recovery in the vicinity of the nucleus is significantly faster than in the remainder of the material.

(iii) In order for a high angle grain boundary to be produced by this rapid recovery, there must be an **orientation gradient** present. This is shown schematically in figure 6.28 which shows two one-dimensional subgrain structures with similar misorientations between the individual subgrains. Recovery of the structure shown in 6.28a produces large subgrains but no high angle boundary (fig 6.28b), whereas the same amount of recovery in the presence of an orientation gradient (fig 6.28c) results in the formation of a high angle grain boundary (fig 6.28d). This important point was first clearly stated by Dillamore et al. (1972). It should be noted that any region with a large orientation gradient will **always** have a high stored energy because of the geometrically necessary dislocations or low angle grain boundaries, which are needed to accommodate the misorientation. As discussed in §5.5.3.4, theory predicts that recovery will be most rapid in regions of large orientation gradient.

Nucleation of recrystallization is therefore considered to be no more than discontinuous subgrain growth at sites of high strain energy and orientation gradient.

The **site** of the nucleus is very important in determining its viability, and it is clear that the general mechanism discussed above can occur at a variety of sites.

6.6.4 Nucleation sites

Recrystallization will originate at inhomogeneities in the deformed microstructure. These regions of inhomogeneity may either be those present before deformation such as second-

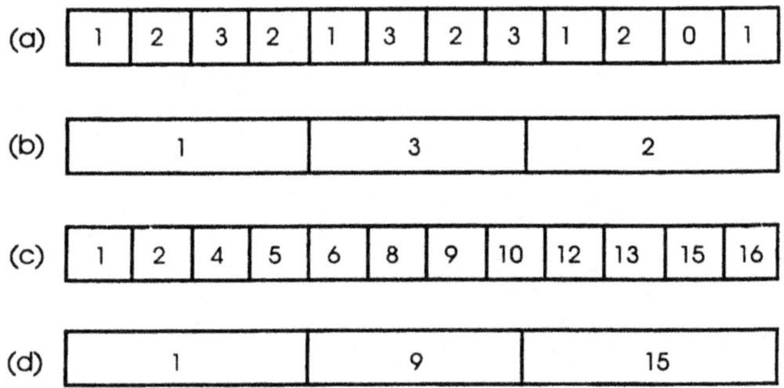

Fig. 6.28. The effect of an orientation gradient on the microstructure after recovery. The numbers represent the orientation of the subgrains with respect to the left edge of the microstructure. a) There is no overall orientation gradient. b) On annealing, no high angle boundary is formed. c) There is an orientation gradient. d) Recovery leads to the formation of a high angle grain boundary.

phase particles (§8.4) or grain boundaries, or else inhomogeneities induced by the deformation as discussed in §2.3.2. As the orientation of the recrystallized grains may depend upon the recrystallization site, the types of site which operate may have a strong effect on the **recrystallization texture** as discussed in chapter 10.

6.6.4.1 Grain boundaries

Nucleation in the vicinity of prior grain boundaries, of new grains of orientations which are not close to those of the parent grains, has been frequently observed (e.g. Beck and Sperry 1950), and is more frequent at larger strains. Evidence of the operation of this mechanism has been obtained by Hutchinson (1989) in iron bicrystals in which the new grains were misoriented by 30° from the parent grains, and an example of grain boundary nucleation in aluminium is shown in figure 6.29.

Little is known about this mechanism of recrystallization or about the orientations of the resulting grains. However, it is known that grain boundaries give rise to inhomogeneity of slip (Ashby 1970, Leffers 1981), and that different combinations of slip systems will operate near grain boundaries (§2.3.5), thereby giving rise to local misorientations in a manner which is probably similar in principle to that which occurs near large second-phase particles (§8.2.4). It is therefore likely that recrystallization originates in the regions of large orientation gradient which will be present near some grain boundaries. Whether or not there is a real distinction between this mechanism and that of SIBM has not been clearly established.

6.6.4.2 Transition bands

A transition band separates parts of a grain which have split during deformation into regions of different orientation (§2.3.6.2). A transition band is therefore a region of large orientation gradient which is an ideal site for recrystallization. Recrystallization at transition bands was

Fig. 6.29. SEM micrograph showing grain boundary nucleation in aluminium. The orientation of the new grain has no simple relationship to the orientations of either of the old grains and is misoriented by 20° and 48° from them, (courtesy A.O. Humphreys).

first reported by Hu (1963) and Walter and Koch (1963) in iron. The crystallographic orientations developed in transition bands are a direct consequence of the slip processes and the strain path and therefore the recrystallized grains tend to have preferred orientations. Because of its importance in determining the recrystallization texture of metals (chapter 10) there have been many extensive studies of this mechanism including those of Bellier and Doherty (1977) and Hjelen et al. (1991) on aluminium, Inokuti and Doherty (1978) on iron and Ridha and Hutchinson (1982) on copper. An example of nucleation at a transition band is shown in figure 6.30.

As shown schematically in figure 6.31, the subgrains in a transition band are often elongated, and Dillamore et al. (1972) proposed a criterion for transition band nucleation, based on this idealised geometry. The large subgrain **A** will grow by the transverse boundaries shrinking if

$$D_r > \frac{4}{3}\left((d_r + d_t) \cdot \left(\frac{4\gamma_t^2}{\gamma_r^2} - 1\right)^{1/2}\right) \tag{6.41}$$

where the subscripts **r** and **t** refer to the directions indicated on the figure.

This shows that nucleation is favoured at elongated subgrains in large orientation gradients (large γ_t/γ_r). A simulation of nucleation in a transition band which is of a similar type to that of figure 6.31 is shown in figure 13.9.

6.6.4.3 Shear bands
Shear bands are thin regions of highly strained material typically oriented at $\sim 35°$ to the rolling plane. They are a result of strain heterogeneity due to instability during rolling and their formation is strongly dependent on the deformation conditions as well as the

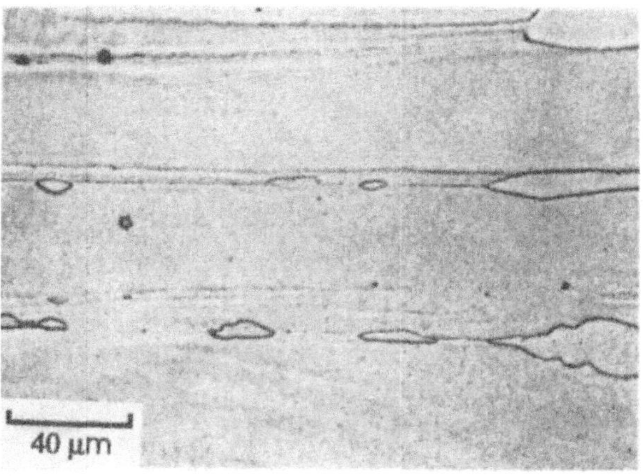

Fig. 6.30. Optical micrograph showing recrystallization at transition bands in a silicon-iron crystal, (Hu 1963).

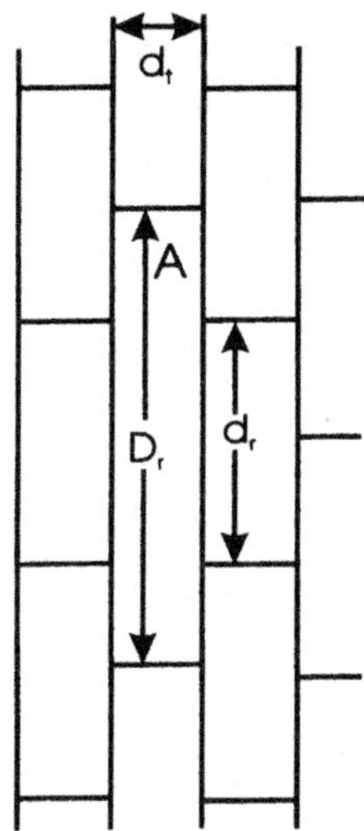

Fig. 6.31. Model for subgrain growth in a transition band, (Dillamore et al. 1972).

composition, texture and microstructure of the material (§2.3). Nucleation of recrystallization at shear bands has been observed in many metals including copper and its alloys, (Adcock 1922, Duggan et al. 1978a, Ridha and Hutchinson 1982, Haratani et al. 1984), aluminium (Hjelen et al. 1991) and steels (Ushioda et al. 1981). A micrograph showing profuse nucleation at shear bands is shown in figure 6.32.

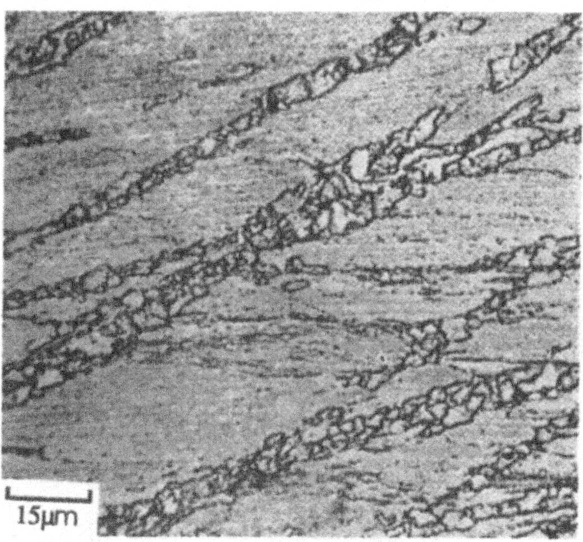

Fig. 6.32. Optical micrograph showing recrystallization occurring at shear bands in copper, (Adcock 1922).

Details of the mechanism of nucleation at shear bands are not known, and the orientations of the grains so formed appear to be very case dependent (Nes and Hutchinson 1989). For example in α-brass (Duggan et al. 1978a) the orientations of the grains were widely scattered, but in aluminium (Hjelen et al. 1991), a strong **S** texture component was found to be nucleated at shear bands.

6.7 ANNEALING TWINS

6.7.1 Introduction

During the recrystallization of certain materials, particularly the face centred cubic metals of intermediate or low stacking fault energy such as copper and its alloys and austenitic stainless steels, annealing twins are formed. Such twins are also found in many intermetallic compounds, ceramics and minerals. These twins take the form of parallel sided lamellae as shown schematically in figure 6.33. In fcc metals the lamellae are bounded by {111} planes or **coherent twin boundaries (CT)** and at their ends or at steps, by **incoherent twin boundaries (IT)** as shown in figure 6.33. The nature of these boundaries is discussed in more detail in §3.4. Twins may form during recovery, primary recrystallization or during grain growth following recrystallization.

6.7.2 Mechanisms of twin formation

Although twins have been extensively studied, the atomistic mechanism of their formation is still not clear. The growth of twins may take place in two configurations as shown in figure 6.34. In 6.34a the growth direction is perpendicular to the coherent twin boundary and in 6.34b it is parallel.

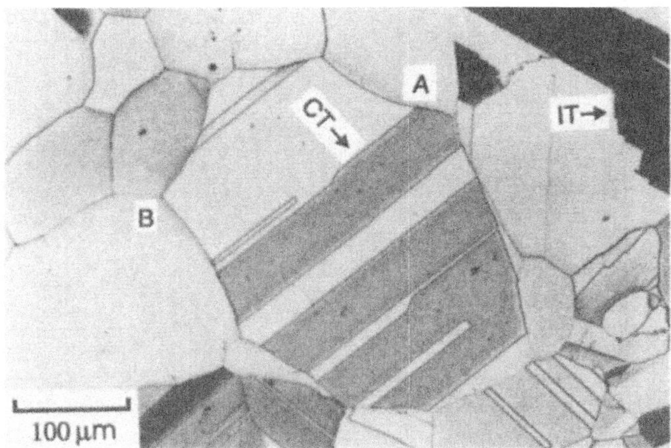

Fig. 6.33. Annealing twins in annealed 70:30 brass. Some coherent (CT) and incoherent (IT) twin boundaries are marked, (courtesy of M. Ferry).

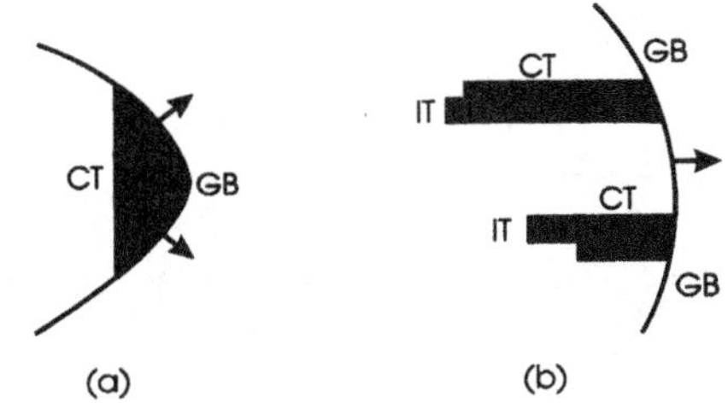

Fig. 6.34. Possible orientations of twins during recrystallization. a) the coherent twin (CT) boundary lies parallel to the recrystallization front. b) The CT boundary is perpendicular to the recrystallization front, (after Goodhew 1979).

6.7.2.1 Twinning by growth faulting

Gleiter (1969c, 1980) proposed that twinning occurred by the movement of ledges on the coherent twin boundary. As the twin propagates by addition of the close packed planes in a sequence such as ABCABC, the start or finish of the twin lamella requires only the nucleation and propagation of a ledge with a low energy growth fault such as to change the sequence e.g. **ABCABC|BACBACBA|BCABC**. Whilst this mechanism is capable of explaining twins propagating as shown in figure 6.34a, it is not consistent with the configuration of figure 6.34b.

6.7.2.2 Twinning by boundary dissociation

Goodhew (1979) found evidence for twin formation by boundary dissociation in gold. He identified the following dissociations:

$$\Sigma 9 \rightarrow \Sigma 3 + \Sigma 3$$
$$\Sigma 11 \rightarrow \Sigma 3 + \Sigma 33$$
$$\Sigma 33 \rightarrow \Sigma 3 + \Sigma 11$$
$$\Sigma 99 \rightarrow \Sigma 3 + \Sigma 33$$

In all these cases a decrease in energy results (Hasson and Goux 1971). A possible mechanism for this is shown schematically in figure 6.35.

It has been suggested that this type of boundary dissociation could occur by the emission of grain boundary dislocations, and variants of this type of mechanism have been suggested by Meyers and Murr (1978) and Grovenor et al. (1980).

6.7.3 Twin formation during recovery

Huber and Hatherly (1979, 1980) reported the formation of **recovery twins** in shear bands, during the annealing of deformed 70:30 brass at $\sim 250°C$, some 30°C below the recrystallization temperature. Recovery twins are $\sim 0.25\mu m$ long and $\sim 0.002\text{-}0.01\mu m$ wide. Only one set of twins developed in a particular region, and these were always associated with

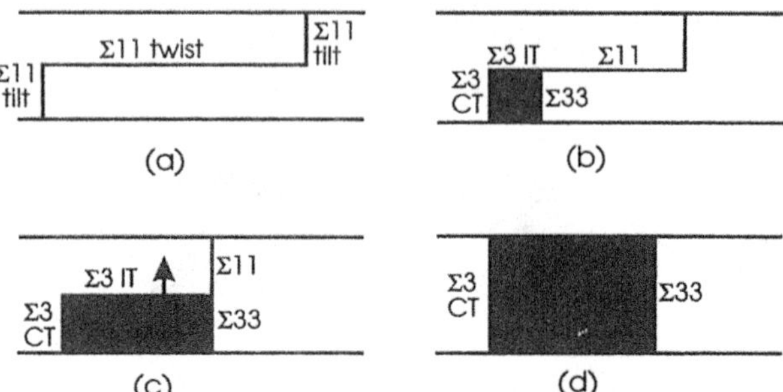

Fig. 6.35. Possible mechanism for the dissociation of a Σ11 boundary. During tilt boundary migration (a), one of the tilt boundaries dissociates (b) leaving a twin behind. The remaining Σ11 twist boundary is then eliminated (c) and the incoherent twin boundary (IT) then migrates to remove the remaining Σ11 boundary (d), (Goodhew 1979).

local orientations of the form $\{110\}<112>$. The twins increased in number and width as annealing continued and there was a gradual loss of dislocations from the regions where they occurred (fig. 6.36). This loss appeared to correspond to the beginning of recrystallization, but it is significant that there was no clearly defined boundary around the developing nucleus.

6.7.4 Twin formation during recrystallization

6.7.4.1 The role of twinning
As twinning produces new orientations that were not present in the deformed microstructure, twinning may play an important part in the development of annealing textures (§10.7.5). An extensive research programme by Haasen and colleagues at Göttingen (e.g. Berger et al. 1988) has shown that multiple twinning in the early stages of recrystallization may be

Fig. 6.36. Recovery twins in 85:15 brass, deformed in torsion and annealed at 240°C, (Huber and Hatherly 1979).

important in determining grain orientations in copper alloys. Their conclusion, based on in-situ high voltage electron microscopy annealing experiments (Berger et al. 1983) that twinning plays an important role in the recrystallization of aluminium is more controversial. Twins are rarely observed in bulk samples of aluminium and the possibility of twin formation during the annealing of thin foils being an artefact cannot be ruled out.

6.7.4.2 Twin selection principles

If multiple twinning were to occur in a random manner then a random texture would result (§10.7.5). However, it is known (e.g. Hatherly et al. 1984) that only a few of the possible twin variants are formed, and therefore twinning during recrystallization cannot occur randomly and some selection principles must apply. Wilbrandt (1988) has reviewed these in detail and concluded that no single principle is capable of explaining all experimental observations. However, the following three factors have been shown to be important.

(i) Boundary energy - There is a large body of evidence which suggests that the lowering of grain boundary energy is a very important factor in determining the twin event and provides stability of the grain against further twinning. The work of Goodhew (1979) discussed above, and of the Göttingen group (Berger et al. 1988) has provided good evidence in support of this principle.

(ii) Grain boundary mobility - The possibility that twinning occurs in a manner such as to provide a more mobile boundary has also been considered. For example during dynamic recrystallization of copper and silver single crystals (Gottstein et al. 1976, Gottstein 1984) (see §11.4), it was found that multiple twinning occurred until a fast growing grain orientation was formed. If this were the overriding factor however, we might expect twin chains to result in progressively more mobile boundaries, whereas it is known that twinning sometimes results in very immobile boundaries (Berger et al. 1988).

(iii) The role of the dislocation arrangement - Evidence that twinning is affected by the dislocation structure has been presented by Form et al. (1980) who found that the twin density increased with the dislocation density. Rae et al. (1981) and Rae and Smith (1981) found that both dislocation structure and the grain boundary orientation were important in determining the ease of twinning.

6.7.5 Twin formation during grain growth

The early work on twin formation was mainly concerned with twin formation during grain growth (Burke 1950, Fullman and Fisher 1951). The postulated mechanism of twinning, which is based on an energy criterion is shown in figure 6.37

In figure 6.37a, grain growth is occurring such that the triple point between grains A,B and C is moving vertically. As growth proceeds, it is assumed that a growth fault may occur at some point leading to the formation of a grain (T) which is twinned with respect to grain A. If the relative orientations of the grains are such that the energy of the boundary AT is lower than that of AC then, because the energy of the coherent twin boundary AT is very low (table 3.2), there may be a reduction in total boundary energy despite the extra boundary area created, and therefore the twin configuration will be stable and grow. This condition, in two dimensions, is:

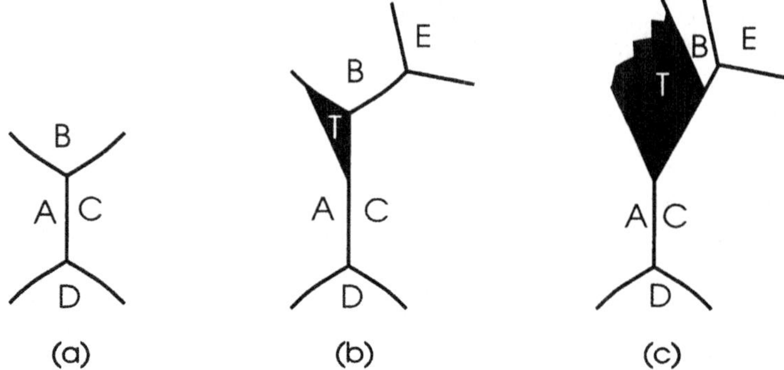

Fig. 6.37. Mechanism for twin formation during grain growth following recrystallization.

$$\gamma_{AT} L_{13} + \gamma_{TC} L_{23} + \gamma_{TB} L_{12} < \gamma_{AC} L_{23} + \gamma_{AB} L_{12} \qquad (6.42)$$

where γ_{ij} is the energy of the boundary between grains i and j and L_{xy} is the distance between points x and y

The growth will be terminated if the triple point ABC reacts with another triple point such as BCE, resulting in a grain configuration with a less favourable energy balance (fig 6.37c). On such a model, the number of twin lamellae should be proportional to the number of triple point interactions, and evidence for this was found by Hu and Smith (1956). The actual atomistic mechanism of twin formation during grain growth is likely to be similar to the mechanisms discussed above for twin formation during recrystallization.

Chapter 7

RECRYSTALLIZATION OF ORDERED MATERIALS

7.1 INTRODUCTION

Although ordered phases exist with a large variety of different lattice structures, many such materials are too brittle to deform significantly and the likelihood of annealing phenomena such as recovery or recrystallization being associated with ordering reactions is limited. However, there are a few cases in which extensive cold working is feasible and others where hot working and dynamic recrystallization are possible; it is with these materials that we are concerned here.

7.2 ORDERED STRUCTURES

7.2.1 Nature and stability

Only two of the possible ordered structures (or **superlattices**) have been studied in detail with respect to annealing behaviour. These are the **L1$_2$** structure associated with A$_3$B alloys such as Cu$_3$Au, Ni$_3$Fe etc., and the **B2** structure typical of AB compounds such as CuZn, FeCo, NiAl etc. The L1$_2$ structure is face centred cubic with the A atoms located on the face centred sites ($\frac{1}{2}$,$\frac{1}{2}$,0 etc.) while the B2 structure is body centred cubic with the B atoms at the body centred site ($\frac{1}{2}$,$\frac{1}{2}$,$\frac{1}{2}$). The B2 structure is subject to a further significant ordering process in which a **D0$_3$** structure is produced. The unit cell consists of eight B2 unit cells arranged as a 2x2x2 cube with alternate cells having A and B atoms at the cell centre. Examples of this structure include the important materials based on Fe$_3$Al.

In some materials the ordered structure is absent above a critical **ordering temperature (T$_c$)**, but in others the ordered state is stable up to the melting point. In many respects materials of the latter type, which are called **permanently ordered materials**, are closely related to chemical compounds. Details of T$_c$ for a number of significant materials are given in Table 7.1.

In materials that are capable of being disordered, the retention of ordering during heating, or its development during cooling, is measured by the **ordering parameter (S)**. S takes

Table 7.1

Ordering Temperature,(T_c) of Significant Materials.

Material	T_c (°C)	Material	T_c (°C)
Cu_3Au	390	Fe_3Al	~540
Ni_3Fe	500	$(Co_{78}Fe_{22})_3V$	950
CuZn	454	Ni_3Al	M.Pt.
FeCo	~725		

values between 1, corresponding to perfect ordering and 0, corresponding to the fully disordered state. However, the extent of ordering below T_c does not follow the same pattern in all materials. In some cases, e.g. FeCo, **S** decreases slowly as the alloy is heated and the disordered state can be retained at room temperature by quenching. In other cases, e.g. Cu_3Au, the disordered state cannot be retained on quenching and there is an abrupt change to high values of **S** at temperatures just below T_c. The two types of behaviour are shown in figure 7.1.

In materials that can be disordered the structure below T_c is characterised by volumes called **domains,** where there is a high degree of order. In a fully ordered material, neighbouring domains, in which the A and B atoms have different positions, are separated by a surface called an **antiphase boundary (APB)** and it is found in practice that such boundaries follow simple crystallographic planes in the parent lattice. APB's are of critical importance to the deformation and annealing behaviour of ordered materials, because they occur not only from the impingement, during ordering, of growing domains but also from the passage of dislocations (see below). In partly disordered materials the domains are separated from each other by disordered regions.

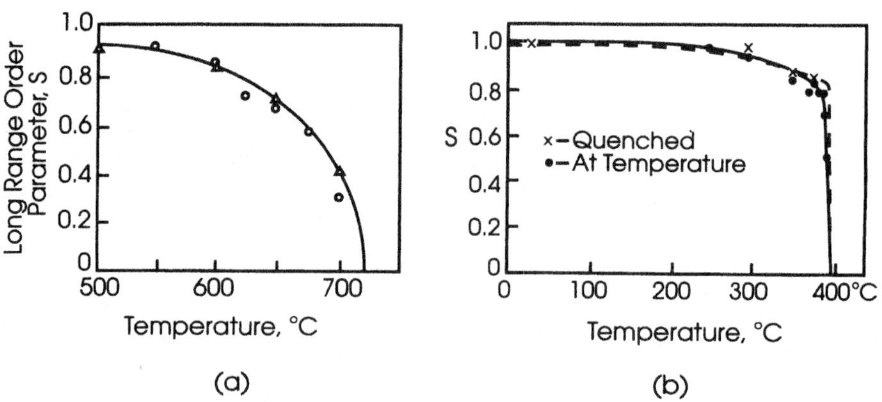

Fig. 7.1. Order parameter, **S**, as a function of temperature for, (a) FeCo (after Stoloff and Davies, 1964), (b) Cu_3Au, (after Hutchinson et al. 1973).

7.2.2 Deformation of ordered materials.

A perfect dislocation in the disordered lattice is merely a partial dislocation in the ordered structure and its movement in the superlattice leaves behind a plane that is an **APB**. In B2 structures, for example, the expected dislocation has a Burgers vector b $= \frac{1}{2}<111>$ and its passage leads to the creation of an **APB**. In order to avoid the energy increase associated with the production of such a boundary, the moving dislocations in an ordered structure are coupled in pairs (**super-dislocations**) so that the second dislocation restores the ordered state. The two dislocations are linked by a small area of antiphase boundary.

The details of the slip processes in ordered B2 alloys have been discussed recently by Baker and Munroe (1990). The commonly observed dislocations have Burgers vectors corresponding to $<100>$, $<110>$ and $<111>$ but only the first and last of these are found at low temperatures. In many alloys (e.g. CuZn), there is a change from $<111>$ slip to $<100>$ slip at temperatures near 0.4-0.5 T_m. However, in others (e.g. FeAl), the temperature at which the transition to $<100>$ slip occurs is also governed by the composition. There is at present, no clear reason for the change from $<111>$ to $<100>$ slip in these materials but the change is not unlike those observed in fcc and bcc metals at elevated temperatures (§2.3.1).

For $<111>$ dislocations in B2 superlattices, the slip planes are $\{011\}$, $\{112\}$ and $\{123\}$ but whether or not the coupled $<111>$ partials can cross slip, depends on the energy of the APB that links them. For large APB energies the partials are close together and recombination and cross slip are easy. If the energy is very low, cross slip of the individual partials becomes possible. For intermediate values cross slip is more difficult and pile ups are to be expected.

A detailed study of the dislocations present in the L1$_2$ alloy $Al_{64}Ti_{28}Fe_8$ deformed in compression, has been made by Lerf and Morris (1991). At room temperature, undissociated $<110>$ dislocations move through the lattice, but $<100>$ dislocations remain pinned. At 500°C. many dislocations dissociate into pairs of $\frac{1}{2}<110>$ partials, thereby constituting mobile super-dislocations on $\{111\}$ planes. At 700°C. the super-dislocations are distributed evenly between the $\{001\}$ and $\{111\}$ planes. Cross slip between octahedral planes is increasingly prevalent at temperatures >500°C.

It has been generally believed that the very limited low temperature ductility of many materials which remain ordered up to the melting point, is attributable in part to the retention of order and hence to the difficulty of dislocation movement. However, ordered materials can be deformed by novel techniques. Jang and Koch (1990) have shown that ball milling is capable of producing complete disorder in Ni$_3$Al powders and Gialanella et al. (1992) have reported similar results with another L1$_2$ phase Zr$_3$Al. Clearly dislocation movement must be involved in such milling, and the X-ray line broadening studies of Gialanella et al. revealed the expected reduction in crystallite size and an increase in lattice strain as the milling time increased. A recent investigation by Dadras and Morris (1993) of a $Fe_{68}Al_{28}Cr_4$ alloy which has a D0$_3$ structure has found that after milling, about 75% of the dislocations present were uncoupled and single, i.e. the D0$_3$ super-dislocations had been dissociated by deformation. An analysis of domain size distribution led these workers to conclude that localized shear bands may have operated. Koch (1991) has reviewed the possible mechanisms involved in deformation during milling.

McKarney and Pierce (1992) have examined the room temperature strength and ductility of Fe_3Al using specimens that had been heat treated to relieve stresses or produce controlled amounts of recrystallization. Maximum ductility (8% elongation) was associated with the stress relieved state; by contrast the fully recrystallized material showed only 3.5% elongation.

7.2.3 Microstructure and deformation texture.

Although there has been much discussion of the mechanism of dislocation movement in ordered materials, there is surprisingly little information about the nature of the microstructure developed during deformation. Amongst the few reports that consider the nature of the inhomogeneities produced, (i.e. the few reports that can be related to the microstructures described in chapter 2), are the recent papers by Ball and Gottstein (1993a,b), Ball et al. (1992), and an earlier paper by Kawahara (1983).

Ball and Gottstein (1993a) rolled a boron-doped $Ni_{76}Al_{24.}$ compound which has an $L1_2$ structure, to >90% reduction at room temperature. Their investigation followed earlier work by Gottstein et al. (1989) which had reported an unusually high value of stored energy (163J/mol) in this alloy. Ball and Gottstein found that at intermediate strains the microstructure consisted of clusters of microbands together with shear bands (fig 7.2); the latter were of the intersecting type normally associated with low SFE alloys (cf. fig 7.2 and 2.6h). No evidence of cell formation was observed at any level of strain. At strain levels below 0.4, very thin ($\sim 0.05\mu$m) microbands were observed, and these authors also reported that at higher strain levels, somewhat thinner shear bands, classified by them as copper type shear bands, appeared between the brass type shear bands formed earlier.

The deformation texture of the material after rolling to 92% reduction is shown in figure 7.3. It was described as a weak copper-type texture consisting of the components, B ($\{110\}<112>$), C ($\{112\}<111>$) and with a broad spread between the Goss orientation $\{110\}<001>$ and S $\{123\}<634>$. This rolling texture is unexpected, because shear bands

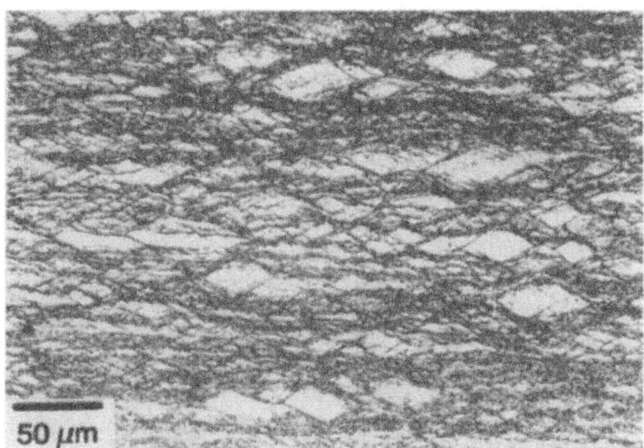

Fig. 7.2. Longitudinal section of 70% cold rolled, B-doped Ni_3Al showing shear bands, (Ball and Gottstein 1993a).

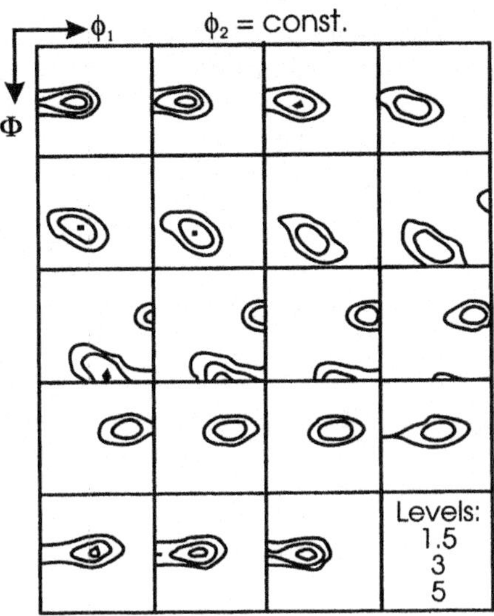

Fig. 7.3. Rolling texture of 92% cold rolled, B-doped Ni₃Al, (Ball and Gottstein 1993a).

of the type shown in figure 7.2a which are common in materials with low-medium stacking fault energy, are normally associated with strong brass-type textures at this level of reduction. Ball and Gottstein detected no evidence of deformation twinning, but it is significant that they reported that annealing twins were common in the new grains developed during recrystallization (see §7.3.1). The development of a "copper type" texture (§2.4.1) in a microstructure resembling that of a low stacking fault energy fcc material (§2.3.3) is unusual and there is an obvious need for further microstructural study of this and other deformable, ordered materials.

Kawahara (1983) examined an FeCo-2V alloy; this material which has a B2 structure can be retained in the disordered condition by quenching. Extensive shear banding occurred in both ordered and disordered specimens after rolling. The first shear bands developed after only 10% reduction, and by 50% reduction the specimens displayed microstructures similar to that shown in figure 7.2a, in which the shear bands were profuse and had the characteristics of the bands normally found in fcc metals with low stacking fault energy. The rolling texture of a specimen rolled to 70% reduction was said to consist of {001}<110>, {112}<110>, {111}<110> and {111}<211> components and to be similar to that of iron-based materials.

A brief reference to the deformation microstructure was also made by Morris and Morris (1991) in their investigation of the B2 alloy, β-brass. After 20% compression of the alloy Cu₅₂Zn₄₈, optical microscopy revealed the presence of numerous "deformation bands".

7.3 RECOVERY AND RECRYSTALLIZATION OF ORDERED MATERIALS

The recovery and recrystallization of ordered structures is complicated by the fact that several quite different processes may be occurring simultaneously in the material, viz. the restoration of the ordered state and those associated with recovery and recrystallization. The kinetics of these processes are quite different and the behaviour is complex. A detailed account of the literature has been given by Cahn (1990) and Cahn et al. (1991) have summarized this as follows:

(i) There is a drastic reduction in grain boundary mobility in the presence of atomic order, which severely retards recrystallization.

(ii) A temperature range may exist where recrystallization does not occur even though it occurs at lower and higher temperatures.

(iii) The pre-recrystallization behaviour depends on the composition of the alloy and the annealing temperature. The deformed alloy may soften, harden, or remain unchanged.

In the following account the $L1_2$ and B2 structures are considered separately.

7.3.1 $L1_2$ structures

In a significant early study Roessler et al. (1963) examined Cu_3Au after rolling to 63% reduction ($\epsilon \sim 1$). Both ordered and previously disordered material were used. It was found (fig 7.4) that in both cases there was a hardness increase when annealing was carried out below the ordering temperature of 390°C.; the increase was greatest for material deformed in the disordered state. The maximum values were associated with an increase in S from 0.50 to 0.85. It is well known that there is a relationship between hardness and domain size and Stoloff and Davies (1964) have shown that the hardness is a maximum at a critical small domain size. The strain-age-hardening shown in figure 7.4 is attributable to rapid domain growth as order returns.

A second classic study of Cu_3Au, concerned in this case with recrystallization, was that of Hutchinson et al. (1973). The disordered alloy was rolled to 90% reduction ($\epsilon \sim 2.5$) and isothermally annealed (fig 7.5). The kinetics of ordering were such that order was fully restored at 330-380°C. and the retardation of recrystallization in the presence of order is clear. Hutchinson et al. established that the retardation could not be attributed to inhibition of nucleation in the presence of order and that the significant factor was inhibition of grain boundary migration. This was believed to be due to the increased diffusion distances associated with boundary migration through an ordered lattice. The recrystallization textures were also determined; for annealing above T_c a very weak texture was found, but a strong copper-type deformation-type texture developed at temperatures below T_c. By comparing their results with those for pure copper and pure gold, Hutchinson et al. estimated that at T_c the retardation factor due to the presence of a superlattice is $\sim 100x$. Cahn and Westmacott (1990) have recently carried out similar experiments with the alloy $(Co_{78}Fe_{22})_3V$ and find an even greater retardation factor of 300x.

The alloy Ni_3Fe, which has $T_c \sim 500°C.$, was found by Vidoz et al. (1963) to harden during annealing in much the same way as Cu_3Au. The effect was greatest at the stoichiometric composition and fell off rapidly as the composition deviated from ideality; off-stoichiometric alloys with $>32\%$ Fe or $>78\%$ Ni softened by recrystallization. In addition to the hardness

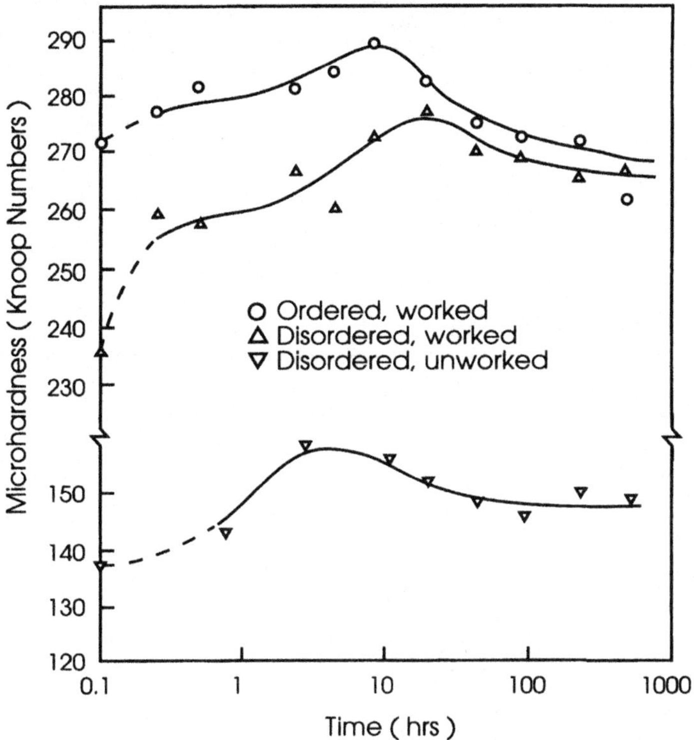

Fig. 7.4. Microhardness of rolled Cu_3Au as a function of annealing time at 288°C, (Roessler et al. 1963).

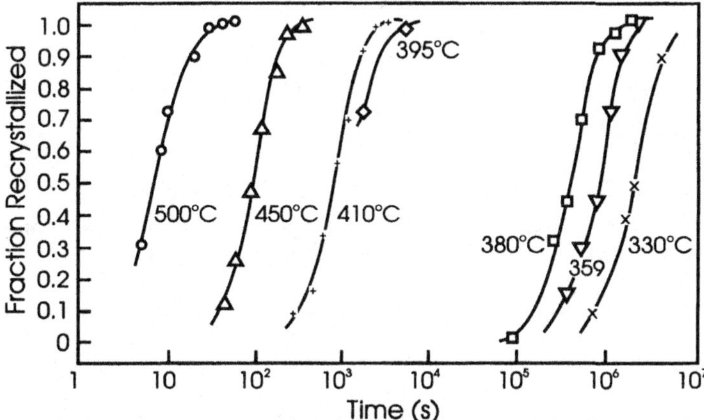

Fig. 7.5. Fraction recrystallized vs annealing time for Cu_3Au reduced 90% by rolling, (Hutchinson et al. 1973).

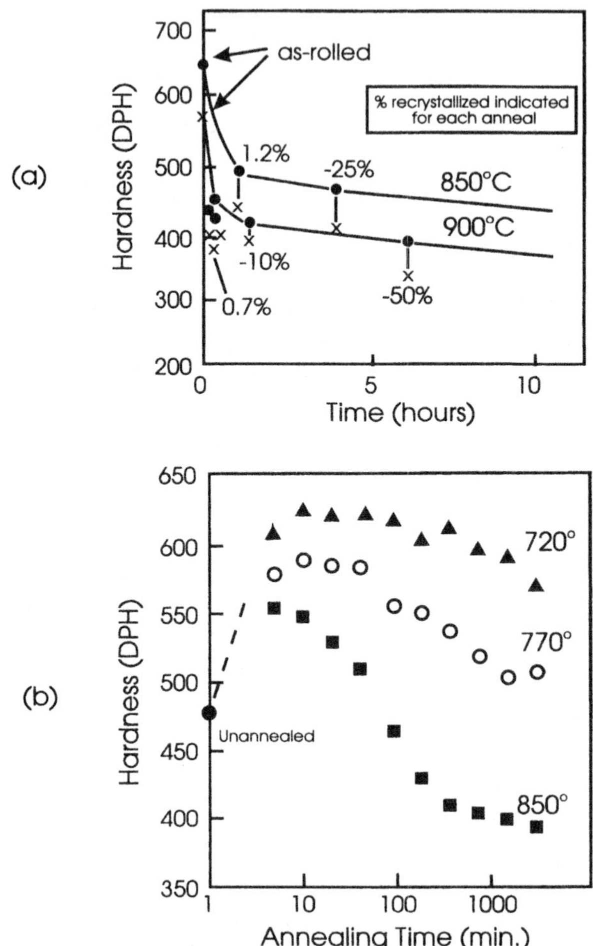

Fig. 7.6. Hardness changes during the annealing of deformed $(Co_{78}Fe_{22})_3V$. (a) Initially fully ordered material rolled to 25% reduction ($\bullet$ =microhardness, X =macrohardness). (b) Initially disordered material rolled to 50% reduction, (Cahn et al. 1991).

increase, the stoichiometric alloy showed a strongly enhanced work hardening rate after annealing; this change is now believed to be due to dislocation drag involving APB tubes.

Considerable attention has been paid to the family of $L1_2$ alloys based on the composition $(FeCoNi)_3V$ (Liu 1984). Cahn et al. (1991) rolled the alloy, $(Co_{78}Fe_{22})_3V$, in both the ordered and disordered conditions to reductions up to 50% ($\epsilon \sim 0.8$) and annealed above and below T_c (910°C.). In this alloy S remains high up to T_c and recrystallization was strongly retarded with a retardation factor of $\sim 300x$ just below T_c. In the case of initially ordered material, both hardness and tensile measurements showed substantial softening prior to, and during the early stages of recrystallization as shown in figure 7.6a. However, initially disordered material showed substantial hardening before and during the early stages of recrystallization as shown in figure 7.6b. The reason for this markedly different behaviour is discussed in §7.5.1

The annealing behaviour of the boron doped Ni_3Al alloy referred to in section 7.2.3 has been examined by Gottstein et al. (1989) and Ball and Gottstein (1993b). In the initial work details of the recrystallization kinetics were determined and recrystallization was reported to be slow. The JMAK exponent was 2.2 and Q_R (the activation energy for recrystallization) was found to be 110kJ/mol which was less than half the value found by Baker et al. (1984). In view of the high value of stored energy, which provides a large driving force for recrystallization, Gottstein et al. attributed the sluggish recrystallization behaviour to a reduced grain boundary mobility in the presence of ordered structures. The evolution of microstructure was said to be similar to that in conventional metals and alloys. Strain induced boundary migration was the significant process at low strain levels while nucleation at microstructural heterogeneities was dominant at higher strains.

In a later investigation, Ball and Gottstein (1993b) investigated the annealing behaviour of material that had been rolled to 70-85% reduction ($\epsilon \sim 1.3$-2). Nucleation was associated with shear bands (figure 7.2) and microbands and equiaxed grains with diameter $0.2\mu m$ developed at these inhomogeneities. Annealing twins were common in the new grains. In the volumes between the bands, nucleation was much slower and the grains that eventually developed were some three times larger. This situation is directly comparable to the case of low zinc brass reported by Carmichael et al. (1982) (fig 6.15 and §6.4.1). In this case the recrystallization texture was reported to be almost random and Ball and Gottstein attributed this to the locally high nucleation rate in the shear bands and microbands and the very low grain boundary mobility of ordered structures. It was maintained that the latter factor effectively ruled out the development of a strong texture by growth selection.

Ball and Gottstein (1993b) also investigated grain growth after recrystallization in the boron-doped Ni_3Al discussed above, in the temperature range 800 to 1150°C. They found that the kinetics were of the form of equation 9.7, with a growth exponent n which varied somewhat with temperature, having a mean value of 3. The activation energy for grain growth was found to be 298kJ/mol, close to that for diffusion of Ni in Ni_3Al.

7.3.2. B2 structures

The annealing behaviour of B2 alloys is different in many respects to that just described, and amongst such materials FeCo has been extensively studied. This alloy has a critical temperature of $\sim 720°C$, and the disordered state can be retained by quenching. On deformation, FeCo work hardens at similar rates in both the ordered and disordered states. The kinetics of re-ordering in FeCo deformed in the disordered state are retarded (Stoloff and Davies 1964) in contrast to the behaviour of alloys such as the $L1_2$ phase Cu_3Au in which acceleration is observed. Studies of grain growth in the industrially important alloy containing 2%V, have shown that the kinetics are described by the normal $D^2 \propto t$ relationship (§9.1). If however, the rate constant is plotted against 1/T the results shown in figure 7.7 are observed (Davies and Stoloff, 1966). A normal Arrhenius relation exists above T_c but at lower temperatures the plot is curved and there is a discontinuous change in the rate constant at T_c.

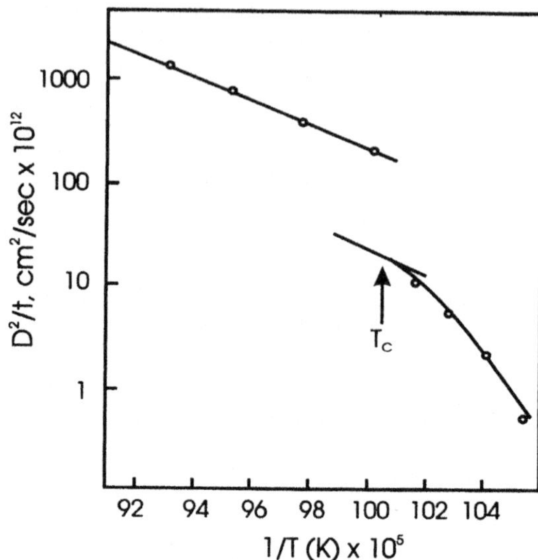

Fig. 7.7. Grain growth rate constant in FeCo-V vs reciprocal temperature, (Davies and Stoloff 1966).

The recrystallization behaviour of FeCo containing small amounts of V or Cr has been studied by Buckley (1979) and Rajković and Buckley (1981). If cold worked material is annealed below T_c then interaction between the ordering and recrystallization reactions leads to the very complex pattern of behaviour summarised below.

(i) Above $\sim 725°$C (T_c) recrystallization is rapid.

(ii) Between 600°C and 725°C the alloy orders rapidly and this is followed by slow recrystallization.

(iii) Between 475° and 600°C, homogeneous ordering occurs at a moderate rate, dislocation recovery takes place but no recrystallization occurs. It is presumed that the driving force for recrystallization is insufficient for grain boundary migration through the superlattice.

(iv) Between 250° and 475° the ordering reaction is slow and partial recrystallization is possible. Ordering is more rapid in the vicinity of the migrating grain boundaries.

The most detailed study of annealing in a B2 alloy is that of Morris and Morris (1991), who examined β brass. As pointed out earlier, the amount of order decreases continuously in this phase as the temperature rises to, and through, T_c and the annealing behaviour (fig 7.8) does not show the marked discontinuity seen in figure 7.5 for Cu_3Au. Electron microscopy showed that recovery of the dislocation structure was rapid well below T_c, but decreased as T_c was approached and this was attributed to short range ordering. The increasing rate of recrystallization as T_c is approached is to be expected, and the activation energy of 145kJ/mol for boundary migration in this range is close to that for diffusion of copper in the ordered

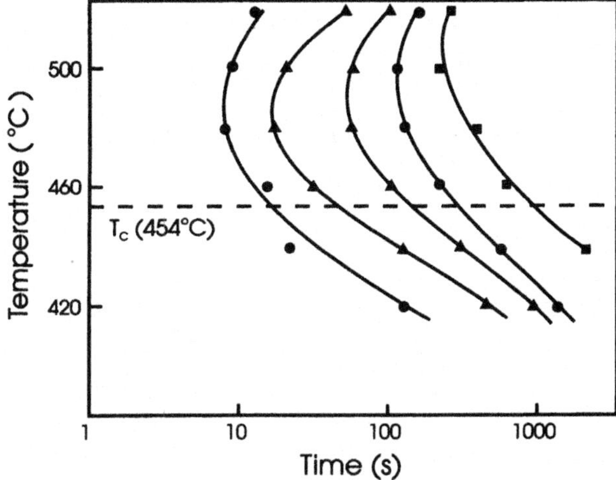

Fig. 7.8. Recrystallization kinetics of 20% compressed, β brass. The curves show contours for recrystallized fractions 0.05 (●), 0.2, 0.4, 0.6, and 0.8 (■), (after Morris and Morris 1991).

structure. The decreasing rate of recrystallization with increasing temperature above T_c is however, surprising.

Recrystallization, as such, has not been studied extensively in Fe_3Al but an early report by Lawley et al. (1961) suggested that complete recovery was possible after small strains. McKarney and Pierce (1992) (see §7.2.2) refer to the properties of stress relieved and recrystallized material.

7.4 DYNAMIC RECRYSTALLIZATION

Evidence of strain induced boundary migration and dynamic recrystallization has been obtained by Baker and Gaydosh (1987) in specimens of FeAl deformed in tension at temperatures between 727°C. and 827°C. Recovery was minimal at all temperatures and it was suggested that the recrystallization effects observed were the result of the retention of large amounts of stored energy. It was suggested also, that dynamic recrystallization and strain induced boundary migration may be common features in ordered materials deformed at high temperatures.

Aretz et al. (1992) have investigated dynamic recrystallization during the hot compression of boron-doped Ni_3Al. They find that dynamic recrystallization occurs at the old grain boundaries, the new grains forming a "necklace microstructure" (§11.2.3). The mechanism of dynamic recrystallization appears to involve progressive lattice rotations in the grain boundary regions. This mechanism is very similar to that found for magnesium (Ion et al. 1982) and some minerals, which is discussed in §11.3.2.

7.5 CONTROVERSIAL AREAS

The discussion above reveals a number of matters relevant to the annealing of ordered structures that are still unclear and in this section some of these will be considered in more detail. The topics involved are related to the property changes taking place prior to recrystallization, the migration of grain boundaries in ordered phases and the development of a domain structure during annealing.

7.5.1 Property changes prior to and during recrystallization

The question to be answered here is: why do some materials such as Cu_3Au harden prior to recrystallization, while others $(Co_{78}Fe_{22})_3V$ soften? Confusion has been heightened by the fact that various studies have used initially ordered alloys, initially disordered alloys or, occasionally both. Figure 7.6 shows the results of hardness tests on $(Co_{78}Fe_{22})_3V$ during isothermal annealing. T_c for this alloy is $\sim 910°C$, and S remains high up to T_c. Material rolled in the fully ordered state softens substantially from the beginning of annealing (fig 7.6a) but initially disordered material shows marked initial hardening (fig 7.6b). The maximum hardness was associated with an **S** value of 0.4-0.5. It is well established that the strength of mixed ordered-disordered materials is a maximum at **S** ~ 0.5, and Cahn (1991a) has interpreted these results in terms of a change in the dislocation structure. It is argued that in the fully ordered state, deformation is based on the movement of super-dislocations and that a change to single dislocation glide occurs at $S < 0.5$. Such an interpretation of the pre-recrystallization behaviour is supported by the work of Jang and Koch (1990) on mechanical disordering of Ni_3Al. Cahn (1991a) suggests that the apparent anomaly in the case of Cu_3Au, which hardened on annealing after deformation in both states, is due to the fact that Roessler et al. (1963) rolled their specimens to 63% reduction ($\epsilon \sim 1$). It is inferred that at this level of strain, and in this material, the value of S would be less than 0.5.

7.5.2 Migration of grain boundaries

There can be no doubt that the rate of grain boundary migration is strongly retarded in ordered materials and it is obvious that the nature of the grain boundary regions is not understood. It is not clear for example, whether the ordered state exists up to the boundary or whether the grain boundary composition is the same as that found within the grains. Local disorder is to be expected at the grain boundaries of an ordered material and it is not surprising that grain boundary migration should be affected. There is considerable evidence to suggest that the major nickel constituent in the $L1_2$ structure Ni_3Fe, is enriched at APB's, but there is an equivalent body of evidence to the contrary (for a discussion of this matter see Baker 1991). Cahn (1990) maintains that enrichment is to be expected at both grain and antiphase boundaries. Grain boundary migration rates in superlattice structures are slow and Cahn (1990) suggests that the presence of a local segregation of "enriched" atoms constitutes an "atmosphere" of atoms that must exert a dragging effect on boundary movement (§4.4.2). The rate of boundary movement would then be governed by the rate of diffusion of the atoms in the atmosphere; if the boundary succeeds in breaking away from the atmosphere the consequent re-ordering of the boundary would again restrict the mobility.

7.5.3 Development of domain structure

Well developed domain structures were reported by Hutchinson et al. (1973) in the recrystallized grains of their annealed specimens of Cu_3Au and this has been confirmed by Cahn and Westmacott (1990). Similar structures are found in FeCo (Cahn 1990). In other materials, and particularly those with high ordering temperatures APB's are extremely rare.

Two points are relevant to discussion of this matter. First, the ordered grains in a permanently ordered material contain few domains. Second, the transformation of a normal disordered structure leads automatically to prominent domains. At first sight, the nucleation of a recrystallized, ordered grain appears little different to the nucleation of a similar grain directly from the melt. In both cases only a single nucleation event is involved and there is no orientation change. Cahn (1990) has suggested two possible reasons for the appearance of the domain structure in the case of recrystallization.

> (i) It has been observed in FeCo that ordering can occur by a process of localized grain boundary movement to produce bulges that contain fine domains (§7.3.2). If nucleation involved the formation of a similar bulged structure the formation of multiple domains would be expected.

> (ii) If the new grain were to nucleate in the disordered state the generation of multiple domains would occur in the normal way.

Of the two, Cahn preferred the latter reason, arguing that, if the boundary of the growing grain had an accompanying disordered region, ordering behind the boundary would involve nucleation of multiple domains.

Chapter 8

RECRYSTALLIZATION OF TWO-PHASE ALLOYS

8.1 INTRODUCTION

Most alloys of commercial importance contain more than one phase, and therefore an understanding of the recrystallization behaviour of two-phase materials is of practical importance as well as being of scientific interest.

The second phase may be in the form of **dispersed particles** which are present during the deformation, or, if the matrix is supersaturated, the particles may precipitate during the subsequent anneal. In addition to conventional alloys which usually contain much less than 5% by volume of the second phase, we will make some reference to **particulate metal matrix composites** which contain large volume fractions of ceramic reinforcement and to **duplex alloys** in which the fractions of the two phases are similar.

Particles have three important effects on recrystallization :

> **(i) The stored energy and hence the driving force for recrystallization may be increased.**

> **(ii) Large particles may act as nucleation sites for recrystallization.**

> **(iii) Particles, particularly if closely spaced, may exert a significant pinning effect on both low and high angle grain boundaries.**

The first two effects tend to promote recrystallization, whereas the last tends to hinder recrystallization. Thus the recrystallization behaviour, particularly the kinetics and the resulting grain size and texture, will depend on which of these effects dominate. Because the size, distribution and volume fraction of second-phase particles in an alloy are affected by composition and thermomechanical processing, second-phase particles may be used to control microstructure and texture in industrial alloys.

In this chapter we will examine the effect of particles on primary recrystallization, and in chapter 12 we will discuss some cases in which the control of recrystallization by second-phase particles is of industrial significance. In most of this chapter we will be dealing with the situation in which a stable dispersion of particles is present during deformation and annealing. The case of concurrent annealing and phase transformation is dealt with as a special case in §8.9.

8.1.1 Particle parameters

As will be discussed in the following sections, the annealing behaviour of a particle-containing alloy is dependent on the **volume fraction**, **size** and **spacing** of the particles, and we should at this stage clarify what is meant by these parameters. The determination and representation of the particle parameters is not a trivial problem, and for further details the reader is referred to Cotterill and Mould (1976) and Martin (1980). However, neither the theories nor the experimental results are currently of sufficient accuracy to warrant the use of very complicated parameters, and wherever possible, we will, for clarity, use the simplest relationships.

PARTICLE SIZE
All multi-phase materials contain particles of a spectrum of sizes and shapes. In modelling the effect of particles it is often necessary to simplify the situation and to assume that the dispersion is of uniform spheres of radius **r**. In comparing experimental data with such a model, a mean value of particle diameter is often used, and for the case of non-spherical but equiaxed particles, an equivalent radius is usually calculated. For particles which are not equiaxed, it is necessary to use more than one parameter to define the particle size and shape.

VOLUME FRACTION
The volume fraction F_V of particles is the other parameter which it is often convenient to measure experimentally or to calculate from a knowledge of the phase diagram. For a random distribution of particles, F_V is also equal to the fractional area occupied by particles on a plane section, and the fractional length occupied by the particles on a random straight line.

INTERPARTICLE SPACING
For a dispersion of uniform spheres of radius **r**, the **number of particles per unit volume** (N_V) is related to the volume fraction by

$$N_V = \frac{3\,F_V}{4\,\pi\,r^3} \tag{8.1}$$

The **number of particles intersecting unit area** (N_S) of a plane surface is $N_V.2r$ and hence

$$N_S = \frac{3\,F_V}{2\,\pi\,r^2} \tag{8.2}$$

If the particles are arranged on a square lattice, then the centre to centre **nearest neighbour spacing on a plane** is

$$\lambda = N_S^{-1/2} \tag{8.3}$$

For randomly distributed particles, the centre to centre **nearest neighbour spacing on a plane** is

$$\Delta_2 = 0.5 \, N_S^{-1/2} \tag{8.4}$$

and the centre to centre **nearest neighbour spacing in the volume** is

$$\Delta_3 = 0.554 \, N_V^{-1/3} \tag{8.5}$$

PARTICLE DISTRIBUTION

The spatial distribution of the particles, i.e. whether it is uniform, random or clustered may be important. However, there is as yet no simple method of quantitative classification (Fridy et al. 1992).

8.2 THE DEFORMED MICROSTRUCTURE

Particles have a large effect on the microstructure developed during deformation, and this in turn affects the recrystallization behaviour. There are three aspects of the deformation structure that are of particular importance in determining the behaviour of the material during subsequent annealing.

(i) **The effect of the particles on the overall dislocation density**. This provides the driving force for recrystallization.

(ii) **The effect of the particles on the inhomogeneity of deformation in the matrix**. This may affect the availability and viability of the sites for recrystallization (§6.6.4).

(iii) **The nature of the deformation structure in the vicinity of the particles**. This determines whether particle stimulated nucleation of recrystallization (PSN) will be possible.

During deformation of an alloy containing particles, the dislocations will bow around the particles as shown in figure 8.1. An analysis of the situation in terms of elementary dislocation theory is as follows. For particles of radius **r** and spacing λ along the dislocation line, the force exerted on each particle (**F**) is given by:

$$F = \tau \, b \, \lambda \tag{8.6}$$

where τ is the applied stress.

If the strength of the particle is less than F, then the particle deforms, otherwise the dislocation reaches the semicircular configuration of figure 8.1b, when the applied stress is given by

$$\tau_0 = \frac{G \, b}{\lambda} \tag{8.7}$$

which is the well known **Orowan stress**.

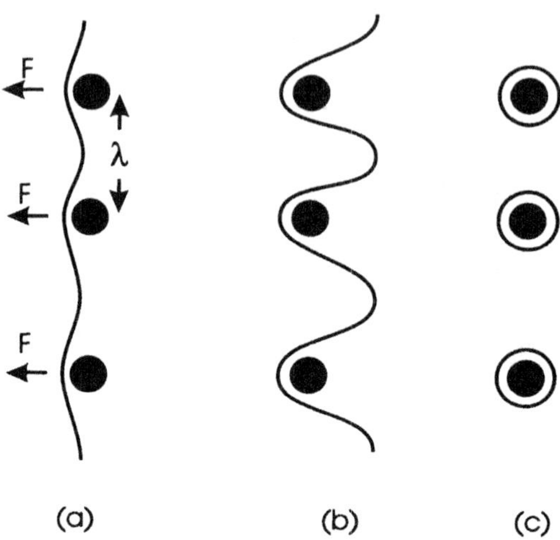

Fig. 8.1. The formation of Orowan loops at second-phase particles.

The dislocation then proceeds to encircle the particle, leaving an Orowan loop as shown in figure 8.1c. A micrograph of such loops in a nickel alloy is shown in figure 8.2. Because of its line tension, an Orowan loop exerts a shear stress on the particle which is given approximately by

$$\tau = \frac{G\,b}{2r} \tag{8.8}$$

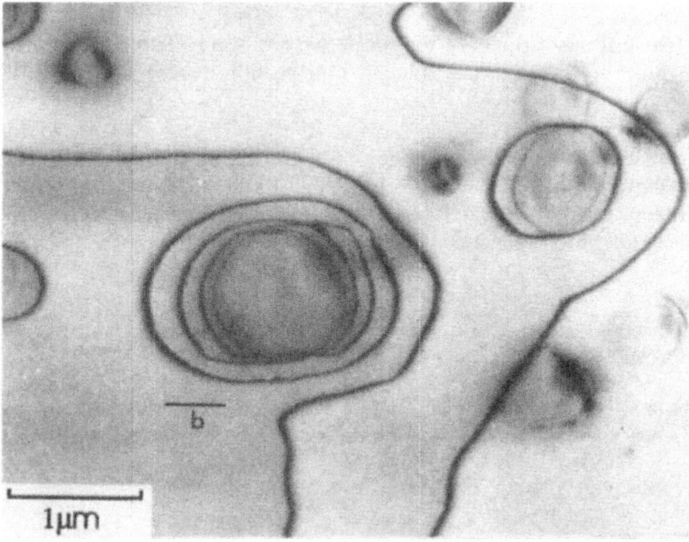

Fig. 8.2. Orowan loops at Ni$_3$Si particles in a Ni-6%Si single crystal, (Humphreys and Ramaswami 1973).

If the particle is strong enough to withstand this stress, it does not deform and the net result of the passage of a matrix dislocation is the generation of extra dislocation at the particle. If the particle deforms either before or after the Orowan configuration is reached, no extra dislocations are generated at the particle.

Detailed discussion of the factors affecting the strength of particles is beyond the scope of this volume, and the reader is referred to the reviews of Martin (1980), Brown (1985), Humphreys (1985) and Ardell (1985) for further details. However it is important to realise that the subsequent deformation behaviour and hence the density and arrangement of dislocations in the deformed material is dependent on whether or not the particles deform.

8.2.1 Dislocation distribution in alloys containing deformable particles

If a particle deforms as shown in figure 8.3b, then its size on the slip plane is effectively reduced by the Burgers vector **b** of the dislocation. As a smaller particle is normally weaker than a larger particle, the slip plane is weakened and subsequent dislocations will tend to move on the same plane, thus concentrating slip into bands as shown in figure 8.3d. This is illustrated in the micrograph of figure 8.4, where shear offsets of $\sim 0.1\mu$m indicate the passage of several hundred dislocations on the same plane.

The slip distribution in alloys containing small particles has been discussed by Hornbogen and Lütjering (1975) and by Martin (1980) as follows.

The yield stress of a crystal containing deformable particles is often given by an equation of the form

$$\tau = C \, F_V^{1/2} \, d^{1/2} \tag{8.9}$$

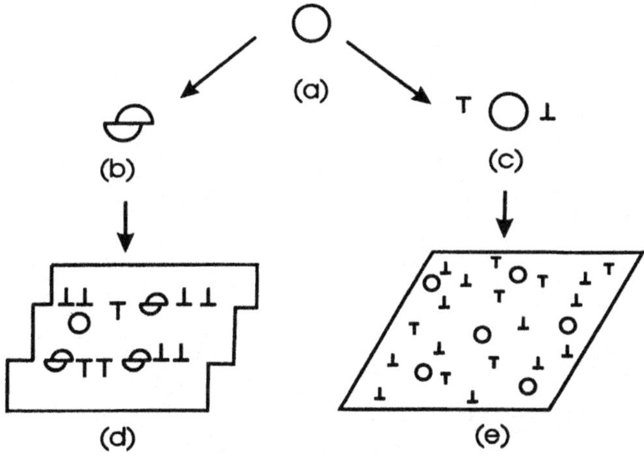

Fig. 8.3. The effect of particle strength on the distribution of slip.

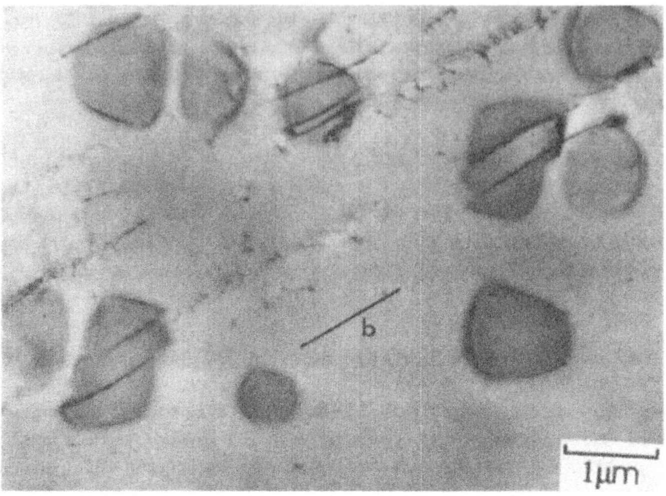

Fig. 8.4. Deformed Ni₃Si particles in a Ni-6%Si single crystal,
(Humphreys and Ramaswami 1973).

where C is a constant dependent upon the particular hardening mechanism which is operative
(e.g. coherency strains). If **n** dislocations shear a particle, the diameter of the particle will
be reduced by **nb**. The stress for further deformation then becomes

$$\tau = C\,F_V^{1/2}\,(d - nb)^{1/2} = C\,F_V^{1/2}\,d^{1/2}\left(1 - \frac{nb}{d}\right)^{1/2} \tag{8.10}$$

Thus the slip plane is softened, so that further slip tends to occur on that plane. This is in
contrast to the case of non-deformable particles, where the dislocation debris left by previous
dislocations (e.g. the Orowan loops of figure 8.2) makes it more difficult for slip to occur
on the same plane, thus tending to make slip relatively homogeneous.

The amount by which a slip plane is weakened by the passage of a dislocation is the
parameter $d\tau/dn$, and differentiation of equation 8.10 gives

$$\frac{d\tau}{dn} = \frac{-\,b\,C\,F_V^{1/2}}{2\,d^{1/2}}\cdot\left(1 - \frac{n\,b}{d}\right)^{-1/2} \tag{8.11}$$

We thus see that the tendency for coarse slip is increased by strong particles, a large volume
fraction and a large value of C.

There are many recorded examples of particle shearing leading to slip concentration. This
is of particular concern in the design of high strength precipitation hardened alloys, in which
the slip concentration may adversely affect the fracture toughness and the fatigue behaviour
(e.g. Polmear 1989). An example of this is found in binary Al-Li alloys in which
precipitates of the ordered δ' phase are formed. These particles deform during deformation,
leading to extensive shear localisation (Sanders and Starke 1982).

As the shape and size of the particles changes during deformation, so the slip distribution may also be a function of strain. Kamma and Hornbogen (1976) found that small platelike particles of Fe_3C in steel deformed, leading to the formation of shear bands. However at high strains, the bands broadened, and the slip became homogeneous. Nourbakhsh and Nutting (1980) showed that in overaged Al-Cu alloys containing large plates of θ', the deformation was initially inhomogeneous as the θ' plates deformed. At strains above 1, the plates disintegrated to give small spherical particles, and as found by Kamma and Hornbogen a more homogeneous slip distribution resulted. The same authors also found that in the underaged alloy, the small GP zones which resulted in slip concentration at lower strains, were dissolved at strains of 5 and the deformation behaviour of the alloy then became similar to that of the solid solution.

We therefore conclude that the slip distribution in alloys containing deformable particles is complex and may alter with strain as changes in the particle size and shape are induced by the deformation.

8.2.2 Dislocation distribution in alloys containing non-deformable particles

8.2.2.1 Dislocation density

If a deforming matrix contains non-deformable particles, then, as was first shown by Ashby (1966, 1970) there is a strain incompatibility between the two phases as shown in figure 8.5. In figure 8.5a, a spherical particle of radius r is contained within an undeformed matrix. On deformation by a shear strain s, then, as shown in figure 8.5b, a deformable particle will comply with the imposed strain. However, if the particle does not deform, then as shown in figure 8.5c, there is a local strain incompatibility. This can be accommodated by the generation of dislocations at the particle-matrix interface and also, if the interface is weak, by the formation of voids. A more detailed consideration of the dislocation generation near the particle is given in §8.2.3. However, the length of dislocation generated is not very sensitive to the details of the mechanism, and it is easily shown that the incompatibility may be accommodated approximately by the generation of n circular prismatic dislocation loops of Burgers vector b and radius r, where

$$s = \frac{n\,b}{2r} \tag{8.12}$$

The length of dislocation per particle is thus

$$L = \frac{4\,\pi\,r^2\,s}{b} \tag{8.13}$$

and the dislocation density (ρ_G) due to this effect is $N_V.L$. Taking N_V as given by equation 8.1, then

$$\rho_G = \frac{3\,F_V\,s}{r\,b} \tag{8.14}$$

These have been termed the **geometrically necessary** dislocations by Ashby.

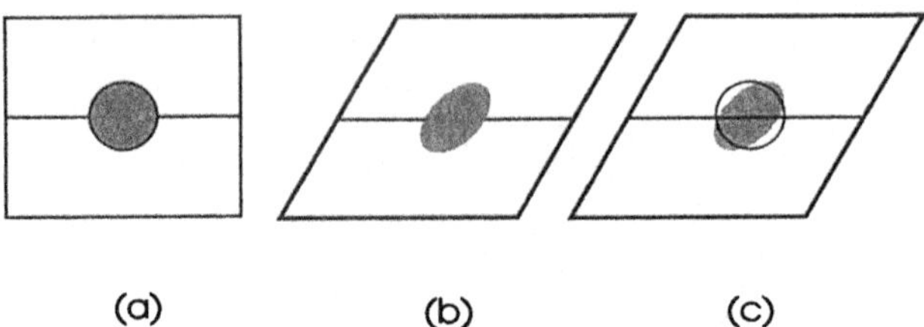

Fig. 8.5. The incompatibility between a deforming matrix and a non-deforming particle.

The total density of dislocations generated in the material will be approximately $\rho_G + \rho_S$, where ρ_S is the density of dislocations which would be generated in the single-phase matrix (see §2.2). However, because dislocations will be lost by recovery during or after the deformation, this is an upper limit. Figure 8.6 shows the calculated density of dislocations ($\rho_G + \rho_S$) generated for a matrix of grain size $100\mu m$ (equations 2.1 and 8.14), as a function of strain and particle content, neglecting the effects of recovery.

Although there is little quantitative data, there is evidence that in particle-containing single crystals or polycrystals deformed to small strains, the dislocation density at low strains is much greater than in single-phase materials of similar composition. For single crystals of copper, oriented for single slip, and containing alumina particles, Humphreys and Hirsch (1976) found that the dislocation density was close to that predicted by equation 8.14. However, analysis of the data of Lewis and Martin (1963) on copper polycrystals containing oxide particles shows that the dislocation densities at tensile strains of 0.08 are some 50 times lower than those predicted. There are no reliable data available for high strains.

There are comparatively few reliable measurements of the stored energy of particle-containing alloys. Adam and Wolfenden (1978) found the stored energy of an Al-Mg-Si

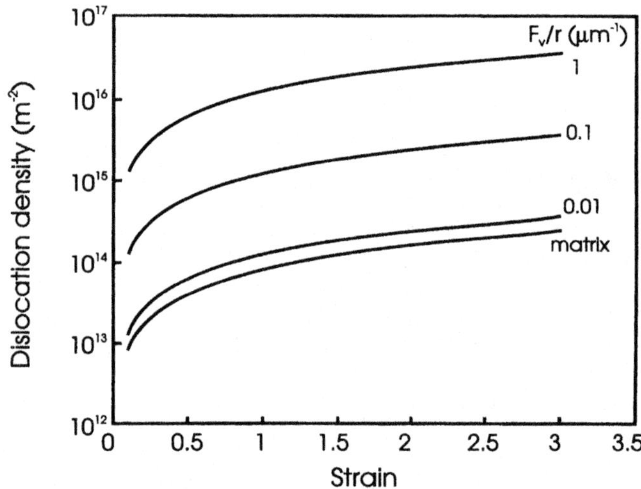

Fig. 8.6. Predicted dislocation densities in particle-containing alloys.

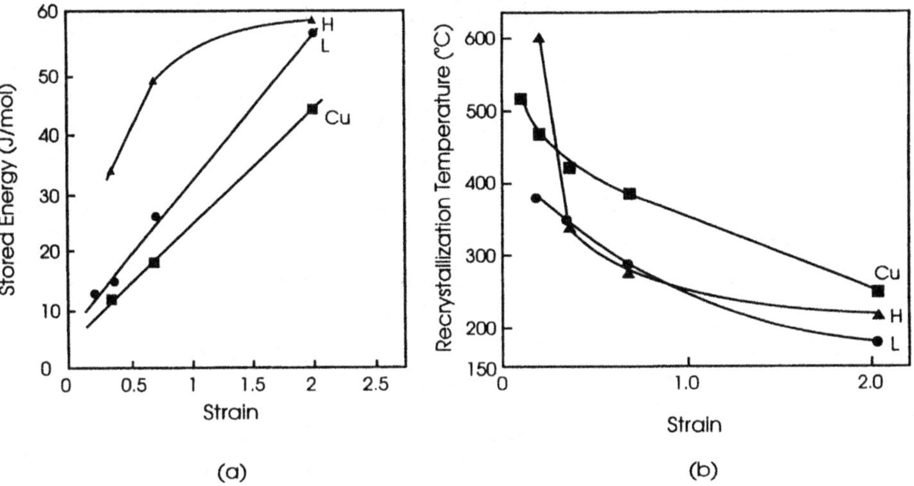

Fig. 8.7. a) The stored energy b) The recrystallization temperature in Cu-Al$_2$O$_3$ alloys compared with that of pure copper. The parameters of the alloys are d=46nm, F_V=1.5x10^{-2} (H) and d=38, F_V=4x10^{-3} (L), (Baker and Martin 1983b).

alloy deformed to a strain of 0.1 to be some ten times that of pure aluminium. Measurements of stored energy in particle-containing copper alloys by Baker and Martin 1983a (fig 8.7), show an increase in stored energy with strain which is in approximate agreement with equation 8.14 at low strains, with some indication that the stored energy saturates at strains larger than ~1. Other measurements of the stored energy of dispersion hardened copper alloys deformed to large strains (Bahk and Ashby 1975 and Chin and Grant 1967) indicate that the stored energy in these alloys is not significantly greater than for pure copper.

As the dislocation density is generally found to be directly related to the flow stress (equation 2.2), some indication of the change in dislocation content with strain may often be inferred from the stress-strain behaviour. From such experiments, there is considerable evidence that the work hardening in two-phase alloys, which is initially high, frequently falls to that of a comparable single-phase alloy after strains of ~0.05 in copper (Lewis and Martin 1963), in steels (Anand and Gurland 1976), and in aluminium alloys (Lloyd and Kenny 1980).

In summary, it is apparent that the large increases in dislocation density predicted by equation 8.14 and figure 8.6 are only obtained for very small strains, and that at the larger strains which are usually of importance for recrystallizing materials, dynamic recovery may reduce the dislocation density to a value little larger than that for a single-phase alloy.

8.2.2.2 Cell and subgrain structures
It has often been reported that particles affect the dislocation cell structures formed on deformation. As discussed in §2.3.2.1, cell and subgrain structures are products of dynamic recovery and will therefore be greatly influenced by any parameter which affects the generation or the recovery of dislocations. Furthermore, particles increase the rate of

dislocation generation (§8.2.2.1), and therefore it is expected that they will affect the rate of dynamic recovery.

A dispersion of small second-phase particles will affect dynamic recovery by impeding the movement of dislocations. Dislocations will only pass strong particles if the local stress exceeds that given by equation 8.7. Therefore although dislocations will be free to move in the regions between particles, they will tend to be held up at particles. In a similar manner, the particles will pin cell or subgrain boundaries (§3.6). Recovery processes which occur on a scale **smaller** than the interparticle spacing (λ) will therefore be relatively unaffected by the particles, but recovery processes involving rearrangement of dislocations on a scale **larger** than the interparticle spacing will be hindered. The interparticle spacing will therefore govern the **maximum** size of the dislocation cells or subgrains. As discussed in §2.3.2, a cell structure is formed at medium to large strains in metals other than those of low stacking fault energy. This cell size (D) is typically 0.5-1μm in many metals and decreases with increasing strain (fig 2.7). If $\lambda < D$ then the cell size in the two-phase alloy will be reduced to $\sim \lambda$. However, if $\lambda > D$ then the particles will have little effect on the cell size.

Observations of cell formation in dispersion strengthened copper with very small interparticle spacing ($\lambda \sim 0.1\mu$m) (Lewis and Martin 1963 and Brimhall et al. 1966) indicate that compared with copper, the dislocation cells are more diffuse, form at larger strains, and have a size comparable with the interparticle spacing. However, in oxide-containing aluminium (Hansen and Bay 1972) and copper (Baker and Martin 1983b), where the interparticle spacing is 0.3 to 0.4μm, the cell size is not significantly different from that of the pure matrix. In a study of the effect of strain on the substructure of two-phase aluminium alloys, Lloyd and Kenny (1980) found that at low strains the cell size was related to the interparticle spacing, but at larger strains, the cell size was reduced and was almost independent of particle size as shown in figure 8.8. As discussed in §2.3.2.1, the determination of cell size is not straightforward, and many early measurements should be regarded with caution. However,

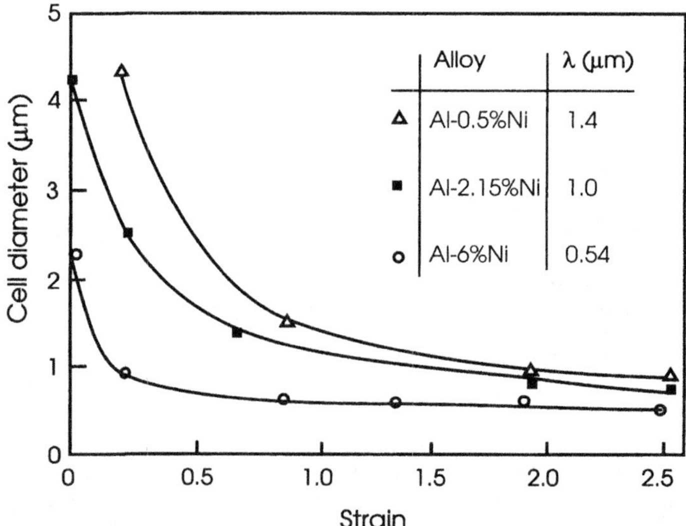

Fig. 8.8. The cell size as a function of strain in Al-Ni alloys, (after Lloyd and Kenny 1980).

these data are consistent with the view expressed above, that the particles exert a pinning effect on the dislocations and on the cell or subgrain boundaries (§3.6) and that the interparticle spacing determines the **maximum** but not the minimum cell size in the material.

Some electron diffraction experiments on copper (e.g. Brimhall et al. 1966) have shown that the lattice misorientation between cells, in alloys with a small interparticle spacing is less than in the single-phase matrix. This is at variance with the results reported for copper (e.g. Baker and Martin 1983b) and aluminum (Hansen and Bay 1972), in which the interparticle spacings were larger and in which the misorientations were found to be similar to those in the matrix alloy. Although the results are limited, it would appear that in cases where the interparticle spacing is sufficiently small to reduce the cell size to below that of the particle-free matrix, there is a corresponding decrease in the cell or subgrain misorientation.

8.2.2.3 Larger scale deformation heterogeneities

There is less evidence as to the effect of non-deformable particles on larger scale heterogeneities in the deformation structure such as **microbands, transition bands** or **shear bands**. During the deformation of single-phase alloys, a grain may, in addition to changing orientation, deform inhomogeneously and divide into regions of different orientation as shown in figure 8.9a and b.

It has been found (Habiby and Humphreys 1993) that sub-micron sized particles have little effect on the formation of either the deformation bands or the transition bands in aluminium. However, there was evidence in the same investigation that large ($>1\mu$m) silicon particles could nucleate deformation bands. An example of this effect is found in the work of Humphreys and Ardakani (1994) who investigated the deformation in plane strain compression of aluminium crystals of the unstable (001)[110] orientation. During deformation, the crystal splits into two orientations which stabilise close to components of the "copper orientation" (112)[11$\bar{1}$] and ($\bar{1}\bar{1}$2)[111], as shown schematically in figure 8.9b. Large non-deforming particles, by altering the operative slip systems close to the particle, cause a small portion of crystal to rotate in the "wrong" sense (fig 8.9c). This then affects adjacent material within the crystal and results in a small deformation band within one of the components which has the orientation of the other component. The microstructure of such a deformation band adjacent to a large particle is shown in figure 8.9d. By this type of mechanism, it is likely that inclusions in any alloy may be responsible for nucleating deformation bands.

Particles may also affect the formation of **shear bands** which as discussed in chapter 2 are the result of instability during deformation. Shear bands often occur at large strains when the work hardening rate in a material becomes small. In alloys containing large volume fractions of non-deformable particles, the initially high work hardening rate is not maintained to high strains and therefore these materials may be prone to the formation of shear bands. Examples of this have been noted for Al-Mg-Si alloys (Liu and Doherty 1986) and for aluminium matrix particulate composites (Humphreys et al. 1990).

8.2.3 Dislocation structures at individual particles

We have seen that second-phase particles affect the density and distribution of dislocations in the matrix, and that this may influence the driving force for recrystallization. In addition, we need to consider the accumulation of dislocations in the vicinity of large particles as this may lead to these regions becoming sites for recrystallization nucleation.

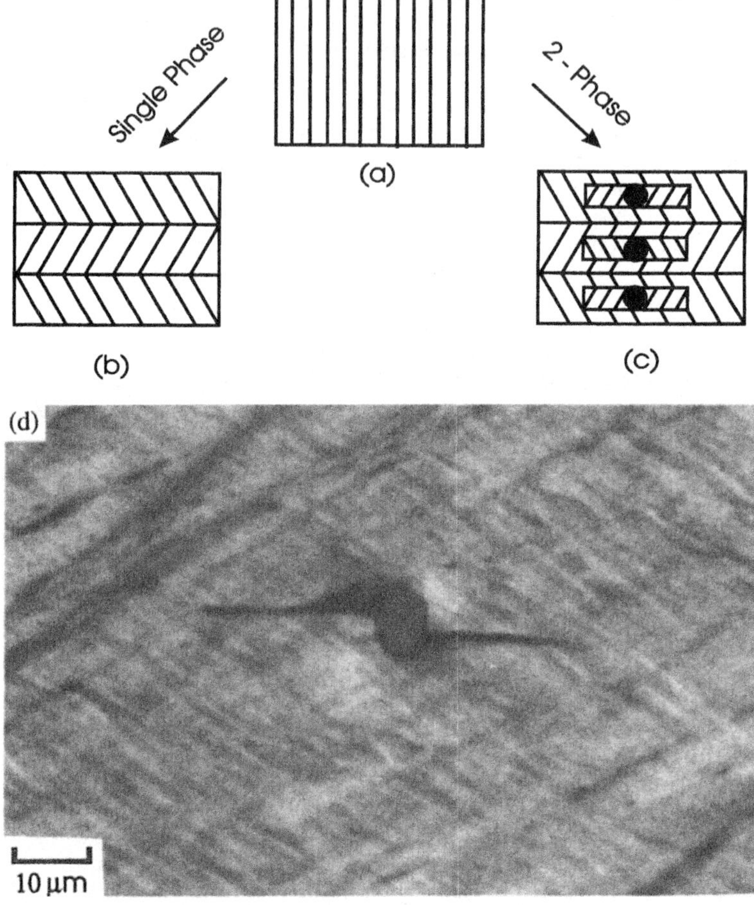

Fig. 8.9. The nucleation of deformation bands as particles. a) Undeformed crystal of unstable orientation. b) Formation of deformation bands in the deformed single-phase crystal. c) Formation of small deformation bands at particles. d) Optical micrograph of deformation bands at a 10μm particle in an Al-Si crystal, (Humphreys and Ardakani 1994).

As discussed in §8.2.1, if the particles are strong enough to resist the passage of dislocations, then Orowan loops may be formed at particles, and the stress on a particle from a single Orowan loop is given by equation 8.8. During continued deformation, more loops will form and the resulting stress will rapidly rise to a level where, if the particle does not deform, then there will be local plastic flow or **plastic relaxation** in the matrix to relieve these stresses. Because of the high stresses associated with them, Orowan loops are very unstable and are rarely observed. The nature of this plastic relaxation has been extensively studied for single crystals deformed to low strains, and to a much lesser extent in single crystals or polycrystals deformed to large strains. We are primarily interested here in those aspects of the deformation structures which are relevant to recrystallization, and for further details of the dislocation mechanisms the reader is referred to the reviews of Brown (1985) and Humphreys (1985).

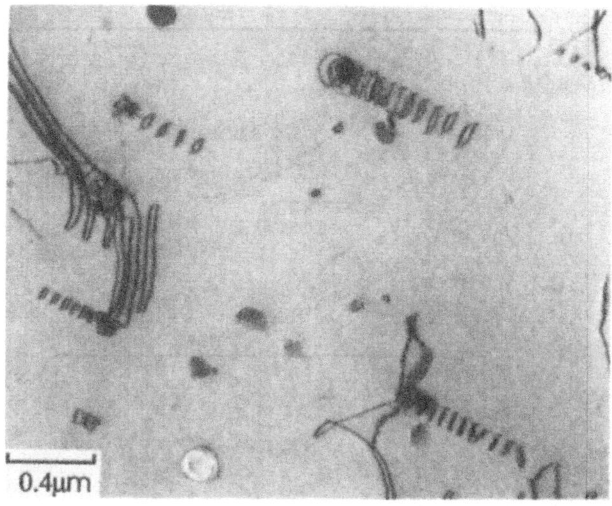

Fig. 8.10. Primary prismatic loops at Al_2O_3 particles in an α-brass crystal, (Humphreys 1985).

For small particles and low strains, plastic relaxation generally involves the generation of prismatic dislocation loops, which are of lower energy than Orowan loops. These may be prismatic loops of the same Burgers vector as the gliding dislocations (primary prismatic loops) or alternatively, secondary prismatic loops of a different Burgers vector may be generated. In deformed single crystals, rows of primary prismatic loops aligned with the particles are found as seen in figure 8.10. In polycrystals, although aligned rows of loops are not generally observed, there is evidence that prismatic loops are formed.

At larger strains and larger particles, more complex dislocation structures are formed, and these are often associated with local lattice rotations close to the particles. Such regions are commonly termed **particle deformation zones**. The form and distribution of the dislocation structures at particles are primarily functions of the strain and the particle size, although other factors such as shape, interface strength and matrix are known to be important (see e.g. Humphreys 1985). The effect of strain and particle size on the relaxation mechanisms in aluminium single crystals is summarised in figure 8.11. The deformation zones formed at large particles are of particular interest as they are the source of **particle stimulated nucleation of recrystallization (PSN)** and may therefore have a strong influence on the grain size and texture after recrystallization.

8.2.4 Deformation zones at particles

8.2.4.1 Single crystals

Detailed measurements of deformation zones at particles have so far been made only on particle-containing single crystals of copper and aluminium, oriented for single slip, and deformed in tension (Humphreys 1979a). It is found that for particles of diameter less than $0.1\mu m$, deformation zones with local lattice rotations are not formed in these metals, and in such cases it is assumed that plastic relaxation occurs by the generation of prismatic loops as discussed above. For particles in the size range 0.1 to $\sim 2.5\mu m$, measurements of lattice

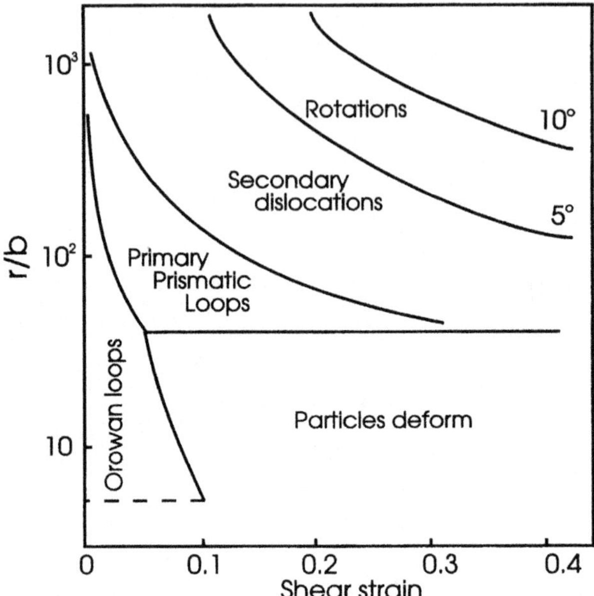

Fig. 8.11. Deformation mechanisms at particles in aluminium as a function of strain and normalised particle radius, (after Humphreys 1979a).

rotation at the particles indicate that plastic relaxation by the formation of both prismatic loops and rotated deformation zones occurs. For larger particles, only lattice rotation occurs.

The local lattice rotations were found to occur about an axis perpendicular to both the primary Burgers vector and the primary slip plane normal. The maximum rotation (θ_{max}) which was found to be close to the particle-matrix interface, and to decrease with distance (x) from the particle as shown in figure 8.12, dropped to a level indistinguishable from that of the matrix at distances of the order of the particle diameter ($x \sim d$). The data were found to fit an empirical equation of the form

$$\tan \theta = \tan \theta_{max} \exp \left(\frac{-c_1 x}{d} \right) \tag{8.15}$$

where c_1 is a constant equal to 1.8.

For particles of diameter greater than $\sim 2.5 \mu m$, where lattice rotations were the only relaxation mechanism, θ_{max} was found to be independent of particle size, and to depend on the shear strain (s) as shown in figure 8.13. The relationship between θ_{max} and s was of the form

$$\theta_{max} = c_2 \tan^{-1} s \tag{8.16}$$

where c_2 is a constant of the order of unity.

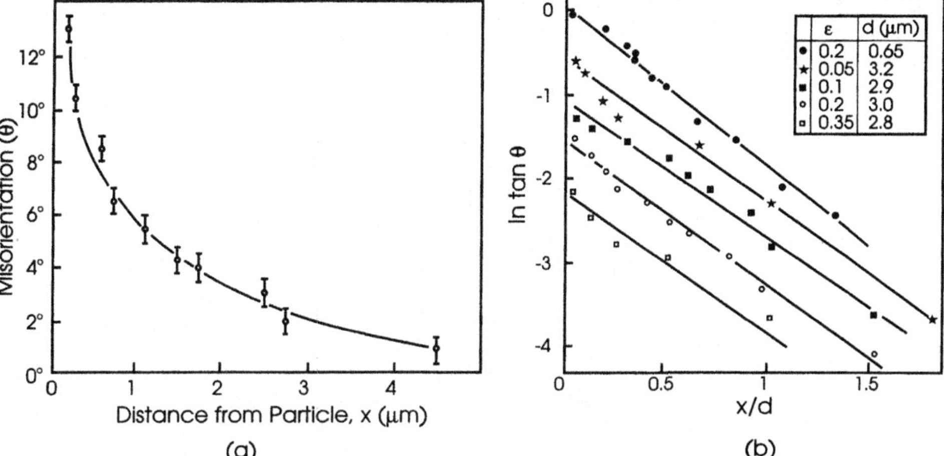

Fig. 8.12. Lattice misorientation at silicon particles in deformed aluminium crystals as a function of distance (x) from the interface. a) Measurements for a single particle. b) Data for various particles and strains plotted according to equation Fig. 8.15. (Humphreys 1979a).

For elongated particles, it was found that the maximum misorientation occurred at the ends of the particle. For particles of diameter less than $\sim 2.5\mu m$, the misorientation (θ_{max}) was found to be a function of both strain and particle size as shown in figure 8.11. For such particles, the maximum misorientation (θ'_{max}) was found to fit the empirical equation

$$\theta'_{max} = 0.8\ \theta_{max}\ (d-0.1)^{1/5} \tag{8.17}$$

where **d** is the particle diameter expressed in microns.

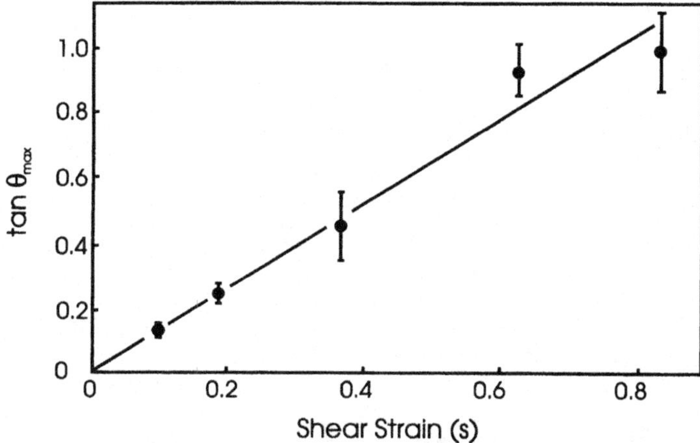

Fig. 8.13. The mean maximum misorientation as a function of shear strain for particles of diameter greater than $2.5\mu m$, (Humphreys 1979a).

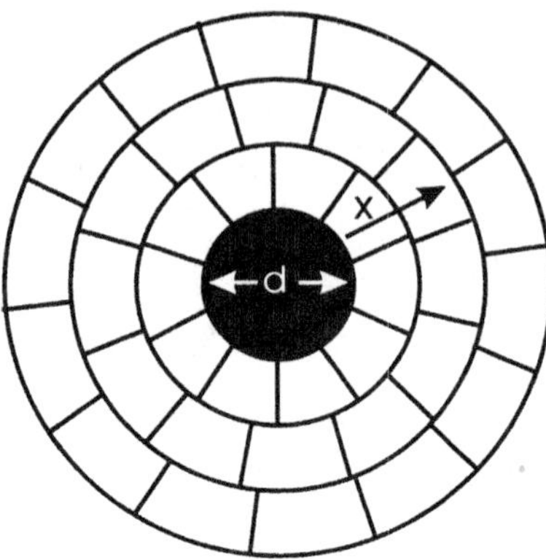

Fig. 8.14. Simplified model of a deformation zone.

The detailed shape of the deformation zones described above was not determined. However, If we assume that it consists of concentric spherical low angle boundaries as shown in figure 8.14, then the distribution of dislocations in the deformation zone is directly related to the orientation distribution as given by equation 8.15.

The orientation gradient at a distance **x** from the surface of the particle is given by

$$\frac{d\theta}{dx} = \frac{-\,c_1 \tan \theta_{max}}{d} \,.\, \cos^2\theta \,.\, \exp\left(\frac{-c_1\, x}{d}\right) \tag{8.18}$$

and the dislocation density at distance x, assuming the low angle boundaries to comprise square networks of dislocations is

$$\rho = \frac{2\, c_1 \tan \theta_{max}}{b\, d} \,.\, \cos^2\theta \,.\, \exp\left(-\frac{c_1\, x}{d}\right) \tag{8.19}$$

It is interesting, but probably fortuitous, that if we take θ_{max} as given by equation 8.16, then the total length of dislocation per particle, obtained by integrating equation 8.19 over the volume of the deformation zone is very close to $\pi d^2.s/b$, the value calculated for the geometric dislocations in equation 8.13.

8.2.4.2 Polycrystals and rolled single crystals

There is little quantitative data for deformation zones in polycrystals or rolled specimens, mainly because of the severe experimental difficulties inherent in such measurements. The deformation zone in a rolled alloy is generally found to be elongated in the rolling direction as shown in figure 8.15. Regions of highly misoriented small subgrains of diameter $<0.1\mu$m are found close to the particle, but further away, the subgrains are elongated and distorted by the presence of the particle (Humphreys 1977, Hansen and Bay 1981).

Few systematic studies of misorientations have been made. However, it is commonly found that even after rolling deformations of up to 90%, the misorientation between the matrix and the deformation zone is typically of the order of 30°- 40° for equiaxed particles (e.g. Gawne

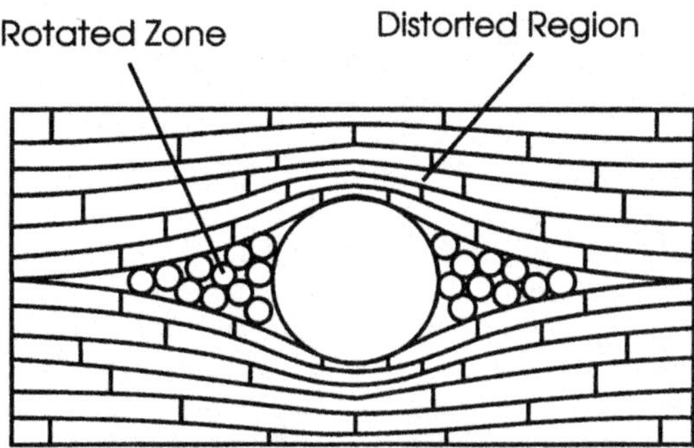

Fig. 8.15. A deformation zone in a rolled polycrystal, (After Porter and Humphreys 1979).

and Higgins 1969, Humphreys 1977, Herbst and Huber 1978, Bay and Hansen 1979), and for elongated fibres (Liu et al. 1989), the misorientations are of a similar magnitude.

The misorientations quoted above, which were obtained from electron diffraction measurements of regions adjacent to and remote from a particle were typically measured as **angle-axis pairs**, defined as the minimum angle of rotation (and the rotation axis) necessary to bring the crystals into register (§3.2). Because of the symmetry of cubic crystals, this is not necessarily the same as the **physical misorientation** which has occurred during deformation, and it should be noted that a mean misorientation of 40° would be measured for **randomly** oriented crystals (§3.2). From the experimental measurements quoted above, we can therefore only deduce that the physical misorientations are at least as large as the measured values, but because of the crystal symmetry they could be considerably larger.

Use of a TEM microtexture technique in which the orientation spread within a small volume adjacent to a particle is measured, allows the physical rotation of the regions adjacent to a

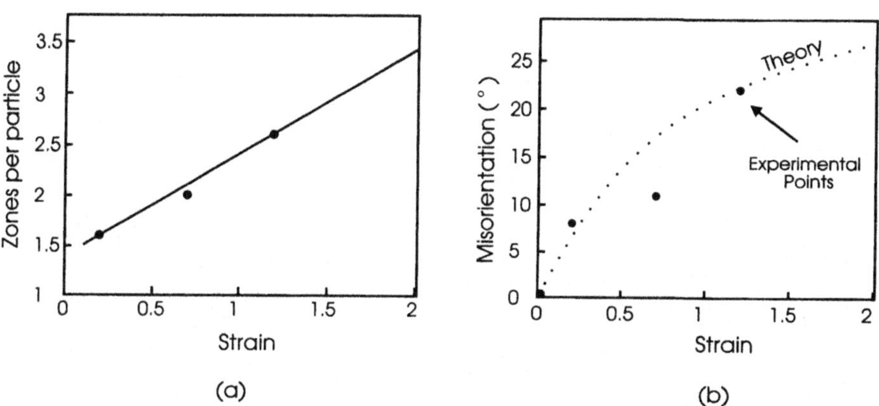

Fig. 8.16. The misorientation at silicon particles (d > 1.5μm) in polycrystalline aluminium deformed in compression. a) The number of deformation zones (>2°) per particle. b) The maximum misorientation of the zones. The theory curve corresponds to equation Fig. 8.22, (experimental data from Humphreys and Kalu (1990).

particle to be determined (Humphreys 1983). Results of the use of this technique to study the misorientation within deformation zones at particles in polycrystalline aluminium deformed in uniaxial compression are shown in figure 8.16. It was found that in many cases more than one deformation zone was formed per particle, and that the magnitude of the rotations increased with both strain and particle size.

8.2.4.3 Modelling the formation of the deformation zone

The formation of a deformation zone at a micron-sized particle in a polycrystal during rolling is a process which occurs at a scale which is difficult to model successfully. The process is too complex to be analyzed in terms of individual dislocation reactions, but as yet, larger scale modelling has not been able to incorporate the subtle fine-scale features of crystal plasticity needed to account for the crystallographic aspects of the deformation.

For the case of deformation on a single slip system, the following model based on the ideas of Ashby (1966) accounts reasonably well for the local lattice rotations. Consider a particle within a crystal as shown in figure 8.17a. During deformation, the particle will be surrounded by Orowan loops (fig 8.17b). The stress may be relaxed and the loop absorbed into the particle-matrix interface if the particle rotates and if a suitable array of secondary dislocations is generated as shown in figure 8.17c. For a shear strain of s, the number of gliding dislocations of Burgers vector b, encountering the particle (of diameter d) is sd/b. If these dislocations are absorbed into the interface and form a low angle boundary, then the spacing of the dislocations in the boundary will be b/s. The misorientation (θ), which equates to the rotation of the particle is then (equation 3.4) approximately equal to s, which is close to the value obtained from figure 8.13 and from equation 8.16.

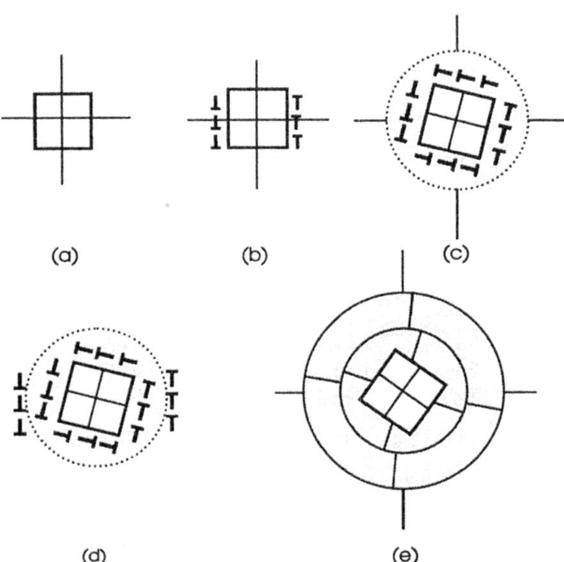

Fig. 8.17. The formation of a deformation zone under conditions of single slip. The secondary dislocations are represented by the shaded region. (After Humphreys and Kalu 1990).

The model of figure 8.17 has been extended (Humphreys and Kalu 1990) so as to predict the size and orientation of the deformation zone. The dislocations formed at the particle will make it more difficult for further primary dislocations to penetrate to the particle surface, i.e. they effectively increase the size of the particle as shown in figure 8.17d. Therefore relaxation of the primary loops results in a further rotation of the volume within the hard (shaded) shell. This process is repeated as deformation continues, resulting in the structure shown in figure 8.17e. The material close to the particle has been rotating throughout the deformation and is therefore misoriented by $\sim s$ with respect to the matrix, whereas the material in the outer skin of the zone has been rotating for only a small part of the total strain and is therefore less misoriented. In this way, an orientation gradient is formed, with the maximum misorientation being close to the particle surface. This model can therefore account for the magnitude of θ_{max} and its dependence on strain as shown in figure 8.13, and can qualitatively account for the variation of θ with distance from the particle shown in figure 8.12.

Humphreys and Kalu (1990) attempted to quantify this model by taking the dislocation density generated at the particle as ρ_G of equation 8.14 and then using a forest hardening model to calculate the dislocation density necessary to prevent penetration by the glide dislocations. They calculated the radius of the deformation zone (Z) as a function of the shear strain (s) at a particle of radius r as:

$$Z^2 = r^2 + \frac{r^3 s}{C b} \tag{8.20}$$

and the rotation at any position R inside the zone as

$$\theta = s - \frac{C b (Z^2 - r^2)}{r^3} \tag{8.21}$$

where C is a constant of value ~ 800

This model fits the experimental data of figure 8.12 and equations 8.16 to 8.19 reasonably well at low to medium strains, but breaks down at large strains when Z becomes unreasonably large.

Although Humphreys and Kalu (1990) extended the model to polycrystalline deformation, and were able to predict the formation of several deformation zones at particles in agreement with the results of figure 8.16a, a model involving significant particle rotations about an axis related to the crystallography of the slip system is unlikely to be applicable to polycrystals deforming on many slip systems, and there is little evidence of the rotation of equiaxed particles during such deformation. Another factor which suggests that this model may be applicable to single slip but not polyslip deformation is the magnitude of the misorientation within the deformation zones. The misorientations (θ_{max}) predicted by equation 8.16 are consistent with the values measured for single crystals shown in figure 8.13, but if this equation is used to calculate misorientations in highly deformed specimens such as those reduced 90% by cold rolling, misorientations much larger than the measured values are predicted.

Humphreys and Ardakani (1994) suggested an alternative approach. The matrix remote from the particle within a grain in a polycrystal is assumed to deform on up to five slip systems

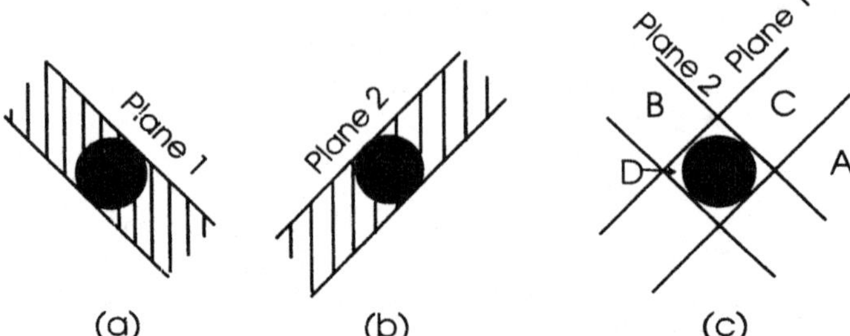

Fig. 8.18. Inhibition of slip in the vicinity of a large particle,
(after Humphreys and Ardakani 1994).

(§2.3.5). The non-deforming particle prevents matrix slip propagating near the particle. Therefore very close to the particle there will be a region which is essentially undeformed. Further from the particle, matrix slip can operate more freely. There is therefore a **strain gradient** from 0 at the particle to ϵ (the matrix strain) far from the particle. The strain gradient will not be uniform or symmetrical about the particle and will depend on the geometry of the imposed strain. The slip blockage or slip shadowing will not be uniform but will vary for the different slip systems. Consider the particle shown in figure 8.18. If a single slip system operates in the matrix with a slip plane 1 as shown in figure 8.18a, then the particle will impede slip within the shaded zone. In a similar manner, a second slip system with plane 2, would be restricted in the shaded region shown in figure 8.18b. If both systems operate, then the situation will be as shown in figure 8.18c. Here we can identify four regions of different slip activity. At A, slip is not affected by the particle and the strain will be the imposed strain ϵ, which comprises that of the two slip systems 1 and 2. i.e. $\epsilon = \epsilon_1 + \epsilon_2$.

At B, slip on system 1 is blocked by the particle and the strain here is given approximately by $\epsilon - k\epsilon_1$, where k is close to 1 near the particle and 0 far from the particle. At C, slip on system 2 is blocked by the particle and the strain here is given approximately by $\epsilon - k\epsilon_2$. At D, both systems are blocked, and the strain is $\epsilon - k_1\epsilon_1 - k_2\epsilon_2$. i.e. the local strain approaches zero close to the particle.

During deformation, the change of orientation of a grain or crystal depends on the strains on the active slip systems, and as both these parameters are seen to vary in the vicinity of the particle, so we expect the orientation close to the particle to vary. With reference to figure 8.18c, in region A, where slip is not blocked, the orientation change of the grain during deformation may be calculated from crystal plasticity theory e.g. using the Taylor model (§2.3.5). In regions B and C, if we assume that the slip activity is similar to that at A except that system 1 or 2 does not operate, the crystal will undergo a different reorientation. In region D, where no slip has occurred, we can presume that no orientation change has occurred. On this model, **the orientation of some of the material near the particle will be close to that of the undeformed grain**, whilst far from the particle the grain reorients according to crystal plasticity theory. In the intermediate regions, varying amounts of slip occur, producing various misorientations.

By using a Taylor analysis, modified to take account of the blockage of slip systems close to the particles, Humphreys and Ardakani (1994) calculated the orientation changes at particles and found some agreement with experimental results for single crystals deformed in plane strain compression. If this model is used to calculate the average misorientation of the deformation zones (θ_{max}) for a large number of randomly oriented grains, then the theoretical curve shown in figure 8.16 is obtained. These data, which compare reasonably well with the experimentally measured misorientations in figure 8.16 approximate to the empirical relationship

$$\theta_{max} = \theta_0 \left(\frac{\epsilon}{1 + \epsilon} \right) \qquad (8.22)$$

where $\theta_0 = 40°$.

It should be emphasized that this is a very simple and approximate model which should be used with caution. Further development of this type of model requires more accurate determination of the slip activity adjacent to the particles by methods such as finite element analysis.

8.2.4.4 Modelling the structure of the deformation zone

In order to understand recrystallization in the vicinity of large particles, it is necessary to know the distribution of dislocations and orientations within the deformation zone. The model for zone formation in single crystals discussed above makes some attempt to predict the distribution of both dislocations and orientations, but is clearly an oversimplification. The model for polycrystals, considers the development of orientation spread within a deformation zone but has not been developed sufficiently to predict the spatial distribution of orientations or dislocations.

At least two other models of the deformation zone have been proposed. Sandström (1980) calculated the dislocation distribution within a zone using a model developed by Argon et al. (1975), in which strain incompatibilities at the surface of a particle are resolved by the punching of prismatic loops into the surrounding matrix (equation 8.14). The local work hardening due to the accumulation of loops hinders their glide, and thus the dislocation density decreases with distance from the particle. The model predicts no orientation gradients within the zone and is therefore of limited applicability.

Ørsund and Nes (1988) propose a deformation zone which is spherical as in figure 8.14, and in which the subgrain size increases linearly outwards from the particle. The misorientation of adjacent subgrains is assumed to be constant and therefore the misorientation decreases with distance from the interface in a manner similar to that shown in figure 8.12. The model was quantified by fitting it to the experimental data of figure 8.12. However, because the distribution of dislocation density in a zone with this geometry is directly related to the orientation gradient, as seen from equations 8.18 and 8.19, this model is effectively a restatement of the empirical equations 8.15 to 8.18.

In the absence of a sound theoretical model for a deformation zone in a deformed polycrystal, we will for the practical purposes of examining the recrystallization behaviour at large particles, adopt the following model.

(i) We will assume the zone to be centred on the particle and spherical as in figure 8.14.

(ii) The average maximum misorientation (θ_{max}) for particles of diameter greater than 2.5μm will be that obtained by the plasticity model and given by equation 8.22. For smaller particles, θ_{max} is reduced according to equation 8.17.

(iii) The distributions of orientation and dislocations will be those found for zones in single crystals and given by equations 8.18 and 8.19. Both are a maximum at the interface and decrease monotonically to values characteristic of the matrix at distances from the interface of $\sim$ d.

8.3 THE OBSERVED EFFECTS OF PARTICLES ON RECRYSTALLIZATION.

Before we discuss the models and mechanisms of recrystallization in particle-containing alloys, we will look briefly at some of the experimental measurements and in particular at the parameters which appear to influence the recrystallization behaviour. It is not our intention to undertake a comprehensive review of the large number of experiments which have been carried out, but to emphasise what appear to be the key points which have emerged from these investigations. For a more detailed survey of the earlier literature, the reader is referred to Cotterill and Mould (1976) and to the reviews of Hansen (1975), Hornbogen and Köster (1978) and Humphreys (1979b).

8.3.1 The effect of the particle parameters

Both the recrystallization kinetics and the final grain size depend strongly on the **particle size** and the **interparticle spacing**, as was first demonstrated by Doherty and Martin (1962). It is difficult to separate the effects of these two parameters because there are few investigations in which they have been independently varied. It has frequently been found that by comparison with a single-phase alloy, recrystallization is retarded or even completely inhibited by closely spaced particles. With a sufficiently small spacing of thermally stable particles, it is possible to preserve the deformed/recovered microstructure up to the melting temperature of the matrix as shown in figure 8.19. The retention of the dislocation substructure at high temperatures provides a strengthening mechanism additional to that of the particles in dispersion hardened alloys for high temperature structural applications (e.g. Kim and Griffith 1988).

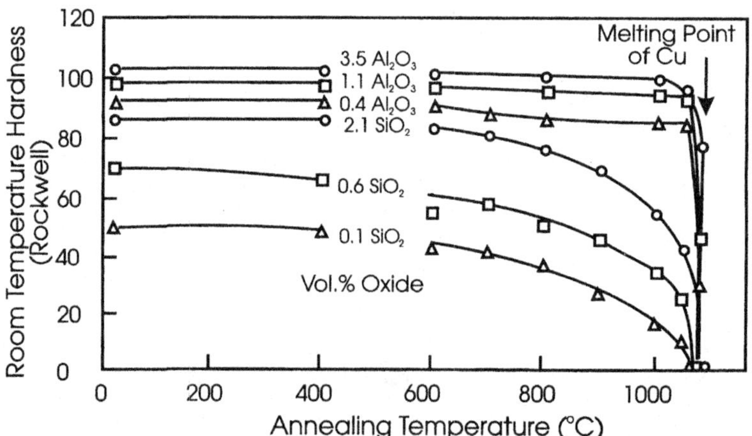

Fig. 8.19. The room temperature hardness of extruded and annealed dispersion-strengthened copper alloys, (after Preston and Grant 1961).

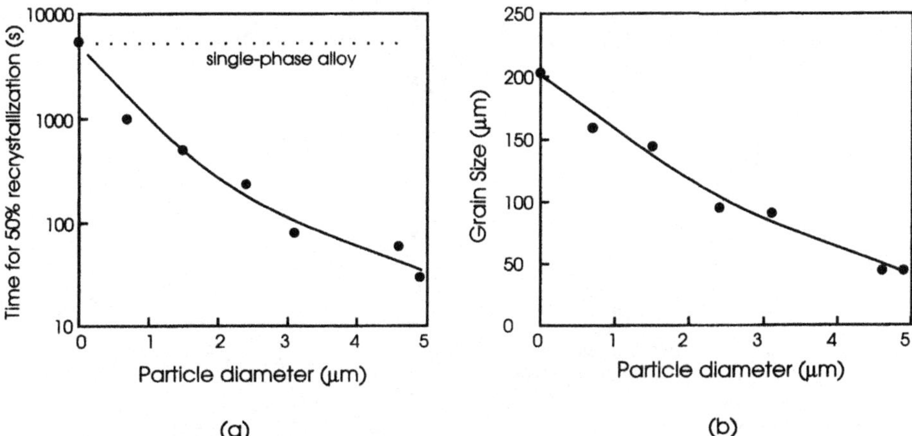

Fig. 8.20. The effect of particle size on recrystallization in Al-Si alloys reduced 50% by cold rolling and annealed at 300°C. a) The time for 50% recrystallization, b) The grain size after recrystallization, (data from Humphreys 1977).

By comparison, alloys containing particles larger than ~1μm and which are widely spaced, show accelerated recrystallization compared to the matrix as shown in figure 8.20a.

By varying the particle size and spacing, Doherty and Martin (1962) showed that acceleration or retardation of recrystallization could be produced in the same alloy as shown in figure 8.21a.

Figures 8.20b and 8.21b show that the final grain size becomes larger as recrystallization becomes more retarded. Figure 8.22 shows a typical recrystallized microstructure in an alloy

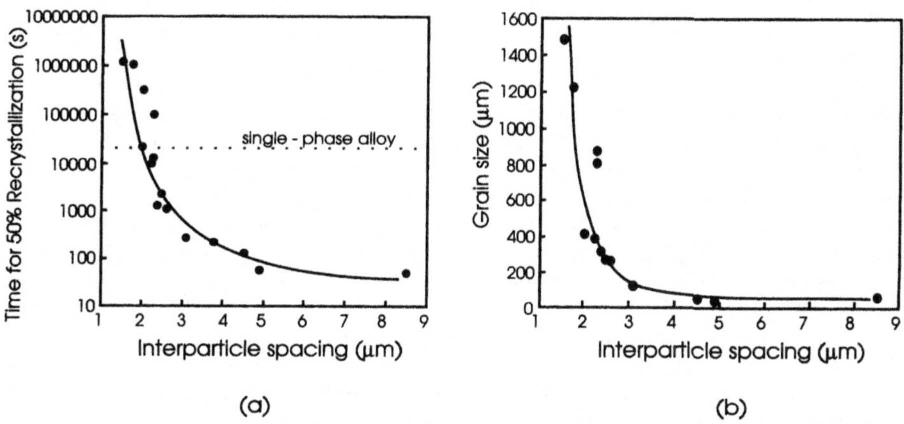

Fig. 8.21. The effect of interparticle spacing on the recrystallization of Al-Cu single crystals reduced 60% by rolling and annealed at 305°C. a) The time for 50% recrystallization, b) The grain size after recrystallization, (data from Doherty and Martin 1964).

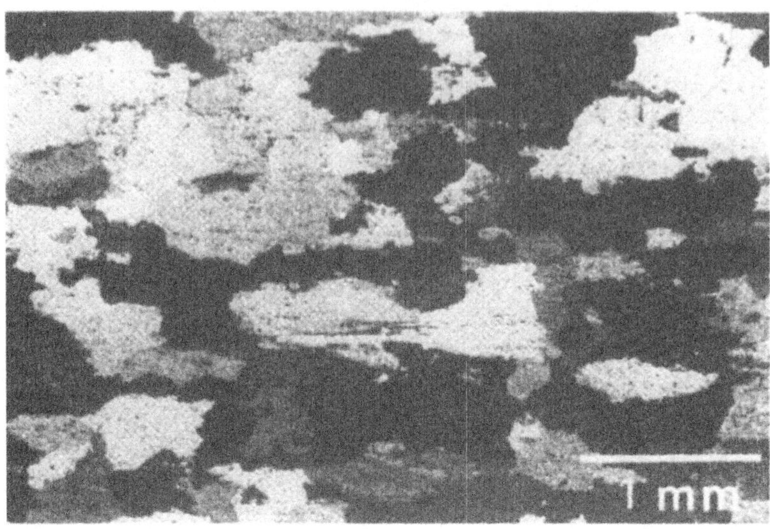

Fig. 8.22. Micrograph of cold rolled and recrystallized Al-Al$_2$O$_3$, (Hansen and Bay 1981).

where recrystallization is retarded by small particles.   The pinning of the grain boundaries by particles leads to a characteristic ragged boundary shape.

Analysis of a number of experimental investigations shows that the transition from accelerated to retarded recrystallization is primarily a function of the volume fraction (F_V) and particle radius (r), and that to a first approximation, retardation of recrystallization occurs when the parameter F_V/r is greater than $\sim 0.2\mu m^{-1}$ (Humphreys 1979b). If F_V/r is less than this then it is often found that recrystallization is accelerated by comparison with the particle-free matrix, even if the particles are too small for particle stimulated nucleation of recrystallization (PSN) to occur. A more recent analysis of published experimental data by Ito (1988) has suggested that a critical value of the interparticle spacing, which is seen from equations 8.2 and 8.3 to be equivalent to $F_V^{1/2}/r$, may be a better indicator of the transition, although the experimental data are not good enough to distinguish between these two possibilities. Acceleration of recrystallization is attributed both to the increased driving force arising from the dislocations generated by the particles during deformation (§8.2.2) and to PSN (§8.4). Accelerated recrystallization in alloys containing small particles is less commonly found in aluminium alloys than in copper alloys (e.g. Hansen 1975). This is probably because the large amount of dynamic recovery in most aluminium alloys removes many of the geometrically necessary dislocations created at the particles. It will be shown in §8.4 that the condition for PSN is that the particle diameter should be greater than $\sim 1\mu m$. The criteria for acceleration/retardation and PSN indicate that the effect of volume fraction and particle size on recrystallization will be as shown schematically in figure 8.23.   The position of the line AB depends on the amount of dynamic recovery in the material and will move to the right for aluminium and to the left for copper. The figure is of course an over-simplification and, as will be discussed later in this chapter, there are other parameters which also have an effect on the recrystallization mechanisms and kinetics.

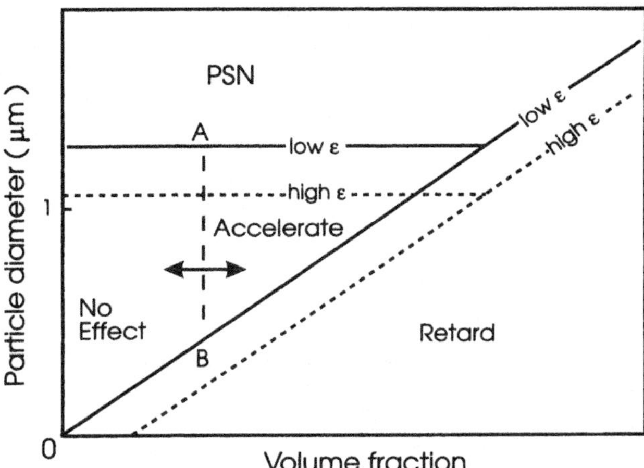

Fig. 8.23. Schematic diagram of the effect of particle size, volume fraction and prior strain on recrystallization kinetics and mechanism.

8.3.2 The effect of strain

There is considerable evidence from research on copper and aluminium alloys that the transition from retarded to accelerated recrystallization is affected by strain, and that an alloy which shows retarded recrystallization after small strains may show accelerated recrystallization after larger strains (fig 8.23). An example of this effect in copper alloys is seen in figure 8.7b. It is explained by an increasing strain raising the driving pressure but not the retarding (pinning) pressure for recrystallization.

8.3.3 The effect of particle strength

8.3.3.1 Coherent particles
Alloys of copper and nickel containing weak equiaxed coherent particles ($F_V/r \sim 2\mu m^{-1}$) which deform at the same rate as the matrix are found to show retarded recrystallization (Phillips 1966). During deformation, the particles become ribbons, but on annealing, they break-up and spheroidise. In such alloys, no geometrically necessary dislocations are produced and therefore there is no increase in the driving force for recrystallization. In addition, the effective interparticle spacing decreases during deformation thus increasing the effectiveness of the particles in pinning boundaries. It is therefore to be expected that this type of alloy should exhibit retarded recrystallization more readily than one containing non-deformable particles.

8.3.3.2 Pores and gas bubbles
Pores or gas bubbles exert a pinning effect on boundaries in a similar manner to second-phase particles, and Bhatia and Cahn (1978) have reviewed their effect on recrystallization and grain growth. As no geometrically necessary dislocations are formed during deformation, pores and gas bubbles always tend to retard recrystallization and these may also be dragged by the moving boundaries as discussed in §9.4.5. A dispersion of pores or gas bubbles has been found to inhibit recrystallization in many metals including platinum

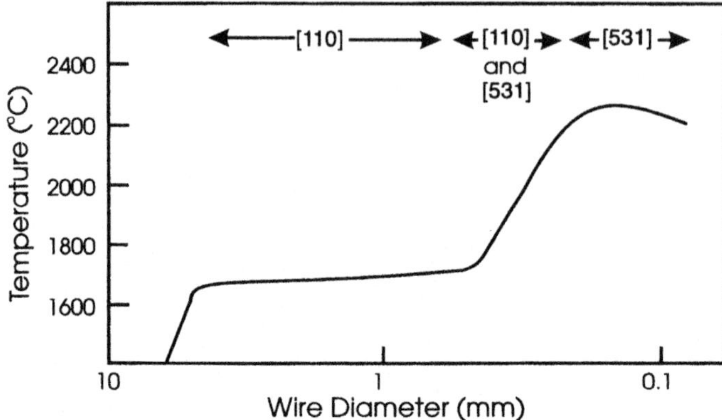

Fig. 8.24. Increase in recrystallization temperature and change in texture, with increasing strain in drawn tungsten wire, (after Thompson-Russell 1974).

(Middleton et al. 1949), aluminium (Ells 1963), copper (Bhatia and Cahn 1978) and tungsten (Farrell et al. 1970). Thomson-Russell (1974) has shown that in doped tungsten, as the strain is increased, the bubble dispersion is progressively refined leading to an increase in recrystallization temperature as shown in figure 8.24. Control of recrystallization in this way is a key part of the processing of tungsten filaments for incandescent light bulbs, as discussed by Dillamore (1978b).

8.3.3.3 Small plate-shaped particles

Even particles of a hard phase will fracture if small enough, because of the increasing stress due to the dislocations (equation 8.8 and fig 8.11), particularly if they are in the form of plates. Kamma and Hornbogen (1976), investigating recrystallization in steels containing carbide plates (200x10nm) found that for rolling reductions of up to 90%, the plates fractured, resulting in the formation of deformation bands. Recrystallization was nucleated at the intersection of these deformation bands, leading to rapid recrystallization. However, at larger strains, the refinement of the dispersion led to more homogeneous deformation, increased particle pinning and to slower recrystallization as shown in figure 8.25.

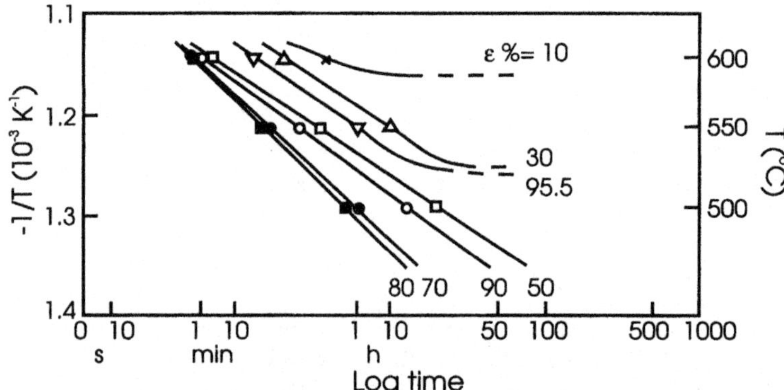

Fig. 8.25. The time to initiate recrystallization as a function of temperature (T) and cold work (ϵ%) in a steel containing a dispersion of 200x10nm carbide plates, (Kamma and Hornbogen 1976).

8.3.4 The effect of microstructural homogenisation

There has been considerable discussion in the literature as to whether or not "homogenisation of slip by a fine dispersion of particles" contributes to the retardation of recrystallization by removing the lattice curvatures necessary for nucleation. This is now thought to be unlikely.

As discussed in §8.2.2, there is evidence that the presence of small, and very closely spaced particles affects the cell or subgrain structure resulting in smaller, less misoriented cells or subgrains. However, it is now clearly established (§6.6.4) that nucleation does not occur at cells or subgrains in regions of uniform orientation but at microstructural heterogeneities associated with large strain gradients. These sites, e.g. grain boundaries, deformation bands, shear bands and deformation zones at large particles, are still present in alloys showing retarded recrystallization (e.g. Chan and Humphreys 1984b). However, their effectiveness is reduced by the pinning effects of particles on both high and low angle grain boundaries, and the recrystallization kinetics are determined mainly by the fine particle dispersion.

We therefore conclude that the effects of particles on recrystallization may be analyzed primarily in terms of particle stimulated nucleation and boundary pinning, and will consider these mechanisms in further detail.

8.4 PARTICLE STIMULATED NUCLEATION OF RECRYSTALLIZATION

PSN has been observed in many alloys, including those of aluminium, iron, copper and nickel and is usually only found to occur at particles of diameter greater than approximately $1\mu m$ (Hansen 1975, Humphreys 1977), although a lower limit of $0.8\mu m$ was inferred from the indirect measurements of Gawne and Higgins (1971) on Fe-C alloys.

In most cases a maximum of one grain is found to nucleate at any particle, although there is evidence that multiple nucleation occurs at very large particles. This is clearly seen in figure 8.26, one of the first micrographs to demonstrate PSN.

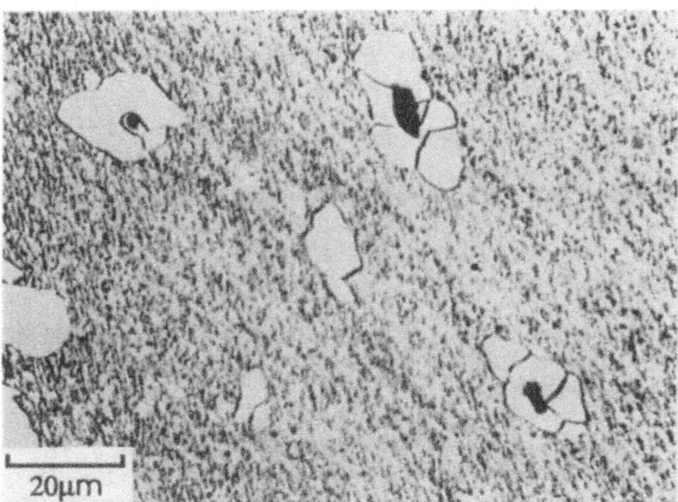

Fig. 8.26. Particle stimulated nucleation of recrystallization at oxide inclusions in iron, (Leslie et al. 1963).

There are four particularly significant aspects of particle stimulated nucleation of recrystallization.

(i) Many **industrial alloys**, particularly those of iron and aluminium, contain large second-phase particles and thus PSN is a likely recrystallization mechanism in many engineering alloys.

(ii) Unlike most other recrystallization mechanisms the nucleation sites are well defined regions of the microstructure which can be controlled by alloying or processing. Thus the number of potential nuclei and hence the **recrystallized grain size** may be controlled.

(iii) The orientations of the recrystallization nuclei produced by PSN will in general be different from those produced by other recrystallization mechanisms. We thus have the potential to exercise some control over the **recrystallization texture** by controlling the amount of PSN (§10.6, §12.2).

(iv) Because the interaction of dislocations and particles is temperature dependent, PSN will only occur if the prior deformation is carried out below a critical temperature or strain rate. This effect is considered in §11.6.4.

8.4.1 The mechanism of PSN

Although specimens which are annealed to produce recrystallization nuclei and which are subsequently examined metallographically, can provide information about the kinetics of recrystallization and the orientation of the new grains, there is little that can be deduced from these specimens about the mechanisms of nucleation. Most of the direct evidence has come from in-situ observations in the HVEM. From a study of recrystallization at particles in rolled aluminium, Humphreys (1977) concluded that :-

(i) Recrystallization originates at a pre-existing subgrain within the deformation zone, but not necessarily at the particle surface.

(ii) Nucleation occurs by rapid sub-boundary migration.

(iii) The grain may stop growing when the deformation zone is consumed.

Later in-situ work (Bay and Hansen 1979), together with work on bulk annealed specimens (Herbst and Huber 1978) has supported these general conclusions.

The electron micrographs of figure 8.27, part of an in-situ HVEM annealing sequence, show the occurrence of particle-stimulated nucleation in aluminium originating in the arrowed region close to the interface.

The kinetics of nucleation, as determined by in-situ HVEM annealing of Al-Si, are illustrated in figure 8.28. The rapid annealing of the deformation zone, shown in figure 8.28a is due to the presence of a large orientation gradient and small subgrain size (§5.5.3.4), compared to the matrix. The drop in the maximum misorientation within a deformation zone shown

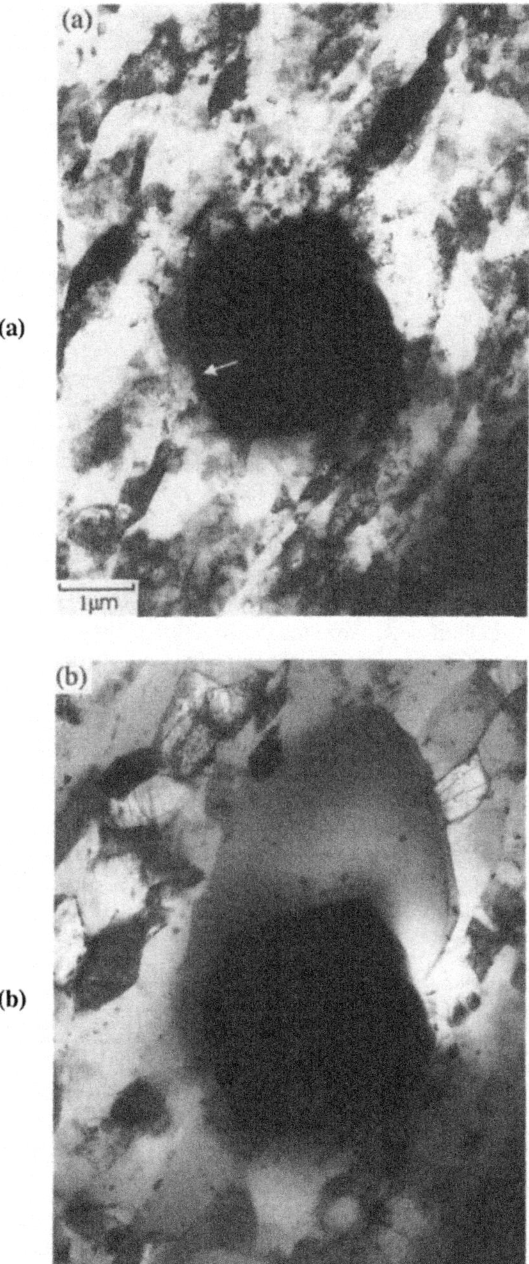

Fig. 8.27. In-situ HVEM annealing sequence of Al-Si. Note the very small subgrain size in the deformation zone above the particle. a) Recrystallization originates in the arrowed region of the deformation zone close to the particle. b) Recrystallization has consumed the deformation zone, (Humphreys 1977).

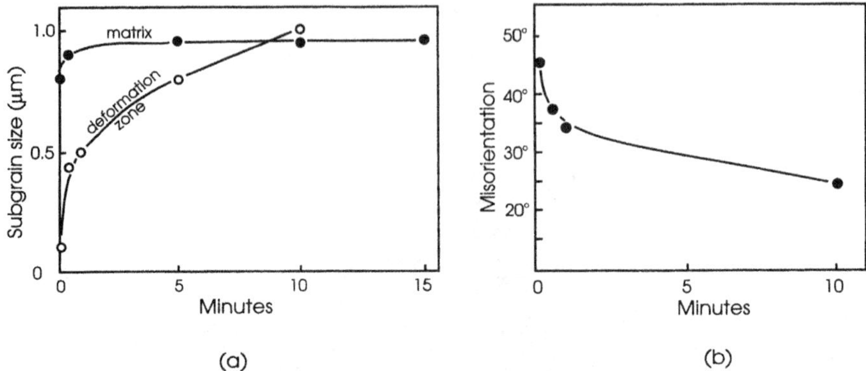

Fig. 8.28. Changes in the subgrain size and misorientation within the deformation zone of Si particles in Al as determined by in-situ HVEM annealing at 250°C. a) Growth kinetics. b) The maximum misorientation within the deformation zone, (Humphreys 1980).

in figure 8.28b is consistent with the observation that the nucleus may not originate in the region of highest misorientation at the interface, but elsewhere in the deformation zone. There is no evidence that PSN occurs by a micro-mechanism different from nucleation of recrystallization in regions of heterogeneous deformation in single-phase materials, and the models of PSN are related to the models of recrystallization at transition bands discussed in §6.6.4.2.

Using the largely empirical model of the deformation zone discussed in §8.2.4 we can examine the occurrence of PSN more quantitatively. The deformation zone has a gradient of both orientation and dislocation density as given by equations 8.18 and 8.19. We will assume that recrystallization originates by the growth of subgrains within the deformation zone, and as shown in figure 8.29, also consider the possibility that this may occur at the particle surface (A), in the centre of the zone (B) or at the periphery (C).

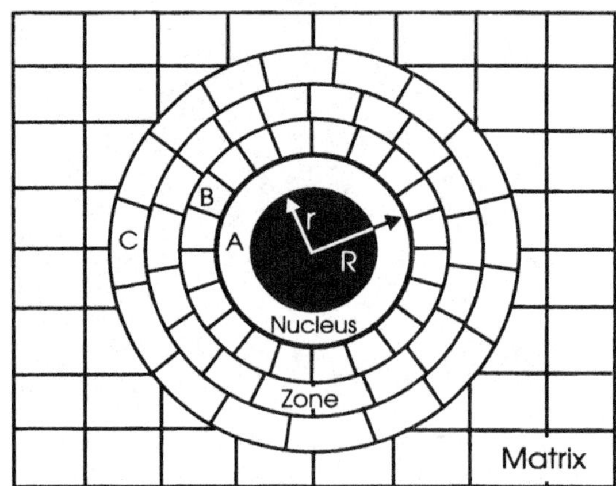

Fig. 8.29. Growth of a nucleus within the deformation zone.

8.4.1.1 Formation of a nucleus within the zone

A necessary criterion for the formation of a nucleus is that the maximum misorientation within the deformation zone (θ_{max}) is sufficient to form a high angle boundary with the matrix, and the conditions for nucleation therefore depend on both the particle size (equation 8.17) and the strain (equation 8.22), and this condition is plotted in figure 8.30b.

From the consideration of subgrain growth in an orientation gradient (§5.5.3.4) and equation 5.38, we see that to a first approximation, the growth velocity of a subgrain in a region of large orientation gradient is independent of subgrain size and is proportional to the magnitude of the orientation gradient. As a subgrain grows in an orientation gradient, it increases its misorientation with neighbouring subgrains. When this misorientation has reached that of a high angle grain boundary (e.g. 10-15°) we can say that a potential recrystallization nucleus has been formed. The size to which a subgrain must grow to achieve a high angle boundary will be inversely proportional to the orientation gradient, and combining these two factors suggests that the **time** for the formation of a nucleus will be approximately inversely proportional to the **square of the orientation gradient**. Therefore the formation of a nucleus close to the particle surface, where the orientation gradient is greatest (fig 8.12 and equation 8.15), will be strongly favoured.

8.4.1.2 Growth of the nucleus

Once a nucleus which is significantly larger than the neighbouring subgrains is formed, then the above model which is based on uniform subgrain growth is less appropriate than a model of the type shown in figure 8.29. The nucleus, which is assumed to have a high angle boundary of energy γ_b, is subjected to two pressures. A retarding pressure (P_C) due to its radius of curvature as given by equation 6.3, and a driving pressure (P_D) due to the dislocation density (ρ) or stored energy (E_D) of the material into which the nucleus is growing, as given by equation 6.2. The nucleus will grow if $P_D > P_C$, i.e

$$D \geq \frac{4\,\gamma_b}{\alpha\,\rho\,G\,b^2} \tag{8.23}$$

where $\alpha \sim 0.5$

The simplest analysis of PSN assumes that the critical condition for PSN is that at which the nucleus has a radius of curvature equal to half the particle diameter (i.e. $R = r$ in fig 8.29) and therefore D in equation 8.23 is replaced by the particle diameter (d). It is also assumed that at this stage the deformation zone has been consumed and that the driving force is provided by the matrix stored energy (E_D). The critical particle diameter (d_g) which will provide a nucleus able to grow beyond the deformation zone is therefore

$$d_g \geq \frac{4\,\gamma_b}{\alpha\,\rho\,G\,b^2} = \frac{4\,\gamma_b}{E_D} \tag{8.24}$$

Calculation of the stored energy or dislocation density in the matrix outside the deformation zone is more difficult as discussed in §2.2.1 and §8.2.2.1, although an estimate may be made by using equation 2.1 and taking the slip distance (L) as the spacing between the large particles. The dislocation density increases with strain and there will therefore be a **critical particle diameter (d_g) which is a function of strain,** below which the nucleus will not grow beyond the deformation zone and which therefore defines the conditions under which PSN

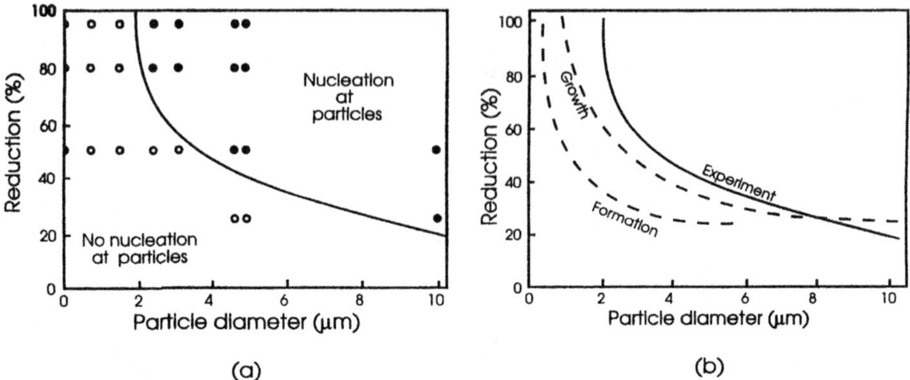

(a)   (b)

Fig. 8.30. a) The effect of rolling reduction and particle size on the occurrence of PSN (Humphreys 1977). b) Comparison of the experimental data with theoretical predictions of nucleus formation and growth.

will occur. The above analysis neglects the driving force provided by the higher dislocation density of the deformation zone, and if we take ρ within the zone as given by equation 8.19, then a more accurate estimate of the growth condition may be calculated.

Experimental measurements of the effects of particle size and strain on the occurrence of PSN in single crystals of Al-Si deformed by cold rolling, are shown in figure 8.30a. It is seen that the critical particle size for PSN (d_g) decreases with increasing strain as expected. In figure 8.30b we compare the experimental data (represented by the solid line of figure 8.30a) with the condition for **growth** calculated from equation 8.23, using the dislocation density in the deformation zone as given by equation 8.19. There is a reasonable fit between experiment and model, although as discussed above there are considerable uncertainties in the analysis. In figure 8.30b we also show the condition for the **formation** of a nucleus as discussed in §8.4.1.1. From a comparison of the criteria for formation and growth it is seen that following low temperature deformation **it is the condition for growth of the nucleus which is the determining factor for PSN**.

8.4.2 The orientations of particle-stimulated grains

As the nuclei are thought to originate within the deformation zones, their orientations will be restricted to those present in these zones (§8.2.4).

8.4.2.1 Experiments on single crystals

Experiments on single crystals of Al-Si deformed in tension (Humphreys 1977, 1980) have established that the orientation of nuclei lie within the orientation spread of the deformation zone. Annealing of lightly rolled crystals of these alloys resulted in a sharp recrystallization texture, which was rotated from the deformation texture by 30-40 degrees about a $<112>$ axis as shown in figure 8.31.

In the few experiments where the initial orientation of the single crystal has been measured before deformation, there is some correlation between the orientation of the PSN nuclei and

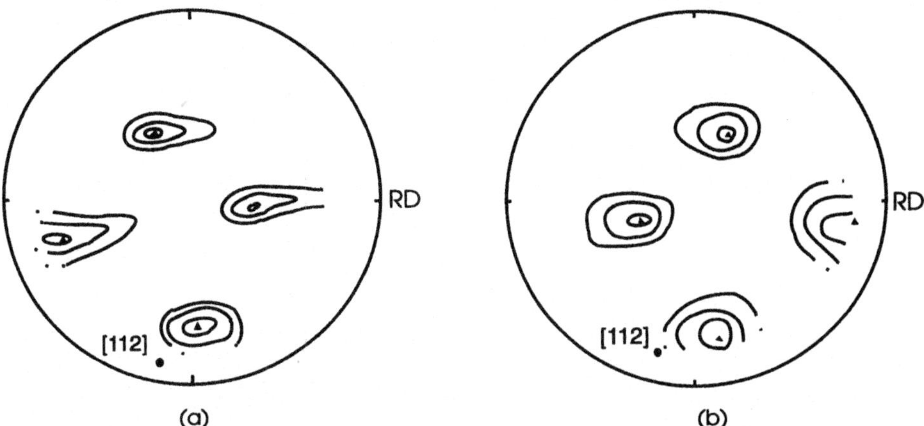

Fig. 8.31. 111 pole figures of (a) the deformation, and (b) the recrystallization texture of a 65% deformed single crystal of Al containing 4.6μm Si particles, (Humphreys 1977).

the initial orientation. Jack et al. (1989) investigated the annealing behaviour following uniaxial compression, of single crystals of [001] orientation, containing 2-3μm Si particles, and found that the orientation of the nuclei tended to lie within 20° of the [001] compression axis. Humphreys and Ardakani (1994) investigated the deformation behaviour in plane strain compression of (001)[110] oriented single crystals of aluminium containing 2-3μm particles of Si. They found that although the crystal reoriented by a 35°[$\bar{1}$10] rotation to a (112)[11$\bar{1}$] orientation during deformation, the orientation of PSN nuclei were within 16° of the initial crystal orientation. The results of both of these investigations are consistent with the deformation model of §8.2.4 in which it was suggested that regions close to the particles have orientations which are similar to those of the initial undeformed material, and in the latter investigation, nuclei orientations consistent with those predicted by the slip-shadowing model discussed in §8.2.4.3 were found.

8.4.2.2 Experiments on lightly deformed polycrystals
As part of a microtexture investigation into deformation zones in polycrystalline Al-Si alloys deformed in compression, Kalu and Humphreys (1988) determined the orientation of deformation zones at particles using the TEM microtexture technique and subsequently annealed the specimens in an HVEM until nucleation occurred. The microtextures were then remeasured and the results are summarised in the inverse pole figure of figure 8.32. The orientations of the nuclei (★) are seen to lie within the deformation zones, and nucleation was found to occur most frequently in the deformation zone associated with slip shadowing of the primary slip system.

8.4.2.3 Heavily deformed polycrystals
Experiments on heavily rolled polycrystals containing large particles have usually shown that the nuclei are either weakly textured or are approximately randomly oriented (e.g. Wassermann et al. 1978, Herbst and Huber 1978, Chan and Humphreys 1984c).

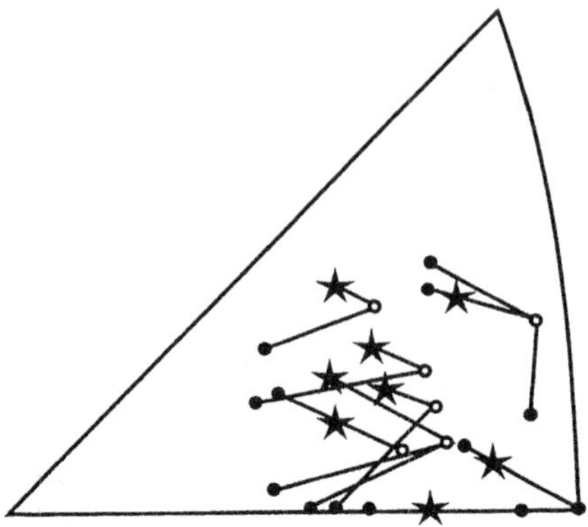

Fig. 8.32. The orientations of deformation zones (solid lines) and nuclei (★) at particles
of Si in compressed polycrystalline Al. The open circles are the matrix orientations,
(Kalu and Humphreys 1988).

In summary, we find that in the case of deformed single crystals, a fairly sharp PSN texture
occurs and this arises from the restricted range of orientations present in the deformation
zones. In the case of heavily deformed polycrystals, each grain, or coherent portion of a
grain will have several deformation zones, which, after large strains may have large
misorientations. The range of orientations for potential nuclei is therefore great and a single
grain will give rise to nuclei spread over a wide range of orientations, thus leading to a
weakening of the texture. **Although the nuclei are not randomly oriented with respect to
the deformation structure, the spread of orientations around what may already be a
weak texture, will give the appearance of random nuclei orientations.**

The arguments presented in §8.4.1 suggest that nuclei will tend to originate in the regions
close to the particle as the orientation gradients and hence the recovery rates are highest in
these regions. Experimental evidence, both microstructural (§8.4.1) and from microtextures
(fig 8.32), suggests that although this is generally observed, nuclei do in some cases originate
elsewhere in the deformation zones. It has been suggested that an alternative annealing
process might involve the growth of a matrix subgrain into the deformation zone (case C of
fig 8.29). This would produce a nucleus which was not highly misoriented with respect of
the matrix. Ørsund and Nes (1988) have presented some evidence of such a process in
deformed Al-Mn alloys. They found that after recrystallizing at high temperatures the
texture was weak, which is consistent with nuclei growing from the core of the deformation
zone, whereas after low temperature annealing, the texture was similar to the deformation
texture, which is consistent with nuclei originating in the outer regions of the deformation
zones. However, as there was little microstructural evidence of PSN originating in the
periphery of deformation zones, these conclusions must be regarded as tentative.

On the basis of the model of §8.2.4, it is predicted that a component of the PSN texture may
be close to that of the undeformed material. Although there is evidence cited above from
single crystal experiments to support this prediction, it has yet to be verified for rolled
polycrystals.

From §8.2.2.3 and figure 8.9, we note that in addition to the deformation zones formed at the particles, small deformation bands may also be associated with the particles. In the work quoted it was found that the orientation of these bands was unsuitable for significant recrystallization (Ardakani and Humphreys 1994). However, in the more general case we might expect recrystallization to originate in such regions.

8.4.2.4 The recrystallization texture

If PSN is the only nucleation mechanism, then the final recrystallization texture will, as discussed above, generally be weak. In alloys in which the number of particles is large, this is often the case. However, in alloys with low volume fractions of particles, it is found that other texture components, are strongly represented in the final texture. This indicates a contribution from grains originating at microstructural features other than particles, a point which is discussed further in §10.6.3.

8.4.3 The efficiency of PSN

An important industrial application of PSN is in controlling the grain size of recrystallized alloys, and in particular, the production of fine grained material. If each particle successfully nucleates one grain (i.e an efficiency of nucleation equal to 1), then the resultant grain size (**D**) will be directly related to the number of particles per unit volume and the grain size will be related to the volume fraction F_V of particles of diameter **d** by:

$$D \approx d\,F_V^{-1/3} \tag{8.25}$$

The grain size predicted using this criterion is shown in figure 8.33.

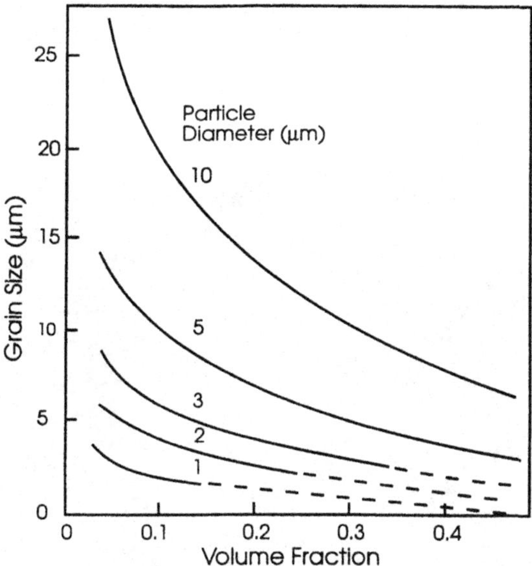

Fig. 8.33. The predicted grain size for a PSN efficiency of 1. The dotted lines show conditions under which pinning effects may prevent discontinuous recrystallization, (Humphreys et al. 1990).

If several grains nucleate at a particle, then an efficiency of greater than 1 may occur. However, as multiple nucleation generally only occurs for particles larger than 5-10μm, this is rarely achieved. If the particle size is close to the minimum (fig 8.30), then the nucleation efficiency is very low (Humphreys 1977, Wert et al. 1981, Ito 1988), and the predicted small grain size is not achieved. As nuclei are essentially in competition with each other, we only obtain a high efficiency if all nucleation events occur simultaneously (site saturation nucleation), and if the grains grow at similar rates. This condition may be met for alloys with large, widely spaced particles, but if the particles are closely spaced, or, if growth of nuclei is affected by a fine dispersion of particles, then the nucleation efficiency may decrease markedly as will be discussed further in §8.7.

It is likely that particles situated in regions where large strains and strain gradients occur, such as at grain boundaries, shear bands etc, may provide more favoured nucleation sites than other particles, thus reducing the overall efficiency of nucleation. There is some evidence of this effect in Al-Mn and Al-Fe-Si alloys (Sircar and Humphreys 1994).

8.4.4 The effect of particle distribution

There is evidence that nucleation occurs preferentially at pairs or groups of particles as shown in figure 8.34. This may occur even if the individual particles are below the critical size for nucleation. This has been detected statistically (Gawne and Higgins 1971), from metallographic observations (Herbst and Huber 1978) and from in-situ annealing (Bay and Hansen 1979). A detailed study of the distribution of recrystallization nuclei in Al-Si specimens containing particles of diameters close to the critical size (Koken et al. 1988), has also shown that under these conditions, nucleation is favoured in sites of particle clustering. The effect of the clustering of nuclei on recrystallization kinetics is discussed in §6.4.1.

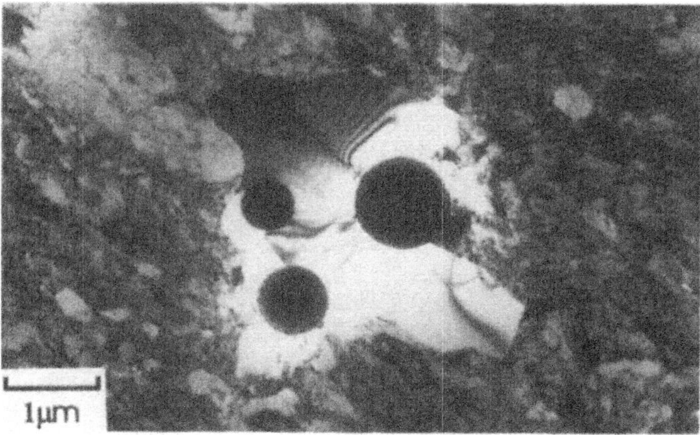

Fig. 8.34. Nucleation of recrystallization at a cluster of SiO_2 particles in nickel during an in-situ HVEM anneal, (Humphreys 1980).

8.5 PARTICLE PINNING DURING RECRYSTALLIZATION (ZENER DRAG)

8.5.1 Nucleation

In Chapter 5, we discussed the ways in which a dispersion of particles might hinder recovery and showed how pinning was particularly effective when the interparticle spacing and subgrain or cell sizes were similar. As the nucleation of recrystallization occurs via recovery, it is clear that particles may inhibit nucleation. There is little quantitative information on the sites for recrystallization nucleation in single-phase alloys, and consequently it is difficult at present to develop a satisfactory model for the effect of particles on recrystallization nucleation in alloys containing only small particles.

The effect of particles on recrystallization nucleation may be discussed in terms of the **particle-limited subgrain size** (§5.6.2). If this is parameter is small enough to prevent the growth of subgrains to a size such that a high angle grain boundary is formed, then nucleation will be suppressed.

If nucleation is taking place in a region of orientation gradient Ω, then the amount of subgrain growth necessary to form a high angle boundary of misorientation θ_m is $\sim \theta_m/\Omega$. The criterion for suppression of nucleation would then, to a first approximation be

$$D_{lim} < \frac{\theta_m}{\Omega} \tag{8.26}$$

where, as discussed in §9.4.2.3, D_{lim} will be given by equation 9.33 for particle volume fractions less than ~ 0.05, and equation 9.37 for larger volume fractions. Thus the criterion for inhibition is

$$\frac{F_V}{r} > \frac{4\,\alpha\,\Omega}{3\,\theta_m} \qquad (F_V < 0.05) \tag{8.27}$$

$$\tag{8.28} \frac{F_V^{1/3}}{r} > \frac{\beta\,\Omega}{\theta_m} \qquad (F_V > 0.05)$$

For alloys containing low volume fractions of particles, using $\alpha=0.35$, $\beta=3$ as discussed in §9.4.2.3, taking θ_m as $15°$ and $\Omega=5°\mu m^{-1}$ (the upper limit of Ω in figure 2.2), then nucleation will be suppressed if $F_V/r > 0.15\mu m^{-1}$. This is in reasonable agreement with the experimental observations discussed in §8.3.1. The orientation gradients will tend to increase with increasing strain and therefore we expect that the critical value of F_V/r or $F_V^{1/3}/r$ necessary to suppress recrystallization will increase with increasing strain, as has been reported (§8.3.2).

8.5.2 Growth

During primary recrystallization, a growing grain in an alloy containing a dispersion of particles is acted on by two opposing pressures, the driving pressure for growth (P_D) as given by equation 6.2 and the Zener pinning pressure (P_Z) arising from the particles, as given by equation 3.21. In the early stages of growth, we should also consider the retarding pressure due to boundary curvature (P_C) (equation 6.3). The net driving pressure for recrystallization (P) is thus

$$P = P_D - P_Z - P_C = \frac{\alpha \rho G b^2}{2} - \frac{3 F_V \gamma_b}{d} - \frac{2 \gamma_b}{R} \qquad (8.29)$$

Recrystallizing grains will not grow unless P is positive and we therefore expect recrystallization growth to be suppressed when F_V/r is greater than a critical value, which will be larger for small driving pressures (i.e. low strains). Qualitatively, this is in agreement with the experimental results discussed in §8.3.1 in that alloys in which F_V/r is large are found to recrystallize more slowly than an equivalent particle-free matrix, and that a transition from retardation to acceleration occurs when a critical strain is exceeded.

Several investigators (e.g. Baker and Martin 1980, Chan and Humphreys 1984b) have found, from analyses of driving and pinning pressures (e.g. equation 8.29), that the Zener pinning pressure (P_Z) is too small to account for the retardation of recrystallization **growth** in their materials. However, the values of F_V/r in these investigations ranged from 0.2 to 0.6μm^{-1}, which are consistent with particle-inhibition of **nucleation** according to equation 8.27.

We therefore conclude that Zener pinning plays a major role in retarding primary recrystallization, and may affect both the nucleation and the growth of the grains. Experiment and theory suggest that the critical parameter in determining the recrystallization kinetics is F_V^n/r, where $\frac{1}{3} < n < 1$, although further work is needed in this area.

8.6 BIMODAL PARTICLE DISTRIBUTIONS

Many commercial alloys contain both large ($>1\mu m$) particles which will act as sites for PSN, and small particles, spaced sufficiently closely to pin the migrating boundaries. The recrystallization of such alloys has been analyzed extensively by Nes and colleagues (e.g. Nes 1976). In such a situation, the driving force (P_D) is offset by the Zener pinning force (P_Z) and the critical particle size for the growth of a nucleus (equation 8.24) now becomes

$$d_g = \frac{4 \gamma_b}{P_D - P_Z} = \frac{4 \gamma_b}{\dfrac{\rho G b^2}{2} - \dfrac{3 F_v \gamma}{2 r}} \qquad (8.30)$$

where F_V and r refer to the small particles.

Thus, as the Zener pinning force increases, the critical particle diameter for PSN increases. As there will be a distribution of particle sizes in a real alloy, this means that less particles are able to act as nuclei, and that the recrystallized grain size will increase.

An alternative analysis of the effects of a fine particle dispersion on PSN may be made by extending that of §8.5.1. The orientation gradient (Ω) is in this case, that at the large particles, and this is given by equation 8.18.

The number of particles capable of acting as nuclei (N_d) is the number of particles of diameter greater than d_g. If other nucleation sites are neglected and if we assume **site saturated** nucleation, then the grain size (D_N), will be given approximately by :-

$$D_N \approx N_d^{-1/3} \qquad\qquad (8.31)$$

The recrystallization kinetics of bimodal alloys in which there is significant Zener pinning are often found to be similar to those of alloys containing a similar dispersion of only small particles (Hansen & Bay 1981, Chan & Humphreys 1984b). The latter authors found an unusual recrystallized "island" grain structure in which small isolated grains from less successful nuclei were engulfed by more rapidly growing grains.

8.7 THE CONTROL OF GRAIN SIZE BY PARTICLES

There is considerable interest in the use of particles to control both the grain size and the texture of alloys. This is of importance in many structural steels, aluminium alloys used for cans, and superplastic alloys, and in chapter 12, we will discuss some cases in more detail. In the present discussion, we show how the effects of particles on recrystallization which were discussed earlier, can be used to control grain size during thermomechanical processing.

The discussion below is largely based on models developed by Nes (1985), Wert and Austin (1986) and Nes and Hutchinson (1989). The main parameter which determines the grain size after annealing an alloy which contains a volume fraction F_V of small particles of mean radius **r** is F_V/r. This is because F_V/r affects the Zener drag (equation 3.21), the number of viable recrystallization nuclei (equations 8.30, 8.31) and also the grain size (D_{lim}) at which normal grain growth will stagnate (§9.4.2). In figure 8.35, we show how the particle dispersion level (F_V/r) is expected to affect the grain size after recrystallization.

The curve **D_N** is the grain size after primary recrystallization by PSN. The number of viable nuclei from large particles or any other sites will decrease as F_V/r increases (equations 8.30, 8.31). At some dispersion level ($F_V/r=B$), the number of nuclei is effectively zero, and recrystallization can not occur. The curve D_{LIM} is the grain size at which normal grain growth will cease, as given by equations 9.33 or 9.37 as appropriate. The point at which these two grain sizes are equal is denoted $F_V/r=A$.

We note three important regimes.

i) $F_V/r < A$: In this regime the grain size after primary recrystallization is determined by the number of available nucleation sites, and if the latter is large, then the former may be very small (equations 8.30, 8.31). However, the grains are unstable with respect to normal grain growth, and if annealed at a sufficiently high temperature then grain growth will occur. In this situation the minimum grain size is achieved when $F_V/r = A$.

ii) $A < F_V/r < B$: In this regime the material recrystallizes to a grain size (D_N) as determined by the number of nuclei. This is above the value of D_{lim} and therefore normal grain growth will not occur. However, if the mean recrystallized grain size is small but the microstructure is not homogeneous, then the suppression of normal grain growth may make the material vulnerable to **abnormal grain growth** if annealed at very high temperatures (§9.5.2).

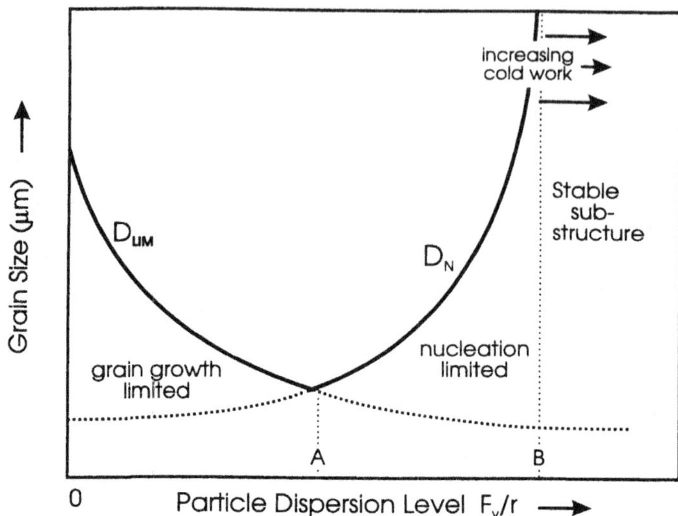

Fig. 8.35. The effect of the particle parameters on the recrystallized grain size.

iii) $F_V/r > B$: In this regime the Zener pinning is now sufficient to suppress discontinuous recrystallization and the particles stabilise the deformed or recovered microstructure as discussed in §5.6.2.

We therefore see that there is an optimum dispersion level, $F_V/r = A$, required to produce a small and stable grain size. A further parameter which affects D_N and hence the critical values of F_V/r is the stored energy of deformation, which, for low temperature deformation is primarily a function of the strain. In figure 8.35, the arrows indicates the effect of strain on the transition between regimes.

8.8 PARTICULATE METAL-MATRIX COMPOSITES

Particulate metal matrix composites (MMCs), such as aluminium alloys containing $\sim 20v\%$ of ceramic particles are of interest as structural materials in automotive and aerospace applications, where a high strength and stiffness combined with a low density is required. One of the advantages of particulate composites is that they can be mechanically processed in the same way as conventional alloys, and the recrystallization behaviour is therefore of interest (e.g. Humphreys at al. 1990, Liu et al. 1991).

In these materials we have the unusual situation of alloys containing both a large volume fraction of particles and also particles which are large enough (typically 3-10μm) for PSN. The particles therefore play an important role in controlling the grain size by particle stimulated nucleation. If each particle produces one recrystallized grain, then the grain size as a function of particle diameter and volume fraction will be as shown in figure 8.33, from which it can be seen that very small grain sizes are achievable in these materials. The grain size of a series of Al-SiC particulate composites as a function of particle size and volume fraction are shown in figure 8.36, and these results are close to the predictions of equation 8.25 and figure 8.33.

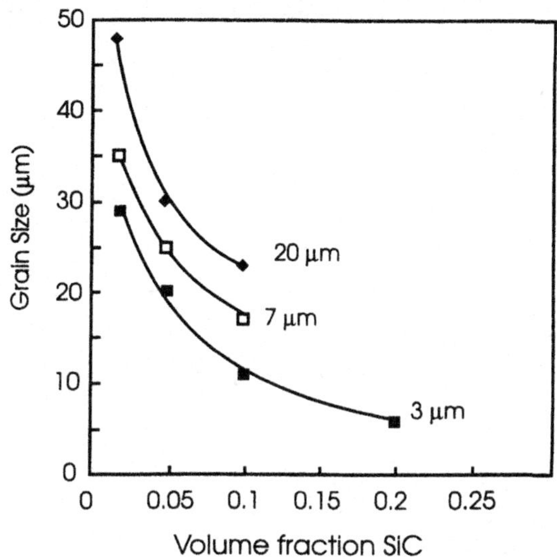

Fig. 8.36. The recrystallized grain size in Al-SiC particulate composites cold rolled 80% and annealed at 600°C, (courtesy M.G. Ardakani).

The recrystallization kinetics of particulate MMCs are usually very rapid compared to the unreinforced matrix (Humphreys et al. 1990, Liu et al. 1989, Sparks and Sellars 1992, Ferry et al. 1992). This is a combination of the large driving force due to the geometrically necessary dislocations (equation 8.14) and the large number of nucleation sites.

In aluminium composites produced by powder metallurgy, there are, in addition to the large ceramic reinforcement particles, small (100nm), inhomogeneously distributed oxide particles. These have several predictable but important effects, particularly in composites containing low volume fractions of ceramic reinforcement.

  i) The recrystallization kinetics are slowed by Zener pinning
  ii) The recrystallized grain size is larger than predicted by equation 8.25
  iii) Elongated subgrains and grains are produced (fig 3.25)

Ferry et al. (1992) have studied the effect of precipitation on the recrystallization of an AA2014-Al$_2$O$_3$ composite and found that as expected for bimodal particle distributions (§8.6), the recrystallization kinetics are determined primarily by the fine precipitate dispersion.

If high volume fractions of large ceramic particles are present then discontinuous recrystallization may be suppressed as found by Humphreys (1988a) and Liu et al. (1991). This has been found to occur when $F_V/r > 0.2$, a similar condition to that found for conventional alloys (§8.3.1), and this is shown as the dotted region in figure 8.33. The resulting microstructures contain a mixture of high angle and low angle boundaries, and the annealing behaviour is an example of the homogeneous annealing phenomena discussed in §5.7.1.

8.9 THE INTERACTION OF PRECIPITATION AND RECRYSTALLIZATION

The mutual interaction of precipitation and recrystallization has been extensively studied by Hornbogen and colleagues in Al-Cu and Al-Fe alloys (e.g. Ahlborn et al. 1969), and has been reviewed by Hornbogen and Köster (1978). Despite its practical importance, there has been little systematic work in this area, and although the outline of the phenomena provided by Hornbogen and colleagues is generally accepted, the details are largely unresolved.

If a supersaturated solid solution is deformed and annealed, then unless recrystallization is complete before precipitation begins, the two processes affect each other. Precipitation on low angle or high angle boundaries will hinder recovery and recrystallization, and dislocations will themselves promote the heterogeneous nucleation of precipitates. The recrystallization behaviour as a function of time and temperature is summarised schematically in figure 8.37.

The temperature T_1 is the solvus temperature, above which, no precipitation occurs. In the absence of precipitation, recrystallization would commence at times given by the line AC. The curve BD shows the onset of recrystallization in the presence of precipitated particles. The dotted curves indicate the onset of precipitation both in the absence and presence of a deformed microstructure. The actual recrystallization behaviour is expected to follow the bold line, and we can identify three regimes.

I. $T > T_1$
      There is no precipitation possible, and so the recrystallization behaviour is that of a solute containing alloy.

II. $T_1 > T > T_2$
      Recrystallization occurs before precipitation, therefore the recrystallization is similar to I.

III. $T < T_2$
      Precipitation occurs before recrystallization. The particles control the recovery rate (see §5.6). Eventually recrystallization may occur. In some cases discontinuous precipitation may occur on migrating high angle grain boundaries.

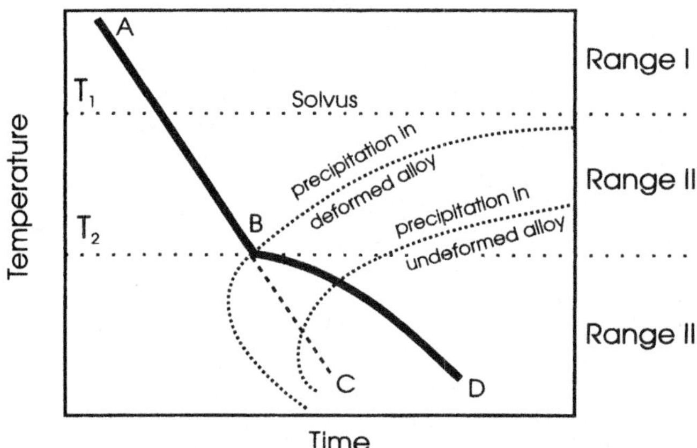

Fig. 8.37. The interaction of recrystallization and precipitation, (Hornbogen and Köster 1978).

8.10 THE RECRYSTALLIZATION OF DUPLEX ALLOYS

Earlier sections of this chapter have discussed the recrystallization of two-phase alloys in which one phase was dispersed in a matrix of the other. There are however many important alloys in which the volume fraction of the two-phases are comparable (e.g. fig 9.14), and it is with the recrystallization of such **duplex alloys** that we are concerned here. Examples of duplex alloys include α/β brasses and bronzes, α/γ steels, α/β titanium alloys. The phase transformations and recrystallization which often occur during the thermomechanical processing of duplex alloys may be used to refine the microstructures, and this technology has been used extensively to produce fine grained duplex alloys capable of undergoing superplastic deformation (Pilling and Ridley 1989). In this section we are primarily concerned with recrystallization following low temperature deformation. The recrystallization of duplex alloys has been reviewed by Hornbogen and Köster (1978) and Hornbogen (1980), and grain growth in duplex microstructures is considered in §9.4.3.2

The α/β brass alloys (e.g. 60:40 brass) provide good examples of recrystallization in coarse duplex microstructures. Some well known early work (Honeycombe and Boas 1948, Clareborough 1950) was carried out on this alloy system, and the later work of Mäder and Hornbogen (1974) provides a basis for the following discussion which illustrates some of the complexities in the behaviour of such systems.

8.10.1 Equilibrium microstructures

The simplest case is that in which the material is annealed at the same temperature at which it was equilibrated before deformation. If the volume fractions of both phases are equal, then the phases are usually constrained to undergo similar deformations. However, if one phase is continuous, then there will be a strain partitioning between the phases dependent on the relative volume fractions, sizes and strength of the phases, and this will influence the recrystallization behaviour. It is usually found that the phases recrystallize independently of each other in a manner that is largely predictable from a knowledge of the recrystallization of the individual phases (e.g. Vasudevan et al. 1974, Cooke et al. 1979).

In duplex α/β brass alloys, the microstructure comprises equiaxed grains of fcc α-brass, which has a low stacking fault energy and whose deformation behaviour is discussed in §2.3; and grains of the ordered bcc (B2) β-brass phase. The annealing behaviour after rolling reductions of less than 40% is shown schematically in figure 8.38. The β phase recovers readily and forms subgrains, whilst the low stacking energy α-phase undergoes little

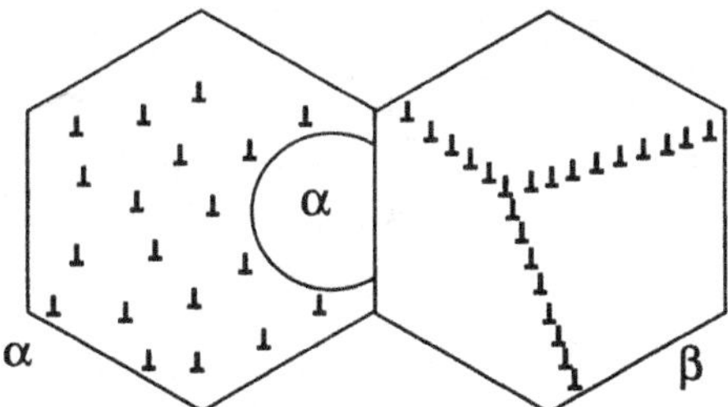

Fig. 8.38. Recrystallization of α/β brass after a low strain, (Mäder and Hornbogen 1974).

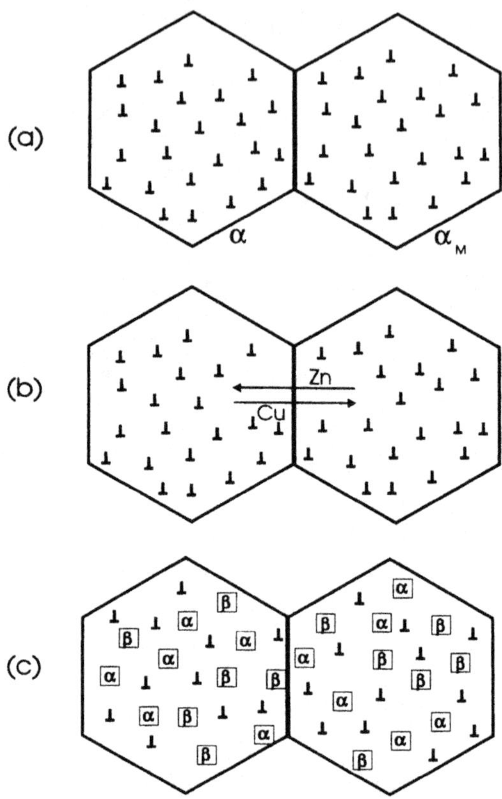

Fig. 8.39. Recrystallization of α/β brass after a large strain. a) The β-phase transforms martensitically to α. b) Interdiffusion occurs. c) Recrystallization and precipitation occur in both phases, (Mäder and Hornbogen 1974).

recovery, but recrystallizes readily, nucleation being mainly at the α/β boundaries. The β-phase generally recrystallizes at longer times, although for lower strains, extended recovery may occur (Mäder and Hornbogen 1974, Cooke and Ralph 1980). As discussed in §7.3.2, the recrystallization kinetics of the β-phase are affected by the ordering reaction.

After rolling reductions of greater than 70%, Mäder and Hornbogen found that the β-phase in a Cu-42%Zn alloy transformed martensitically into a distorted fcc phase (α_M) during deformation. On annealing, diffusion occurred between the two phases (fig 8.39) tending to equalise their compositions, and recrystallization and precipitation occurred in all grains (fig 8.39c), leading to the formation of a fine-grained duplex microstructure.

8.10.2 Non-equilibrium microstructures

If the annealing is carried out at a temperature different from that at which the material was originally stabilized, then phase transformations will occur during the anneal, and complex

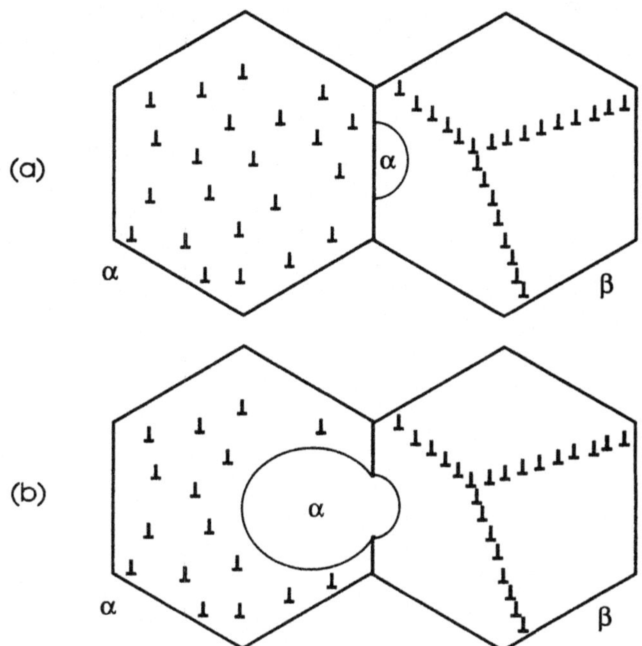

Fig. 8.40. Recrystallization of α/β brass at a temperature different from that of the pre-anneal. a) The volume fraction of β is reduced by growth of α into the β grains. b) The new α region nucleates recrystallization in the α grains, (Mäder and Hornbogen 1974).

interactions between precipitation and recrystallization may result (§8.9), and much of the early work on α/β brasses was carried out in such conditions. The application of combined recrystallization and phase transformation to the control of microstructure in duplex materials has been discussed for iron and copper alloys by Hornbogen and Köster (1978), and for titanium alloys by Williams and Starke (1982) and Flower (1990), and will not be considered in detail here. However, the α/β brass system (Mäder and Hornbogen 1974) may again be used to illustrate the principles.

The necessary phase changes can take place either by movement of the original α/β interfaces or by precipitation within the phase whose volume fraction is being reduced. For example in figure 8.40a in which the fraction of β phase is being reduced, α nucleates at the phase boundary and grows into the deformed β phase. These grains subsequently act as recrystallization nuclei and consume the deformed α grains (fig 8.40b). If the driving force is increased, then heterogeneous nucleation of α inside the β grains may occur.

Chapter 9

GRAIN GROWTH FOLLOWING RECRYSTALLIZATION

9.1 INTRODUCTION

Compared with primary recrystallization, the growth of grains in a recrystallized single-phase material might appear to be a relatively simple process. However, despite a large amount of theoretical and experimental effort, many important questions remain unanswered. The theoretical basis for understanding grain growth was laid down over 40 years ago in the classic papers of Smith (1948, 1952) and Burke and Turnbull (1952), and the apparent conflict of theory with experiment prompted several other theoretical models over a period of some 30 years. The application of computer simulation techniques (Anderson et al. 1984) provided a fresh approach to the problem and the interest and controversy surrounding the computer simulations gave a stimulus to the subject. This has resulted in renewed theoretical and experimental interest in grain growth as may be seen from the international conference on the subject held recently (Abbruzzese and Brozzo 1992) and the comprehensive review of developments in the theory of grain growth by Atkinson (1988).

Although primary recrystallization often precedes grain growth, it is of course not a necessary precursor, and the contents of this chapter are equally relevant to grain growth in materials produced by other routes, such as casting or vapour deposition. In this chapter we are primarily concerned with the kinetics of grain growth and the nature and stability of the microstructure; the factors affecting the mobilities of the grain boundaries are discussed in chapter 4.

9.1.1 The nature and significance of grain growth

When primary recrystallization, which is driven by the stored energy of cold work, is complete, the structure is not yet stable, and further growth of the recrystallized grains may occur. The driving force for this is a reduction in the energy which is stored in the material in the form of grain boundaries. The driving pressure for grain growth is some two orders of magnitude less than that for primary recrystallization (§1.3.2), and is typically $\sim 10^{-2}$ MPa. Consequently grain boundary velocities will be slower than during primary recrystallization and boundary migration will be much more affected by the pinning effects of solutes and second-phase particles.

The technological importance of grain growth stems from the dependence of properties, and in particular the mechanical behaviour, on grain size. In materials for structural application at lower temperatures, a small grain size is normally required to optimise the strength and toughness. However, in order to improve the high temperature creep resistance of a material, a large grain size is required. Examples of the application of the control of grain growth which are considered in chapter 12, include the processing of silicon-iron transformer sheet and the development of microstructures for superplastic materials. There is also considerable interest in grain growth in thin oxide and semiconductor films for electronic applications as discussed in §9.5.4. A good understanding of grain growth is therefore a prerequisite for control of the microstructures and properties of metals and ceramics during solid state processing.

Grain growth may be divided into two types, **normal grain growth** and **abnormal grain growth or secondary recrystallization**. During normal grain growth, the microstructure changes in a rather uniform way, there is a relatively narrow range of grain sizes and shapes, and the form of the grain size distribution in usually independent of time and hence of scale (fig 9.1a). During abnormal grain growth, a few grains in the microstructure grow and consume the matrix of smaller grains and a bimodal grain size distribution develops. However, eventually these large grains impinge and normal grain growth may then resume (fig 9.1b).

9.1.2 Factors affecting grain growth

The main factors which influence grain growth, and which will be considered later in this chapter include:

Temperature
Grain growth involves the migration of high angle grain boundaries and the kinetics will therefore be strongly influenced by the temperature dependence of boundary mobility as discussed in §4.3.1. As the driving force for grain growth is usually very small, significant grain growth is often found only at very high temperatures.

Solutes and particles
Although grain growth is inhibited by a number of factors, grain boundary pinning by solutes (§4.3.3, §4.4.2) and by second-phase particles (§3.6) are particularly important.

Specimen size
The rate of grain growth diminishes when the grain size becomes greater than the thickness of a sheet specimen. In this situation the grains are curved only in one direction rather than two, and thus the driving force is diminished. The grain boundaries, where they intersect the surface, may also develop grooves by thermal etching, and these will impede grain growth.

Texture
A strongly textured material inevitably contains many low angle boundaries, and there is therefore a reduced driving force for grain growth.

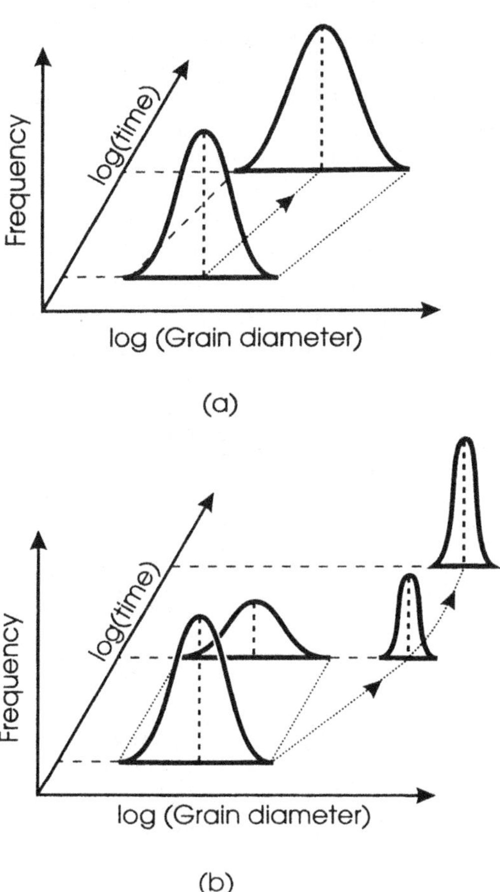

Fig. 9.1. Schematic representation of the change in grain size distribution during a) Normal grain growth and b) Abnormal grain growth, (After Detert 1978).

9.1.3 The Burke and Turnbull analysis of grain growth kinetics

Burke (1949) and Burke and Turnbull (1952) deduced the kinetics of grain growth on the assumption that the driving pressure (P) on a boundary arises only from the curvature of the boundary. If the principal radii of curvature of a boundary of energy γ_b are $\mathbf{R_1}$ and $\mathbf{R_2}$ then

$$P = \gamma_b \left(\frac{1}{R_1} + \frac{1}{R_2} \right) \qquad (9.1)$$

If the boundary is part of a sphere of radius $\mathbf{R}$, then $\mathbf{R} = \mathbf{R_1} = \mathbf{R_2}$ and

$$P = \frac{2\gamma_b}{R} \qquad (9.2)$$

Burke and Turnbull then made the following assumptions:

(i) γ_b **is the same for all boundaries**.

(ii) The radius of curvature (R) is proportional to the mean radius ($\bar{R}$) of an individual grain, and thus

$$P = \frac{\alpha \gamma_b}{\bar{R}} \tag{9.3}$$

where α is a small geometric constant.

(iii) The boundary velocity is proportional to the driving pressure P (equation 4.1), and to $d\bar{R}/dt$. i.e. $d\bar{R}/dt = c.P$, where c is a constant.

Hence,

$$\frac{d\bar{R}}{dt} = \frac{\alpha c_1 \gamma_b}{\bar{R}} \tag{9.4}$$

and therefore

$$\bar{R}^2 - \bar{R}_0^2 = 2\alpha c_1 \gamma_b t$$

which may be written as

$$\bar{R}^2 - \bar{R}_0^2 = c_2 t \tag{9.5}$$

where $\bar{R}$ is the mean grain size at time t, $\bar{R}_0$ is the initial mean grain size and c_2 is a constant.

This **parabolic growth law** is expected to be valid for both 2-D and 3-D microstructures, although, according to equation 9.1 the constant c_2 will be different for the two situations. In the limit where $\bar{R}^2 \gg \bar{R}_0^2$ then

$$\bar{R}^2 = c_2 t \tag{9.6}$$

Equations 9.5 and 9.6 may be written in the more general form

$$\bar{R}^n - \bar{R}_0^n = c_2 t \tag{9.7}$$

$$\bar{R} = c_2 t^{1/n} \tag{9.8}$$

The constant **n**, often termed the **grain growth exponent** is, in this analysis equal to **2**.

9.1.4 Comparison with experimentally measured kinetics

The use of equations 9.5 and 9.6 to describe grain growth kinetics was first suggested empirically by Beck et al. (1949). These authors found that **n** was generally well above 2 and that it varied with composition and temperature. It is significant that very few measurements of grain growth kinetics have produced the grain growth exponent of 2 predicted by equations 9.4 or 9.5, and values of **1/n** for a variety of metals and alloys as a function of homologous temperature are shown in figure 9.2. The trend towards lower values of **n** at higher temperatures has been reported in many experiments.

Data for zone-refined metals in which the impurity levels are no more than a few ppm are shown in table 9.1. The values of **n** range from 2 to 4, with an average of 2.4±0.4. Grain growth kinetics have been extensively measured in ceramics, and compilations of the data (Anderson et al. 1984, Ralph et al. 1992) reveal a similar range of grain growth exponents as is shown in table 9.2.

Much effort has been expended in trying to explain why the measured grain growth exponents differ from the "theoretical" value of 2 given by the Burke and Turnbull analysis, and the earlier explanations fall into two categories:

(i) The boundary mobility (M) varies with the boundary velocity
The boundary mobility, as discussed in §4.1.3, may under certain circumstances be a function of boundary velocity, in which case the linear dependence of velocity on driving force (equation 4.1), which is assumed in the Burke and Turnbull analysis will not apply. An example of this is the case of solute drag on boundaries (§4.4.2). Figure 4.27 shows that the velocity is not linearly proportional to the driving pressure except for very low or very high boundary velocities. However, the shape of these curves would only predict a higher grain growth exponent if conditions were such that the boundary changed from breakaway to solute drag behaviour during grain growth, and it is unlikely that these conditions would be commonly met.

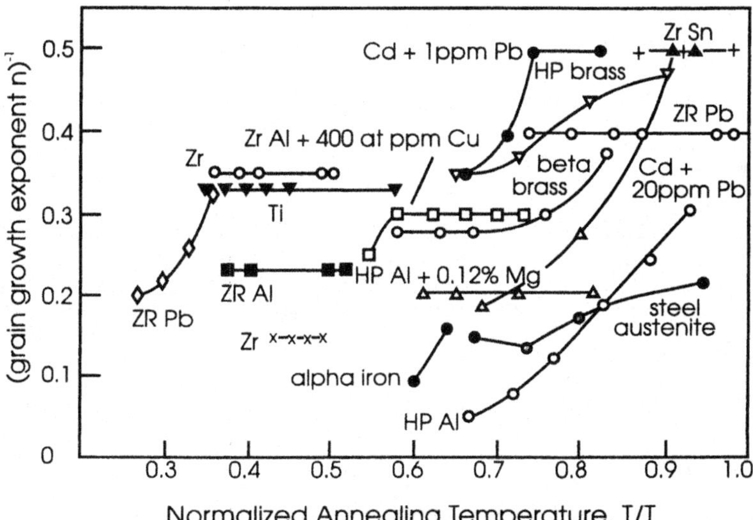

Fig. 9.2. The temperature dependence of the grain growth exponent **n** for isothermal grain growth in a variety of materials, (Higgins 1974).

Table 9.1

**Grain growth exponents for isothermal grain growth in high purity metals
(from Anderson et al. 1984)**

Metal	Exponent n	Reference
Al	4	Gordon & El Bassyoumi (1965)
Fe	2.5 (varies with T)	Hu (1974)
Pb	2.5	Bolling and Winegard (1958)
Pb	2.4	Drolet & Gallibois (1968)
Sn	2.3	Drolet & Gallibois (1968)
Sn	2	Holmes and Winegard (1959)

Table 9.2

**Grain growth exponents for isothermal grain growth in ceramics
(from Anderson et al. 1984)**

Ceramic	Exponent n	Reference
ZnO	3	Dutta & Spriggs (1970)
MgO	2	Kapadia & Leipold (1974)
MgO	3	Gordon et al. (1970)
CdO	3	Petrovic & Ristic (1980)
$Ca_{0.16}Zr_{0.84}O_{1.84}$	2.5	Tien & Subbaro (1963)
UO_2	3	Kingery & Francois (1965)

As discussed in §4.3.1.2, there is evidence of changes in grain boundary structure and mobility at very high temperatures, even in metals of very high purity. It is conceivable that this could account in some cases for transitions to lower n values, although there is no direct evidence for this.

Grain growth in ceramics has been extensively investigated, and in many cases the measured values of **n** have been ascribed to particular mechanisms of boundary migration. For example Brook (1976) lists eleven proposed mechanisms for boundary migration with growth exponents between 1 and 4.

(ii) There is a limiting grain size
An alternative empirical analysis of the data has been given by Grey and Higgins (1973). They propose that equation 4.1 is replaced by

$$v = M (P - C) \tag{9.9}$$

where C is a constant for the material.

If this relationship is used in the Burke and Turnbull analysis then equation 9.4 becomes

$$\frac{d\bar{R}}{dt} = c_1 \left(\frac{\alpha \, \gamma_b}{\bar{R}} - C \right) \tag{9.10}$$

When, as a result of grain growth, the driving pressure P falls to the value C, there is no net pressure, and grain growth ceases at a **limiting value**. Grey and Higgins show that this form of equation° accounts quite well for the kinetics of grain growth in several materials. The term C is similar to the Zener pinning term which accounts for a limiting grain size in two-phase materials (§9.4.2), and Grey and Higgins suggest that the physical origin of C may be solute clusters which are unable to diffuse with the boundary.

However, there is little evidence for $n=2$ even in very pure materials, and it has been suggested that one or more of the underlying assumptions of the Burke and Turnbull analysis may be incorrect. Although we will later conclude that there is little evidence to suggest that the Burke and Turnbull result is in serious error, a great deal of work has gone into producing more refined models of grain growth which address not only the kinetics but also the grain size distribution.

9.1.5 Topological aspects of grain growth

The Burke and Turnbull analysis assumes that the mean behaviour of the whole array of grains can be inferred from the migration rate of part of one boundary and does not consider the interaction between grains or the constraints imposed by the space-filling requirements of the microstructure. This aspect of grain growth was first addressed by Smith (1952) who discussed grain growth in terms of grain topology and stated that **"Normal grain growth results from the interaction between the topological requirements of space-filling and the geometrical needs of surface tension equilibrium"**.

From the introduction to grain topology in §3.5, we note that in a two-dimensional grain structure, the only stable arrangement which can fulfil both the space-filling and boundary tension equilibrium requirements is an array of regular hexagons as shown in figure 9.3, and any other arrangement must inevitably lead to grain growth. For example, if just one 5-sided polygon is introduced (fig 9.4a), then it must be balanced by a 7-sided one to maintain the average number of edges per grain at 6 (§3.5.1) as shown in figure 9.4a. In order to maintain the 120° angles at the vertices, the sides of the grains must become curved. Grain boundary migration then tends to occur in order to reduce the boundary area, and the boundaries migrate towards their centres of curvature (fig 9.4b). Any grain with more than six sides will tend to grow because its boundaries are concave and any grain with less than six sides will tend to shrink as it has convex sides. The shrinkage of the 5-sided grain in figure 9.4a leads to the formation of a 4-rayed vertex (fig 9.4b) which decomposes into two

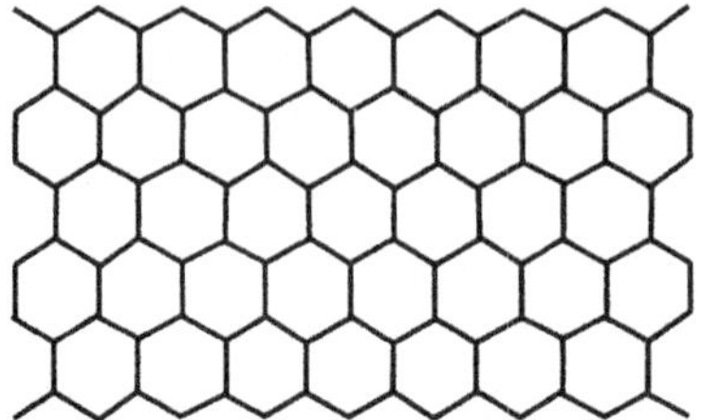

Fig. 9.3. A 2-dimensional array of equiaxed hexagonal grains is stable.

3-rayed vertices and the grain becomes 4-sided (fig 9.4c). A similar interaction allows the grain to become 3-sided (fig 9.4e) and to eventually disappear leaving a 5-sided grain adjoining a 7-sided grain (fig 9.4f).

Von Neumann (1952) proposed on the basis of surface tension requirements, that the growth of a 2-D cell of area A with N sides is given by

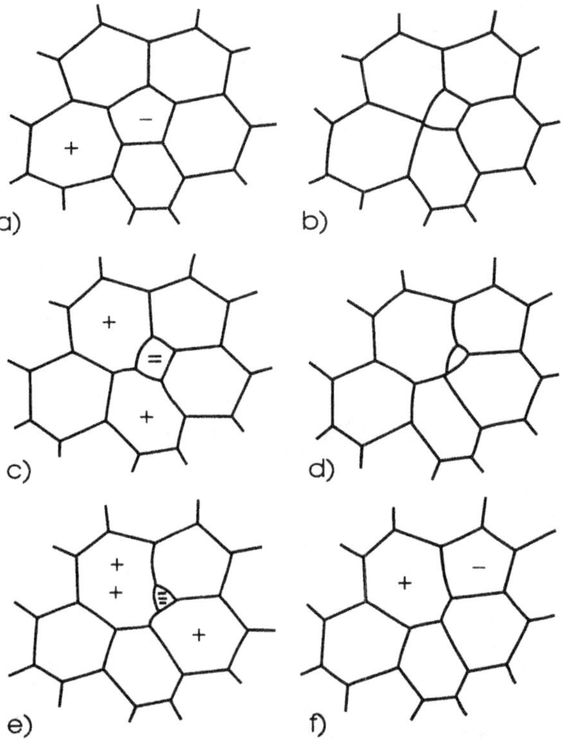

Fig. 9.4. Schematic diagram of growth of a 2-dimensional grain structure. a) A grain of less than or more than 6 sides introduces instability into the structure. b) - f) Shrinking and disappearance of the 5-sided grain, (Hillert 1965).

$$\frac{dA}{dt} = c\,(N - 6) \tag{9.11}$$

or, if written in terms of grain radius (R),

$$\frac{dR}{dt} = \frac{c\,(N - 6)}{2\,R} \tag{9.12}$$

Rivier (1983) later showed that **Von Neumann's law** is actually a geometric result and not due to surface tension.

In three dimensions, no regular polyhedron with plane sides can fill space and have its sides at the appropriate angles to balance the boundary tensions. As discussed in §3.5, the nearest approach is the Kelvin tetrakaidecahedron (fig 3.14) but the angles are not exact and the boundaries must become curved to obtain equilibrium at the vertices. Therefore **grain growth is inevitable in a 3-dimensional grain structure**.

9.2 THE DEVELOPMENT OF THEORIES AND MODELS OF GRAIN GROWTH

A complete theory of grain growth must take into account both the topological space-filling requirements discussed by Smith and the kinetics of local boundary migration as discussed by Burke and Turnbull. In this section we review the various attempts that have been made to improve on the theory discussed in §9.1.3. We should note that all the theories discussed assume that the rate limiting factor in grain growth is the migration of the boundaries. However, grain growth also involves the migration of grain vertices, and the possibility that the mobility of these defects might become rate limiting under certain circumstances should not be ignored (Galina et al. 1987, Palumbo and Aust 1992).

9.2.1 Introduction

Because the geometry of an array of grains is very similar to that of a soap froth, and the stability and evolution of the latter is of interest in its own right, this analogy has been widely studied. For example Smith (1952) suggested, on the basis of data from soap froth experiments that the distribution of grain sizes and shapes should be invariant, and that the grain area should be proportional to time. In both systems the driving force for growth is the reduction in boundary energy. The analogy between the two cases is very close, and similar growth kinetics and grain size distributions are found as discussed by Weaire and Rivier 1984, Atkinson 1988 and Weaire and Glazier 1992. However, the mechanisms of growth are different because in froths, growth occurs by gas molecules permeating through the cell membranes in order to equalise the pressures. There are other significant differences, and there are therefore limits to the extent that soap froth evolution can be used as an analogy for grain growth.

Theories and models of grain growth may be divided into two general categories, **deterministic** or **statistical**. Deterministic models are based on the premise that the behaviour of any grain in the assembly is dependent upon the behaviour of **all** the other grains. If the geometry of the whole grain assembly is known, then topological constraints are automatically accounted for, and by the application of relatively simple local rules such as equation 9.4, the evolution of the grain structure is predicted. This very powerful approach requires extensive computing power if reasonable sized grain assemblies are to be studied, and deterministic models will be further discussed in §9.2.4.

Statistical models are based on the assumption that the behaviour of the whole grain assembly can be calculated by generalising the behaviour of a small part of the microstructure. The Burke and Turnbull analysis is an example of such a model, and the later refinements which predict grain size distributions and which approximate the topographical constraints are discussed in §9.2.2 and §9.2.3.

The development of statistical theories of grain growth has a long and complex history and the merits of different approaches are still hotly debated. The essential problem is that the theories need to reduce the topological complexity of the real grain structure and its effect on the driving forces for grain growth to some average value, with a manageable number of parameters, for the particular grain under consideration. In assessing the theories we will concentrate on their predictions of the kinetics of grain growth and of the grain size distributions, and the comparison of these with experiment.

9.2.2 Early statistical theories

The majority of statistical grain growth theories fall into the category of **mean field theories** which determine the behaviour of a single grain or boundary in an environment which is some average representation of the whole assembly. The theories may be divided into two groups. During grain growth, the larger grains grow and the smaller grains shrink and statistically the grains can be considered to move in grain size-time-space under the action of a force which causes a drift in the mean grain size. Such models, typified by the theories of Feltham and Hillert are commonly known as **drift models**. Another approach, taken by Louat, is to consider the grain faces to undergo a random walk in the grain size-time-space. Grain growth then formally becomes a diffusion-like process, and this is often known as the **diffusion model**.

9.2.2.1 Feltham and Hillert's theories
Hillert (1965) developed a statistical theory of grain growth which was based on the assumption that the grain boundary velocity is inversely proportional to its radius of curvature. He used previous analyses of the Ostwald ripening of a distribution of second-phase particles to obtain the relationship

$$\frac{dR}{dt} = c\,M\,\gamma_b\left(\frac{1}{R_{crit}} - \frac{1}{R}\right) \tag{9.13}$$

where $c = 0.5$ for a 2-D array and 1 for a 3-D array. R_{crit} is a critical grain size which varies with time according to

$$\frac{d\,(R_{crit}^2)}{dt} = \frac{c\,M\,\gamma_b}{2} \tag{9.14a}$$

$$\frac{dR_{crit}}{dt} = \frac{c\,M\,\gamma_b}{4\,R_{crit}} \qquad\qquad (9.14b)$$

A grain such that $R < R_{crit}$ will shrink, and one with $R > R_{crit}$ will grow. Hillert showed that topographic considerations resulted in the mean grain radius $\bar{R}$ being equal to R_{crit}, and therefore equation 9.14 predicts parabolic grain growth kinetics of the form of equations 9.4 and 9.5.

Hillert also solved equation 9.13 to obtain the grain size distribution $f(R,t)$. As shown in figure 9.6a, this is much narrower than the log-normal distribution which is close to that found experimentally (§9.2.4 and fig 9.6b). In Hillert's grain size distribution the maximum was $1.8\bar{R}$ for a 3-D array and $1.7\bar{R}$ for a 2-D array. He argued that if the initial grain size distribution contained no grains larger than $1.8\bar{R}$ then normal grain growth would result and the distribution would adjust to the predicted $f(R,t)$. However, if grains larger than $1.8\bar{R}$ were present then he predicted that abnormal grain growth would result, although this latter conclusion has been shown to be incorrect (§9.5.1)

Hillert's result is very similar to that obtained by Feltham (1957), who started from the assertion that the normalised grain size distribution was log-normal and time invariant. He obtained

$$\frac{dR^2}{dt} = c\ln\left(\frac{R}{\bar{R}}\right) \qquad\qquad (9.15)$$

where c is a constant. Setting $R = R_{max} = 2.5\bar{R}$ he obtained parabolic growth kinetics.

9.2.2.2 Louat's random walk theory

Louat (1974) argued that boundary motion can be analyzed as a diffusional process in which sections of the boundary undergo random motion. This will lead to grain growth because the process of grain loss by shrinkage is not reversible. The theory predicts parabolic grain growth kinetics and an invariant Rayleigh grain size distribution (fig 9.6a) which is close to that found experimentally (§9.2.4). Although the theory has been criticised by a number of authors on the grounds that it lacks a strong physical basis for its assumptions, it has recently been defended and further developed by its author (Louat et al. 1992). Pande (1987) has developed a statistical model which combines the random walk element of Louat's theory with the radius of curvature approach of Hillert.

9.2.3. The incorporation of topology

9.2.3.1 Defect models

Hillert (1965) proposed an alternative approach to two-dimensional grain growth based on the topological considerations discussed in §9.1.5. Within an array of six-sided grains, the introduction a 5-sided and 7-sided grain as shown in figure 9.4a, constitutes a stable defect. As growth occurs and the 5-sided grain ultimately disappears (fig 9.4b-e), then the 5-7 pair defect moves through the structure (fig 9.4f). Hillert argued that the rate of growth depended on the time taken for the defect to move (i.e. for a 5-sided grain to shrink) and on the number of such defects in the microstructure. If the latter remained constant, which is a reasonable assumption for a time invariant grain distribution, then a parabolic growth rate similar to that for his statistical theory (equation 9.14) results. Morral and Ashby (1974) extended this model to 3-D by introducing 13 or 15-sided grains into an array of 14-sided polyhedra.

9.2.3.2 The Rhines and Craig analysis

Rhines and Craig (1974) emphasised the role of topology in grain growth. They argued that when a grain shrinks and disappears as shown in figure 9.4, then not only must this volume be shared out between the neighbouring grains, but because the topological attributes (shape, faces, edges, vertices etc) of the neighbours also alter, this will in turn influence grains which are further away. They introduced two new concepts, the **sweep constant** and the **structural gradient**.

They defined a **sweep constant** Θ, as the number of grains lost when the grain boundaries in the specimen sweep out unit volume of material. They argued that Θ will remain constant during grain growth. An alternative parameter Θ^*, the number of grains lost when the boundaries sweep through a volume of material equal to that of the mean grain volume, was suggested by Doherty (1975). Clearly both these parameters cannot remain constant during grain growth, neither can be measured directly by experiment, and the matter is unresolved. The second parameter introduced by Rhines and Craig was the dimensionless **structural gradient** ς, which is the product of the **surface area per unit volume (S_v)** and the **surface curvature per grain (m_v/N_v)**. i.e.

$$\varsigma = \frac{m_v S_v}{N_V} \tag{9.16}$$

where N_v is the number of grains per unit volume and

$$m_v = \int_{S_v} \frac{1}{2}\left(\frac{1}{r_1} + \frac{1}{r_2} \right) dS_v \tag{9.17}$$

where r_1 and r_2 are the principal radii of curvature.

Rhines and Craig argued that ς should remain constant during growth and as shown in figure 9.5a, found ς to be constant in their experiments. However, Doherty (1975) suggested an alternative structure gradient $\varsigma^* = m_v/N_v$, which is the mean curvature per grain.

The Rhines and Craig analysis, using Doherty's modifications as suggested by Atkinson (1988) is as follows. The mean pressure P on the boundaries is

$$P = \frac{\gamma_b m_v}{S_v} \tag{9.18}$$

and the mean boundary velocity

$$v = MP = \frac{M \gamma_b m_v}{S_v} \tag{9.19}$$

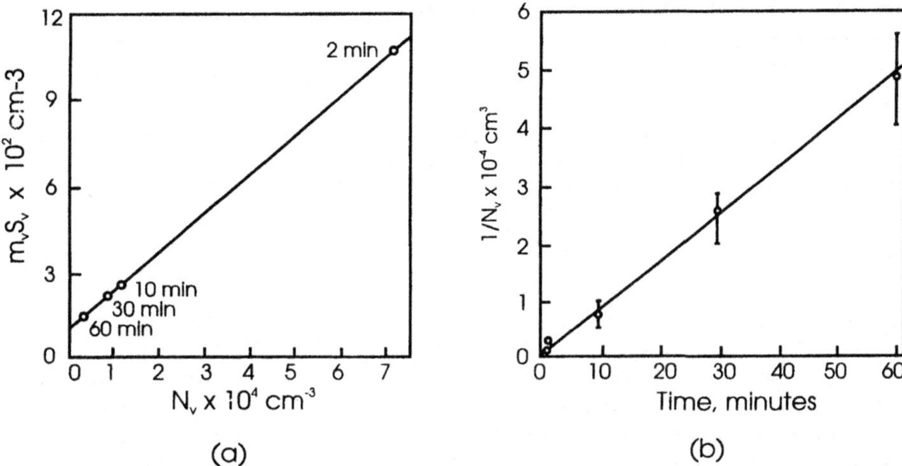

Fig. 9.5. a) Plot of $m_v S_v$ against N_v for grain growth in aluminium, b) Plot of mean grain volume $(1/N_v)$ against annealing time for aluminium, (after Rhines and Craig 1974).

The volume swept per second per unit volume of specimen is $v.S_v$, and if Θ^* grains are lost per unit volume, for each $\bar{V}$ (where $N_v = \bar{V}^{-1}$), then the rate of loss of grains will be

$$\frac{dN_v}{dt} = \frac{\Theta^* v S_v}{\bar{V}} = \Theta^* M \gamma_b m_v N_v \qquad (9.20)$$

For each grain lost, per unit volume, there is a net increase in volume of $\bar{V}$ which on average is distributed over the remaining N_v grains. Therefore

$$\frac{d\bar{V}}{dt} = \frac{dN_v}{dt} \cdot \frac{\bar{V}}{N_v} = \frac{\Theta^* M \gamma_b m_v}{N_v} \qquad (9.21)$$

If Θ^*, M, γ_b, and m_v / N_v $(=\varsigma^*)$ are constant with time then equation 9.21 can be integrated to give

$$\bar{V} = \frac{\Theta^* M \gamma_b m_v t}{N_v} + \bar{V}_0 = ct + \bar{V}_0 \qquad (9.22)$$

where $\bar{V}_0$ is the mean grain volume at $t=0$.

As shown in figure 9.5b, Rhines and Craig (1974) found such a relationship in aluminium of 99.99% purity. A linear dependence of $\bar{V}$ on time implies that the grain radius is growing as $t^{1/3}$, i.e. **the grain growth exponent n of equation 9.7 is equal to 3**. As emphasised by Rhines and Craig (1974), this is an important difference from the value of **2** predicted by most other theories.

The linear dependence of $\bar{V}$ on time is only predicted in the above analysis if ς^* remains constant, whereas the results shown in figure 9.5b shown a constancy of ς. Doherty (1975) has suggested that this could be reconciled if in equation 9.18, the driving pressure **P** was replaced by **(P-c)**, in accord with equation 9.9.

The Rhines and Craig experiments were carried out by the very time-consuming method of serial sectioning of specimens rather than by analysis of 2-D sections as is commonly done. They found that conventional 2-D analysis of their specimens resulted in the grain radius being proportional to $t^{0.43}$, i.e. **n=2.3**, which is rather close to that commonly found (table 9.1). They concluded from the discrepancy between their 2-D and 3-D measurements that **N_v can only be determined from 3-D methods such as serial sectioning and cannot be inferred from measurements on a 2-D section**. There has been widespread discussion of this point, and the analysis of 3-D microstructures produced by computer simulation (Anderson et al. 1984, Srolovitz et al. 1984a) has been of help. These have shown that there will only be a serious discrepancy between 2-D and 3-D metallographic methods if the structure is **anisotropic**, and the difference in the grain growth exponents measured by Rhines and Craig (1974) in 2-D and 3-D therefore implies that their microstructure was anisotropic. There is of course a serious implication in this for all experiments based on 2-D sections and the work of Rhines and Craig emphasises that grain shape anisotropy must be determined if 2-D measurements are to be used for comparison with theory.

This seminal work left a number of unanswered questions about theoretical and experimental aspects of grain growth kinetics. In particular, it was not clear whether a grain growth exponent of 3 would be the result of any model based on topographic considerations or whether it is specific to this particular analysis. As discussed by Atkinson (1988), it is not obvious that the microstructure will be in topographic equilibrium, particularly at low temperatures, and if this is the case then the local topographic constraints discussed in 9.2.3a may be more appropriate. Kurtz and Carpay (1980) proposed a detailed statistical theory of grain growth, which placed an emphasis on topographic considerations, and was essentially an extension of the Rhines and Craig model. However, unlike Rhines and Craig, Kurtz and Carpay predicted a parabolic (**n=2**) grain growth relationship.

Because of the difficulty of serial sectioning methods, experimental confirmation of $\bar{V} \propto t$ is limited to the work of Rhines and Craig (1974), and as shown by the results in tables 9.1 and 9.2, it would be inadvisable to place too much reliance on a study of a single material, in particular aluminium, in which boundary mobility is known to be very sensitive to small amounts of impurity (§4.3.3). However a grain growth exponent of ~ 3 would not be inconsistent with many investigations of high purity metals.

9.2.3.3 The Abbruzzese-Heckelmann-Lücke model

These authors have developed a 2-D statistical theory of grain growth (Lücke et al. 1990, Abbruzzese et al. 1992, Lücke et al. 1992). A key element of their approach is the introduction of topological parameters relating the number of sides of a grain ($\bar{n}_i$) to its size (r_i). From experimental measurements they find that

$$\bar{n}_i = 3 + 3\, r_i \tag{9.23}$$

and that the mean size ($\bar{r}_n$) of grains with **n** sides is given by

$$n = 6 + \frac{3\,(\bar{r}_n - 1)}{\xi^2} \qquad\qquad (9.24)$$

where ξ is a correlation coefficient equal to 0.85.

Although this **special linear relationship** is derived from experiments and has not been proved analytically, the authors expect it to be generally applicable to equiaxed real grain structures. The parabolic grain growth kinetics predicted on this model are similar to those of the Hillert theory i.e. equation 9.14.

9.2.3.4 Other recent statistical theories

The discussions above show that although statistical theories of normal grain growth in pure single-phase polycrystals have been developed over a period of some 45 years, there is no general agreement as to the correct solution, or even if this type of approach can ever yield a satisfactory solution. New theories and modifications of the old theories are still being produced at an alarming rate, and for example at a recent conference on grain growth (Abbruzzese and Brozzo 1992), **no fewer than 10 statistical models** were discussed. One of the main problems in reaching agreement is the question of how to represent the topology of the grain structure both accurately and with a manageable number of parameters.

9.2.4 Deterministic theories

If instead of considering the behaviour of an "average" grain, we consider the growth and shrinkage of **every grain in the assembly** then many of the topological difficulties of the statistical models are bypassed. Hunderi and Ryum (1992b) discuss the relative merits of **statistical** and **deterministic** models of grain growth and illustrate this with an analysis of grain growth in one dimension, developed from the earlier deterministic model of Hunderi et al. (1979). These authors pointed out that the statistical theories described in §9.2.3 do not allow for the fact that a grain of a particular size can grow in an environment where it is surrounded by smaller grains, but will shrink if surrounded by larger grains, i.e. R_{crit} in equation 9.13 varies with position. They proposed a **linear bubble model** in which a bubble **i** makes contact with a number of other bubbles **i-n** to **i+n**, where **n** depends on the relative sizes of the bubbles. The pressure difference between bubbles of different sizes leads to the transfer of material between the bubbles and sets of coupled equations are solved to predict the bubbles size distribution and the grain growth kinetics, which are found to be close to those for parabolic growth predicted by Hillert. Extension of such an analytical model to large numbers of grains in 3-D is however, not currently practicable. The most promising deterministic models of grain growth in recent years have been those based on computer simulation of grain growth using the **equation of motion** or **Monte-Carlo** simulation methods, details of which are discussed in §13.2.

9.2.4.1 Equation-of-motion computer simulation.

In this approach, which has mainly been applied to 2-D at present, a starting grain structure is specified, and this microstructure is then allowed to equilibrate by allowing the boundaries and vertices to move according to specific equations. For example, a boundary is adjusted to allow the angles at the triple points to be 120° and then the boundary is allowed to move by an amount proportional to its radius of curvature, this cycle being repeated over all boundaries a large number of times. This is essentially a quantification of the logical arguments which suggested the sequence shown in figure 9.4. The key feature of this type of approach is that once the initial microstructure is constructed, the laws of motion formulated, and procedures for dealing with vertex contact and grain switching (fig 13.9) specified, then no further assumptions are needed regarding topology, and the grain growth behaviour is thus truly deterministic.

A number of different authors have developed such methods (§13.2.3), and these are reviewed by Anderson (1986) and Atkinson (1988). The slightly different physical principles used have resulted in rather different grain size distributions and growth kinetics, although the latter are generally close to parabolic. These simulations have also shown that the initial growth kinetics are very sensitive to the starting grain structure. In order to make a significant contribution to our understanding of grain growth, this approach, which is still in its infancy, needs to be developed in 3-D, and preliminary work in this area has been demonstrated by Nagai et al. (1992).

9.2.4.2 Monte-Carlo computer simulation

The Monte-Carlo simulation technique, the principles of which are discussed in §13.2.2, has been extensively used to study grain growth. Early tests of a 2-D model (Anderson et al. 1984) showed that the shrinkage of a large isolated grain of area A, followed the relationship

$$A - A_0 = -c\,t \qquad (9.25)$$

where A_0 is the grain size at $t=0$ and c is a constant.

This leads to a parabolic relationship between grain size and time similar to that predicted by many theories (equation 9.5) and shows that the simulation leads to a linear dependence of boundary velocity on driving pressure (i.e. equation 4.1).

Significantly however, the growth of a 2-D grain structure such as is shown in figure 13.4, was found after an initial transient, to give a grain growth exponent of **2.44**. This differs significantly from the value of **2** predicted by the statistical theories discussed in §9.2, and is closer to the experimentally measured values (tables 9.1 and 9.2). Analysis of the computer-generated microstructures suggested that deviation from the exponent of 2 was due to topological effects. In particular, the movement and rotation of vertices was found to result in the redistribution of curvature between adjacent boundaries, and it was suggested by Anderson et al. (1984) that this lowered the local driving pressures for boundary migration, and was responsible for the higher grain growth exponent. These simulations indicated the importance of the random motion of boundaries which was first considered by Louat (1974) and also emphasised the importance of the local environment of a grain. Extension of these simulations to 3-D (Anderson et al. 1985) resulted in a grain growth exponent of **2.81** as compared to the Rhines and Craig (1974) prediction of **3**. These results therefore supported the Rhines and Craig (1974) view that grain growth exponents larger than **2** were a consequence of topological factors.

However, more recent simulations (Anderson et al. 1989a) run for longer times with larger arrays have indicated that the earlier results **did not represent steady state grain growth and were influenced by the starting grain structure**. The more recent results have revealed a growth exponent of **2.04** in 2-D and **2.12** for a 3-D simulation, exponents which are very close to the $n=2$ parabolic kinetics which are predicted on most theories. Anderson concludes that the asymptotic long-time growth exponent is **2**, although it should be noted in view of the misleading results from the earlier simulations and the fact that the 3-D simulations are limited to a lattice of only 100^3 points that the simulation may not yet be optimised.

The grain size distributions obtained by the Monte-Carlo simulations have been analyzed by Srolovitz et al. (1984a) and Anderson et al. (1989a). The grain size distribution function, expressed in terms of $R/\bar{R}$ is found to be time invariant and close to experimental measurements as shown in figure 9.6. The grain size distribution determined from 2-D sections of the 3-D grain structure is closest to the Rayleigh distribution suggested by Louat (1974).

9.2.5 Which theory best accounts for grain growth in an ideal material?

The question should now be considered as to which theory or model, if any, is close to accounting for grain growth in pure single-phase materials. That there is no obvious answer is clear from the large number of approaches that are actively being pursued. Ryum and Hunderi (1989) have given a very clear analysis of the statistical theories of grain growth and shown that all make questionable assumptions.

We will briefly consider a number of basic questions:

(i) Is a grain growth exponent of 2 inevitable?
The discussions in this section have shown that the vast majority of theories predict parabolic kinetics. The main dissent from this is the work of Rhines and Craig (1974) which predicted $\mathbf{n=3}$. However, their analysis is not universally accepted, and the extension of their work by Kurtz and Carpay (1980) predicted $\mathbf{n=2}$. Indications from early Monte-Carlo simulations that a higher value of n was associated with topographic factors (Anderson et al. 1984) have proved to be an artefact of the model, as later simulations (Anderson et al. 1989a) have found $n \sim 2$. In a recent review of the theory of the coarsening behaviour of **statistically self similar structures**, i.e. those in which the structure remains geometrically similar in a statistical sense, Mullins and Vinals (1989) conclude that for curvature driven growth, such as occurs during grain growth, an exponent of $\mathbf{n=2}$ is inevitable.

At the present time, the evidence is strongly in favour of n=2 being the prediction of theory for an ideal single-phase material in which the boundary velocity is proportional to driving pressure and boundary energies are isotropic.

(ii) Can grain size distributions be used to prove or disprove a theory?
As discussed earlier in this section, the various theories predict different grain size distributions, although it is usually predicted that the grain size distribution expressed in terms of normalised grain size $(R/\bar{R})$ will remain invariant during growth.

The most commonly predicted distributions are:

(i) Hillert's (1965) distribution

$$f(R) = c_1^{\beta}\left(\frac{\beta R}{(2-R)^{2+\beta}}\right) \cdot \exp\left(\frac{-2\beta}{(2-R)}\right) \qquad (9.26)$$

where β is the dimensionality of the distribution

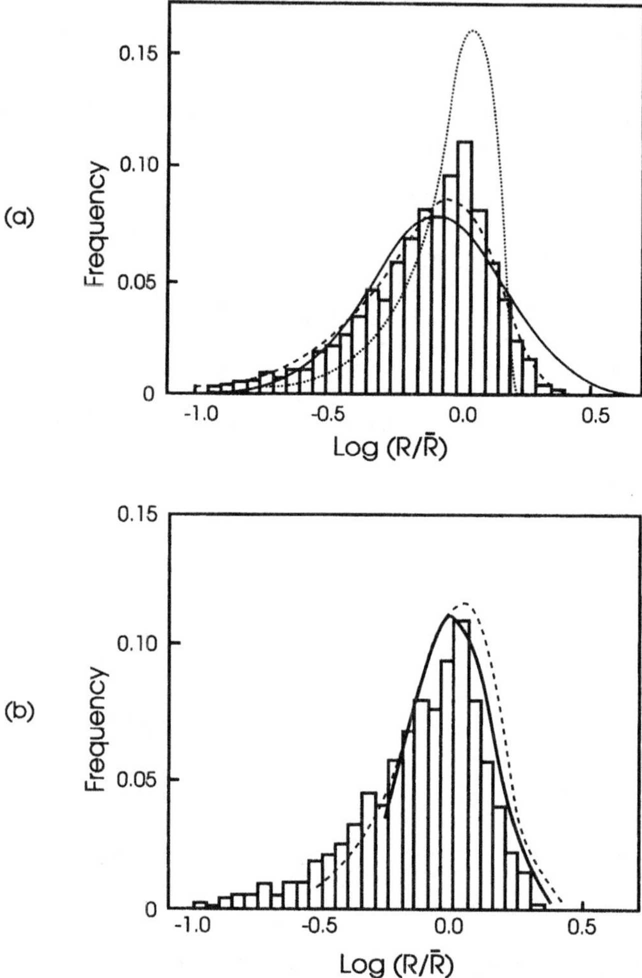

Fig. 9.6. Histogram of the grain size distributions from 2-D Monte-Carlo simulations compared with: a) Theoretical distributions - log-normal (Feltham 1957), Hillert (1965) dotted, and Rayleigh (Louat 1974) dashed. b) Experimental data - for aluminium (Beck 1954) and MgO (Aboav and Langdon 1969), dashed line. (After Srolovitz et al. 1984a).

(ii) Rayleigh distribution (Louat 1974, Anderson et al. 1989a)

$$f(R) = c_2 \, R \, \exp\!\left(-c_3 R^2\right) \tag{9.27}$$

(iii) Log-normal distribution (Feltham 1957)

$$f(R) = c_4 \exp\left(-(\ln R)^2\right) \tag{9.28}$$

As shown in figure 9.6, the experimental data appear to be closest to the Rayleigh distribution, with the narrow Hillert distribution giving the worst fit. More recent reviews of experimental distributions (Pande 1987, Louat et al. 1992) have confirmed this, and the results of later Monte-Carlo simulations (Anderson et al. 1989a) have produced data consistent with the Rayleigh distributions. In comparing experiment with theory it should be noted that as discussed in §9.2.3.2, measurements from 2-D sections will only reflect the 3-D grain distribution if the grain structure is isotropic. Although there have been numerous correlations of measured and predicted distributions, Frost (1992) has shown that the grain size distribution predicted by individual statistical models can be varied by very small adjustments to the models and cannot therefore be used to prove the validity of a model.

(iii) Why is a grain growth exponent of 2 rarely measured?

As discussed in §9.1.4 and summarised in tables 9.1 and 9.2, grain growth exponents of **2** are rarely found experimentally and average values are close to **2.4**. If we accept that theory predicts $n=2$, then we must conclude that higher measured exponents are a consequence of the materials used not being ideal i.e. not consistent with the basic assumptions about the material which are incorporated in the models. We should recall that the driving pressure for grain growth is very small (§9.1) and therefore any small deviations from an "ideal material" may have a very large effect on kinetics. There are several important parameters of the material which may lead to loss of ideality and have an influence on the kinetics:

(a) **The initial grain structure is not equiaxed or is far from the steady state grain size distribution**. Changes in grain size distribution during growth are known to affect experimentally measured grain growth kinetics (e.g. Takayama et al. 1992, Matsuura and Itoh 1992) and have been shown to produce large exponents during the early stages of growth in Monte-Carlo simulations (Anderson et al. 1984, 1989a).

(b) **The presence or development of a texture**. This would result in the occurrence of non-uniform boundary energies and mobilities thereby invalidating equations 9.3 and 9.4 which form the basis of grain growth theory.

(c) **The presence of very small amounts of a second-phase or other pinning defect**. As discussed in §9.1.4 this could account for high values of **n**.

These last two important factors are considered in more detail in the following sections.

9.3 GRAIN ORIENTATION AND TEXTURE EFFECTS IN GRAIN GROWTH.

9.3.1 Kinetics

9.3.1.1 Experimental measurements

The rate of grain growth may be affected by the presence of a sharp crystallographic texture (Beck and Sperry 1949). This arises at least in part from a large number of grains of similar orientation leading to more low angle (i.e. low energy and low mobility) boundaries (§4.2). Thus the driving force (equation 9.3) and hence the rates of growth are reduced. The texture may also alter during grain growth thereby affecting the kinetics (Distl et al. 1982,

Heckelmann et al. 1992). The evolution of textures during grain growth is discussed in §10.5, and an example of the complex grain growth kinetics which are found when there are concurrent texture changes is seen in figure 10.15b.

9.3.1.2 Theories

Novikov (1979) modified a statistical model of grain growth to include variations of boundary energy and showed how the presence of texture would affect the kinetics of grain growth. Abbruzzese and Lücke (1986) and Eichelkraut et al (1988) have incorporated the effects of texture into Hillert's statistical model of grain growth. They argued that if a material contained texture components A,B,C...etc., then the uniform boundary energies (γ_b) and mobilities (M) in equations 9.13 and 9.14 should be replaced by specific values relating to the texture components, (γ_b^{AB}, M^{AB} etc.) and that the critical radius R_{crit} in equation 9.13 would be different for each group of boundaries. They then calculated the grain growth of each texture component and showed that this might have a very strong effect on the grain size distribution and growth kinetics. Figure 9.7 shows the predicted change in these parameters for a material containing two texture components, A and B. The initial relative mean grain sizes are $\bar{R}_B/\bar{R}_A = 0.8$, and the mobilities are $M^{AB}/M^{AA} = 5$ i.e. the mobility of A grains growing into B grains is five times that of A grains growing into A. It may be seen that in this case the A component is strongly depleted during growth (fig 9.7a,b), that the grain size distribution is drastically altered, and that the relationship between $\bar{R}$ and t is complex (fig 9.7c). There is some evidence of agreement between the predictions of the model and experimental measurements (Abbruzzese and Lücke 1986), although too much reliance should not be placed on quantitative agreement because as discussed in §4.3.2, the relationship between the mobility of a grain boundary and the misorientation is often not known.

9.3.1.3 Computer modelling

Grest et al. (1985) extended their Monte-Carlo simulation studies of grain growth to include variable boundary energies. They set the boundary energies according to the Read-Shockley relationship (equation 3.6) and varied the energy range by altering θ_m, the angle at which the energies saturate according to this equation. The simulation is therefore close to representing the annealing behaviour of a microstructure comprising a mixture of both high angle and low angle boundaries. However, as the mobilities of the boundaries were not varied, the low angle (low energy) boundaries did not have the correspondingly low mobilities expected in a real material (§4.2). They found that the introduction of variable boundary energy led to an increase in the growth exponent (**n**) to **4**. The number of low angle boundaries increased during the anneal, compared with the constant energy simulation and the grain size distribution was broader. This simulation may be compared with the simulation of recovery shown in figure 5.20, in which the low angle boundaries had both low energies and low mobilities and in which a large growth exponent **n** and a decrease in boundary misorientation were also found.

9.3.2 The effect of grain growth on grain boundary character

9.3.2.1 Mesotexture

As automated techniques for the rapid determination of crystallite orientations have become widely available (see Appendix), there has been a growing interest in the effect of grain

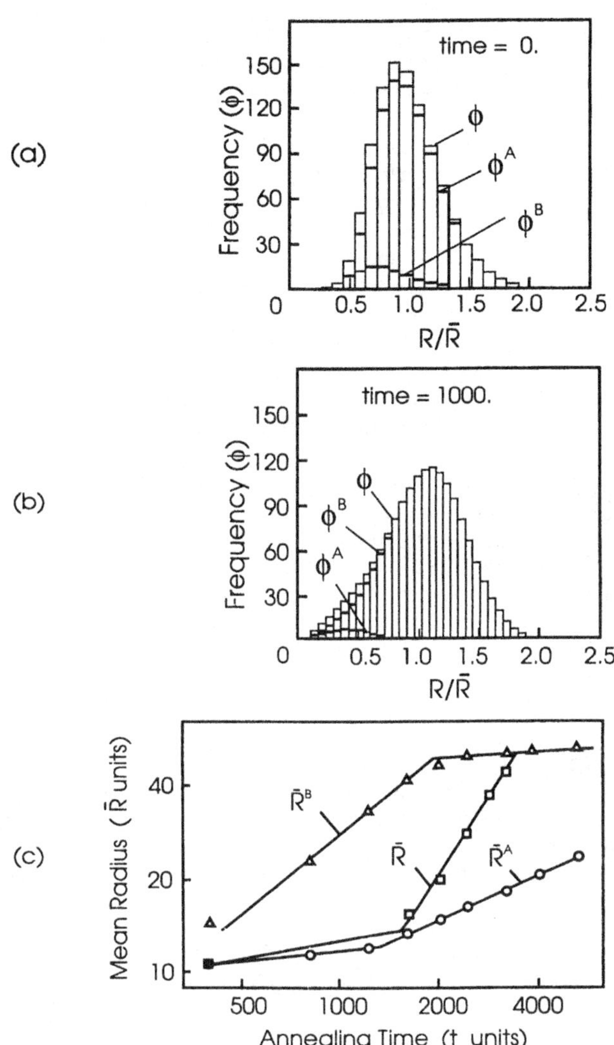

Fig. 9.7. Predicted grain growth in a microstructure containing texture components A and B. a) Initial grain size distribution, b) Grain size distribution after growth c) Growth kinetics, (after Abbruzzese and Lücke 1986).

growth and other annealing processes on **grain boundary character** (e.g. Watanabe 1992a,b) and the term **mesotexture** has been used to describe the distribution of grain boundary character (Randle 1992). The interest in this parameter arises from the fact that certain properties such as creep failure or susceptibility to intergranular corrosion may depend on the nature of a grain boundary. Low angle boundaries or special (low Σ) boundaries may be beneficial in these circumstances and therefore the possibility of controlling or **engineering** suitable grain boundary distributions is of some practical importance.

The relationship between macrotexture and mesotexture is complex and a material which does not have a strong macrotexture may possess a strong mesotexture (e.g. a large fraction of low Σ boundaries). A mesotexture would be expected to affect grain growth kinetics by altering the mobility and energy of the boundaries compared to a structure comprising random boundaries. Information as to the effect of grain growth on boundary character is as yet very incomplete and is generally confined to a consideration of only the grain misorientation and not the boundary plane.

9.3.2.2 The frequency of special boundaries

In nickel, Furley and Randle (1991) reported an increase in $\Sigma=3$ (twin) boundaries and a decrease in $\Sigma=5$ boundaries during grain growth, and Randle and Brown (1989) found an increase in both low angle ($\Sigma=1$) and other low Σ boundaries during the annealing of austenitic stainless steel. Pan and Adams (1994) reported a greatly increased number of special boundaries ($\Sigma=1,3,9$) in Inconel 699. From an analysis of results from a variety of materials, Watanabe et al. (1989) suggest that the frequency of CSL boundaries is proportional to $\Sigma^{-1/3}$. In most cases, because the material has been thermomechanically processed, it is difficult to determine whether the mesotexture is a result of primary recrystallization, grain growth, or both. Some of the clearest evidence as to the change of boundary character during grain growth has been obtained by Watanabe et al. (1989) who examined the misorientations of boundaries after grain growth and abnormal grain growth in an Fe-6.5%Si alloy which was annealed after rapid solidification. After a short annealing time (fig 9.8a) the distribution of misorientation is close to the random value (fig 3.2), but at long times (fig 9.8c) it shifts markedly towards lower misorientations and a strong {100} texture develops. The frequency of low Σ and low angle boundaries was also found to increase markedly as grain growth proceeded as shown in figure 9.8d.

9.3.2.3 Interpretation of the data

It was argued in §5.5.3.5 that in a microstructure with a distribution of boundary energies, the boundary tensions would lead to a decrease in the total amount of high energy boundary on annealing. Such behaviour has been observed in several materials (§9.3.2.2) and has been shown to occur in both Monte-Carlo (Grest et al. 1985) and vertex (Humphreys 1992b) computer simulations.

In order to illustrate semi-quantitatively, the complexities of the problem we have run a vertex simulation with an initial microstructure containing five generic types of grain boundary:

1. Random high angle boundaries of constant energy and mobility.
2. Low energy boundaries with low mobility (typified by LAGBs or $\Sigma3$ twins)
3. Low energy boundaries with high mobility (typified by low Σ boundaries in pure metals)
4. High energy boundaries with low mobility
5. High energy boundaries with high mobility

High or low energies or mobilities were given values of x5 or 1/5 respectively that of the random boundaries. It was found that the distribution of boundary types changed during grain growth as shown in figure 9.9.

It is seen that there is a tendency for the proportion of low energy boundaries to increase and for high energy boundaries to decrease during grain growth. However, the boundary mobility plays a significant role, and it is interesting that the largest increase is predicted for

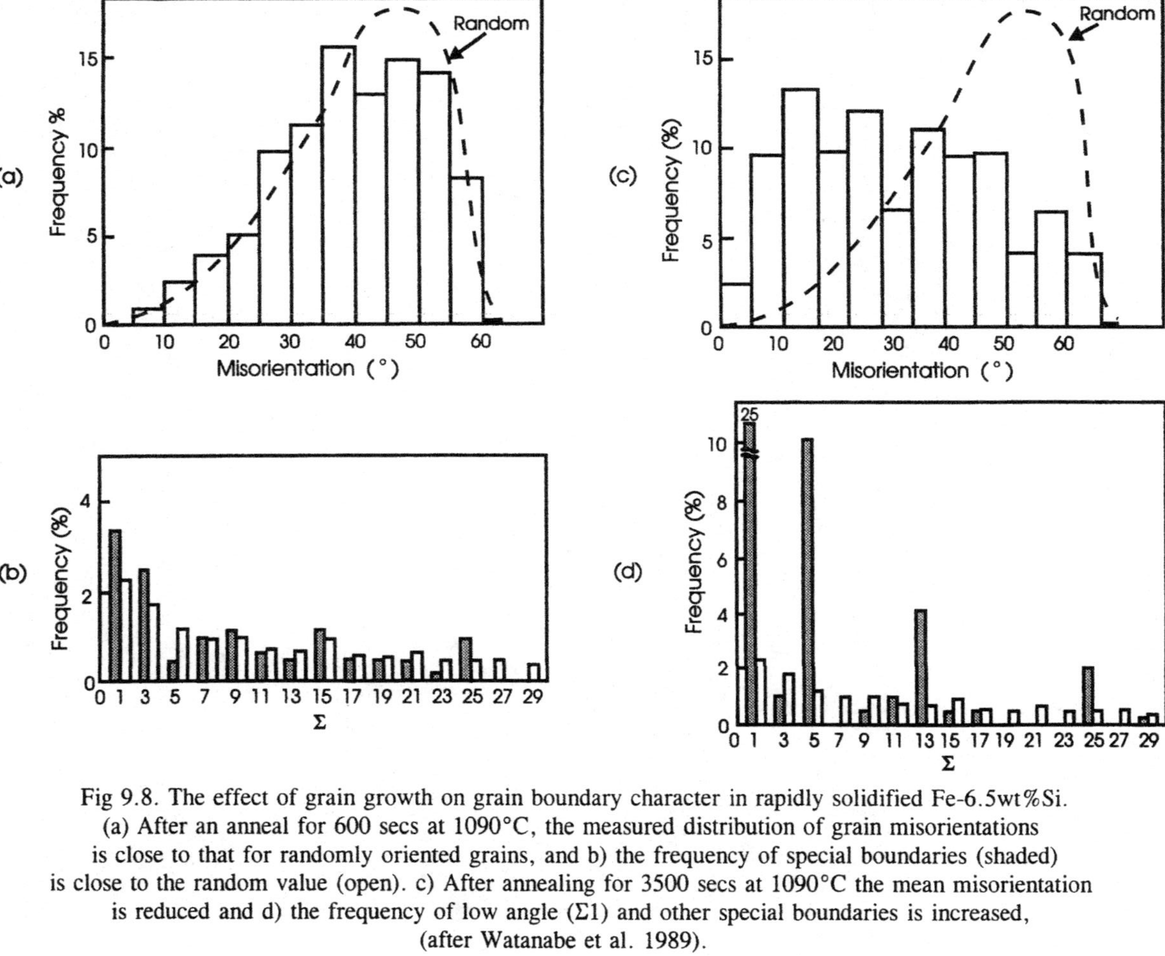

Fig 9.8. The effect of grain growth on grain boundary character in rapidly solidified Fe-6.5wt%Si. (a) After an anneal for 600 secs at 1090°C, the measured distribution of grain misorientations is close to that for randomly oriented grains, and b) the frequency of special boundaries (shaded) is close to the random value (open). c) After annealing for 3500 secs at 1090°C the mean misorientation is reduced and d) the frequency of low angle ($\Sigma 1$) and other special boundaries is increased, (after Watanabe et al. 1989).

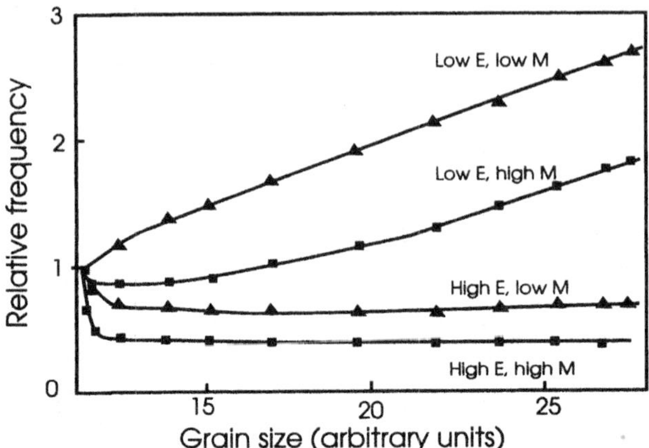

Fig. 9.9. Computer simulation of the change in boundary type during grain growth. The boundary frequencies are expressed as multiples of the frequency of a "random" boundary.

low energy/low mobility boundaries such as LAGBs ($\Sigma = 1$) and other low Σ boundaries, which is in accord with the experimental results of figure 9.8. If it is assumed that the development of the mesotexture during grain growth is controlled primarily by the boundary energies and mobilities, then there will be difficulties in modelling this for real materials because, as discussed in chapter 4, these parameters are often not known and are strongly dependent on factors such as purity. There is already evidence that the mesotexture is dependent on solute (Furley and Randle 1991, Palumbo and Aust 1990).

We conclude that there is experimental and theoretical evidence that bulk and local orientation effects are important during grain growth, and that these affect both the kinetics of grain growth and the microstructure. There is growing evidence that the **ideal materials with isotropic and unchanging boundary energy which are addressed in the standard theories of grain growth simply do not exist**, and the most fruitful developments in the theory of grain growth in single-phase materials are likely to be those which take a more realistic account of orientation effects during grain growth.

9.4 THE EFFECT OF SECOND-PHASE PARTICLES ON GRAIN GROWTH

It was shown in §3.6 that second-phase particles exert a strong pinning effect (**Zener pinning**) on boundaries, with the pinning pressure being determined primarily by the size, volume fraction, interface and distribution of the particles. Because the driving pressure for grain growth is extremely low, particles may have a very large influence both on the kinetics of grain growth and on the resultant microstructure. It should be noted that at the high temperatures at which grain growth occurs, the dispersion of second-phase particles may not be stable. We will first deal with the behaviour of a material containing a stable dispersion of particles and at the end of this section consider the cases where the particles themselves form, coarsen or migrate during the process of grain growth. It should be noted that the quantitative relationships presented below are based generally on the interaction of boundaries with a random distribution of non-coherent equiaxed particles. For other types of particle the retarding pressure due to the particles may be different as discussed in §3.6.1 and appropriate modifications to the theory should therefore be made.

Although the discussion below is presented in terms of **normal grain growth**, many of the considerations are also relevant to the **growth of a subgrain structure during recovery**. Those aspects of the theory which are particularly relevant to subgrain growth are further discussed in chapter 5.

9.4.1 Kinetics

A dispersion of stable second-phase particles will reduce the rate of grain growth because the driving pressure for growth (P in equation 9.3) is opposed by the pinning pressure (P_z) due to the particles (equation 3.21). If this is incorporated into a simple grain growth theory such as the Burke and Turnbull (1952) analysis (§9.1.3), then the rate of grain growth becomes

$$\frac{dR}{dt} = M(P - P_z) = M\left(\frac{\alpha\,\gamma_b}{R} - \frac{3\,F_v\,\gamma_b}{2\,r}\right) \tag{9.29}$$

This predicts a growth rate which is initially parabolic, but which subsequently reduces and eventually stagnates when $P = P_z$.

Hillert (1965) extended his theory of normal grain growth to include the effects of particle pinning on the kinetics of grain growth and on the grain size distribution. The pinning pressure due to the particles results in a modification to the growth rate for single-phase materials (equation 9.13)

$$\frac{dR}{dt} = c\,m\,\gamma_b\left(\frac{1}{R_{crit}} - \frac{1}{R} \pm \frac{z}{c}\right) \tag{9.30}$$

where $c = 0.5$ for 2-D and 1 for 3-D and $z = 3F_v/4r$

For grains in the size range $1/R \pm z/c$, the net pressure for boundary migration will be zero and therefore no growth or shrinkage will occur. Grains larger or smaller than this will shrink or grow, but at a reduced rate. Hillert suggests that the mean growth rate will be given by

$$\frac{d\bar{R}^2}{dt} = \frac{c}{2}\cdot M\,\gamma_b\left(1 - \frac{z\bar{R}}{c}\right)^2 \tag{9.31}$$

which predicts a more gradual retardation of growth rate than equation 9.29. Hillert (1965) also points out that the grain size distribution will be affected by particle pinning. Abbruzzese and Lücke (1992) have extended this model and showed that the width of the grain size distribution during normal grain growth should be reduced by particle pinning, for which there is some experimental evidence (Tweed et al. 1982).

9.4.2 The particle-limited grain size

It was shown many years ago by Zener (1948) that when the pressure on a boundary due to particle pinning equalled the driving pressure for grain growth, growth would cease and a **limiting** grain size would be reached. The existence of a limiting grain size is of great practical importance in preventing grain growth during high temperature heat treatment of industrial alloys.

9.4.2.1 The Zener limit

In the situation which was first considered by Zener, the grain boundary is considered to be macroscopically planar as it interacts with the particles and therefore the pinning pressure (P_z) is given by equation 3.21. The driving pressure for growth (P) arises from the curvature of the grain boundaries, and is given by equation 9.3. Grain growth will cease when $P = P_z$. i.e.

$$\frac{\alpha \, \gamma_b}{R} = \frac{3 \, F_v \, \gamma_b}{2r} \tag{9.32}$$

If the mean grain radius is taken to equal the mean radius of curvature (R) then we obtain a limiting grain size

$$D_z = \frac{4 \, \alpha \, r}{3 \, F_v} \tag{9.33}$$

Setting $\alpha = 1$ (some authors use other values), results in the well known **Zener limiting grain size**

$$D_{Zener} = \frac{4 \, r}{3 \, F_v} \tag{9.34}$$

It has long been recognised that this is only an approximate solution, and numerous alternative approaches have been attempted. Hillert (1965) derived a limiting grain size from equation 9.30 by equating $1/R_{crit}$ with z/c for the case when R is large, giving a limiting grain radius for the 3-D analysis of $4r/3F_v$, i.e. a grain diameter of twice that of equation 9.34 (α in equation 9.33 equal to 0.5). Gladman (1966) developed a geometric model consisting of tetrakaidecahedral grains and considered the effect of particles on the growth and shrinkage of the grains. He concluded that the limiting grain size was given by

$$D_G = \frac{\pi \, r}{3 \, F_v} \left(\frac{3}{2} - \frac{2}{Z} \right) \tag{9.35}$$

where Z is the ratio of the maximum grain size to the average grain size, a parameter which is not readily calculated, but which is expected to lie between 1.33 and 2. Hillert's (1965) grain growth theory gives $Z = 1.6$, and Gladman has suggested $Z = 2$. Using this value, Gladman's model, which has the same dependence on F_v and r as equation 9.33, predicts a limiting grain size which is smaller than that of equation 9.34, corresponding to equation 9.33 with $\alpha = 0.375$.

Numerous refinements to the Zener treatment have been carried out (Louat 1982, Hellman and Hillert 1975, Hillert 1988) and **these generally predict a limiting grain size which is**

of a similar form to equation 9.33 with 0.25 < α < 0.5, i.e. considerably smaller than that predicted by equation 9.34.

9.4.2.2 Comparison with experiment
There are surprisingly few experimental results available to test these relationships accurately in materials containing low volume fractions of particles, and in particular to verify the dependence of D_Z on F_V. Gladman (1980) compared measurements of the limiting grain size (D_Z) from a number of investigations with his theory and found reasonable agreement. Tweed et al. (1982) in a very detailed and widely quoted metallographic study of the recrystallized grain sizes and size distributions in three aluminium alloys containing very low volume fractions of Al_2O_3, found limiting grain sizes which in some cases corresponded to very small values of α in equation 9.33. However the data showed considerable scatter and it is difficult to reconcile the trend of the data with any model of limiting grain size.

The limiting grain sizes obtained by Koul and Pickering (1982) after grain growth of iron alloys containing volume fractions of $\sim 5 \times 10^{-3}$ of carbide particles are shown in figure 9.10. It may be seen that although the data are limited and scattered, there is fair agreement with the theoretical line corresponding to equation 9.33 with α=0.37, i.e. Gladman's model.

9.4.2.3 Particle-boundary correlation effects
The analyses discussed in §9.4.2.1 assumed that the pinning pressure (P_z) was that for a macroscopically planar boundary. However, as discussed in §3.6.2.2, when the interparticle spacing is similar to the grain size, this assumption is not valid, and non-random correlation of particles and boundaries must be taken into account. This will be particularly important in materials with large volume fractions of particles. In this situation, a limiting grain size (D_{Z1}) is found by equating the driving pressure (equation 9.3) with the pinning pressure as given by equation 3.24.

$$D_{Z1} = \left(\frac{8\,\alpha}{3\,F_v} \right)^{1/2} . r \sim \frac{1.6\,\alpha^{1/2}\,r}{F_v^{1/2}} \tag{9.36}$$

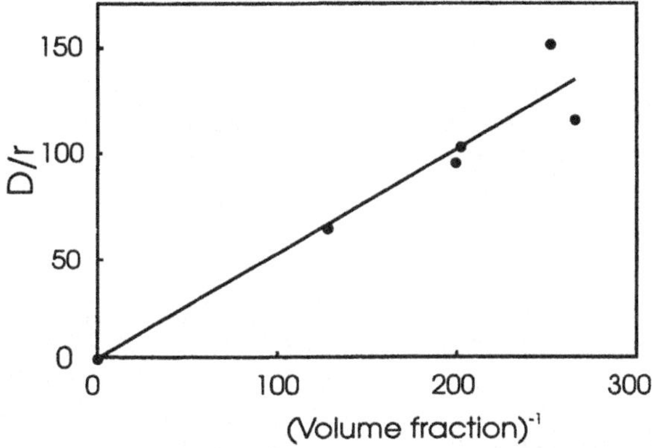

Fig. 9.10. The particle-limited grain size in carbide-containing Fe-Ni-Cr alloys, (data from Koul and Pickering 1982).

The limiting grain size given by equation 9.36 is very similar to that suggested by Anand and Gurland (1975) for subgrain growth (fig 5.29 and §5.6.2.1). It is however likely that D_{Z1} which represents the situation when all particles are on boundary corners but not all boundary corners are occupied by particles (fig 3.20a), represents a lower bound for the limiting grain size and that grain growth will actually continue to occur until all grain corners are pinned, i.e the grain size D_C of equation 3.25 (fig 3.20b)).

$$D_{zc} = D_c \sim N_v^{-1/3} = \frac{\beta r}{F_v^{1/3}} \qquad (9.37)$$

where β is a small geometric constant.

Various authors (Hellman and Hillert 1975, Hillert 1988, Hunderi and Ryum 1992a) have discussed this type of relationship, and Hillert (1988) suggests that $\beta = 3.6$.

The experimental measurements of limiting grain size in alloys containing large volume fractions of particles ($F_v > 0.05$) are very scattered (Hazzledine and Oldershaw 1990, Olgaard and Evans 1986) and although there is some indication that the limiting grain size is inversely proportional to F_v^n, where $n < 1$, the results are by no means conclusive. If equation 9.37 is assumed to hold, then analyses of the results of Hellmann and Hillert (1975) on steels and the data for Al-Ni of figure 9.13 give β as 3.3 and 3.4 respectively. The subgrain data of Anand and Gurland (1975) which were analyzed by them in terms of equation 9.36 also fit equation 9.37 with $\beta = 2.7$.

Figure 9.11 shows the predicted variation of the limiting grain sizes D_Z and D_{ZC} with volume fraction according to equations 9.33 and 9.37 using values of the geometric constants ($\alpha = 0.35$, $\beta = 3$) which are consistent with experiment. It may be seen that the curves cross at a volume fraction of 0.06. At volume fractions less than this the particle-correlated limit (D_{ZC}) is unstable with respect to growth, and the uncorrelated limit (D_Z) is appropriate, whereas at higher volume fractions the limit will be D_{ZC}. In practice there will be a gradual transition between correlated and uncorrelated pinning as considered in §3.6.2.2, and this has been discussed by Hunderi and Ryum (1992a). The actual volume fraction at which the transition from D_Z to D_{ZC} occurs is not yet firmly established either theoretically or experimentally, but theoretical estimates (Hillert 1988, Hazzledine and Oldershaw 1990, Hunderi & Ryum 1992a) put it within the range $0.01 < F_v < 0.1$, which is consistent with figure 9.11.

We conclude that the effective limiting grain size (D_{lim}), will be the larger of D_Z or D_{ZC}, the solid line in figure 9.11, and that the transition will occur at a volume fraction of ~ 0.05.

9.4.2.4 Computer simulation

Extensive computer simulation of particle-controlled grain growth has been carried out using the Monte-Carlo technique (§13.2.2). The initial work (Srolovitz et al. 1984b) which was a 2-D simulation found a limiting grain size proportional to $F_v^{-1/2}$. However, it has since been shown (Hillert 1988, Hazzeldine and Oldershaw (1990) that the situation in 3-D is quite different and that the **2-D simulations of particle-limited grain growth are not applicable to grain growth in a 3-D microstructure.** Three dimensional Monte-Carlo simulations have been carried out more recently (Anderson et al. 1989b, Hazzledine and Oldershaw 1990).

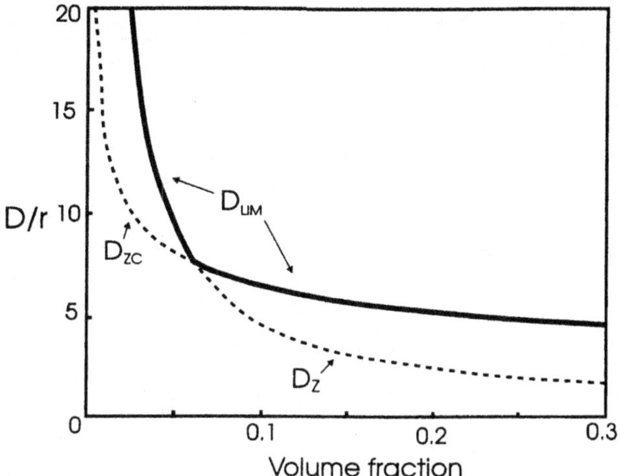

Fig. 9.11. The predicted variation with particle size and volume fraction of the particle-boundary correlated limiting grain size (D_{ZC}) for $\beta=0.3$, and the non-correlated limiting grain size (D_Z) for $\alpha=0.35$.

These produce a limiting grain size in accordance with equation 9.37, although their value of β (~ 9) is considerably larger than that found experimentally (§9.4.2.3). Currently, 3-D Monte-Carlo simulations are limited to a lattice of 100^3 units, with a particle diameter equal to one unit. Therefore for volume fractions less than ~ 0.05, as may be seen from equation 9.37, the limiting grain size approaches the size of the array, and accurate results cannot be obtained. The simulations are therefore currently limited to the range in which D_{ZC} is expected to apply and can neither verify nor contradict the prediction that the limiting grain size (D_{lim}) changes from D_Z to D_{ZC} at a critical volume fraction.

9.4.3 Particle instability during grain growth

We have assumed so far, that the second-phase particles are stable during grain growth. There are many situations in which this is not so, and in this section we examine the consequences of instability of the second-phase during grain growth for three important cases.

9.4.3.1 Precipitation after grain or subgrain formation

In some situations the second-phase particles may be precipitated after the formation of a grain or subgrain structure. In this situation the particles are unlikely to be distributed uniformly and will form preferentially on the boundaries. The pinning pressure due to the particles will therefore be greater than for a random particle distribution (Hutchinson and Duggan 1978) as discussed in §3.6.2.3. If it is assumed that all the particles are on boundaries and randomly distributed on these boundaries, then the pinning pressure is given by equation 3.33 and by equating this with the driving pressure given by equation 9.3, the limiting grain size is

$$D_{ZP} = r \left(\frac{8\,\alpha}{F_v} \right)^{1/2} \tag{9.38}$$

It is expected that in practice this situation will be more applicable to the **growth of subgrains during recovery** (§5.6.2.3) than to **grain growth**, although no quantitative tests of equation 9.38 have yet been reported.

9.4.3.2 Coarsening of dispersed particles during grain growth

An important situation is one in which grain growth has stagnated due to the particle dispersion, but the particles coarsen (Ostwald ripening). In this situation the grain size (D) is equal to D_{lim} as defined in §9.4.2.3, and the rate of growth is controlled by the rate of change of particle size. Thus

$$\frac{dR}{dt} = c \frac{dr}{dt} \qquad (9.39)$$

where the constant c is $2\alpha/3F_V$ for low volume fractions and $\beta/2F_V^{1/3}$ for large volume fractions.

The rate of particle coarsening will depend on the rate controlling mechanism (Hillert 1965, Gladman 1966, Hornbogen and Koster 1978, Ardell 1972, Martin and Doherty 1976). If the particle growth is controlled by volume diffusion (diffusivity $=D_s$) then

$$\bar{R}^3 - \bar{R}_0^3 = c_1 D_s t \qquad (9.40)$$

whereas for particle growth controlled by diffusion along the grain boundaries (diffusivity$=D_b$), which is commonly found for large volume fractions then

$$\bar{R}^4 - \bar{R}_0^4 = c_2 D_b t \qquad (9.41)$$

Figure 9.12 shows the kinetics of grain growth in an Al-6wt%Ni alloy which contains a volume fraction of 0.1 of NiAl$_3$ particles of initial diameter 0.3μm. The grain growth kinetics, which are controlled by the particle coarsening are seen to be in accord with equation 9.40 over a wide temperature range.

Figure 9.13 shows the relationship between the size of the grains and the second-phase particles in the same alloy system. The grain size is found to be proportional to the particle size as would be predicted by equation 9.37. From the slope of the line, the constant β in the equation is found to be 3.4.

Coarsening of the particles by grain boundary diffusion will of course only affect those particles which are on grain boundaries, and other particles will coarsen more slowly. However, as grain growth occurs the boundaries will lose some particles and acquire others, and therefore the overall particle coarsening rate may be uniform, although the rate constant in equation 9.40 should include a correction factor to account for the fraction of the time a particle is not attached to a boundary.

9.4.3.3 Coarsening of duplex microstructures

In a duplex alloy in which the volume fractions of the two phases are comparable then the rate of coarsening of the structure is controlled by interdiffusion between the phases over distances which are of the order of the size of the phases. In this situation the coarsening rate is predicted to be dependent on the coarsening mechanism and morphology and volume

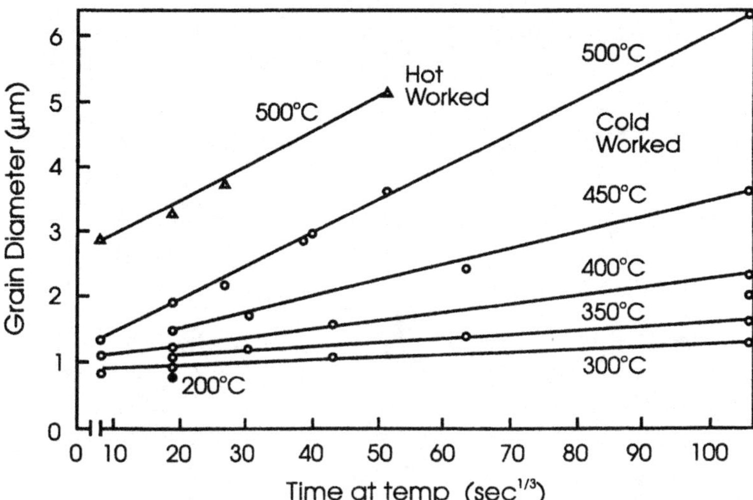

Fig. 9.12. The kinetics of grain growth when controlled by particle coarsening for an Al-6wt%Ni alloy containing a volume fraction of 0.10 of $NiAl_3$ particles, (Morris 1976).

fractions of the phases (e.g. Ardell 1972, Martin and Doherty 1976). A detailed analysis of such phase growth is beyond the scope of this book, but typically, it is expected that the growth exponent will be ~3 for bulk diffusion control (equation 9.40) and ~4 for interface control (equation 9.41). Grewel and Ankem (1989, 1990) have studied grain growth in a variety of titanium-based alloys containing a wide range of volume fractions, such as the microstructure shown in figure 9.14 and find that, as shown in figure 9.15, the growth exponent is not particularly sensitive to volume fraction. However, it is found that as the temperature is raised from 700°C to 835°C the growth exponent **n** in Ti-Mn alloys decreases

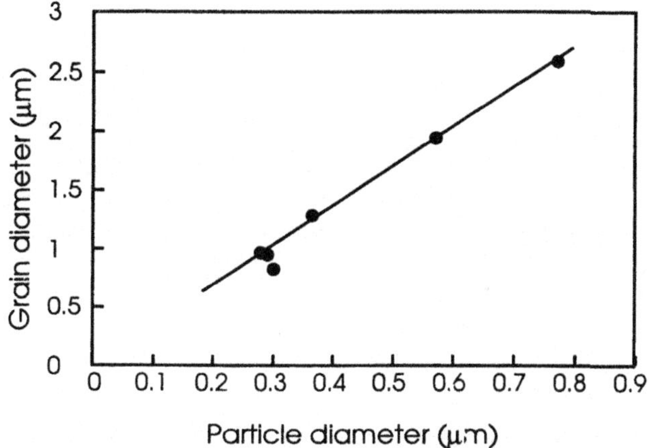

Fig. 9.13. The relationship between grain size and particle diameter for an Al-6wt%Ni alloy containing a volume fraction of 0.10 of $NiAl_3$ particles, (Humphreys and Chan 1996).

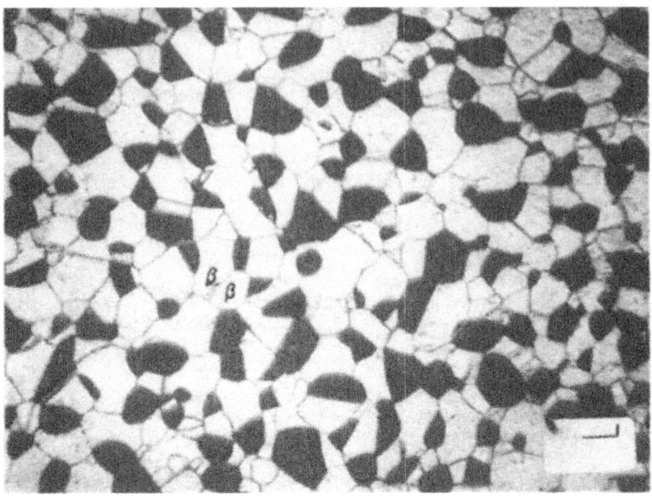

Fig 9.14. Microstructure of a Ti-6%Mn alloy containing 44% of α-phase (dark) and 56% of β-phase, (Grewel and Ankem 1990).

from 3.6 to 3.1, indicating a change in growth mechanism from interface to bulk diffusion control with increasing temperature.

Although Higgins et al. (1992) report a growth exponent of ~ 4 for duplex Ni-Ag alloys, which is consistent with interface diffusion-controlled coarsening as discussed above, they find that the rate constants are several orders of magnitude greater than predicted by the theory of Ostwald ripening. They suggest that the non-spherical shape of the particles (e.g. fig 9.14) and the resulting curvature provide a driving force for **particle migration and coalescence**.

9.4.4 Grain rotation

Randle and Ralph (1987) found that in a nickel-based superalloy in which grain growth was inhibited by the second-phase particles, there was a much higher frequency ($\sim 50\%$) of low Σ boundaries than in the same alloy in which precipitation had not occurred. It was suggested by the authors that this might be the result of grains rotating to produce lower energy boundaries (c.f. subgrain rotation in §5.5.4). There is as yet no direct experimental evidence nor any quantitative theoretical treatment to support this proposed mechanism, and as discussed in §9.3.2, a high frequency of special boundaries may arise as a result of grain growth by boundary migration.

9.4.5 Dragging of particles by boundaries

Equation 9.29 shows that grain growth in a particle-containing material occurs when the driving pressure (P) is greater than the pinning pressure (P_z), and in the subsequent discussions it has been assumed that growth ceases at a limiting grain size when these two pressures are equal. However, the pressure exerted by a boundary on a particle may, at high

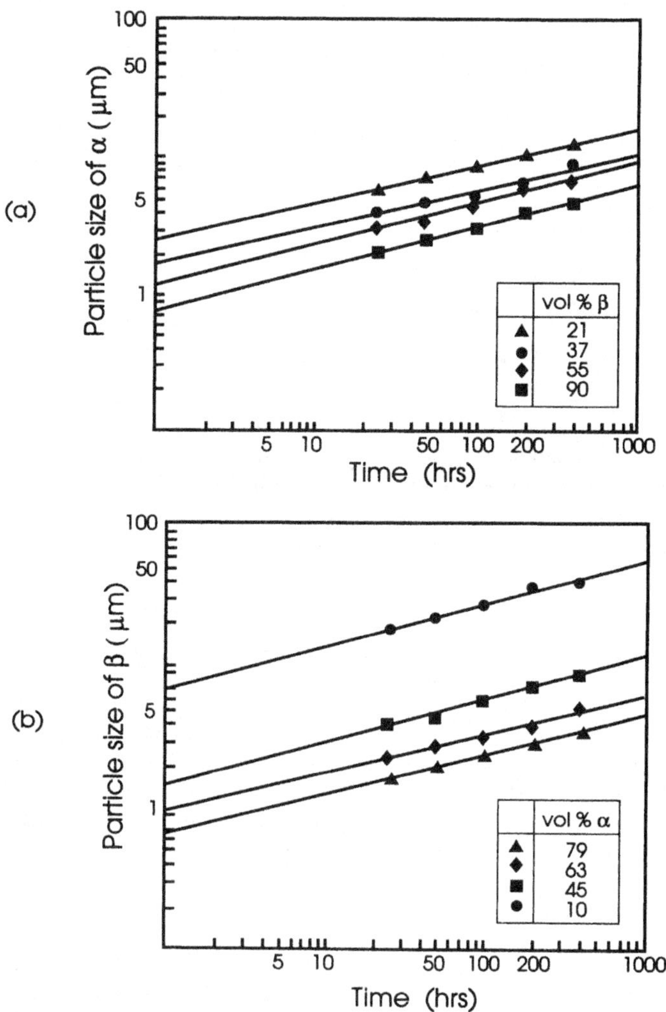

Fig 9.15. Growth kinetics of the phases in a series of Ti-Mn alloys containing different volume fractions at 973K. a) α-phase   b) β-phase, (Grewel and Ankem 1989).

temperatures be sufficient to drag the particle through the matrix, and in these circumstances boundary migration will continue at a rate determined by the mobility of the particle in the matrix (M_p) (Ashby 1980, Gottstein and Schwindlerman 1993).

The effect of driving pressure on boundary velocity is shown schematically in figure 9.16. For low driving pressures, ($P < P_z$) the boundary moves together with the particles and the velocity is given by

$$v = \frac{M_p P}{N_s} \tag{9.42}$$

where N_s is the number of particles per unit area of boundary and M_p is the particle mobility.

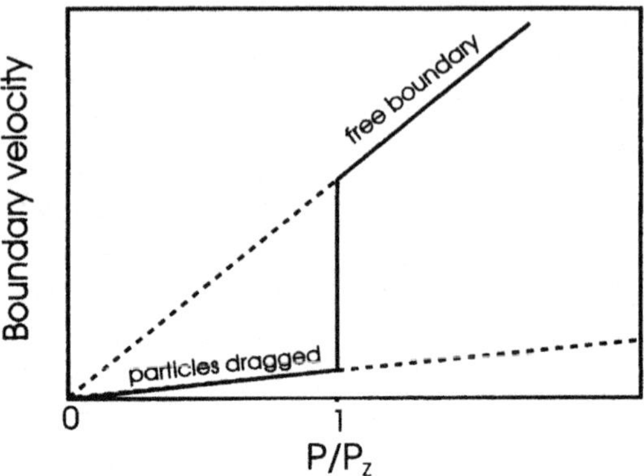

Fig. 9.16. The grain boundary velocity as a function of the driving pressure (P). At low driving pressures the boundary velocity is controlled by the dragging of particles, but when the driving pressure exceeds the pinning pressure due to particles (P_z), the boundary breaks free from the particles.

However, when $P > P_z$ the boundary breaks away from the particles and velocity is given by $M(P-P_z)$, where M is the intrinsic mobility of the boundary (i.e. equation 9.29).

There are several mechanisms by which the particles can move through the matrix (Ashby 1980), including the migration of atoms (matrix or particle) by diffusion through the matrix, through the particle or along the interface. Ashby (1980) has shown that in general, coupled diffusion of both matrix and particle atoms is required. The resulting particle mobilities are low and are generally negligible for stable crystalline particles. However, for gas bubbles or amorphous or liquid particles, measurable particle velocities may occur at very high temperatures (Ashby 1980).

Ashby and Centamore (1968) observed particle drag in copper containing various oxide dispersions annealed at high temperatures. As particles are swept along with the boundary, the number of particles on the boundary (N_s) increases as is seen in figure 9.17, and therefore the boundary velocity decreases according to equation 9.42. The mobilities of amorphous SiO_2, GeO_2 and B_2O_3 were all measurable at high temperatures, but significantly no migration of stable, crystalline Al_2O_3 particles was detectable. This phenomenon is therefore unlikely to be important in alloys containing stable crystalline particles. However, if such a mechanism were to occur in an engineering alloy, the resulting increased concentration of particles at grain boundaries would be expected to lead to increased brittleness and susceptibility to corrosion.

9.5 ABNORMAL GRAIN GROWTH

In the previous section we discussed the uniform growth of grains following recrystallization. There are however, circumstances when the microstructure becomes unstable and a few grains may grow excessively, consuming the smaller recrystallized grains (fig 9.1b). This

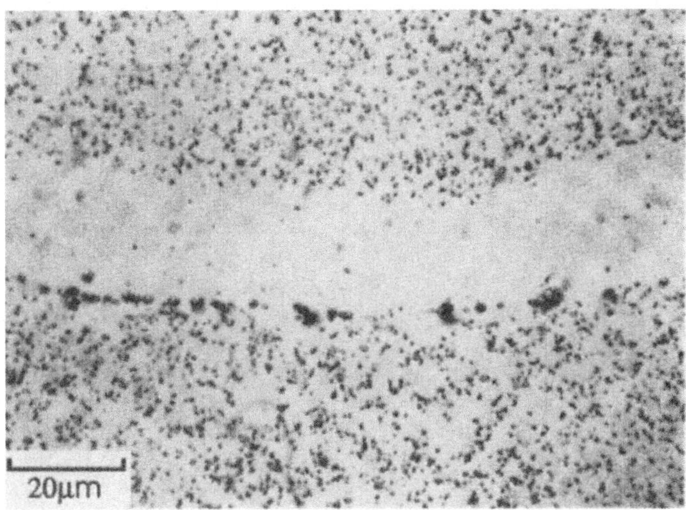

Fig. 9.17. Particle dragging by a boundary in copper containing amorphous SiO_2 particles. The particle-free band behind the migrating boundary and the increased particle concentration at the boundary are clearly seen, (Ashby and Palmer 1967).

process, which may lead to grain diameters of several millimetres or greater is known as **abnormal grain growth.** Because this discontinuous growth of selected grains has similar kinetics to primary recrystallization and has some microstructural similarities, as shown in figure 9.18, it is sometimes known as **secondary recrystallization.** Abnormal grain growth is an important method of producing large grained materials and contributes to the processing of Fe-Si alloys for electrical applications (§12.4). The avoidance of abnormal grain growth at high temperatures is an important aspect of grain size control in steels and other alloys. Dunn and Walter (1966) give an extensive review of abnormal grain growth in a wide variety of materials.

9.5.1 The phenomenon

The driving force for abnormal grain growth is usually the reduction in grain boundary energy as for normal grain growth. However, in thin materials an additional driving force may arise from the orientation dependence of the surface energy (§9.5.4). Abnormal grain growth originates by the preferential growth of a few grains which have some special growth advantage over their neighbours, and the progress of abnormal grain growth may be described in some cases by the JMAK kinetics of equation 6.17 (Dunn and Walter 1966).

An important question to consider is whether or not abnormal grain growth can occur in an "ideal grain array". i.e. one in which there are no impurities and the boundary energy is constant. As was shown by Thompson et al. (1987), the answer to this question can be deduced from Hillert's (1965) theory of grain growth. From equation 9.13, we see that the rate of growth of a grain of radius R is proportional to $(1/R_{crit} - 1/R)$ where R_{crit} is the radius of a grain that will neither grow nor shrink, which was shown by Hillert to equal the mean

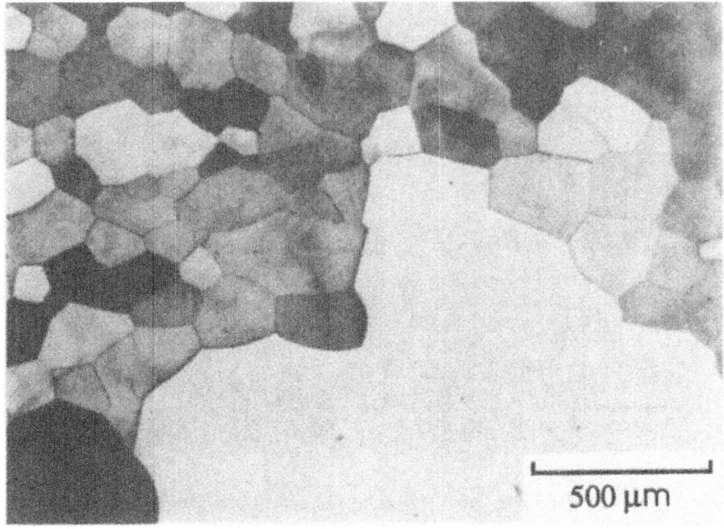

Fig. 9.18. Abnormal grain growth in Fe-3% Si annealed at 1373K, (Detert 1978).

grain radius of the assembly ($\bar{R}$). If we consider a large grain of radius R in the structure, then its growth rate relative to that of the normally growing grains is

$$\frac{d}{dt}\left(\frac{R}{\bar{R}}\right) = \frac{1}{\bar{R}^2}\left(\bar{R}\frac{dR}{dt} - R\frac{d\bar{R}}{dt}\right) \qquad (9.43)$$

Thus the large grain will grow faster than the normally growing grains and lead to abnormal grain growth if

$$\bar{R}\frac{dR}{dt} - R\frac{d\bar{R}}{dt} > 0 \qquad (9.44)$$

Substituting for dR/dt and d$\bar{R}$/dt from equations 9.13 and 9.14 (putting $\bar{R} = R_{crit}$), then it is seen that although $\bar{R}$dR/dt = Rd$\bar{R}$/dt when R=2$\bar{R}$, the condition of 9.44 is never achieved. **Thus a very large grain will always grow more slowly than the average grain and will eventually rejoin the normal size distribution. Therefore abnormal grain growth cannot occur in an "ideal grain assembly".** A similar result has also been obtained by Monte-Carlo simulation of abnormal grain growth (Srolovitz et al. 1985). Abnormal grain growth can therefore only occur when normal grain growth is inhibited, unless the abnormally growing grain enjoys some advantage other than size over its neighbours. The main factors which lead to abnormal grain growth - **second-phase particles**, **texture** and **surface effects**, are considered below.

9.5.2 The effect of particles

The discussion below, which is an extension of the analyses of §9.4.2, is based on similar principles and comes to broadly similar conclusions to the earlier work of Hillert (1965) and Gladman (1966).

9.5.2.1 Conditions for abnormal grain growth

If we consider an assembly of grains of mean size D_M whose growth is prevented by the particles, then this grain structure provides the driving force for abnormal grain growth. If the boundaries are of energy γ_b, then the driving pressure for abnormal grain growth (c.f. equation 2.7) is

$$P_m = \frac{3\,\gamma_b}{D_M} \qquad\qquad (9.45)$$

This is opposed by the particle pinning pressure (P_z of equation 3.21) and therefore the condition under which a **very large grain** will grow, is $P_m > P_z$ or

$$D_M \leq \frac{2\,r}{F_V} \qquad\qquad (9.46)$$

If the grain size following primary recrystallization is less than the particle-limited grain size, then we can take D_M to be D_{lim} which is the larger of D_Z (equation 9.33) or D_{ZC} (equation 9.37), i.e. the solid line in figure 9.11.

In figure 9.19, the critical value of D_M is compared with D_{lim} and it is seen that for all likely volume fractions the condition for abnormal grain growth is satisfied and therefore it is predicted on the basis of the analysis that **abnormal grain growth will be viable in an alloy in which normal grain growth has stagnated due to particle pinning**.

Equation 9.46 is based on the propagation of a planar boundary (infinitely large grain), whereas in practice the "nucleation" of abnormal grain growth (i.e. growth of a favoured large grain) may be the limiting factor. The above analysis may be modified to take into account the finite size of the abnormally growing grain, by adding a retarding force due to the radius of curvature of the grain ($c\gamma_b/R_a$), where R_a is the radius of curvature of the abnormal grain, and c is a constant often taken as 2, so that the condition for its growth becomes

$$P_m > P_z + \frac{c\,\gamma_b}{R_a} \qquad\qquad (9.47a)$$

or

$$D_M < \frac{2\,r}{F_v}\left(1 - \frac{2\,c}{3\,x}\right) \qquad\qquad (9.47b)$$

where $x = D_a/D_M$ and $D_a = 2R_a$

The condition for the abnormal growth of a grain equal to **twice** that of the array ($x=2$) is shown as the broken line in figure 9.19. It is seen that growth of a grain of this size is now

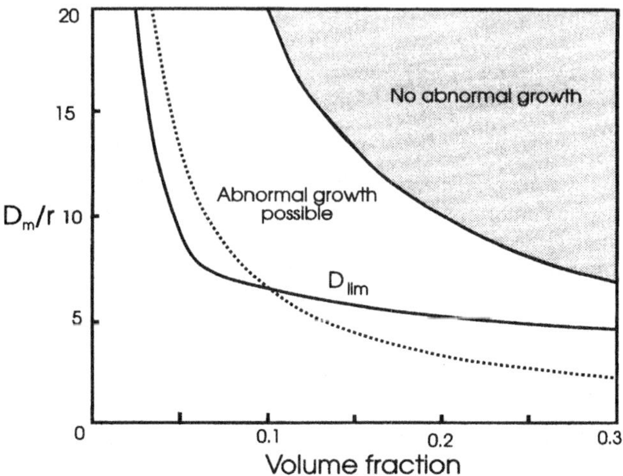

Fig. 9.19. A planar boundary (infinitely large grain) can propagate into a matrix of particle-pinned grains of size D_M if D_M is within the shaded region, and a grain of diameter $2D_M$ can grow if D_M is below the dotted line. The line D_{lim} is the value of D_M corresponding to the particle-limited grain size.

only possible if the particle volume fraction is less than ~ 0.1. Although this analysis is not rigorous and is sensitive to the values of the constants α, β and c in equations 9.32, 9.37 and 9.47, it does indicate that whereas abnormal grain growth may be viable in most particle-containing alloys, its nucleation may be the critical factor in alloys containing larger volume fractions, and this will be sensitive to the grain size distribution.

9.5.2.2 Experimental observations
Abnormal grain growth has been reported for a large number of alloys containing particle volume fractions of between 0.01 and 0.1, and details may be found in Dunn and Walter (1966), Mould and Cotterill (1976) and Detert (1978). However, there is very little evidence in the literature as to the precise conditions necessary to induce or prevent this phenomenon.

The fact that abnormal grain growth does not occur more readily in particle-containing alloys may be due to a number of factors.

(i) In many cases the grain size produced during primary recrystallization is significantly greater than the particle-limited grain size, and therefore according to the analysis above, abnormal grain growth may not be possible. As pointed out by Hillert (1965), this is a very effective way of suppressing abnormal grain growth. As discussed in §8.3.1, the grain size produced by primary recrystallization is a complex function of the particle parameters and the thermomechanical processing route, and in alloys in which the ratio F_V/r is large, particle pinning often results in a large grain size after primary recrystallization. As an example, consider the recrystallization data shown in figures 8.20 and 8.21 for particle-containing aluminium alloys which exhibit accelerated and retarded recrystallization. Data for two specimens from each investigation (the extreme data points of each figure), are summarized in table 9.3. It can be seen that particularly for the material showing retarded recrystallization (code 1), the measured grain size is very much larger than the limiting grain

Table 9.3

Conditions for abnormal grain growth in some particle-containing aluminium alloys

Code	Alloy	F_V	d (μm)	Grain size (μm)	D_M (μm)	D_Z (μm)	Reference
1	Al-Cu	0.053	0.56	1483	11	2.5	Doherty and Martin (1964)
2	Al-Cu	0.03	1.01	68	67	15.7	Doherty and Martin (1964)
3	Al-Si	0.0012	0.7	160	583	114	Humphreys (1977)
4	Al-Si	0.01	4.9	44	490	16.3	Humphreys (1977)

size (D_Z as given by equation 9.33 with $\alpha = 0.35$). If the measured grain sizes are compared with D_M, the critical grain size for abnormal grain growth (equation 9.46), it may be seen that abnormal grain growth will not be possible in material 1, is marginal in material 2 and is highly likely in materials 3 and 4.

(ii) The occurrence of abnormal grain growth may be limited by "nucleation" rather than growth considerations. Strong evidence for this comes from the large number of experimental observations which show that abnormal grain growth is particularly likely to occur as the annealing temperature is raised and as the particle dispersion becomes unstable. This is clearly shown in the classic early work of May and Turnbull (1958) on the effect of MnS particles on abnormal grain growth in silicon iron. Gladman (1966, 1992) has discussed the unpinning of boundaries under conditions of particle coarsening, with particular reference to steels in which the control of abnormal grain growth is of great technological importance, and the grain growth behaviour of some steels is shown in figure 9.20.

Plain carbon steels without grain refining additives show a coarsening of the grain size by normal grain growth. However, if conventional grain growth inhibitors (AlN) are present (fig. 9.20a), then small grains persist up to a temperature of ~ 1050°C at which temperature coarsening of the particles allows abnormal grain growth to occur, producing a very large grain size. At temperatures of ~ 1200°C, most of the particles have dissolved and normal grain growth then results. However, if a coarser dispersion of more stable particles (e.g. oxides or titanium carbonitrides) is present (fig 9.20b) then all grain growth is inhibited to very high temperatures.

According to the analysis of §9.5.2.1, a coarsening of the particle dispersion or a lowering of the volume fraction should not in itself make abnormal grain growth more likely. Therefore what is probably occurring is local destabilization of the grain structure by removal or weakening of critical pinning points (Gladman 1966), leading to local inhomogeneous grain growth. The resulting broader grain size distribution then enables the "nucleation" of abnormal grain growth.

9.5.2.3 Computer simulation

Monte-Carlo simulations of abnormal grain growth in the presence of particles have been carried out in 2-D (Srolovitz et al. 1985) and 3-D (Doherty et al. 1990). In 2-D, although

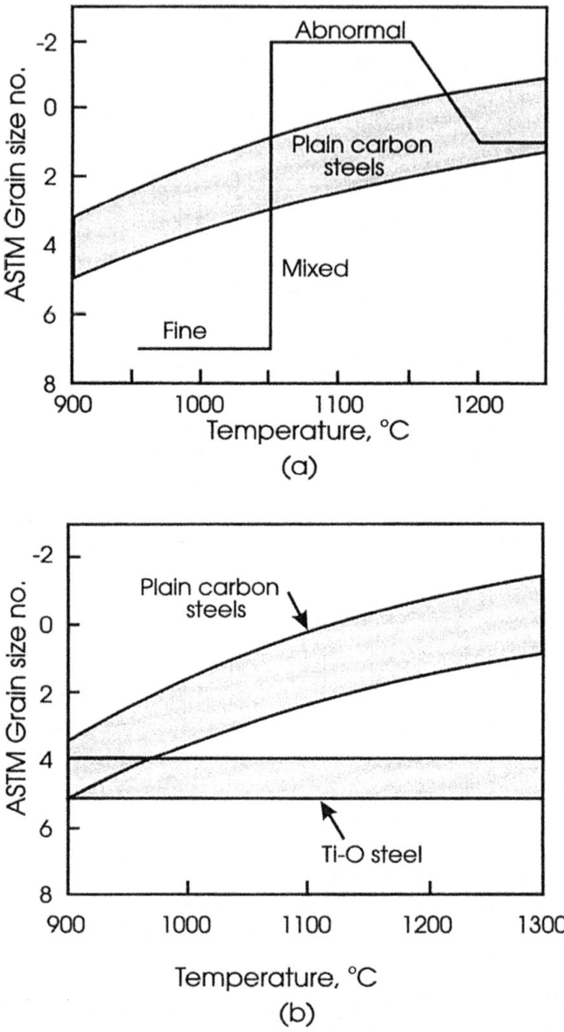

Fig. 9.20. Abnormal grain growth in steels. a) TiN grain growth inhibitors are very effective in preventing normal grain growth, but their coarsening and dissolution lead to abnormal grain growth at high temperatures. b) Coarse, insoluble, non-metallic inclusions are less effective in preventing normal grain growth, but their stability prevents abnormal grain growth, (Gladman 1992).

normal grain growth stagnated due to particle pinning, abnormal grain growth could not be induced. However, in the 3-D simulations, large artificially-induced grains were found to be stable and to consume the smaller pinned grains, which is in keeping with the analytical models discussed above and with the experimental observations. This result again emphasises the necessity of using 3-D simulations.

9.5.3 The effect of texture

The inhibition of normal grain growth by texture (§9.3) may lead to the promotion of abnormal grain growth, and a review of the extensive work in this field is given by Dunn and Walter (1966) and Cotterill and Mould (1976). Perhaps the best example of this is the intensification of the Goss texture in silicon iron (§12.4). In many cases normal grain growth is restricted by a number of factors such as particles and free surfaces in addition to texture, in which case it is difficult to quantify the role of texture in anomalous grain growth.

If a **single strong texture component** is present in a fine-grained recrystallized material, then abnormal grain growth commonly occurs on further annealing at high temperatures. This has been clearly demonstrated for the cube texture in aluminium (Beck and Hu 1952), copper (Dahl and Pawlek 1936, Bowles and Boas 1948, Kronberg and Wilson 1949), nickel alloys (Burgers and Snoek 1935) and silicon iron (Dunn and Koh 1956).

Abnormal grain growth is possible in such a situation because within a highly textured volume the grain boundaries have a lower misorientation and hence a lower energy and mobility than within a normal grain structure. If some grains of another texture component are present, then these introduce boundaries of higher energy and mobility which may migrate preferentially by a process which is therefore closely related to primary recrystallization of a material containing subgrains (chapter 6).

Rollett et al. (1989b) determined the conditions for the abnormal growth of a grain which had a high mobility by extending Hillert's (1965) model in a similar manner to that discussed in §9.5.1. Abnormal grain growth is possible only if the abnormal grain grows faster than the average grain assembly. The general condition for this is given by equations 9.43 and 9.44. Using equation 9.13, we take the rate of growth of the abnormal grain of radius R, which is assumed to have a high angle boundary with the textured matrix as

$$\frac{dR}{dt} = c\,M_b\,\gamma\left(\frac{1}{\bar{R}} - \frac{1}{R}\right) \tag{9.48}$$

where c is a constant, and M_b is the mobility of a high angle boundary.

Using equation 9.14, we take the rate of growth of the textured grain assembly as

$$\frac{d\bar{R}}{dt} = \frac{c\,M_s\,\gamma}{4\,\bar{R}} \tag{9.49}$$

where M_s is the mobility of a low angle boundary.

Substituting into equation 9.44 we obtain the condition for abnormal grain growth as

$$\bar{R}\,\gamma\,M_b\left(\frac{1}{\bar{R}} - \frac{1}{R}\right) > \frac{R\,\gamma\,M_s}{4\,\bar{R}} \tag{9.50}$$

Letting $Q = M_b/M_s$ and $X = R/\bar{R}$, then abnormal grain growth occurs if

$$\frac{X}{4\,Q} + \frac{1}{X} < 1 \tag{9.51}$$

This relationship, plotted in figure 9.21 shows that abnormal grain growth becomes viable for smaller grains as Q increases. It is also seen that there is a maximum value of the ratio X, which increases almost linearly with Q, beyond which the abnormal grain will not grow. It is difficult to put an exact value on Q, but a value of $M_b/M_s \sim 10$ is not unreasonable, which suggests that a grain only 1.03 times larger than that of the mean of the assembly would be viable for abnormal grain growth, and would grow to some 39 times larger than the grain assembly.

9.5.4 Surface effects

It has long been recognised that abnormal grain growth is often easier in thin sheets than in bulk materials. This phenomenon is of particular importance in the production of sheet Fe-Si alloys (Dunn and Walter 1966) as is discussed in §12.4. The control of grain size and texture in the thin polycrystalline films used for electronic applications provides a more recent example of the importance of surface effects in grain growth (Abbruzzese and Brozzo 1992, Thompson 1992). Abnormal grain growth may occur in thin sheets if normal grain growth is inhibited by the free surfaces or by particles or texture as discussed above, and is particularly favoured if the texture leads to a significant variation of surface energy among the grains.

9.5.4.1 Surface inhibition of normal grain growth
When grains are of the order of the thickness of the sheet, then the grains are significantly curved only in one direction, and from equation 9.1, the driving pressure is seen to fall

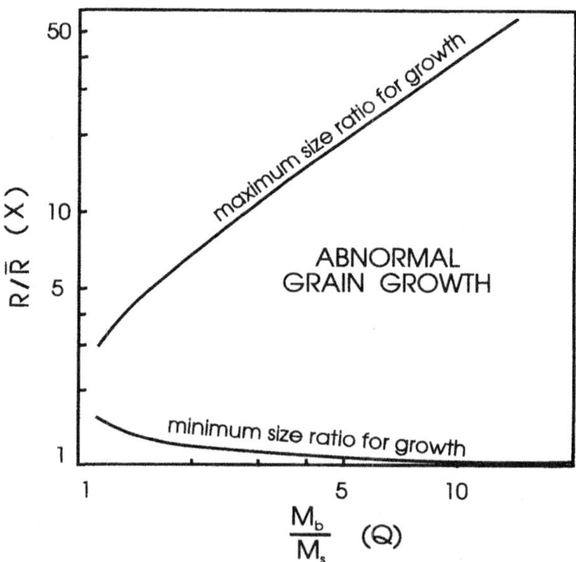

Fig. 9.21. The conditions for abnormal grain growth in a strongly textured material. The minimum critical size of a grain for abnormal growth and the maximum size to which it will grow are shown as a function of the parameter $Q = M_b/M_s$.

towards half that of a grain within a 3-D assembly. Thermal grooving at the junction between the boundary and the free surface, due to the balance between the surface tension and grain boundary tension may also exert a pinning effect on the boundary. The classic work on thermal grooving is due to Mullins (1958) and this treatment has been extended by Dunn (1966), Frost et al. (1990) and Frost et al. (1992). An application of Frost's computer model to the simulation of grain growth in thin metallic strips is shown in figure 13.7. The following is a simplified treatment of the phenomenon.

Consider a single grain in a sheet of thickness S, whose boundaries are pinned by thermal grooving as shown in figure 9.22. The angle θ_g at the thermal groove as shown in figure 9.22b is determined by the balance between the forces arising from the boundary energy γ_b and the surface energy γ_{sur} and is given by

$$\theta_g = \arcsin\left(\frac{\gamma_b}{2\gamma_{sur}}\right) = \frac{\gamma_b}{2\gamma_{sur}} \tag{9.52}$$

The boundary can be pulled out of the groove if, at the surface, it deviates from the perpendicular by an angle of $\sim\theta_g$, which corresponds to a radius of curvature of

$$R_2 = \frac{S}{2\theta_g} \tag{9.53}$$

which can be equated with a pinning pressure due to thermal grooving of

$$P_g = \frac{\gamma_b^2}{S\gamma_{sur}} \tag{9.54}$$

This is opposed by the pressure P (equation 9.3) due to the in-plane curvature R_1, (fig 9.22a). Thus, no grain growth will occur if $P_g \geq P$, and therefore the limiting grain size for normal grain growth (D_L) is given as

$$D_L = \frac{S c_1 \gamma_{sur}}{\gamma_b} \tag{9.55}$$

where c_1 is a small constant.

Mullins (1958) proposes that $c_1 = 0.8$, and computer simulations by Frost et al. (1992) give $c_1 \sim 0.9$. Typically, it is found that $\gamma_{sur} \sim 3\gamma_b$, suggesting a limiting size for normal grain growth in thin films of 2 to 3 times the thickness, which is consistent with experimental measurements in metals (Beck et al. 1949) and semiconductor films (Palmer et al. 1987).

9.5.4.2 Abnormal grain growth in thin films

There is considerable evidence that abnormal grain growth can readily occur in thin films or sheets in which normal grain growth has stagnated for the reasons discussed above (Beck et al. 1949, Dunn and Walter 1966, Palmer et al. 1987, Thompson 1992). That this abnormal grain growth is **not driven by boundary curvature** is clear from observations which show that the direction of motion may be in the opposite direction to that predicted by the curvature (Walter and Dunn 1960b), and from measurements which show that the boundary velocity is constant with time (Rosi et al. 1952, Walter 1965). Computer simulations of 2-D

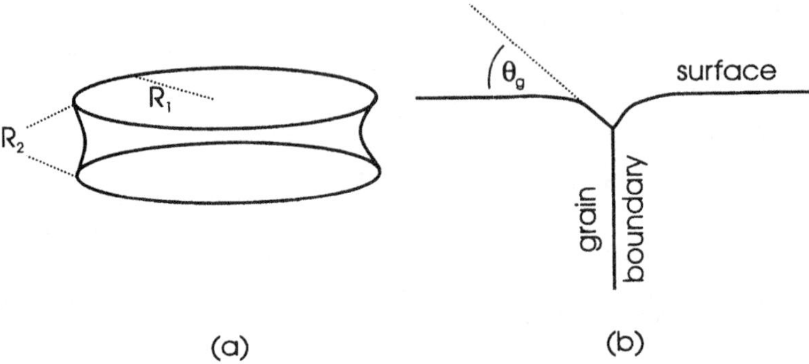

Fig. 9.22. Thermal grooving of a thin specimen. a) The shape of an isolated grain in a thin sheet. b) Thermal grooving at the surface.

grain growth are particularly effective for studying abnormal grain growth in thin films (Srolovitz et al. 1985, Rollett et al. 1989b, Frost and Thompson 1988).

The driving pressure for abnormal grain growth comes from the orientation dependence of the surface energy. If the difference in surface energy between two adjacent grains is $\Delta\gamma_{sur}$ then the driving pressure for boundary migration is $P_s = 2\Delta\gamma_{sur}/S$. As this has to overcome the drag due to thermal grooving the growth velocity will be

$$v = M(P_s - P_g) = \frac{M}{S} \cdot \left(2\Delta\gamma_{sur} - \frac{\gamma_b^2}{\gamma_{sur}} \right) \qquad (9.56)$$

The velocity of abnormal grain growth is thus predicted to be inversely proportional to the sheet thickness, which is consistent with experiments on metal sheet (Foster et al. 1963) and semiconductor thin films (Palmer et al. 1987).

The condition that abnormal grain growth should occur is given from equation 9.56 as

$$2\Delta\gamma_{sur} > \frac{\gamma_b^2}{\gamma_{sur}} \qquad (9.57)$$

or

$$\frac{\Delta\gamma_{sur}}{\gamma_{sur}} > c_2 \left(\frac{\gamma_b}{\gamma_{sur}} \right)^2 \qquad (9.58)$$

where c_2 is a small constant which is given by Mullins (1958) as 1/6 and Frost et al. (1992) as $\sim 1/4$.

Taking $\gamma_{sur}/\gamma_b = 3$, then equation 9.58 indicates that abnormal grain growth will occur if $\Delta\gamma_{sur} > 0.02\gamma_{sur}$, i.e. only a few percent difference in surface energy is required to promote abnormal grain growth in thin specimens.

Surface energy variations between grains are a consequence of grain orientation. Therefore the analysis above, which takes account of the variation of **surface energy** due to orientation but not the variation of the **grain boundary energy** due to texture, must be an oversimplification.

The surface energy of a grain is of course very dependent on the surface chemistry, and it has long been known that abnormal grain growth in Fe-Si alloys (§12.4) is affected by the atmosphere (Detert 1959, Walter and Dunn 1959, Dunn and Walter 1966). Recent work on grain oriented, silicon steel sheet (§12.4) has shown the importance of small alloying additions in determining surface energies.

9.5.5 The effect of prior deformation

There has been interest in the effects of small strains on grain growth and abnormal grain growth (Riontino et al. 1979, Randle and Brown 1989). The observation that a small prior plastic strain may prevent normal grain growth and promote the onset of anomalous grain growth has long been known (§1.2.1.3) and is best interpreted in terms of **primary recrystallization by strain induced boundary migration** (§6.6.2) rather than as a grain growth phenomenon.

Chapter 10

RECRYSTALLIZATION TEXTURES

10.1. INTRODUCTION.

The recrystallization textures developed when deformed metals are annealed have been the subject of extensive study. There are two reasons for this: first, and most important practically, these textures are largely responsible for the directionality of properties observed in many finished products and second, their origin is a source of much scientific interest. With respect to the latter, and despite frequent claims to the contrary, present understanding of the origin of recrystallization textures is little more than qualitative in nature. There are many reasons for this and most of them have already been discussed in earlier chapters. They are related to the nature of the nucleation event, the orientation dependence of the rate of nucleation in inhomogeneities of various type and orientation environment and to the nature, energy and mobility of the boundaries between grains of various orientations. The answer to some of these problems will probably come from the recent development of techniques capable of determining microtextures which enable a correlation to be made between local orientation and microstructure.

We begin by noting that there are several important matters which are often neglected in discussions of texture evolution.

(i) **The effect of subsequent grain growth.** The development of an annealing texture does not cease when recrystallization is complete. It continues during grain growth and the final texture is not necessarily representative of the texture present when primary recrystallization is complete. Much of the immense amount of material available in the literature refers to textures that have been altered by growth.

(ii) **The effect of strain.** Much of the data, and particularly that for fcc metals, refers to the annealing of very heavily rolled materials, e.g. 95-99% reduction by rolling. Such extreme deformation is not generally typical of industrial practice.

(iii) **The starting texture.** Although a great deal of attention has been given to the relationship between the deformation texture and the recrystallization texture, there is increasing evidence that the orientation of the material before deformation plays a significant role in determining the recrystallization texture.

(iv) **The effect of purity.** The purity of a particular metal may strongly influence the recrystallization texture. In many cases very small amounts of second elements may, by affecting the boundary mobility, change the annealing texture completely, e.g. 0.03% phosphorus in copper.

10.2. RECRYSTALLIZATION TEXTURES IN FCC METALS.

The range of recrystallization textures seen in fcc metals and alloys is much wider and more complex than is the case for the corresponding deformation textures, and in many cases the origin of the textures is still a matter of debate.

10.2.1. Rolled sheet

10.2.1.1 Copper and its alloys

As discussed in §2.4.1, the rolling texture of pure copper consists basically of a tube in orientation space which includes as major components the orientations {123}<634> (S) and {112}<111> (C), together with a less significant {110}<112> (B) component (table 2.7). In highly alloyed metals of low stacking fault energy such as 70:30 brass the B orientation is dominant and there is additionally a less important {110}<001> (G) component. The intermediate alloys have textures in which there is a more or less gradual transition from one extreme to the other. A similar, simple transition is not found in the recrystallized metals.

In the case of heavily rolled copper, and other fcc metals and alloys of medium to high stacking fault energy, the recrystallization texture consists of a strong cube texture

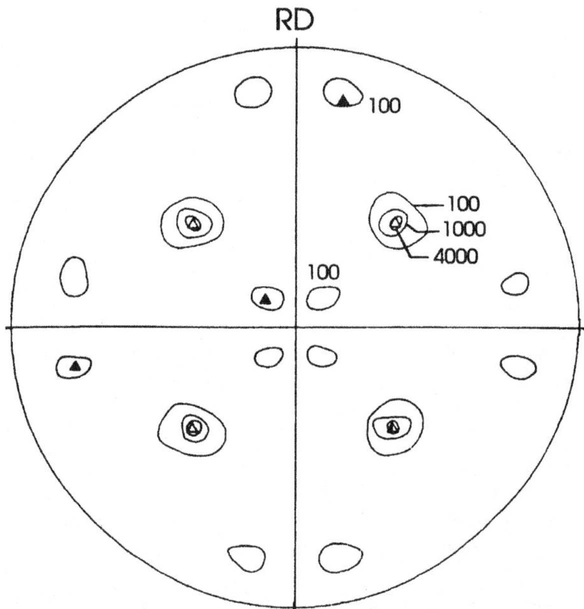

Fig. 10.1. 111 pole figure of 97% cold rolled copper, annealed at 200°C (intensity in arbitrary units). Open symbols indicate the ideal "cube" orientation {001}<100> and filled symbols {122}<212> the "twin" orientation, (Beck and Hu 1966).

{001} < 100 > (fig 10.1). If the stacking fault energy is sufficiently low, minor components that correspond to the twins of this orientation are also present. The strong cube texture of copper is eliminated by quite small amounts of many alloying elements, e.g. Al 5%, Be 1%, Cd 0.2%, Mg 1.5%, Ni 4.2%, P 0.03%, Sb 0.3%, Sn 1%, Zn 4% (Barrett and Massalski 1980). It will be recognized that for the most part these alloying elements reduce the stacking fault energy and change the nature of the deformation texture. A similar change in deformation texture occurs if copper is rolled at low temperatures (-196°C.) and as would be expected, the cube texture is then absent after annealing.

There have been many investigations of the copper-zinc alloys. The recrystallization texture of 70:30 brass (fig 10.2) is usually reported to be {236} < 385 >, but it will be shown later that this can be significantly altered by variation of the annealing temperature. The strong cube texture of copper is lost rapidly as the alloy content increases and in the 3% zinc alloy is very weak. Similarly the {236} < 385 > texture is virtually absent from alloys containing less than 15% zinc. In the intermediate range of 3-15%Zn a series of complex orientations is developed. These annealing textures have been investigated extensively by Lücke and his colleagues, (see for example Schmidt et al. 1975, Virnich and Lücke 1978, Hirsch 1986), using ODF techniques. Their results for this and several other alloy systems are summarized in figure 10.3. It is found that stacking fault energy is the dominant parameter and if the results are normalised with respect to stacking fault energy then consistent behaviour is observed for a range of alloys. In figure 10.3 the volume fractions for the cube and {236} < 385 > texture components are plotted as functions of γ_{RSFE}, the reduced stacking fault energy defined in equation 2.9.

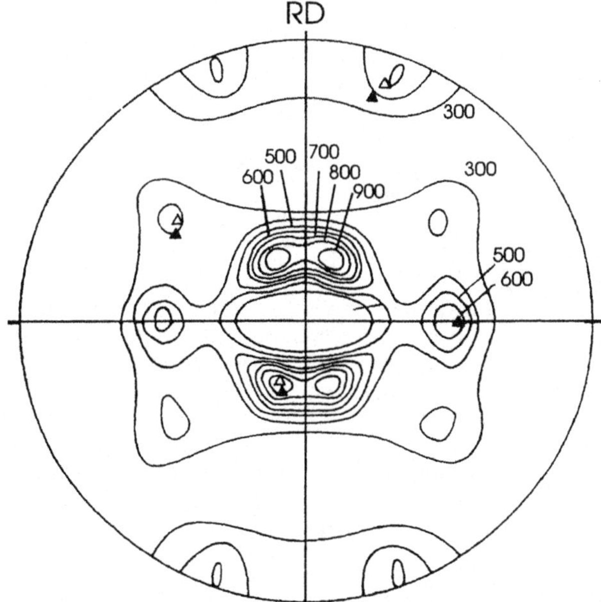

Fig. 10.2. 111 pole figure of 95% cold rolled 70:30 brass, annealed at 340°C, (intensity in arbitrary units). Opens symbols indicate ideal {225} < 374 > and filled symbols {113} < 211 > orientations, (Beck and Hu 1966).

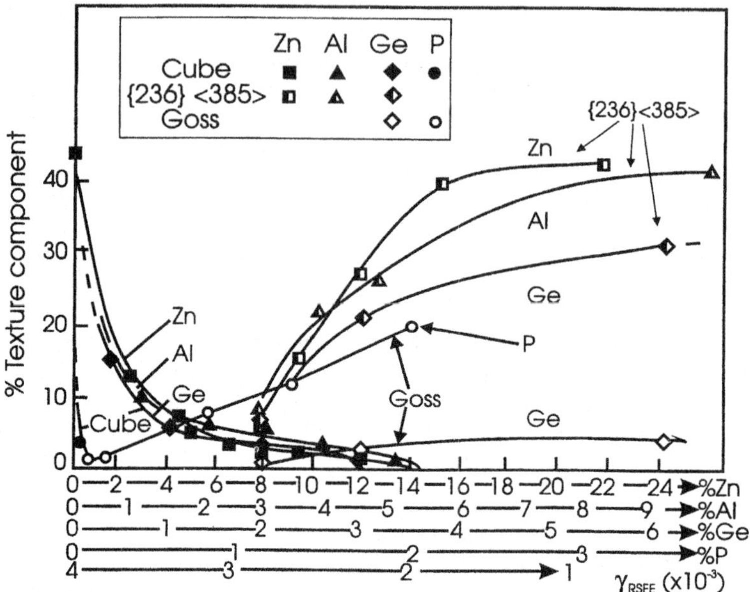

Fig. 10.3. Volume fraction of major texture components ({100] $<001>$; {236} $<385>$; {110} $<001>$) in recrystallized copper based alloys, (Lücke 1984).

It will be seen that there is a marked change in the nature of the recrystallization texture at $\gamma_{RSFE} \sim 3.10^{-3}$ corresponding to ~8%Zn, 3%Al or 2%Ge. At about this value the {236} $<385>$ component begins to develop in the alloys and thereafter rises rapidly to a near maximum value at $\gamma_{RSFE} \sim 2.2.10^{-3}$. At lower values the cube texture is the strongest component although its strength is significant only for low alloy contents. Typical ODFs for the system Cu-Zn are given in figure 10.4, and the volume fractions of the various texture components are shown in figure 10.5. The Euler angles for these and other recrystallization texture components are given in Table 10.1.

There have been many investigations of the effects of microstructure and processing variables such as prior grain size and texture, strain, time, temperature of annealing and rate of heating on these basic textures. It is not possible to review all of these and only a few specific examples are discussed. In the case of the cube texture it has been known for more than 50 years that this texture is reduced by:

(i) Small rolling reductions ($\leq 50\%$) with intermediate annealing,
(ii) A large grain size at the beginning of the penultimate rolling stage, and
(iii) A low final annealing temperature.

An example of the effects of prior grain size, rolling reduction and annealing temperature in 70:30 brass is found in the work of Duggan and Lee (1988). After annealing for 1 hour at 300°C (low temperature) the texture changed from the normal {236} $<385>$ (fig 10.6a) to {110} $<110>$ (fig 10.6b) as the prior grain size increased, but the magnitude of the change was strain dependent and non-existent at 97% reduction. Annealing at high

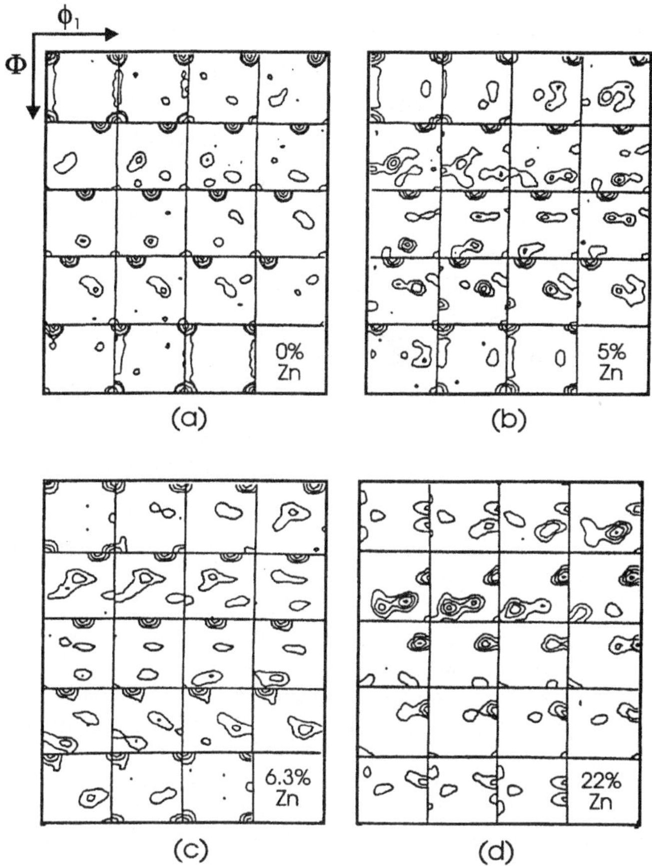

Fig. 10.4. Recrystallization textures of rolled and annealed copper-zinc alloys.
a) Cu b) 5%Zn c) 6.3%Zn d) 22%Zn, (Virnich and Lücke 1978).

temperatures (600°C) led to a $\{113\}<332>$ texture for small grain sized material and high reductions (fig 10.6c).   After low reductions the $\{110\}<112>$ and more general $\{110\}<hkl>$ orientations were produced. The $\{110\}<112>$ texture was present to higher strain levels as the grain size increased.   It seems probable that the high temperature $\{113\}<332>$ texture involved grain growth, and in the 97% rolled material, where grain growth was limited by the specimen thickness and the short annealing times, the texture was $\{236\}<385>$.   For all grain sizes investigated (30-3000μm) the texture of heavily rolled material was $\{236\}<385>$ after annealing at 300°C, and $\{113\}<332>$ for 600°C annealing.   These results illustrate the importance of grain growth effects on annealing textures.

ODF data from Yeung et al. (1988) have confirmed the importance of annealing temperature. Specimens of 70:30 brass of 100μm grain size, were cold rolled and annealed in a salt bath at 300° and 900°C. The volume fractions of some of the main texture components measured

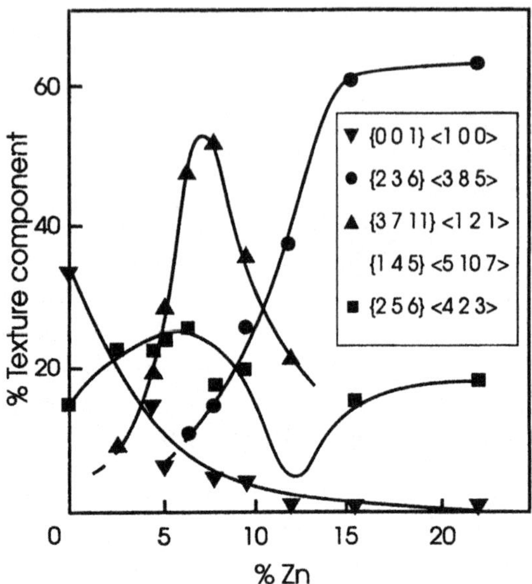

Fig. 10.5. Volume fraction of texture components in copper-zinc alloys,
(Virnich and Lücke 1978).

from the ODF data are summarised in table 10.2. Nucleation was found to occur in all of the various types of deformation inhomogeneities present in the microstructure after annealing at 900°C, and for 95% rolled material the volume fraction of the $\{236\}<385>$ orientation was reduced from 35% at 300°C to 9.4% at 900°C. The major components after the high temperature annealing were characteristic of the rolling texture, i.e. 32% by volume of a component with orientation between B and S and 16.8% of one that was developed by an ND rotation of the cube component. The changes were attributed to the absence of recovery during the rapid heating to the high annealing temperature, which resulted in recrystallization

Table 10.1

Components of recrystallization textures in fcc metals

Notation	Miller indices	Euler angles		
		ϕ_1	Φ	ϕ_2
Cube	$\{001\}<100>$	0	0	0
-	$\{236\}<385>$	79	31	33
Goss (G)	$\{011\}<100>$	0	45	0
S	$\{123\}<634>$	59	37	63
P	$\{011\}<122>$	70	45	0
Q	$\{013\}<2\bar{3}1>$	58	18	0
R (Aluminium)	$\{124\}<211>$	57	29	63

Table 10.2

**Volume percentages of some texture components in recrystallized 70:30 brass
after the rolling reductions shown and subsequent annealing
(Yeung et al. 1988).**

Component	Volume% of component after annealing					
	T=300°C			T=900°C		
	80% red	87% red	95% red	80% red	87% red	95% red
{001}<100>	2.4	3.1		0.5	3.3	
Brass-Goss					24.0	5.9
{236}<385>			35.2			9.4
ND rotat. Cube						16.8
Brass-S						31.8
Others	20.0	34.7	20.9	54.3	29.0	7.1
Random	77.6	62.2	43.9	45.2	43.7	29.0

nucleating uniformly over all components of the deformation microstructure. When coupled
with the rapid boundary migration rate associated with high temperature annealing, the
retention of rolling texture components is to be expected.

Table 10.2 illustrates well a little recognized aspect of recrystallization textures. It is
commonly asserted that such textures are sharp and well defined, particularly after large
strains. However, as shown in the table, the volume fraction of random and other minor
components is high in all cases.

10.2.1.2 Single phase aluminium alloys
When annealed after cold rolling, high purity aluminium develops the characteristic cube
texture found in medium-high SFE materials. As in the case of copper, the texture is
strengthened by high rolling reductions and high annealing temperatures.

The presence of magnesium in solid solution reduces recovery, increases the strength of the
alloy and leads to the formation of shear bands during cold rolling. On annealing,
recrystallization occurs at the shear bands, and texture components such as {011}<100>
(Goss), {011}<122> (P) and {013}<231> (Q) occur (Hirsch 1990, Lücke and Engler
1992). It is known that the formation of cube texture is suppressed by shear bands (Ridha
and Hutchinson 1982), and this effect, combined with the orientations associated with the
shear bands results in a weakening of the cube texture in these alloys.

Because very small amounts of elements such as iron and silicon, which are present in
aluminium of commercial purity, result in the formation of second-phase particles either
before or during the recrystallization anneal, **almost all** commercial aluminium alloys are
multiphase, and **the amount of research on genuinely single-phase aluminium alloys is
therefore very limited**.

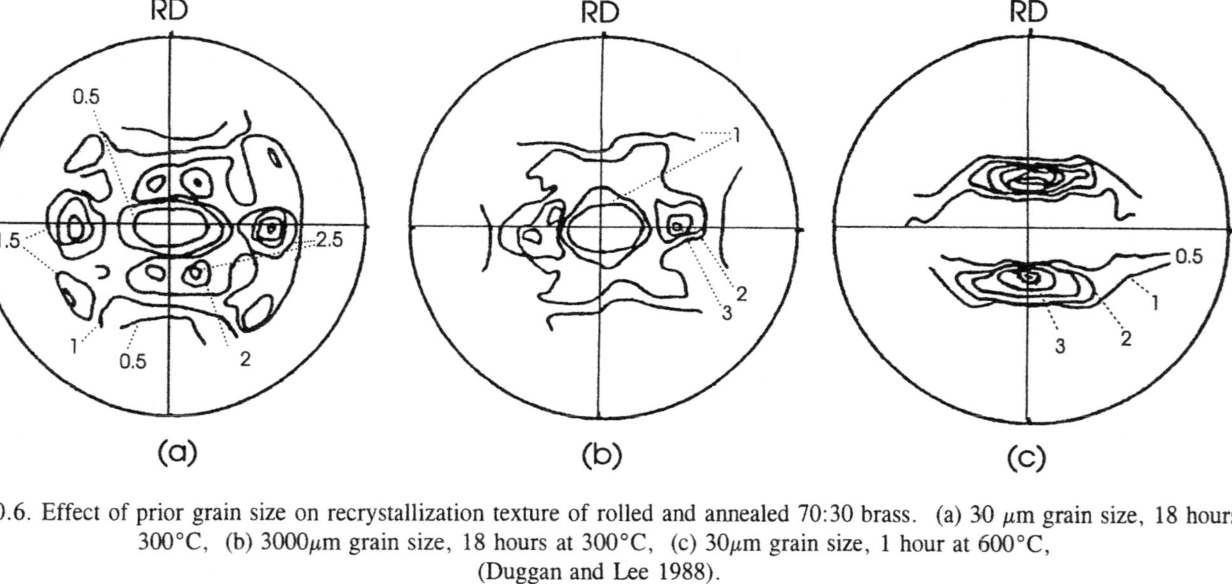

Fig. 10.6. Effect of prior grain size on recrystallization texture of rolled and annealed 70:30 brass. (a) 30 μm grain size, 18 hours at 300°C, (b) 3000μm grain size, 18 hours at 300°C, (c) 30μm grain size, 1 hour at 600°C, (Duggan and Lee 1988).

10.2.2 Fibre textures

The recrystallization textures of copper-based wires, and other products which have fibre textures are normally similar to the corresponding deformation textures. In the case of copper, low temperature annealing of wires leads to fibre textures based on $<100>$ or $<112>$, but high temperature annealing leads to mixed $<111>$-$<112>$ type textures.

10.3 RECRYSTALLIZATION TEXTURES IN BCC METALS.

Because annealed low carbon steel sheet is a product of major industrial importance the optimization of properties is of continuing interest. Texture, grain size, composition, second phase dispersions and processing at each of the several production stages are all important and their respective roles in selected products are discussed in detail in chapter 12. In this section we are concerned only with basic data about texture that is generally typical of bcc metals and as before we begin with rolled sheet.

As discussed in chapter 2, the rolling texture of iron and low carbon steel consists of two prominent orientation spreads. One of these is a fairly complete fibre texture with the $<111>$ fibre axis parallel to the sheet normal; the major components in this fibre texture have $<110>$, $<112>$ and $<123>$ aligned with the rolling direction. The second is a partial fibre texture with a $<110>$ fibre axis parallel to the rolling direction and intensity maxima at $\{001\}<110>$, $\{112\}<110>$ and $\{111\}<110>$ (§2.4.2). The most prominent changes in the second of these fibres during annealing are an increase in the intensities of the $\{110\}$ components and an accompanying decrease in the strength of the $\{100\}$ orientations. With respect to the $<111>$ ND fibre the situation is less reproducible and the strength of $\{111\}<uvw>$ components may either decrease or increase; there is, however an initial decrease in all cases. Notwithstanding this the $\{111\}<uvw>$ components are invariably the strongest in both rolled and annealed sheet. These changes are summarized in figure 10.7, which also shows that texture evolution does not cease with the completion of recrystallization, and the texture continues to change during grain growth (see also §10.5).

Emren et al. (1986) have made a detailed study of texture development in vacuum degassed, aluminium-killed and interstitial-free steels and in Armco iron. The extent of rolling, the time and temperature of annealing and the heating rate were all varied. The 110 pole figures for each material after 80% rolling and annealing at 700°C were basically similar; figure 10.8 refers to an interstitial free steel. The corresponding ODF is shown in figure 10.9. It is clear that the recrystallization texture is basically similar to the rolling texture shown in figure 2.29 and that the recrystallization texture can be described by reference to the α and γ fibres used earlier for the rolled state (§2.4.2).

The major effect of recrystallization is the reduction in strength of the α fibre; this fibre lies in the $\phi_1=0°$ section at constant $\phi_2=45°$ and extends from $\{001\}<110>$ to $\{110\}<110>$. The γ fibre which extends from $\{111\}<110>$ to $\{111\}<112>$ and lies in the section $\Phi=55°$ with $\phi_2=45$ constant is relatively unchanged. Examination of the development of texture during annealing at 700°C of a vacuum degassed steel (fig 10.10), shows a loss of orientations near $\{112\}<110>$ and $\{001\}<110>$ and an increase in the strength of the $\{111\}<112>$ component. The $\{111\}<110>$ component remains constant.

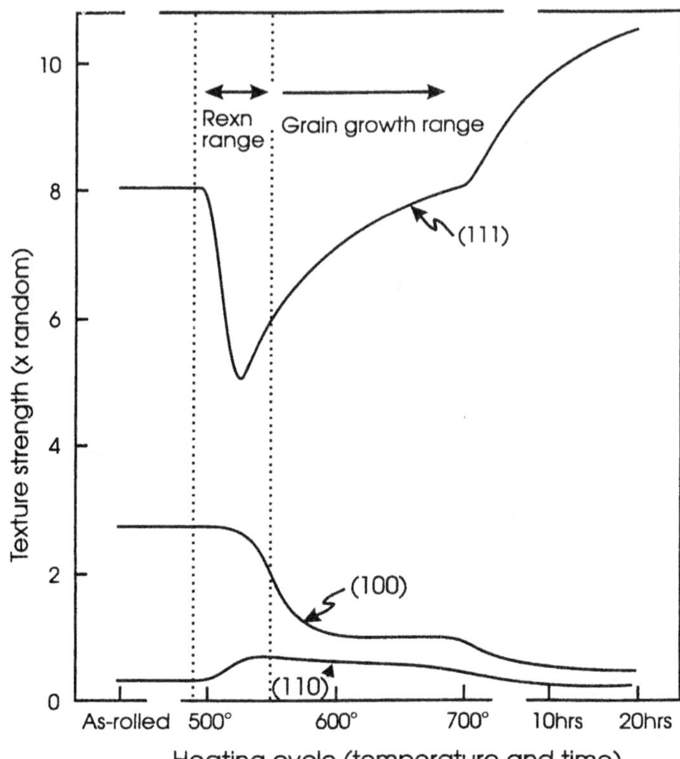

Fig. 10.7. Variation of important texture components during box-annealing of rolled, low carbon steel (Hutchinson 1984; adapted from Michalak and Hu 1978).

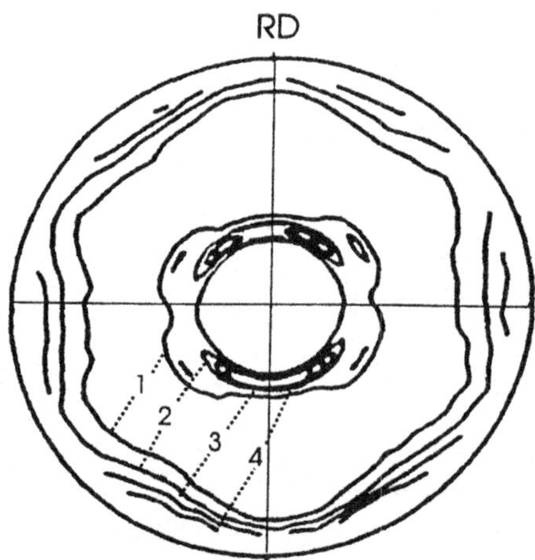

Fig. 10.8. 110 pole figure for recrystallized, interstitial free steel, prior reduction 80%, (Emren et al. 1986).

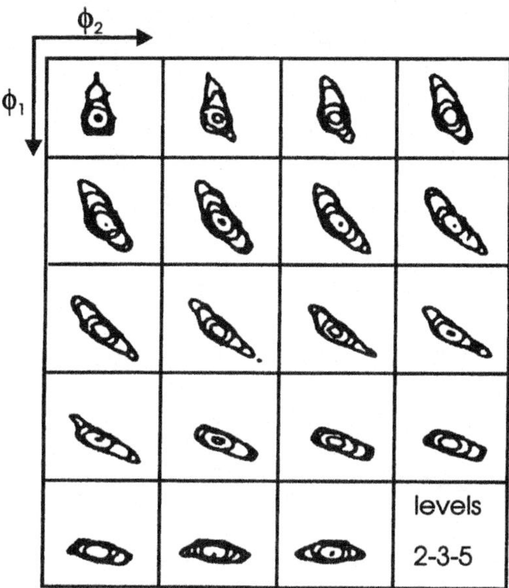

Fig. 10.9. ODF for recrystallized, interstitial free steel, prior reduction 80%, (Emren et al. 1986).

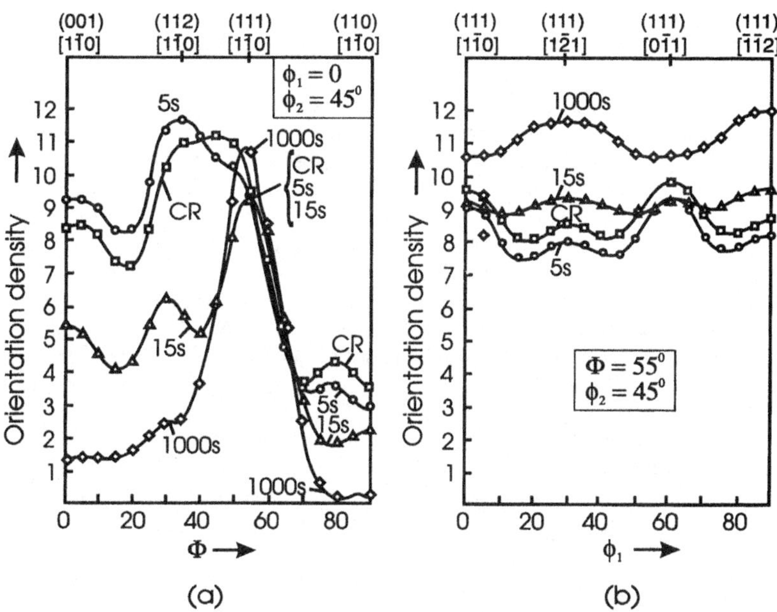

Fig. 10.10. Orientation density f(g) along α and γ fibres for different stages of recrystallization in 85%, cold rolled steel after annealing at 700°C, (a) Along α fibre (b) along γ fibre, (Emren et al. 1986).

There is a significant difference in the nature of the recrystallization textures for bcc and fcc metals. The rolling textures for both can best be described by prominent fibres and in bcc metals these fibres are largely retained in the recrystallization texture. In contrast, the recrystallization textures of rolled fcc metals are better described by individual peak-type components which are not always prominent in the deformation texture.

The industrially important iron-silicon, electrical steels will not be discussed in this chapter, and a detailed account of the generation of the strong $\{110\}<001>$ (Goss) texture typical of this material is given in §12.4.

10.4 RECRYSTALLIZATION TEXTURES IN HEXAGONAL METALS

The early studies of recrystallization textures in cph metals reported that, in general, the rolling textures were retained (Barrett and Massalski 1980) and this is largely true of the rather limited number of recent studies. ODF results from Inoue and Inakazu (1988) for rolled and recrystallized titanium showed that the major component after annealing within the α phase field was $(02\bar{2}5)[2\bar{1}\bar{1}0]$ irrespective of the extent of rolling or the annealing temperature. This orientation is related to the major component of the rolling texture, $(\bar{1}2\bar{1}4)[10\bar{1}0]$ by a $30°[0001]$ rotation. However, the intensity increased considerably as the reduction or the annealing temperature was increased. When annealing was carried out at 1000°C, i.e. within the β phase field, the texture was modified by variant selection associated with the phase transformation. Figure 10.11 shows selected sections of the ODFs for material recrystallized in the α and β temperature ranges.

10.5. TEXTURE DEVELOPMENT DURING GRAIN GROWTH

The textures developed during recrystallization are not necessarily permanent and in many cases changes occur during subsequent grain growth. In some cases the components initially present may be replaced by quite different components, whilst in others an initially rather diffuse texture may be retained and strengthened. The more dramatic changes of the former type are usually associated with abnormal grain growth (Beck 1954). Amongst the best known examples is that associated with abnormal grain growth in rolled copper sheet, where the cube texture is replaced by new orientations generated by a $30°\text{-}40°<111>$ rotation (Bowles and Boas 1948, Kronberg and Wilson 1949). Other examples include silver and 70:30 brass in which the $\{236\}<385>$ texture typical of low SFE metals reverts to the rolling texture $\{110\}<112>$, and Al-3%Mg alloys in which an initial cube texture is replaced by components near $\{013\}<231>$ and $\{124\}<211>$ (Heckelmann et al 1992).

However the more general result, particularly for commercial materials, is that texture sharpening occurs rather than texture change (Hutchinson and Nes 1992). Figure 10.12 shows that the strength of the cube texture in a number of fcc metals and alloys is increased during grain growth and figure 10.13 demonstrates a similar strengthening of the $\{111\}$ texture components in a bcc, low carbon steel. Abnormal grain growth, which would be necessary for major changes in texture generally does not occur in commercial alloys because they usually contain sufficient amounts of dispersed phases to suppress it (§9.5.2). Additionally, the differences in boundary energy and mobility between the texture components present may be insufficient to promote abnormal grain growth (§9.5.3).

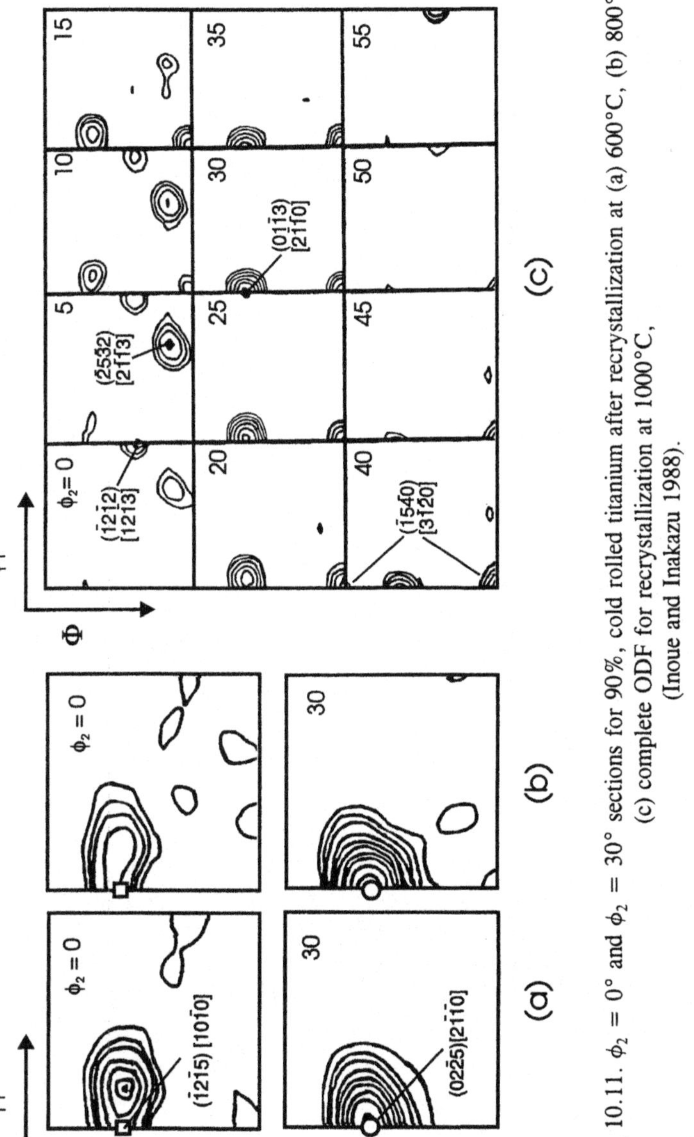

Fig. 10.11. $\phi_2 = 0°$ and $\phi_2 = 30°$ sections for 90%, cold rolled titanium after recrystallization at (a) 600°C, (b) 800°C, (c) complete ODF for recrystallization at 1000°C, (Inoue and Inakazu 1988).

                    Recrystallization

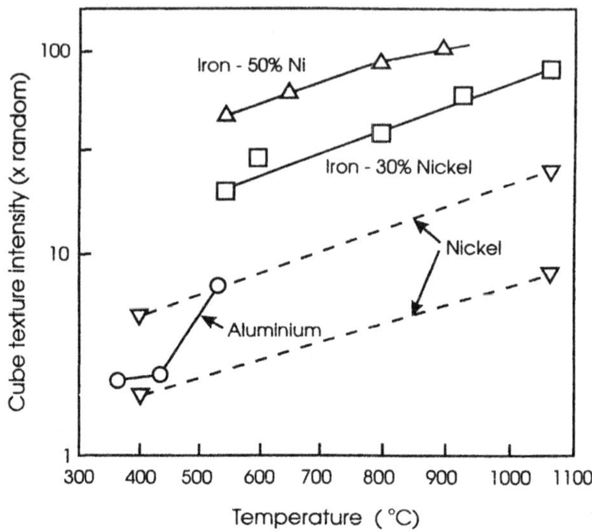

Fig. 10.12. Strengthening of cube texture with grain growth for fcc metals and alloys, (Hutchinson and Nes 1992).

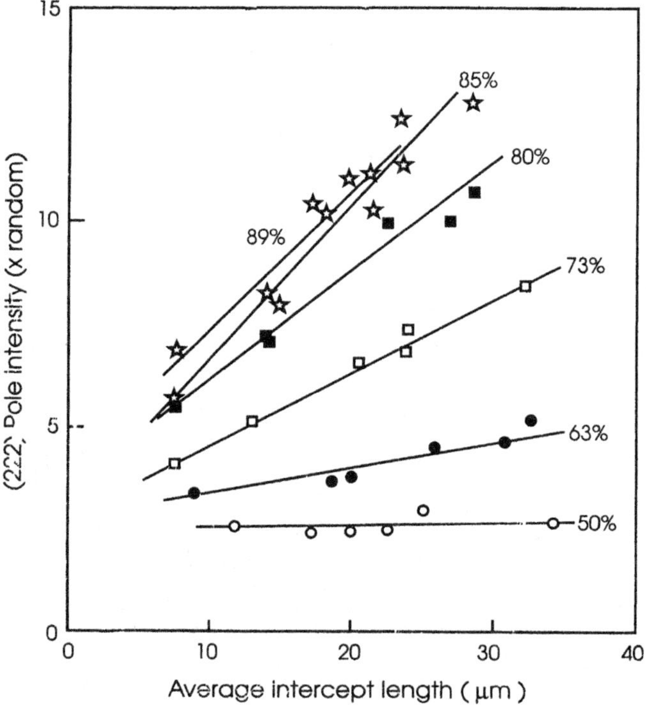

Fig. 10.13. Strengthening of {111} texture with grain growth in low carbon steel with different rolling reductions, (Hutchinson and Nes 1992).

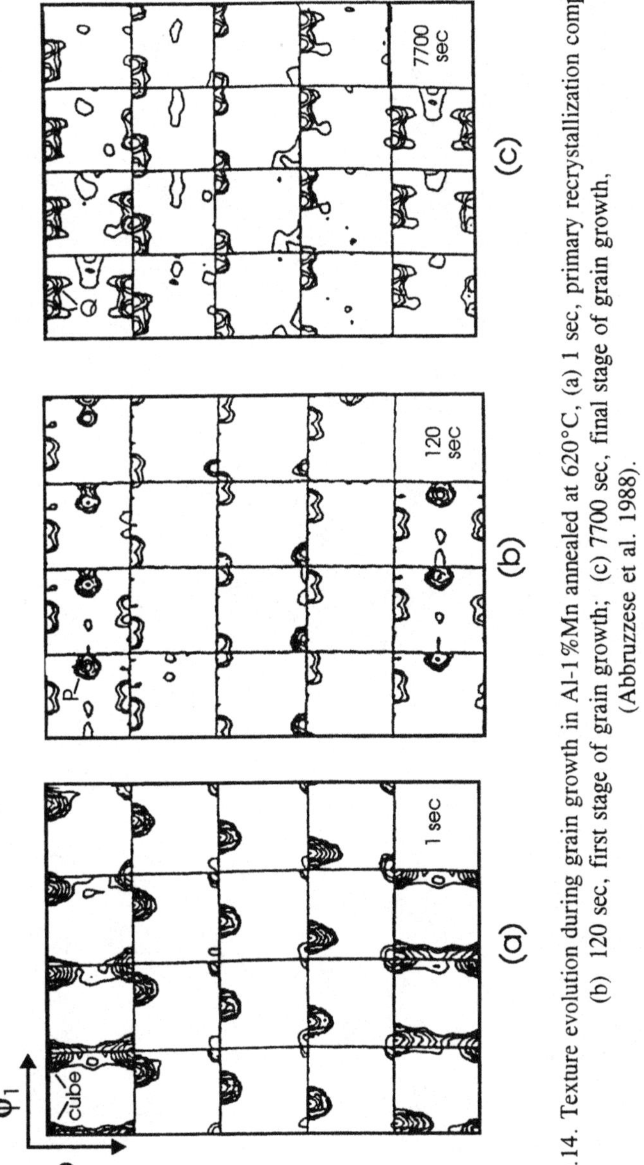

Fig. 10.14. Texture evolution during grain growth in Al-1%Mn annealed at 620°C, (a) 1 sec, primary recrystallization completed; (b) 120 sec, first stage of grain growth; (c) 7700 sec, final stage of grain growth, (Abbruzzese et al. 1988).

Current theories of grain growth after primary recrystallization, including the influence of texture, are discussed in chapter 9, and we are concerned here only with the evolution of texture during grain growth. Several models for grain growth in textured materials have been proposed as discussed in §9.3 (Abbruzzese and Lücke 1986, Eichelkraut et al. 1988, Bunge and Dahlem-Klein 1988). Abbruzzese et al. (1988), have compared model predictions with experimental results from Al-1%Mn which was rolled to 95% reduction and annealed at 620°C. As shown in figure 10.14a the primary recrystallization texture has a strong $\{001\}<100>$ component with a prominent spread generated by a rotation about the rolling direction. After further annealing for 120 seconds the cube component is completely lost and an $\{011\}<122>$ component, which was previously insignificant, predominates (fig 10.14b). At this stage a third component, $\{013\}<231>$, can be detected and as annealing continues this replaces the $\{011\}<122>$ component (fig 10.14c). Figure 10.15 shows that the experimental results are in good agreement with the model used during the early stages of annealing; the lack of agreement at longer times was attributed to the fact that the mean grain size approached the sheet thickness. Weiland et al. (1988) have reported somewhat similar results from the same alloy and have shown by EBSP that the different components have different grain sizes.

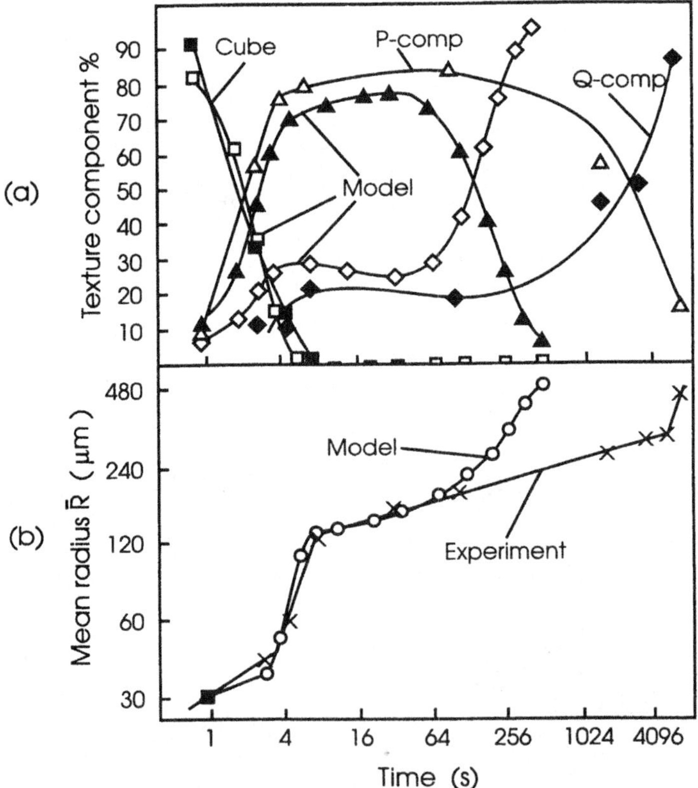

Fig. 10.15. Texture controlled grain growth in Al-1%Mn as a function of annealing time.
(a) Volume fractions of components (filled points are experimental data).
(b) Grain radii, (Abbruzzese et al. 1988).

It is obvious that the nature of the texture changes taking place during grain growth is very complicated and as yet, relatively unexplored. Although theories to explain these changes are being developed, it should be noted that the orientation dependence of grain boundary energies and mobilities on which they are based, are often unknown (§4.3.2).

10.6. RECRYSTALLIZATION TEXTURES IN MULTIPHASE ALLOYS.

10.6.1 Introduction

In addition to the size and spacing of the particles, the recrystallization textures formed when multi-phase alloys are annealed depend particularly on whether the second phase is present during deformation, or whether precipitation occurs on annealing (see Hornbogen and Kreye 1969, Hatherly and Dillamore 1975, Hornbogen and Köster 1978, Humphreys and Juul Jensen 1986, Juul Jensen et al. 1988, Humphreys 1990).

The particles will affect the recrystallization texture in two ways. First, it was shown in chapter 8 that large ($>1\mu m$) pre-existing particles are favoured nucleation sites during recrystallization and that highly misoriented deformation zones associated with them lead to a wide range of nucleus orientations. Second, if a dispersion of closely spaced particles is present then pinning (Zener drag) of low or high angle boundaries affects, not only the recrystallization kinetics, but also the final grain size and texture.

Humphreys and Juul Jensen (1986) and Juul Jensen et al. (1988) have collated data on the recrystallization textures of a number of alloys, and some selected examples, shown in table 10.3 will be used to illustrate the effects of particles on texture.

Table 10.3.
Major components of the recrystallization texture for some
particle-containing alloys.(from Humphreys and Juul Jensen 1986).

Material; Reduction	Particles		Recrystallization Texture
	F_V x 100	Diameter (μm)	
Al(99.9965); 90%	0	-	Cube
Fe-AlN ;70%	0.06	0.017	Rolling
Al-Al$_2$O$_3$;30-90%	0.4	0.1	Deformation
Cu-SiO$_2$;70%	0.5	0.23	Rolling + Twins
Al-FeSi; 90%	0.5	0.2-7	Cube, Rolling, Random
Al-Si; 90%	0.8	2	Cube, rolling, Random
Al-Ni; 80%	10.0	1	Rolling, Random
Al-SiC; 70%	20.0	10	Random

10.6.2 The influence of second-phase particles

10.6.2.1 Particle stimulated nucleation

The contribution of particle-stimulated nucleation to recrystallization and the orientations of the nuclei, has been considered in chapter 8, and here we are concerned mainly with the effects on the final recrystallization texture. It seems certain that the nuclei originate in the deformation zones and their orientations are therefore restricted to those present in these zones. The situation is summarised as follows:

(i) For small deformations, or in particle-containing single crystals, the range of PSN orientations may be very restricted and PSN may in this situation result in a sharp texture (fig 8.31).

(ii) In lightly deformed polycrystals, PSN may originate in several deformation zones at a single particle (fig 8.32). Although the new orientations have a crystallographic relationship with the deformed matrix, which is a function of the slip activity (§8.2.4), this will inevitably lead to a spread of orientations around those of the deformed matrix.

(iii) Several investigations have shown that the nuclei developed in heavily deformed polycrystalline materials are either randomly oriented or weakly textured (e.g. Herbst and Huber 1978, Wassermann et al. 1978. Chan and Humphreys 1984c). Each grain, or each coherent volume within a grain, will have several deformation zones and each of these will have a large misorientation range. A single grain in the initial material will therefore give rise to a wide range of orientations and the texture will be weakened. It is important to understand that the nuclei are not randomly oriented with respect to the deformation texture, and their apparent randomness is derived from the wide range of orientations present both in the matrix and in the deformation zones around the particles.

(iv) Habiby and Humphreys (1993) have shown that even if PSN grains are highly misoriented from the matrix grains, a weakened rolling texture rather than a random texture would result. Starting with a computer-generated texture typical of the rolling texture of an aluminium alloy, they generated a "recrystallization texture" by allowing each orientation in the rolling texture to generate new grains misoriented by up to 45° about random axes from the parent grain. The resultant texture was a weak rolling texture with peak intensity reduced from 5.5 to 3.1xR.

(v) There is a possibility, discussed in §8.4.2.4, that nucleation may occur at the small deformation bands formed adjacent to some particles (fig 8.9).

The annealing behaviour of particulate metal-matrix composites is an extreme example of texture control by PSN (Humphreys 1990, Bowen and Humphreys 1990, Bowen et al. 1991). In many such materials the volume fraction of large ($>3\mu$m) particles is much larger than in conventional alloys and may exceed 20%. PSN readily occurs at the large particles and the strength of the recrystallization texture (a weak rolling texture) is reduced as the volume fraction of particles is increased, becoming effectively random for volume fractions in excess of 0.1.

10.6.2.2 Fine particle dispersions

As was discussed in §8.3, a dispersion of closely-spaced small ($<1\mu$m) particles may have a strong influence on the recrystallization kinetics and the grain size, and the effects depend on whether the particles are present before or after the deformation. Although the recrystallization texture is likely to be influenced by particle size, strength and spacing, there is little data on well characterised material and it is not yet possible to clarify the effects of these parameters.

Small deformable particles. The presence of a dispersion of small deformable particles such as those formed during the age-hardening of e.g. Al-4%Cu, affects the homogeneity of deformation (§8.2.1), and often leads to the formation of shear bands. In this case the recrystallization texture is very similar to that of the solid solution alloys discussed in §10.2.1.2 (Lücke and Engler 1992).

Small non-deformable particles. If the particles are non-deformable and too small for PSN, or if the particles are precipitated during the recrystallization anneal, then their effect is mainly one of boundary pinning (§3.6). Humphreys and Juul Jensen (1986) have analyzed the published data for such alloys, and they find that, in general a texture similar to that of the deformed material is retained after recrystallization (see table 10.3). For example in bcc alloys, the {111} components are often strengthened, and in both fcc and bcc alloys the recrystallization texture is often stronger than the deformation texture, although the balance between the components may be altered. As discussed in §8.5, for small particle spacings discontinuous recrystallization may not be possible and the annealing mechanism is then one of recovery, although coarsening of the particle dispersion may lead to **extended recovery** (§5.7). In either of these cases it is to be expected that the annealing texture will be similar to the deformation texture. The reason for the retention of the rolling texture for alloys of larger interparticle spacing and which undergo normal discontinuous recrystallization (e.g. $Cu-SiO_2$ in table 10.3) is not entirely clear. It appears that a strong cube texture is rarely formed, and particles may be responsible for suppressing the formation or viability of the cube sites, thus enabling other components including the retained rolling components to dominate the texture.

10.6.3 Competition between sites

The orientations of PSN nuclei in a heavily rolled polycrystal are relatively random (§8.4.2, §10.6.2.1). However, except in metal-matrix composites containing high volume fractions of large particles (table 10.3), the recrystallization textures in alloys in which PSN is occurring are seldom close to random. This indicates that grains from other sites and with different orientation are also formed and that these contribute to the final texture. It is difficult to draw general conclusions about the types of site which are competing with PSN, but it may be seen from table 10.3 that for low particle volume fractions, both cube and rolling components are present (e.g. Al-Fe-Si, Al-Si), and for high particle volume fractions, rolling components are formed (e.g. Al-Ni). As not all particles which are large enough for PSN actually produce grains (§8.4.3), the possibility of nucleation at particles in particularly favourable sites (e.g. grain boundaries) should also be considered.

There has been considerable discussion of texture development in heavily rolled Al-Fe-Si which normally contains $\sim 0.5\%$ of $1\text{-}7\mu m$ particles (see also §10.6.4.1). Juul Jensen et al. (1985) have found that the first grains nucleate at the particles with random orientations while cube oriented grains are nucleated later. The cube grains grow rapidly and the final microstructure consists of small grains ($< 10\mu m$), usually associated with large particles or clusters of particles, together with large grains ($> 100\mu m$). The cube oriented grains are

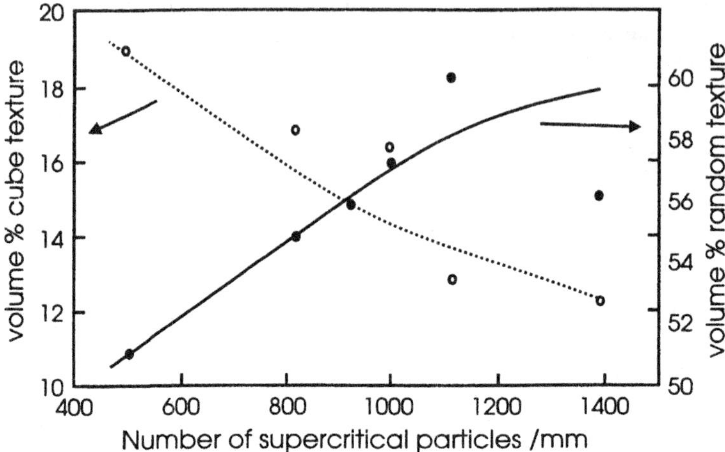

Fig. 10.16. Correlation between recrystallization texture and the number of constituent particles that can act as PSN sites in hot-rolled AA3004, (Bolingbroke et al. 1993).

amongst the largest and the randomly oriented grains amongst the smallest, although the reason for the rapid growth of the cube oriented grains is not clear, and in some cases (Nes and Solberg 1986) the growth advantage of the cube grains is less. It has been suggested that the cube grains grow by virtue of a preferential location compared to the PSN grains (Nes and Solberg 1986) but this has been disputed by Juul Jensen and Hansen (1986). Recent experiments by Ardakani and Humphreys (1994) on the recrystallization of particle-containing (001)[110] crystals of Al-Si have shown very clearly that PSN grains may, although nucleated at an early stage, grow more slowly than other grains under conditions of low strain and low particle volume fraction.

An important example of competition between nuclei from different types of site is found in the non-heat-treatable Al-Fe and Al-Mn alloys in which it is important to control the texture in order to optimise the deep drawing properties (§12.2). When such alloys are hot-rolled and subsequently annealed, the main contributions to the recrystallization texture are the cube component and an essentially random component arising from PSN. As shown in figure 10.16, the strength of the cube component in AA3004 decreases as the number of supercritical sized particles increases. During hot deformation, the critical particle size for PSN is a function of the deformation conditions (§11.6.4), and therefore the amount of cube texture is controlled not only by the particle size distribution, but also the deformation temperature and strain rate.

10.6.4 Some industrial aluminium alloys

The complex effects of solutes and precipitates on annealing textures may be illustrated by a consideration of some examples of industrial aluminium alloys. The most important example is the Al-Mg-Mn alloy AA3004, which is extensively used for beverage cans, and a discussion of the recrystallization of this alloy, in the context of its commercial processing, is given in §12.2.

10.6.4.1 Commercial purity aluminium (AA1xxx) and other Al-Fe-Si alloys
Aluminium of commercial purity, typically has a combined iron and silicon content of more than 1wt%, and the recrystallization behaviour of this type of material is deceptively complex, the annealing textures depending in particular on the heat treatment prior to cold rolling, the Fe/Si ratio and the annealing temperature.

The role of iron.
Small amounts of iron may change the annealing texture from almost pure cube to a strong retained rolling (R) texture. Recrystallization may originate at cube sites (§10.7.6) and sites such as shear bands or grain boundaries where orientations close to the deformation texture (R) may be formed. For very low iron concentrations, the cube texture predominates. However, very small amounts (<100ppm) of iron in solid solution have a large effect on recrystallization kinetics (fig 6.9) and greater amounts may lead to precipitation on the recrystallizing front. The consequent solute drag and/or precipitation lead to slower growth of the dominant cube grains and allow the R component to develop, thereby reducing the strength of the cube component.

The importance of the annealing temperature is well illustrated by the work of Hirsch and Lücke (1985) with an Al-0.007%Fe alloy. The 95% cold-rolled alloy was annealed at 280°C, 360°C and 520°C and figure 10.17 shows ODF results for the as-rolled material and for each annealing temperature. With the exception of a small amount of cube texture the rolling texture (R) was retained at 360°C (fig 10.17c), whereas the results for higher and lower temperatures were quite different. In both cases the cube texture was the major component observed and the strength of the R component was substantially reduced. This was interpreted by the authors as being due to the precipitation of an iron-rich phase at 360°C which hindered boundary migration. At the lower and higher temperatures where precipitation occurs after, or before recrystallization is complete, the effect on boundary migration is less drastic and a stronger cube texture develops.

The combined role of iron and silicon.
The presence of silicon in addition to iron leads to the formation during casting of a stable α-Al-Fe-Si phase in the form of plates or rods of maximum dimension up to $\sim 10\mu$m. Consequently, the level of iron in solid solution is reduced, the exact amount being dependent on the homogenising and cooling conditions. In addition to the recrystallization behaviour discussed above, PSN at the Al-Fe-Si particles may also occur, leading to a random texture component, with a consequent weakening of the R component in particular.

Inakazu et al. (1991) have made a detailed study of the effects of small amounts of iron and silicon on the recrystallization texture of drawn aluminium. Both <100> and <111> texture components were obtained, but the balance depended on the annealing temperature. At temperatures above 450° the <111> deformation component was strengthened, but below 300°C the <100> component increased. The results were interpreted in terms of precipitation during the recrystallization anneal.

The recrystallization of commercial purity aluminium and related alloys is a very good example of how the final annealing texture is the result of competition between grains originating at different sites (§10.6.3), and shows how the balance between the components is affected by small changes in composition or annealing temperature. Oscarsson et al. (1991) have investigated the effect of the deformation temperature on the texture after recrystallization in commercial Al-Fe-Si alloys. They show that in addition to texture effects arising from changes in solute, the deformation temperature affects the amount of PSN (see §11.6.4) and hence the "random" component of the recrystallization texture.

10.6.4.2 Aluminium-lithium alloys

There is interest in the use of strong, low density lithium-containing alloys such as AA8090 in the aerospace industry. After deformation, these alloys have a pronounced brass texture

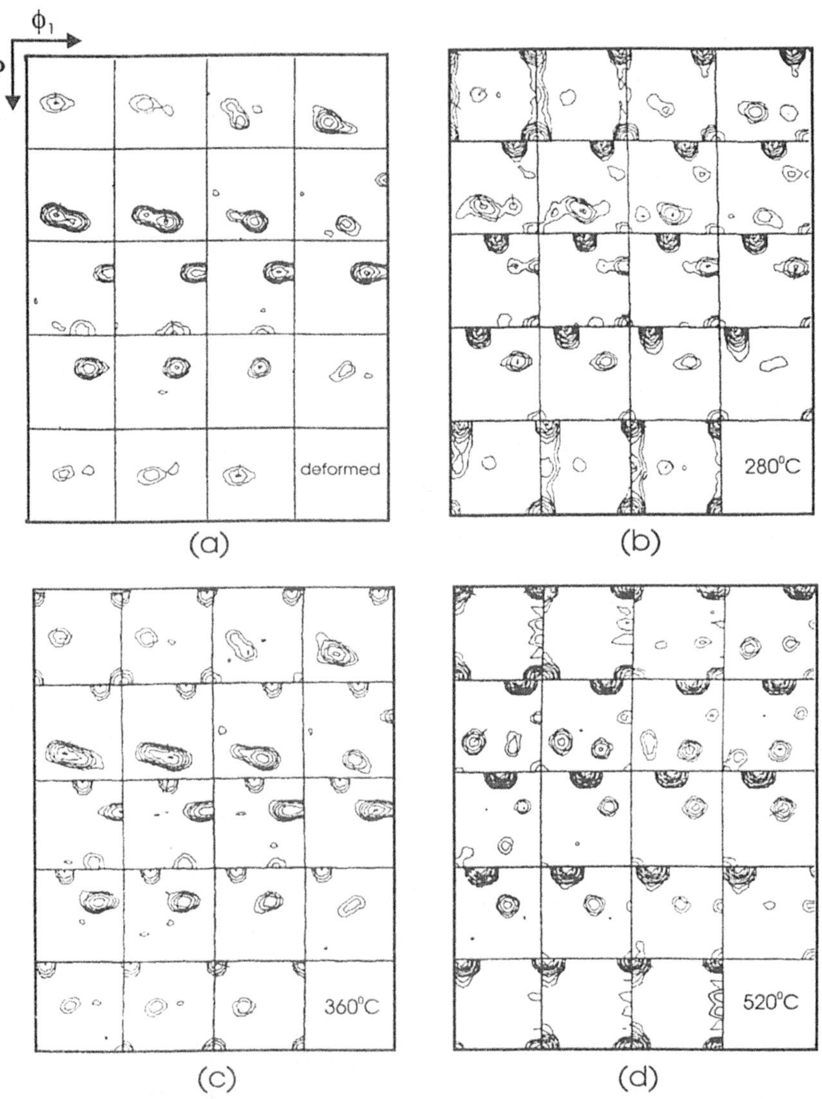

Fig. 10.17. Effect of annealing temperature on the recrystallization texture of 95% cold rolled, Al-0.007%Fe alloy. (a) As rolled, (b-d) annealed at (b) 280°C, (c) 360°C, (d) 520°C, (Lücke 1984).

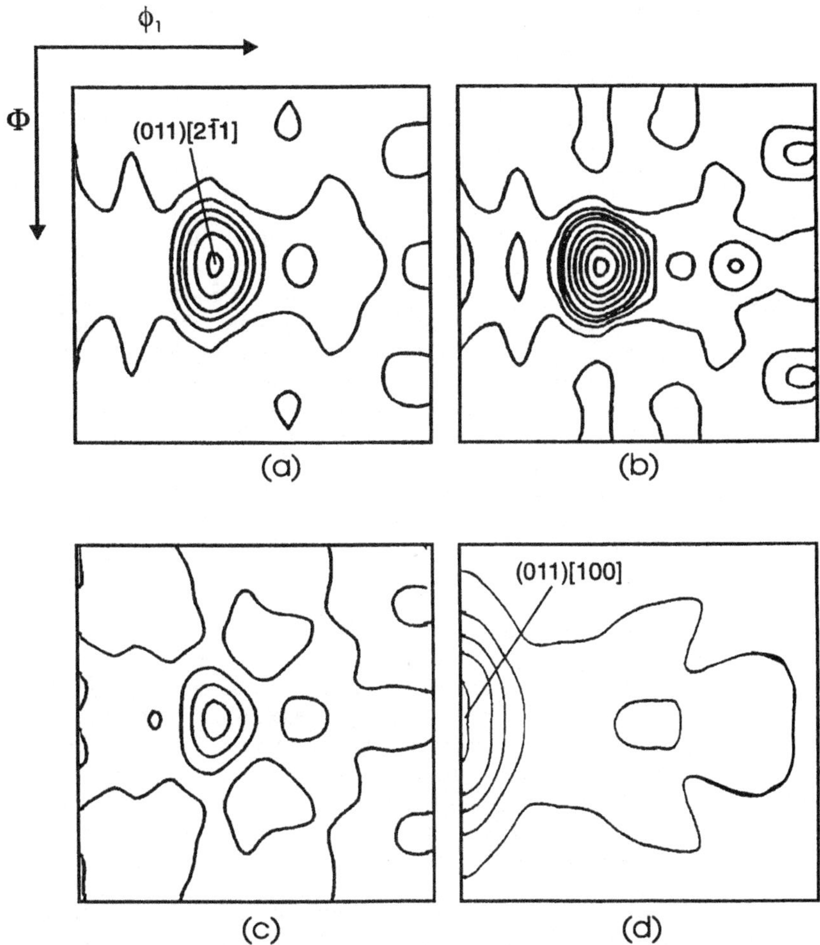

Fig. 10.18. Textures of rolled and annealed Al-Li sheet; $\phi_2=0$ sections of the ODF. (a) As rolled (b-d) annealed at 530°C, (b) Heating rate 60°C/min. (c) Heating rate 250°C/min. (d) commercially produced with heating rate 3000°C/min, (Bowen 1990).

component as shown in figure 10.18a, and there is a strong through-thickness texture gradient. On annealing, the recrystallization behaviour is affected by both solid solution and particle effects, in particular the pinning due to small stable particles of $ZrAl_3$. If the alloy is heated slowly to the solution treatment temperature of 530°C, then recovery rather than recrystallization occurs and the limited boundary migration results in a sharpening of the deformation texture (fig 10.18b). However, an increased heating rate results in less recovery and thus a greater driving force for recrystallization, producing a microstructure of small equiaxed grains with a strong Goss component. The Goss component increases and the brass component decreases as the heating rate is increased (Fig 10.18 c,d).

10.7. THEORIES OF RECRYSTALLIZATION TEXTURES.

10.7.1 Historical background

The nature of the recrystallization texture is determined primarily by two factors:

(i) The orientations of the new grains.

(ii) The relative nucleation and growth rates of these grains.

The manner in which these factors operate to produce a unique recrystallization texture from a particular deformation texture and microstructure has been the subject of controversy for more than 50 years and for the whole of that time two major theories have been strongly advocated. It has been claimed that the recrystallization texture has its origin in either the preferred nucleation of grains with a particular orientation (**oriented nucleation theory**) or the preferred growth of grains of specific orientations from a more randomly oriented array of nuclei (**oriented growth theory**). Most of the experimental work related to the controversy has been concerned with fcc metals and the early work was mainly carried out on copper and its single-phase alloys. The importance of the cube texture in the thermomechanical processing of commercial, multiphase aluminium alloys (§12.2) has resulted in an increased activity in this area in recent years.

It was pointed out in §6.6 that the nuclei from which the recrystallized grains originate are small regions which pre-exist in the deformed state. i.e. the orientations of the nuclei are already present in the deformed structure. With the single exception of changes that may be introduced by twinning there is no way that other orientations can develop. To the early workers faced with a recrystallization texture that was apparently quite different to the deformation texture from which it developed this presented a major problem, and several models were proposed to explain how a new orientation might be generated. Most of these are no longer considered tenable and will not be discussed here; details of this early work may be found in Beck and Hu (1966).

The oriented nucleation theory began with the assertion by Burgers and Louwerse (1931) that the preferred nuclei in lightly compressed aluminium single crystals consisted of crystal "fragments" that were more heavily deformed than the bulk of the crystal. A formal crystallographic theory was devised which associated the "fragments" with local lattice curvatures and involved a $<112>$ rotation. Subsequently, Barrett (1940) found that the recrystallized grains in a single crystal specimen were related to the deformed matrix by a $45°<111>$ rotation. Results of this type led to speculation that the observed rotations were associated with a maximum growth rate, and as experimental methods improved, the value of the angular rotation was refined and **the oriented growth theory** was established. Considerable further evidence for the high mobility of boundaries with certain orientation relationships has been obtained and this is discussed in §4.3.2.

An early alternative oriented nucleation theory was based on the **inverse Rowland transformation** (Verbraak and Burgers 1957, Verbraak 1958). This so called "martensitic" theory relied on a formal crystallographic analysis (the Rowland transformation) which describes a mechanism capable of deriving two twin-related lattices of type $\{112\}<111>$ from a single $\{100\}<001>$ lattice. The martensitic model assumes that the two neighbouring twin-related {112}<111> copper components of the rolling texture jointly undergo an inverse Rowland transformation to generate an {001}<100> cube texture nucleus. The necessary shear requires the presence of <112> partial dislocations and the analysis requires therefore that the stacking fault energy be sufficiently low to allow the formation of the partials. If these are absent, as in aluminium, the process cannot occur and the development of the cube texture in that metal would require a different mechanism. For this and other reasons the theory never attracted strong support and a later attempt by Verbraak (1975) to revive interest was not successful.

10.7.2. Oriented growth.

The current theory of oriented growth is based on observations of specific rotation relationships associated with rapid grain boundary migration. Much of the experimental work has been carried out by Lücke and his associates, and the crystallographic relationships found for various materials are given in table 4.3. The most important of these is the 40°<111> relationship that is used to relate the deformation and annealing textures of fcc metals (e.g. Ibe and Lücke 1966, Lücke 1984). Although discussions of oriented growth have generally stressed the role of high mobility boundaries, it should be recognised that any orientation dependence of boundary mobility may be important, and Juul Jensen (1995) has emphasised that boundaries of particularly low mobility (e.g. low angle boundaries) may have an important influence on the growth of grains during recrystallization.

There are several difficulties in accepting the general applicability of the oriented growth theory.

10.7.2.1 The number of active rotation axes.
Only a small number of the possible rotation axes are actually active and for any particular axis, rotation generally occurs only in one sense.

Many single crystal studies have been made in this area. Consider the orientations {112}<111>, {123}<634> and {110}<112> which form the basis of all fcc rolling textures. Single crystals of the first type retain that orientation during rolling to 80% reduction, and when recrystallized, orientations derived from 6 of the 8 possible 40° <111> rotations are found (Köhlhoff et al. 1981). {123}<634> crystals are a little less stable during rolling, but only one of the possible rotations occurs during recrystallization (Lücke et al. 1976). The {110}<112> orientation is very stable during rolling and all 8 rotations occur on annealing, albeit with some scattering (Köhlhoff et al. 1981).

A similar situation is found for polycrystalline materials as can be illustrated by considering the cube texture, {001}<100>. Oriented growth theory maintains that this orientation is related to the {123}<634> (S) orientation present in the rolling texture of copper and aluminium by a near 40°<111> rotation. However only one of the eight possible rotations of each variant of S yields the cube texture and there is at present no way of predicting which.

These results can only be interpreted on the basis that the number of nuclei available is limited, i.e. oriented growth does not occur independently of oriented nucleation.

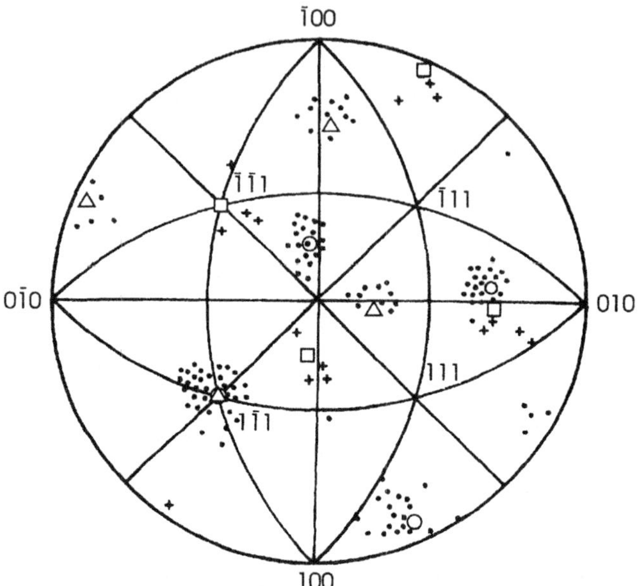

Fig. 10.19. Orientation relationship between artificially nucleated grains and the deformed matrix of aluminium crystals. The matrix is plotted in standard projection.
● <111> poles of new grains close to a 40°[1$\bar{1}$1] relationship.
+ <111> poles of new grains close to a 40°[$\bar{1}\bar{1}$1] relationship.
○ Ideal orientation for 40°[1$\bar{1}$1] clockwise rotation
△ Ideal orientation for 40°[1$\bar{1}$1] anticlockwise rotation
□ Ideal orientation for 40°[$\bar{1}\bar{1}$1] anticlockwise rotation
(Yoshida et al. (1959).

10.7.2.2 The precision of high mobility relationships

Although definite orientation relationships between deformed and recrystallized textures are often quoted and used to support the oriented growth theory, there are several factors that suggest that precise relationships do not exist during recrystallization.

(i) The spread in both the deformation and annealing textures is such that it is unrealistic to select a single orientation relationship to relate them.

Figure 10.19 shows the orientations of recrystallized grains artificially nucleated in an aluminium single crystal. The orientation of the deformed crystal is plotted in standard projection. It was asserted that the results demonstrated a 40° [1$\bar{1}$1] relationship with the deformed crystal although **"in most cases the scatter....around the ideal orientation was rather large"**. The common [1$\bar{1}$1] directions are in fact up to 16° apart and [1$\bar{1}$1] for the annealed grains is spread over an angle of approximately 28°. Although there is clearly an orientation relationship based at or near <111>, involving a rotation of ~40°, the scatter is considerable.

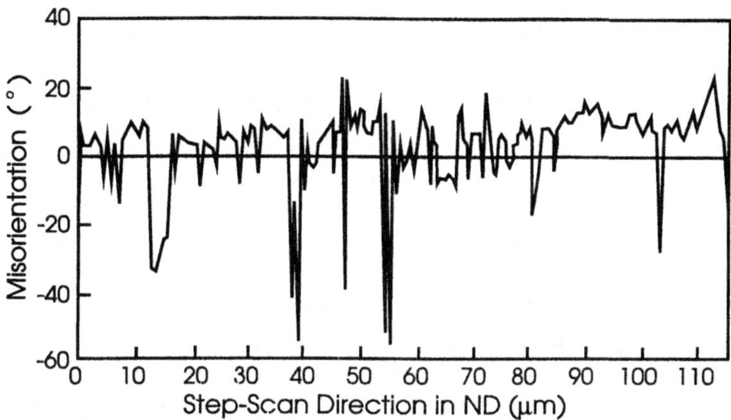

Fig. 10.20. The variation in orientation as determined by a TEM step scan in the ND direction, in a $\{112\}<111>$ aluminium crystal deformed to a strain of 1 in channel die compression, (Hjelen et al. 1990).

(ii) Experiments on single crystals (§4.3.2.2) suggest that for the high mobility $40°<111>$ orientation relationship, tilt boundaries migrate much faster than twist boundaries. However the recrystallized grains in polycrystals are usually equiaxed.

(iii) Although the reasons for the high mobility of a low Σ or CSL boundary can be broadly understood in the case of boundaries migrating in perfect material (§4.4.4.1), it is difficult to see how the necessary ordered boundary structure can be maintained during recrystallization.

(iv) Because of the rapid local orientation variations in the deformed microstructure, a precise relationship between a growing grain and the adjacent deformed matrix cannot be defined.

As discussed in chapter 2, during deformation a single grain breaks up into regions of different misorientation as cells or subgrains and other heterogeneities such as deformation bands develop. The orientation of a deformed grain therefore varies over small distances. Such orientation changes have been measured in both single crystals (Hjelen et al. 1990) and polycrystals (Hjelen et al. 1991), and an extreme example is shown in figure 10.20, in which orientation changes of more than $20°$ are found over distances of a few micrometres.

In summary, we see that although there is clear evidence of oriented growth in single crystals and lightly deformed polycrystals where an orientation relationship between the recrystallizing grain and the deformed matrix may exist and be maintained during growth, it is now clear that during the recrystallization of highly deformed polycrystals, any orientation relationships will not only be very localised, but will be transient as a grain grows into new surroundings. Whether, during the later stages of growth, the existence of favourable orientation relationships over only part of the boundary of a growing grain will lead to a faster growth rate for the grain, remains to be clarified.

10.7.3. Oriented nucleation.

The realisation that the nucleus of a recrystallized grain must have an orientation that is close
to or identical with that of the volume element in the deformed structure from which it grew
has existed for many years. Attention was refocussed on this by the development of electron
microscope techniques that allowed orientation determinations from very small volumes of
material and by advances in the theoretical analysis of texture development during
deformation. Many of the developments in this field have been stimulated by the need to
find the nuclei responsible for the cube texture. Although it is universally accepted that
oriented nucleation occurs, many of the details of how nuclei of particular orientations
develop remain a matter of debate.

10.7.3.1 The Dillamore-Katoh model

The benchmark work was the paper by Dillamore and Katoh (1974) who calculated the
rotation paths during the compression of polycrystalline iron, and showed that a transition
band centred on <411> would be formed when the rotation paths of neighbouring volumes
diverged (§2.3.6.2) as shown in figure 2.19a. They argued that this would provide a very
favourable site for the nucleation and growth of a recrystallized grain and pointed out that
a <411> fibre texture was indeed prominent in the recrystallization texture of compressed
iron as shown in figure 2.19c. Inokuti and Doherty (1978) found <411> oriented nuclei
in transition bands after annealing compressed iron samples.

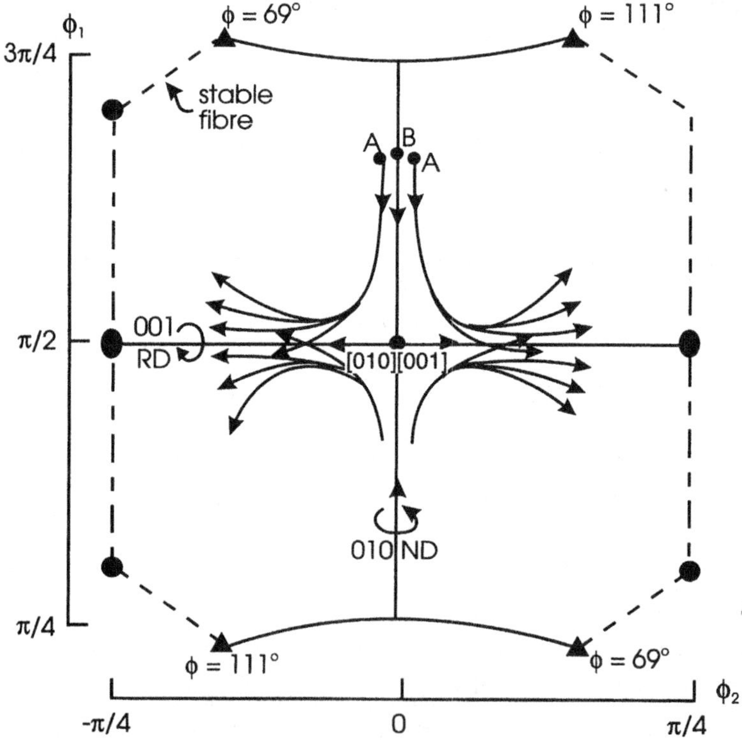

Fig. 10.21. Section through orientation space at $\Phi = 90°$ showing schematically the slip
rotations in the vicinity of the cube orientation in fcc metals, (Dillamore and Katoh 1974).

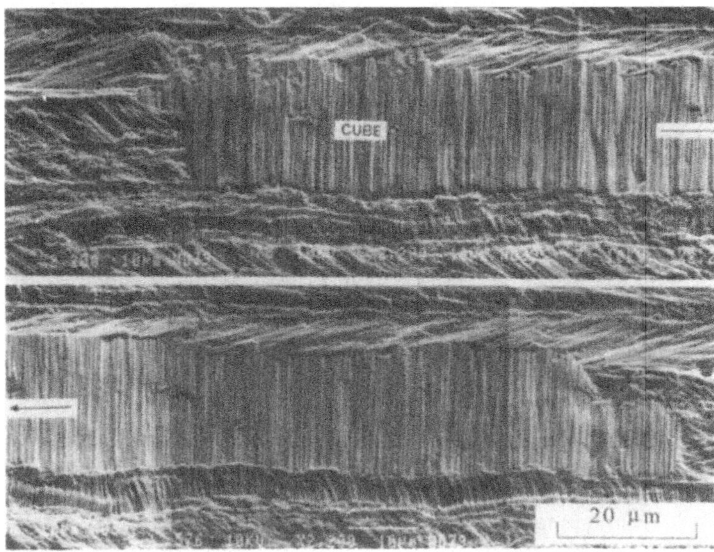

Fig. 10.22. Long, thin, cube oriented grain developing in 95%, cold rolled copper after annealing for 100 sec at 140°C. Longitudinal section, cut at 45° to RD to expose cube oriented grains as {110}<001>, (Duggan et al. 1992).

Another important result came from their analysis of rolling texture development in copper. In this case it was predicted that similar transition bands would form in which the central region of the curved lattice contained the cube orientation as shown in figure 10.21. This shows that certain crystals will approach the cube orientation by rotation around the normal direction and then subsequently diverge by rotation about the rolling direction, eventually arriving at a stable orientation outside the plane of the figure. A small initial curvature, indicated by A-B-A will rapidly become larger, but the centre point B, linking the two diverging orientations will continue to rotate into and remain at the cube orientation.

If nucleation were to occur at this site, which, because of its large orientation gradient is likely, then the cube texture should be found. The first experimental support for this was provided by Ridha and Hutchinson (1982) who found long narrow regions of precisely the predicted type in cold rolled copper. This observation has been confirmed many times since in both copper and aluminium by SAD and EBSP experiments (e.g. Hjelen et al. 1991). Somewhat related observations have also been made by the use of an orientation-sensitive deep etching technique (Köhlhoff et al. 1988b) and a typical microstructure is shown in figure 10.22.

10.7.3.2 Other models

Although it has been widely accepted that the Dillamore-Katoh model provides a satisfactory explanation of the appearance of cube-oriented volumes in the deformed material, other models have subsequently been proposed. Amongst these the recent work of Lee et al. (1993) is significant. These workers assert that the cube-oriented volumes (nuclei) in cold rolled copper are formed by deformation banding and maintain that the Dillamore-Katoh model corresponds to a deformation band in which all regions of the grain involved have the same homogeneous strain and that this is the macroscopic strain (i.e. the model corresponds to the first of the two deformation bands defined by Chin (1969) and discussed in §2.3.6.1).

Lee et al. (1993) have used a simple one-dimensional model, that considers only one banding mode, to examine the formation of deformation bands in the presence of heterogeneous strain and have concentrated on the formation of cube-oriented volumes and the orientation environment in which they occur. It was found that cube-oriented volumes appeared more profusely than in the Dillamore-Katoh model, and from a much wider range of starting orientations. These cube-oriented volumes were similarly related to neighbouring volumes by a 30-40° <111> rotation. However the angular relationship deteriorated at high reductions and they suggested that a unified theory was needed. In this the cube volumes are generated by both types of deformation banding at low-medium rolling reductions with the Dillamore-Katoh model more significant at high reductions. The net result is an increased number of suitably oriented nuclei.

Akef and Driver (1991) point out that the Dillamore-Katoh analysis was based on the use of the Taylor model, i.e. the analysis assumes full constraints (§2.3.5). It is well known that this leads to ambiguities in the slip rate particularly for high symmetry orientations like {001}<100>. Furthermore the analysis referred to the case of bcc iron undergoing pencil glide, {hkl}<111> and by reversing the signs of the rotations Dillamore and Katoh extended it to the case of fcc metals and {111}<110> slip. Akef and Driver question the validity of this transposition and also the use of the Taylor analysis. They have analyzed the rotation paths and rotation rates for crystals originally oriented near {001}<100> and conclude that the behaviour of their single crystal specimens can be predicted more accurately by a simple model of deformation band development in which the normal strain components are equal to the macroscopic strains and the shear strain components are assumed to be unconstrained.

10.7.4 The relative roles of nucleation and growth

The relative roles of nucleation and growth require the examination of microstructure and local texture (microtexture) of both the deformed and recrystallizing material. As emphasised by Duggan et al. (1993), there is a fundamental difficulty in interpreting the results of experiments based on examination of partially recrystallized samples, because in such circumstances, the creation of a growing grain destroys what might have been significant evidence relative to both the nucleus and its immediate environment. One possible solution, that of in-situ HVEM, can only examine small volumes and the interpretation of the results is also open to question (§5.5.4.1). Microtexture investigation of deformed samples, followed by examination of partly recrystallized material, is currently the most promising approach, and techniques such as EBSP or deep etching (see Appendix), which allow large areas to be examined, provide the types of result which are leading to an improved understanding of the origin of recrystallization textures.

The recent work of Hjelen et al. (1991) provides support for oriented nucleation. Directionally cast aluminium was rolled, and EBSP was used to examine the microstructure of rolled and partly recrystallized samples. In agreement with the Dillamore-Katoh model, nucleation was found to occur in transition bands and preferentially in bands that were present in volumes of the deformed matrix that had either the {112}<111> (C) orientation or an ND-rotated C orientation. It was also found that deformation heterogeneities in which the core structure of the heterogeneity had a $40° < 111 >$ orientation relationship with the surrounding matrix were common and amongst the microstructural features involved were transition bands and shear bands. Nucleation occurred frequently at these sites. Because $\{112\} < 111 >$ and $\{001\} < 100 >$ do not have a common $< 111 >$ axis, it was argued that an oriented growth theory based on a $40° < 111 >$ relationship cannot be applicable to the development of the cube texture in this case and that the observed $40° < 111 >$ relationship was due to nucleation. They also suggested that there was no direct evidence in the literature to support the hypothesis that $40° < 111 >$ grains, when uniformly distributed in space, have a growth rate exceeding that of grains of other orientations.

It is often difficult to compare growth rates of grains of different orientation within the same deformed grain, because in many cases only grains of one orientation are found. However, recent work of Ardakani and Humphreys (1994) has shown clearly that in (001)[110] single crystals of the alloy Al-0.05%Si, deformed to 75% reduction in thickness, $40° < 111 >$ grains grew faster than those of other orientations which were nucleated at large second-phase particles, in contradiction to the conclusions of Hjelen et al. (1991).

The same investigation also showed that, in Al-0.8%Si crystals, which contained a larger volume fraction of particles giving rise to PSN, the balance between nucleation and growth control of recrystallization depended on the prior strain in the following manner:

(i) After reductions of 50% or less, $40° < 111 >$ grains grew most rapidly.

(ii) After a deformation of 75%, grains with a $20° < 012 >$ relationship grew most rapidly

(iii) After a reduction of 90%, the PSN grains were found to grow at similar rates to all the other grains.

As in all single crystal studies the range of orientations present in the deformed microstructure and therefore available for the nuclei was limited. However, these results clearly show that associations between growth rate and orientation environment may occur under certain circumstances.

A rigid polarisation between theories of ORIENTED GROWTH and ORIENTED NUCLEATION is now considered to be untenable. The orientations of the nuclei are restricted to those present in the deformed material, and there is no evidence that nuclei of random orientations ever occur. There is also undisputed evidence that under certain circumstances, some orientation relationships are associated with a high boundary mobility.

The final recrystallization texture is determined by both the range of nuclei orientations available and the subsequent growth advantage enjoyed by any grain. Both of these factors will depend on the material and the processing parameters and the relative importance of each will vary from case to case.

It is now recognised that any recrystallization nucleus in a heavily deformed polycrystal will be surrounded by a wide spread of orientations and that these surroundings will change during growth, and this must be taken into account in any model of recrystallization. Much of the discussion about the origin of recrystallization textures is now centred on the orientation dependence of recrystallization mechanisms at the micron level (e.g. §10.7.6.1), where a distinction between "nucleation" and the "early stages of growth" is meaningless. This again emphasises the artificiality of the "oriented nucleation" versus "oriented growth" arguments.

10.7.5. The role of twinning.

Annealing twins are common in the recrystallized grains of low to medium SFE metals (§6.7). The formation of an annealing twin provides the only means of generating an orientation different to those present in the deformed metal prior to recrystallization. It is not surprising therefore, that the role of twinning in the generation of recrystallization textures has been actively investigated. In recent years such studies have been carried out extensively by Gottstein and colleagues at Aachen and by Haasen, Berger and colleagues at Göttingen (see for example Gottstein 1984, Berger et al. 1988, Haasen 1993).

There are twelve possible twinning elements ($\{112\} < 111 >$) in fcc metals and if all these were to operate successively during annealing, the strength of the texture would be rapidly reduced. Gottstein (1984) has pointed out that only six generations of twinning would be needed to achieve a random texture, and this is illustrated in figure 10.23. However, most recrystallization textures are sharp and clearly defined and it follows that unrestricted twinning does not occur.

Some selection principles must apply, and a preferred sequence of twinning events (**twin chain**) must occur. The factors responsible for selecting a particular twin variant are still in dispute (§6.7.4.2), but the two most likely are the development of a high mobility boundary (Gottstein 1984) or a reduction of grain boundary energy (Berger et al. 1988).

An extensive investigation by Berger et al. (1988) found that whereas recrystallized grains in aluminium were related to the matrix by approximately $40° < 111 >$, no such relationship existed for copper. In experiments with tensile deformed $< 100 >$ and $< 111 >$ crystals of copper (99.9998%) the favoured relationships were $20° < 100 >$, $50° < 111 >$, $130° < 430 >$ and $145° < 541 >$. The orientations produced by these rotations are all related by twin chains of various lengths. The $145° < 541 >$ orientation relationship is twin-related to those for $20° < 100 >$ and $50° < 111 >$ by three link chains and to that for $130° < 430 >$ by a single link chain. All of these rather complex relationships are close to coincident site lattices as shown in table 10.4

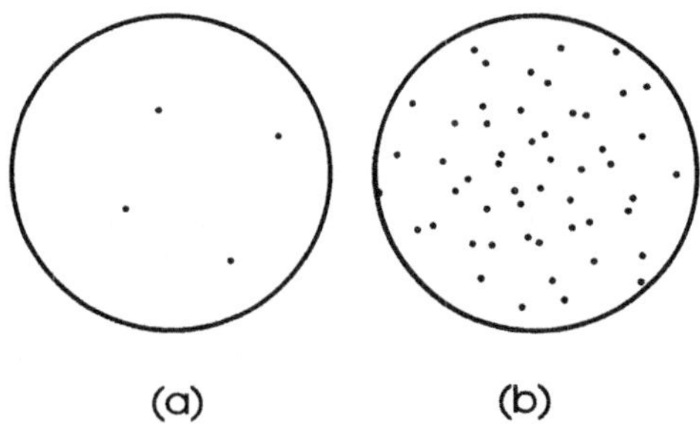

(a)   (b)

Fig. 10.23. Effect of repeated random twinning on texture (a) 111 poles of initial orientation (b) After 3 generations of twins.

Table 10.4.

Orientation relationships in twin chains*.
(Berger et al. 1983)

Orientation relationship	Nearest coincidence relationship
20° <100>	22.6° <100>; Σ=13a
50° <111>	46.8° <111>; Σ=19b
130° <430>	129.8° <540>; Σ=25b
145° <541>	145.7° <541>; Σ=23

The coincidence relationships shown are not necessarily those of minimum rotation angle.

Berger et al. (1988) went on to extend these observations to the development of the annealing texture in rolled polycrystalline copper and pointed out that similar orientation relationships existed. The cube texture is twin-related to two of the major components of the rolling texture ({112}<111> and {110}<112>) by a 130°<430> rotation and to the third component {123}<634> by a 50°<111> rotation. These are very significant results but some caution is needed in extrapolating the single crystal results to explain the origin of recrystallization textures in rolled polycrystals. The crystals were only lightly deformed (20% elongation for <100> crystals and 4% elongation for <111> crystals). The microstructure was, therefore, very homogeneous and none of the major inhomogeneities of deformation that act as nucleation sites in heavily rolled material would have been present. A second concern, the precision of the rotations involved, has already been referred to, and in this case the rotations 130°<430>, 50°<111> and 20°<100> lead respectively to 8°, 5° and 4° deviation from the ideal.

10.7.6. The cube texture.

No account of the origin of recrystallization textures would be complete without further reference to this famous texture (fig 10.1). It can be produced in many fcc metals and alloys of medium-high SFE and the precision and sharpness are remarkable. In addition to copper and aluminium the cube texture is found in gold and nickel, in iron-nickel alloys containing more than 30% nickel and in some ternary alloys of iron, nickel and copper (Barrett and Massalski 1980). The conflicting experiments and different theories referred to in this section show that this very old problem is still the subject of controversy. This has been fuelled to some extent by the very considerable industrial significance attached to the development of the cube texture during the processing of deep drawn aluminium alloy beverage cans (§12.2).

10.7.6.1 The origin of the cube texture

Neither pole figure nor ODF data of cold rolled material give any indication of a significant {100}<001> component and therefore the origin of the cube texture has been difficult to explain. However, not a large amount of cube-oriented material is needed to provide the necessary recrystallization nuclei and the theoretical analysis of Dillamore and Katoh (1974) (see §2.3.6.2 and §10.7.3.1) showed that cube orientations could develop in transition bands as shown in figure 10.21.

Under such conditions a cube nucleus would have available in the neighbouring microstructure a large misorientation gradient appropriate for rapid growth. Subsequently long thin cube oriented regions were detected in the microstructure of heavily rolled copper by Ridha and Hutchinson (1982) and shown to develop into cube-oriented grains. Ridha and Hutchinson also suggested that the rapid development of the cube nuclei from these sites would be aided by the dislocation structure within the cube bands. They argued that the predominant slip systems active for material of the cube orientation (001)[100] would involve only the Burgers vectors $a/2[101]$ and $a/2[\bar{1}01]$, which are orthogonal to one another. As there is little interaction between dislocations of orthogonal Burgers vectors, rapid and extensive recovery is expected to take place in the cube-oriented subgrains, thus giving them an added advantage in forming recrystallization nuclei.

Despite this apparent success, the origin of the cube texture is still hotly debated. The most persistent proponents of oriented growth have been Lücke and Hirsch and their colleagues at Aachen. Amongst the modifications that have been made to oriented growth theory, the concept of the compromise texture has had considerable attention, (see for example, Lücke 1984, Lücke and Engler 1992). It is argued that the cube grains have the best chance of growing because this orientation provides the best compromise towards meeting the $40° < 111 >$ condition with respect to all components of the deformation texture. In the case of the S component the agreement is nearly exact for all four components; the worst situation is found for the C component where the deviation from the proposed relationship is 23° for $30° < 111 >$ and 30° for $40° < 111 >$.

Hjelen et al. (1991) found that cube oriented grains in aluminium nucleated mainly from transition bands located in $\{112\} < 111 >$ C regions. Because there is no $40° < 111 >$ relationship between these orientations and the growth rates of the cube and other grains are similar, they concluded that preferred growth was not significant in the development of the cube texture. Similar conclusions have been reached by Doherty et al. 1993.

In an elegant, and novel experiment, Duggan et al. (1993) used the Köhlhoff etching technique to determine the orientation environment in rolled copper that favoured the development of successful cube oriented grains. The deformed microstructure was searched for cube oriented volumes and the orientations of the surrounding regions were determined. Analysis of the results indicated that about 60% of the cube oriented regions had either

Table 10.5

**Orientation relationships for cube oriented volumes
(Duggan et al. 1993).**

Common Axis	Cold Rolled (no. of volumes)	Partially Annealed (no. of volumes)
$(15\text{-}60°) < 110 >$	34	25
$(25\text{-}40°) < 111 >$	36	1
Others	48	22
Total	118	48

$<111>$ or $<110>$ common with their adjacent regions (table 10.5). The specimen was then lightly annealed to produce isolated cube oriented grains and the orientations of the volume elements immediately adjacent to any unrecrystallized cube volumes determined. Table 10.5 indicates that during partial annealing, virtually all of the deformed cube-oriented regions that had at least one $(25\text{-}40°)<111>$ related neighbour (such regions were generally close to the **S** orientation) had been consumed by the growing recrystallized cube grains. This is interpreted as a special case (which they term "microgrowth selection") of oriented growth, in which the local texture determines the viability of the nucleus. Cube-oriented nuclei grow into adjacent regions which have a $40°<111>$ relationship, and the larger combined volumes subsequently grow into the deformed microstructure.

The observation that cube bands adjacent to **S** components nucleate preferentially compared to cube bands next to other orientations has been confirmed by Vatne and Nes 1994 and Samajdar and Doherty 1995 in aluminium. Although it is accepted that the cube nuclei gain an advantage relative to other nuclei because of their **local environment** as discussed above, the precise reason for this is not yet clear. The factors which may be important include:

- A large driving pressure provided by the higher stored energy of adjacent S components.
- A local growth mobility advantage of the migrating high angle boundary.
- A lower energy of the migrating high angle boundary.

10.7.6.2 The effect of high temperature deformation

In commercial aluminium alloys it is generally found that the cube texture developed on recrystallization is stronger after hot rolling than after cold rolling (Hirsch and Engler 1995). Although this can be partly explained by a reduction of the random components of the texture that are derived from particle stimulated nucleation there is now considerable evidence that other factors also promote the formation of the cube texture in recrystallized, hot rolled aluminium alloys. Amongst these the stability of the cube orientation is important. Sanders et al (1986) demonstrated that in cold rolled, cube textured polycrystalline material this orientation was unstable at large reductions and this has been confirmed by Maurice and Driver (1993) for cube oriented single crystals deformed in plane strain compression. However, they found the cube orientation to be relatively stable at temperatures above 400°C and there is evidence that this may be due to the occurrence of non-octahedral slip at high temperatures (§2.3.1). Whether or not this factor explains the stability of the cube orientation in hot rolled, polycrystalline materials has yet to be confirmed.

Weiland and Hirsch (1991) and Bolingbroke et al. (1994, 1995) have examined the factors affecting the development of the cube texture in hot rolled aluminium alloys and find that the strength of this component is increased by an increased proportion of cube texture prior to rolling, higher rolling temperature and lower strain rate.

Investigation of the microstructure of hot rolled, polycrystalline alloys has shown that the cube oriented elements are retained as long, ribbon-like bands. The experimental evidence of Vatne and Nes (1994) and Samajdar and Doherty (1995) indicates strongly that the recrystallized, cube texture grains originate from these bands and that the strength of the cube component in fully recrystallized material is due to rapid and prolific nucleation from the bands and not to any growth advantage of the cube orientation. The reason for favoured nucleation at the bands is not entirely clear but the larger size of the cube subgrains and their greater perfection may be important. In this respect the earlier work of Ridha and

Hutchinson (1982) on cold rolled polycrystalline copper in which similar cube oriented, long bands were formed (§10.7.6.1) may be relevant.

Finally it must be pointed out once more that as in the case of cold rolled material (§10.7.6.1), the environment of the nucleating subgrains is most important and similar arguments to those given there must also apply to the nucleation in hot deformed material.

Chapter 11

RECOVERY AND RECRYSTALLIZATION DURING AND AFTER HOT DEFORMATION

The softening (restoration) processes of recovery and recrystallization may occur **during** deformation at high temperatures. In this case the phenomena are called **dynamic recovery** and **dynamic recrystallization** in order to distinguish them from the **static annealing** processes which occur during post-deformation heat treatment and which have been discussed in earlier chapters. The static and dynamic processes have many features in common, although the simultaneous operation of deformation and softening mechanisms leads to some important differences. Although dynamic restoration processes are of great industrial significance, they are not well understood because they are very difficult to study experimentally and to model theoretically.

Dynamic recovery and dynamic recrystallization form part of the much larger subject of **hot working.** This is a very large and important area which cannot be covered in detail here. Some aspects of the historical development of the understanding of dynamic restoration are discussed by McQueen (1981) and Tegart (1992).

Dynamic recovery and dynamic recrystallization occur during metalworking operations such as **hot rolling, extrusion** or **forging**. They are important because they lower the flow stress of the material, thus enabling it to be deformed more easily and they also have an influence on the texture and the grain size of the worked material. Dynamic recrystallization may also occur during creep deformation (Gifkins 1952, Poirier 1985), the main difference between **hot working** and **creep** being the strain rate. Hot working is generally carried out at strain rates in the range of 1-100 s^{-1}, whereas typical creep rates are below 10^{-5} s^{-1}. Nevertheless, in many cases similar atomistic mechanisms occur during both types of deformation. Dynamic recrystallization also occurs during the natural deformation of minerals in the Earth's crust and mantle, and is therefore of interest to Structural Geologists (e.g Nicolas and Poirier 1976, Poirier 1985).

11.1 SOFTENING BY DYNAMIC RECOVERY

In metals of high stacking fault energy, such as aluminium and its alloys, α-iron and ferritic steels, dislocation climb and cross-slip occur readily (§2.2.2). Dynamic recovery is therefore rapid and extensive at high temperatures and is usually the only form of dynamic restoration which occurs. The stress-strain curve in this case is typically characterised by a rise to a plateau followed by a constant or steady state flow stress as shown in figure 11.1.

During the initial stages of deformation there is an increase in the flow stress as dislocations interact and multiply. However, as the dislocation density rises, so the driving force and hence the rate of recovery increases (§5.3), and during this period a microstructure of low angle boundaries and subgrains develops. At a certain strain, the rates of work hardening and recovery reach a dynamic equilibrium, the dislocation density remains constant and a **steady-state flow stress** is obtained as seen in figure 11.1. During deformation at strain rates larger than ~ 1 s^{-1} the heat generated by the work of deformation cannot all be removed from the specimen and the temperature of the specimen rises during deformation. This may then cause a reduction in the flow stress as straining proceeds. In modelling the high temperature deformation behaviour as discussed below, it is very important that such effects are taken into account.

11.1.1 Constitutive relationships

At temperatures where thermally activated deformation and restoration processes occur, the microstructural evolution will be dependent on the **deformation temperature (T)** and **strain rate ($\dot{\epsilon}$)** in addition to the **strain (ϵ)**. The strain rate and deformation temperature are often incorporated into a single parameter - **The Zener-Hollomon parameter (Z)**, which is defined as:

$$Z = \dot{\epsilon} \exp\left(\frac{Q}{RT}\right) \qquad (11.1)$$

where Q is an activation energy

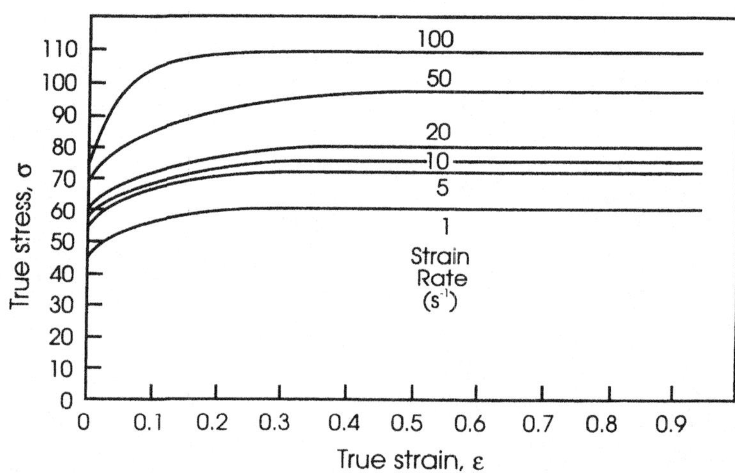

Fig. 11.1. Stress-strain curves for Al-1%Mg at 400°C, (Puchi et al. 1988).

For the purposes of analytical or computer modelling of hot working operations it is necessary to express the relationship between flow stress, temperature, strain and strain rate algebraically. If the flow stress follows a mechanical equation of state, i.e. it is dependent only on the **instantaneous** values of T, ϵ and $\dot{\epsilon}$ and not on their **history**, then the relationship between these parameters may be expressed by relatively simple empirical equations (e.g. Jonas et al. 1969, Frost and Ashby 1982). It is found that during hot work such a mechanical equation of state is closely followed, and during steady state deformation, the relationship between the flow stress (σ) and Zener-Hollomon parameter is often found to be

$$Z = c_1 \sinh (c_2 \, \sigma)^n \qquad (11.2)$$

where c_1, c_2 and n are constants.

Equation 11.2 may also be expressed in the form

$$\sigma = \frac{1}{c_2} \ln \left[\left(\frac{Z}{c_1} \right)^{\frac{1}{n}} + \left(\left(\frac{Z}{c_1} \right)^{\frac{2}{n}} + 1 \right)^{1/2} \right] \qquad (11.3)$$

At low values of stress, equation 11.3 reduces to a power relationship of the form

$$\dot{\epsilon} = c_3 \, \sigma^m \exp \left(\frac{-Q_1}{R\,T} \right) \qquad (11.4)$$

where c_3, m and Q_1 are constants.

and at high stress values it simplifies to

$$Z = 0.5^n \, c_1 \exp (n \, c_2 \, \sigma) = c_4 \exp (c_5 \sigma) \qquad (11.5)$$

Thus Z is seen to be closely related to the flow stress and hence to the dislocation density (equation 2.2). The Zener-Hollomon parameter is particular convenient for discussions of hot working processes in which the temperature and strain rate are generally known, whereas the flow stress may not be measurable.

A detailed consideration of the mechanical properties during hot deformation, the constitutive relationships and the microstructures developed during high temperature deformation is outside the scope of this book and further details may be found in Jonas et al. (1969), Roberts (1984, 1985) and Sellars (1978, 1986, 1990, 1992a, 1992b).

11.1.2 Microstructural development

The basic mechanisms of dynamic recovery are dislocation climb and glide, which result in the formation of low angle boundaries as also occurs during static recovery (§5.4). However, the applied stress provides an additional driving pressure for the movement of low angle boundaries and those of opposite sign will be driven in opposite directions (Exell and Warrington 1972). This leads to some annihilation of dislocations in opposing boundaries and Y-junction boundary interactions (§5.5.3.1) and these enable the subgrains to remain approximately equiaxed during the deformation. Some subgrain growth occurs, and within the subgrains the processes of work hardening and recovery lead to the continual formation of low angle boundaries and to a constant density of unbound or "free" dislocations within the subgrains. After a strain of typically 0.5 to 1, the subgrain structure often appears to achieve a steady state. In some cases, for example pure aluminium, the misorientation between subgrains is found to remain constant at a few degrees when the steady-state is reached (McQueen and Jonas (1975). However, in other materials it is not yet established as to whether or not the subgrain misorientation actually saturates or if it continues to increase with strain. The microstructural changes occurring during dynamic recovery are summarised schematically in figure 11.2.

Although the dislocation and subgrain structures often remain approximately constant during steady-state deformation, the original grain boundaries do not migrate significantly and the grains continue to change shape during the deformation. This of course means that although the flow stress may remain constant, **a true microstructural steady state is not achieved** during dynamic recovery. During deformation under conditions of low Z, the grain boundaries migrate locally in response to the boundary tensions of the substructure and to local dislocation density variations, and become serrated with a wavelength which is closely related to the subgrain size as shown in figure 11.3.

11.1.3 Homogeneity of deformation

As was discussed in chapter 2, the microstructures developed during low temperature deformation are usually very inhomogeneous and these regions of inhomogeneity are frequently sites for the origin of recrystallization during subsequent annealing. It is usually found that as the temperature of deformation increases, the deformation becomes more

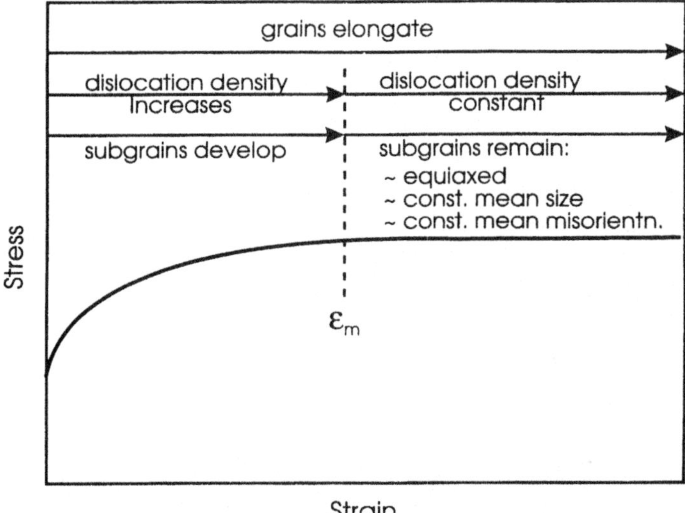

Fig. 11.2. Summary of the microstructural changes which occur during dynamic recovery, (after Sellars 1986).

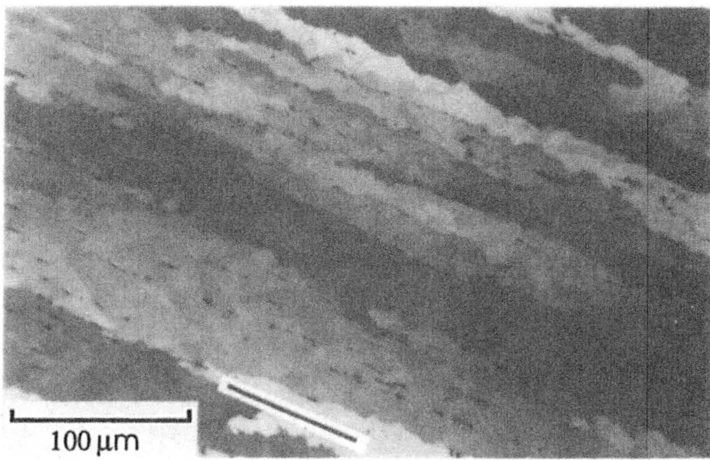

Fig. 11.3. Grain boundary serrations in TD-plane section of commercial purity aluminium, deformed at 400°C, $\epsilon = 1$, $\dot{\epsilon} = 0.25$. (courtesy K.D. Vernon-Parry).

homogeneous (e.g. Drury and Humphreys 1986, Hansen & Jensen 1991). This is generally interpreted in terms of an increase in the number of operating slip systems, which gives rise to homogeneous deformation more typical of the ideal "Taylor type" plasticity (§2.3.5) than that which occurs at lower temperature. During deformation at high temperatures, slip systems which do not operate at low temperatures because of a high Peierls-Nabarro stress may become active. This is particularly important in non-metals and non-cubic materials, (Ion et al. 1982) and even in face centred cubic metals there is some evidence that non-octahedral slip may occur at high temperatures (see §2.3.1). The microstructures after high temperature (low Z) deformation often consist of subgrains whose orientations tend to alternate about a mean, with little overall orientation gradient within a grain.

11.1.4 The effect of the deformation conditions

The mechanisms of dislocation generation and recovery discussed above, operate over a wide range of temperatures and strain rates. However, it should be noted that at very low strain rates and high temperatures (i.e. creep conditions), other mechanisms such as Herring-Nabarro and Coble creep may be important, although they will not be considered here. At low temperatures and high strain rates (high Z) the dislocation generation (work hardening) factor is dominant whereas at high temperatures and low strain rates (low Z) the dynamic recovery dominates. Therefore after hot deformation, the microstructure will be dependent on both strain and Zener-Hollomon parameter.

It has been found for steels and for aluminium that the high temperature flow stress is related to the density of dislocations **within** the subgrains (ρ_i) by the relationship (Castro-Fernandez et al. 1990)

$$\sigma = c_1 + c_2 G b \rho_i^{1/2} \tag{11.6}$$

where c_1 and c_2 are constants

This is similar to the relationship between the flow stress and overall dislocation density found at low temperatures (equation 2.2). Figure 11.4a shows this relationship for an Al-Mg-Mn alloy.

If subgrains are formed then it is usually found (Sherby and Burke 1967) that the high temperature flow stress is inversely proportional to the mean subgrain diameter (D) (equation 5.30) and an example of this is shown in figure 11.4b. It has been shown (Takeuchi and Argon 1976) that if equation 5.30 is expressed in terms of normalised stress and subgrain size and a dimensionless constant K, then K is constant for any particular class of material.

$$\frac{\sigma}{G} \cdot \frac{D}{b} = K \qquad (11.7)$$

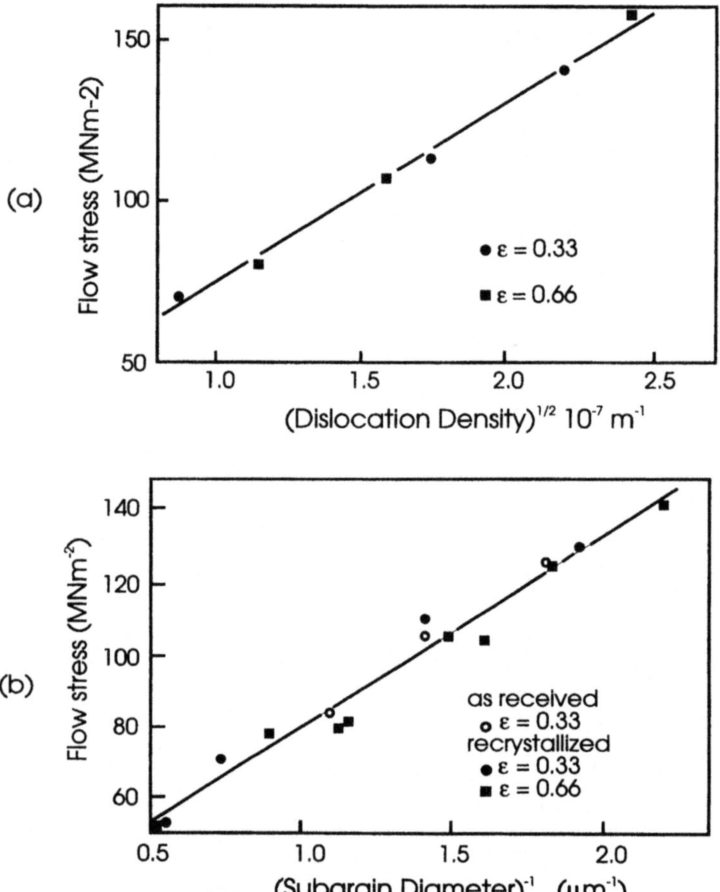

Fig. 11.4. The relationship between the high temperature flow stress and microstructure in Al-1%Mg-1%Mn. a) The dislocation density within subgrains (Castro-Fernandez et al. 1990). b) The subgrain size, (Castro-Fernandez and Sellars 1988).

Derby (1991) has analyzed subgrain data for a range of metals and minerals and these are plotted according to equation 11.7 in figure 11.5. K has a value of ~10 for fcc metals and ~25-80 for ionic crystals of the NaCl structure.

At constant flow stress (or Z), equations 11.6 and 5.30 imply a unique relationship between the subgrain size and the dislocation density within the subgrains of the form

$$\rho_i^{1/2} = c_3 D^{-1} \tag{11.8}$$

Such a relationship is obtained in theoretical models of subgrain formation during deformation (e.g. Holt 1970, Edward et al. 1988), and experimental investigations in aluminium, (Castro-Fernandez et al. 1990), copper (Straker and Holt 1972) and ferritic steels (e.g. Barrett et al. 1966, Urcola and Sellars 1987), have reported values of c_3 between 10 and 20.

11.2 DYNAMIC RECRYSTALLIZATION

In metals in which recovery processes are slow, such as those with a low or medium stacking fault energy (copper, nickel and austenitic iron), dynamic recrystallization may take place when a critical deformation condition is reached. A simplified description of the phenomenon of dynamic recrystallization is as follows. New grains originate at the old grain boundaries, but, as the material continues to deform, the dislocation density of the new grains increases, thus reducing the driving force for their further growth, and the recrystallizing

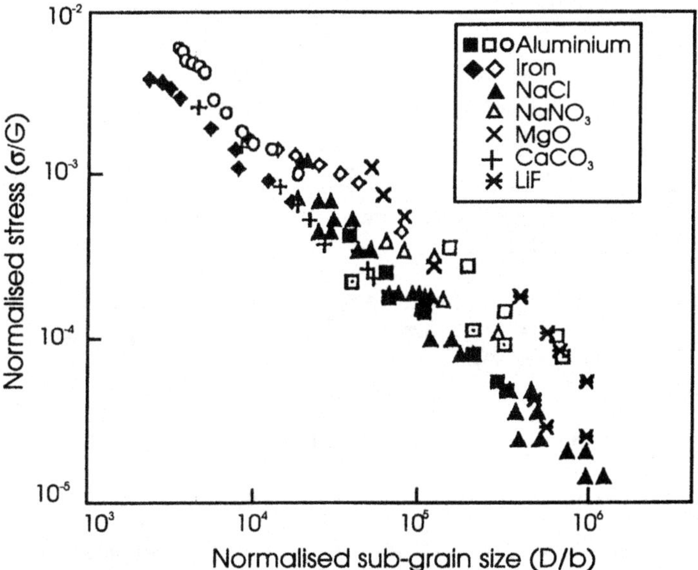

Fig. 11.5. The relationship between subgrain size and high temperature flow stress, (after Derby 1991).

grains eventually cease to grow. An additional factor which may limit the growth of the new grains is the nucleation of further grains at the migrating boundaries.

We should note that in addition to this type of dynamic recrystallization, which is sometimes termed **discontinuous dynamic recrystallization**, there are other mechanisms which produce high angle grain boundaries during high temperature deformation, and which may be considered to be types of dynamic recrystallization. These phenomena are discussed in §11.3.

11.2.1 The characteristics of dynamic recrystallization

The general characteristics of dynamic recrystallization are as follows:

(i) As shown in figure 11.6, the stress-strain curve for a material which undergoes dynamic recrystallization generally exhibits a broad peak that is different to the plateau, characteristic of a material which undergoes only dynamic recovery (fig 11.1). Under conditions of low Zener-Hollomon parameter, multiple peaks may be exhibited at low strains, as seen in figure 11.6.

(ii) A critical deformation (ϵ_C) is necessary in order to initiate dynamic recrystallization. This is somewhat before the peak (σ_{max}) of the stress strain curve. For a range of testing conditions, σ_{max} is uniquely related to the Zener-Hollomon parameter (Z).

(iii) ϵ_C decreases steadily with decreasing stress or Zener-Hollomon parameter, although at very low (creep) strain rates the critical strain may increase again (Sellars 1978).

(iv) The size of dynamically recrystallized grains (D_R) increases monotonically with decreasing stress. Grain growth does not occur and the grain size remains constant during the deformation.

(v) The flow stress (σ) and D_R are almost independent of the initial grain size (D_0), although the kinetics of dynamic recrystallization are accelerated in specimens with smaller initial grain sizes.

(vi) Dynamic recrystallization is usually initiated at pre-existing grain boundaries although for very low strain rates and large initial grain sizes, intragranular nucleation becomes more important.

11.2.2 The nucleation of dynamic recrystallization

Dynamic recrystallization originates at high angle boundaries. These may be the original grain boundaries, boundaries of dynamically recrystallized grains or boundaries created during straining (e.g. those associated with deformation bands). Bulging of grain boundaries is frequently observed as a prelude to dynamic recrystallization, and it is usually assumed that a mechanism closely related to strain induced grain boundary migration (§6.6.2) operates.

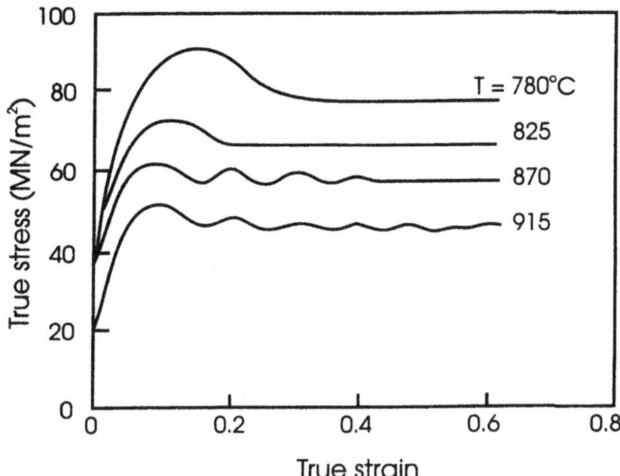

Fig. 11.6. The effect of temperature on the stress-strain curves for 0.68%C steel, deformed in axisymmetric compression, $\dot{\epsilon}=1.3\times10^{-3}\text{s}^{-1}$, (Petkovic et al. 1975).

Several models of dynamic recrystallization have been proposed. Regardless of the details of the mechanism of nucleation, the condition for the growth of a dynamically recrystallized grain is thought to depend on the distribution and density of dislocations (both in the form of subgrains and free dislocations §11.1.4), which provide the driving force for growth. Figure 11.7, based on the model of Sandström and Lagneborg (1975), shows schematically the dislocation density expected in the vicinity of a migrating boundary. The boundary at A is moving from left to right into unrecrystallized material which has a high dislocation density ρ_m. As the boundary moves, it reduces the dislocation density in its wake to around zero by recrystallization. However, the continued deformation raises the dislocation density in the new grain, so that it builds up behind the moving boundary and reaches ρ_x at a distance x behind the boundary, tending towards a value of ρ_m at large distances. The following very approximate analysis, which does not take dynamic recovery into account is based on the approach proposed by Sandström and Lagneborg (1975) and Roberts and Ahlblom (1978).

The migrating grain boundary moves under a driving pressure of $\sim \rho_m Gb^2$ arising from the dislocation density difference across the boundary and from equation 4.1, the resulting velocity is given as

$$\frac{dx}{dt} = M \rho_m G b^2 \qquad (11.9)$$

If the mean slip distance of the dislocations is L, then the rate of increase of dislocation density behind the migrating boundary due to the continuing deformation is found by differentiating equation 2.1.

$$\frac{d\rho}{dt} = \frac{\dot{\epsilon}}{bL} \qquad (11.10)$$

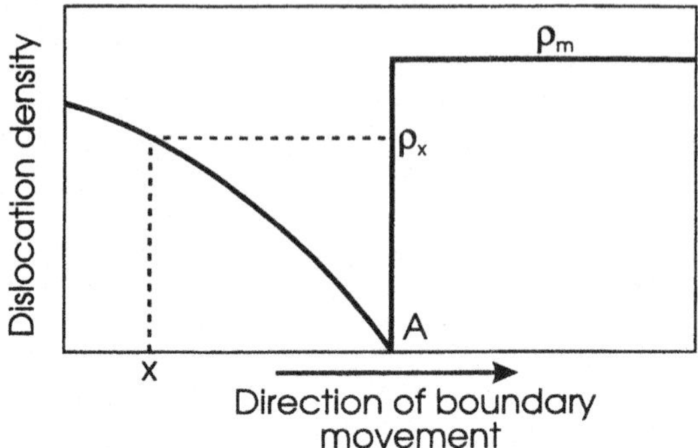

Fig. 11.7. Schematic diagram of the dislocation density at a dynamic recrystallization front, (after Sandström and Lagnegorg 1975).

and hence

$$\frac{d\rho}{dx} - \frac{\dot{\epsilon}}{MLGb^3\rho_m} \tag{11.11}$$

and

$$\rho = \frac{\dot{\epsilon}\,x}{MLGb^3\rho_m} \tag{11.12}$$

The dislocation density behind the moving boundary reaches the value of ρ_m when $x = x_c$ and hence

$$x_c = \frac{MLGb^3\rho_m^2}{\dot{\epsilon}} \tag{11.13}$$

If nucleation is assumed to occur by a bulge mechanism, then the critical condition for the formation of a nucleus of diameter x_c would (equation 6.40) be

$$E > \frac{2\gamma_b}{x_c} \tag{11.14}$$

Where E is the stored energy, which is taken to be a fraction K of $\rho_m Gb^2$. Using equation 11.13, then the condition for nucleation becomes

$$\frac{\rho_m^3}{\dot{\epsilon}} > \frac{2\gamma_b}{KMLGb^5} \tag{11.15}$$

The terms on the right hand side of equation 11.15 may be regarded as approximately constant at a particular temperature, and thus the condition for the nucleation of dynamic recrystallization is that a critical value of $\rho_m^3/\dot{\epsilon}$ must be achieved. In materials such as aluminium and pure iron, recovery occurs readily and this parameter, strongly dependent on dislocation density, never reaches the critical value. Therefore only dynamic recovery occurs. However in materials of lower stacking fault energy such as copper, nickel and stainless steel, recovery is slow and the dislocation density increases to the critical value necessary for dynamic recrystallization to occur.

11.2.3 Microstructural evolution

Dynamic recrystallization generally starts at the old grain boundaries as shown schematically in figure 11.8a. New grains are subsequently nucleated at the boundaries of the growing grains (fig. 11.8b), and in this way a thickening band of recrystallized grains is formed as shown in figure 11.8c. If there is a large difference between the initial grain size (D_0) and the recrystallized grain size (D_R), then a "**necklace**" structure of grains may be formed (fig 11.8b-c), and eventually the material will become fully recrystallized (11.8d). A micrograph of copper which has undergone partial dynamic recrystallization is shown in figure 11.9.

Unlike static recrystallization, the mean size of the dynamically recrystallized grains does not change as recrystallization proceeds as shown in figure 11.10

In some cases, microstructural evolution is more complicated than that described above, because the onset of dynamic recrystallization may lead to changes in deformation mechanism. For example, if dynamic recrystallization results in a very small grain size, then subsequent deformation may occur preferentially by the mechanism of **grain boundary sliding**. This will tend to occur if deformation by dislocation glide and climb is particularly difficult, and has been reported in α-brass (Hatherly et al. 1986) in which the low stacking

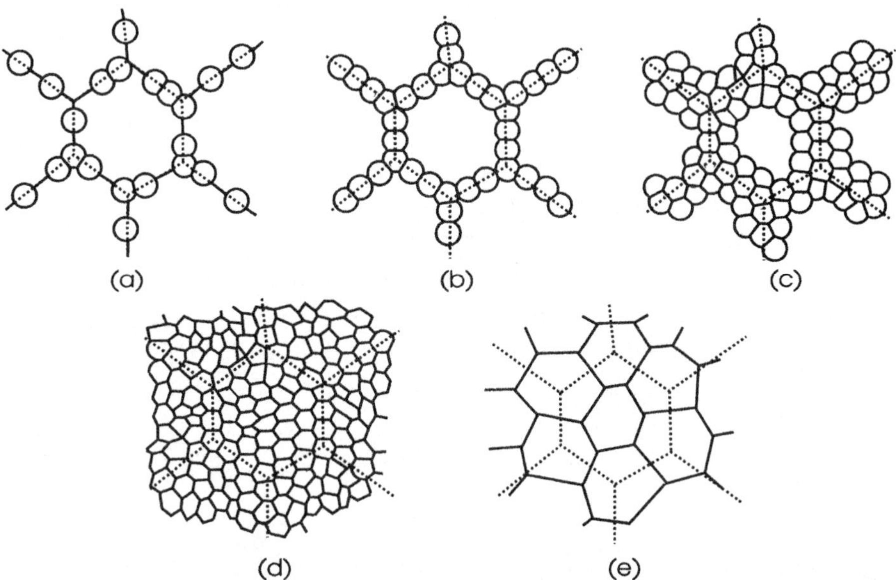

(a)   (b)   (c)

(d)   (e)

Fig. 11.8. The development of microstructure during dynamic recrystallization. a-d) Large initial grain size. e) small initial grain size.

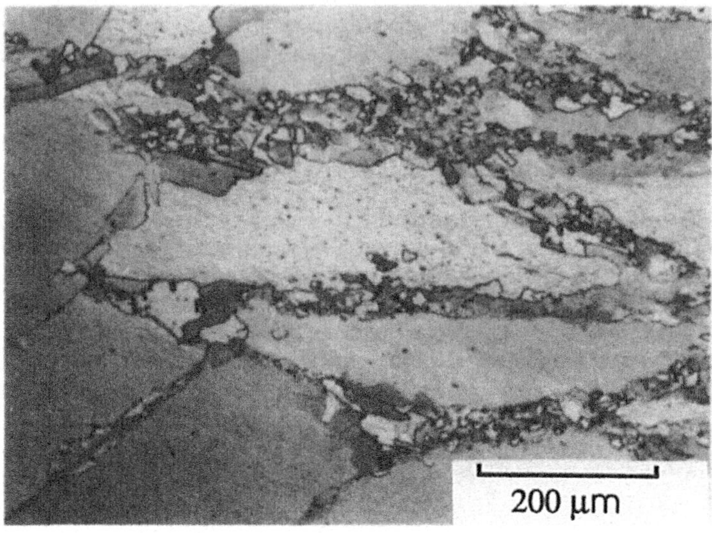

Fig. 11.9. Dynamic recrystallization in polycrystalline copper at 400°C, $\dot{\epsilon}=2x10^{-2}$, $\epsilon=0.7$, (Ardakani and Humphreys 1992).

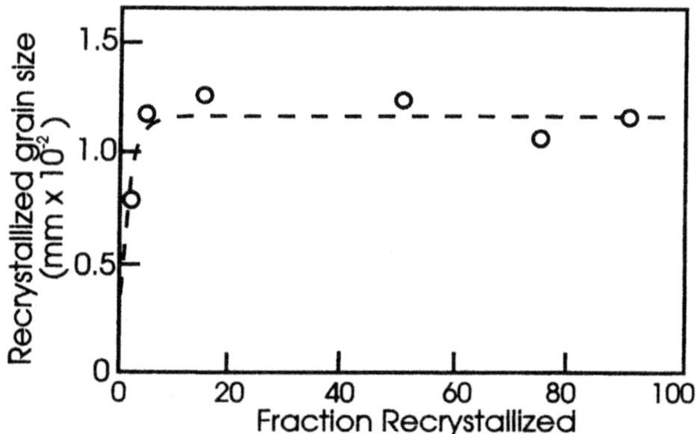

Fig. 11.10. Relationship between dynamically recrystallized grain size and fraction recrystallized in nickel deformed at 880°C, $\dot{\epsilon}=5.7x10^{-2}s^{-1}$, (Sah et al. 1974).

fault energy hinders climb, in magnesium (Drury et al. 1989) where there are few slip systems, and in dispersion hardened copper (Ardakani and Humphreys (1992) in which dislocation motion is hindered by second-phase particles.

11.2.4 The steady state grain size

Experimental observations generally show that the steady state grain size (D_R) during dynamic recrystallization is a strong function of the flow stress and is only weakly dependent on deformation temperature. The empirical relationship is often given as

$$\sigma = K D_R^{-m} \tag{11.16}$$

where $m < 1$ and K is a constant. Twiss (1977) examined the relation between the mean grain size and flow stress of a number of materials and proposed a universal relationship which can be expressed in normalised form as

$$\left(\frac{\sigma}{G}\right) \cdot \left(\frac{D_R}{b}\right)^n = K_1 \tag{11.17}$$

where $n = 0.8$ and $K_1 = 15$, which is of a rather similar form to the subgrain relationship of equation 11.7. Derby (1991) has found that such a relationship holds for a very wide range of materials of metallurgical and geological significance as shown in figure 11.11, and has shown that the data in figure 11.11 are bounded by locii of the form

$$1 < \left(\frac{\sigma}{G}\right) \cdot \left(\frac{D_R}{b}\right)^{2/3} < 10 \tag{11.18}$$

Dynamic recrystallization is a continuous process of deformation, nucleation of grains and subsequent migration of grain boundaries leaving new dislocation-free grains which then deform further. Although the details of dynamic recrystallization cannot be directly observed in metals, these processes have been observed to occur in transparent crystalline materials (§11.5).

For a steady state to occur there must be a dynamic balance between the nucleation of new grains and the migration of the boundaries of the previously nucleated grains. Derby and

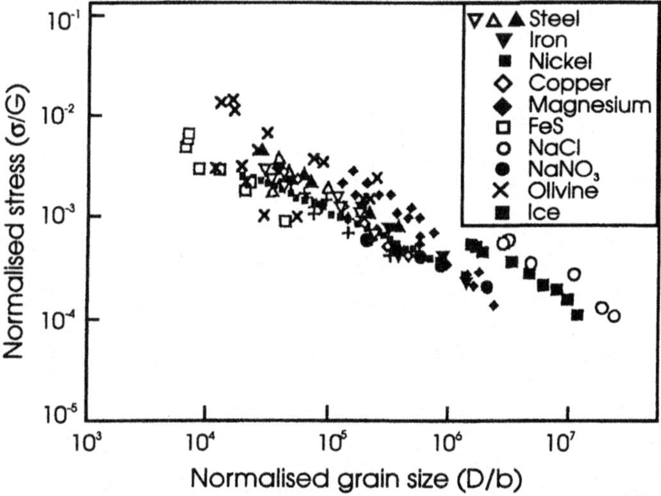

Fig. 11.11. The relationship between dynamically recrystallized grain size and high temperature flow stress, (after Derby 1991).

Ashby (1987) have analyzed this balance as follows. In the time taken for a moving boundary to sweep out a volume equivalent to the mean steady state grain size (D_R), the nucleation rate ($\dot{N}$) should be sufficient for one new nucleation event to occur in each equivalent volume averaged over the microstructure. If nucleation is confined to pre-existing grain boundaries then this condition may be expressed as

$$\frac{C\,D_R^3\,\dot{N}}{\dot{G}} \approx 1 \tag{11.19}$$

where C is a geometric constant of the order of 3.

The steady state grain size will therefore be dependent on the ratio of the nucleation and growth rates. Derby and Ashby (1987) argue that as the steady state grain size depends on the ratio of the nucleation and growth rates, both of which may have a similar temperature dependence, then it is to be expected that the steady state grain size will be only weakly dependent on temperature, as is found in practice. If the size of the dynamically recrystallized grains (D_R) is taken to be equal to x_c as given by equation 11.13, and we assume a relationship between flow stress and dislocation density of the form of equation 2.2 and a relationship between flow stress and strain rate as given by equation 11.4 with $m=5$, then equation 11.13 may be written in the form $\sigma = \mathbf{K'}/\mathbf{D}$, which is close to equation 11.17. Derby and Ashby (1987) have used a more sophisticated version of this analysis to calculate the steady state grain size starting from equation 11.19 and arrive at a relationship close to that of equation 11.17.

11.2.5 The flow stress during dynamic recrystallization

As shown in figure 11.6, the stress strain curves of a dynamically recrystallizing material may be characterised by a single peak or by several oscillations. Luton and Sellars (1969) have explained this in terms of the kinetics of dynamic recrystallization. At low stresses, the material recrystallizes completely before a second cycle of recrystallization begins, and this process is then repeated. The flow stress, which depends on the dislocation density, therefore oscillates with strain. At high stresses, subsequent cycles of recrystallization begin before the previous ones are finished, the material is therefore always in a partly recrystallized state after the first peak, and the stress strain curve is smoothed out, resulting in a single broad peak.

Sakai and Jonas (1984) have suggested that the shape of the stress strain curve depends primarily on the ratio of the recrystallized and starting grain sizes (D_0/D_R). If (D_0/D_R) > 2 then the microstructure develops as shown in figure 11.8a-d, the material is only partly recrystallized except at very high strains, and a smooth curve with a single peak results. However, if D_0/D_R < 2 then the new grains all develop at about the same time because there are enough sites (i.e. old boundaries) for recrystallization to be complete in one cycle as shown in figure 11.8e. This fully recrystallized and softened material then undergoes further deformation, hardens, and then recrystallizes again. As this cycle is repeated, an oscillatory stress strain curve results. The shape of the stress-strain curve therefore depends on the deformation conditions (Z) and on the initial grain size. Figure 11.12 illustrates schematically the relationship between the stress-strain behaviour and these parameters.

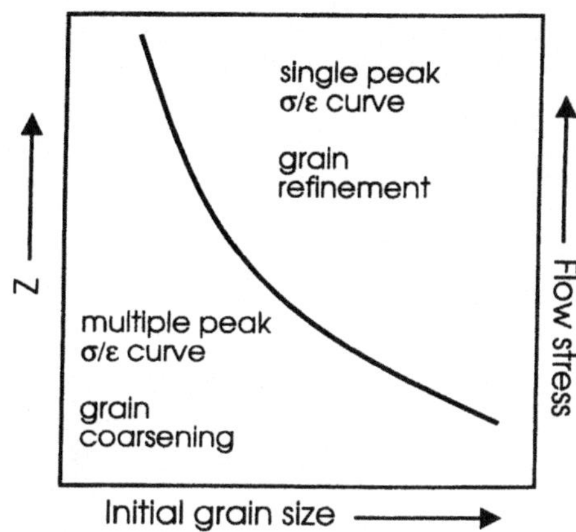

Fig. 11.12. The conditions for multiple and single peak dynamic recrystallization, (after Sakai et al. 1983).

11.3 OTHER MECHANISMS OF DYNAMIC RECRYSTALLIZATION

The dynamic recrystallization discussed in §11.2 is the normal discontinuous dynamic recrystallization which occurs for example in metals of low stacking fault energy. However, in recent years it has become apparent that under certain conditions a microstructure of high angle grain boundaries may evolve in ways other than the nucleation and growth of grains at pre-existing boundaries as discussed above, and in this section we examine some of these alternative mechanisms of dynamic recrystallization. Although such processes generally fall into the overall phenomenological classification of **continuous dynamic recrystallization**, it must be recognised that within this category there are several quite different types of mechanism which need to be considered separately.

11.3.1 Geometric dynamic recrystallization

As shown in figure 11.3, grain boundaries develop serrations during dynamic recovery, and the wavelength of these serrations is similar to the subgrain size. If the material is subjected to a large reduction in cross section, for example by hot rolling or hot compression, then, as the original grains become flattened, the size of these serrations will become comparable with the grain thickness as shown schematically in figure 11.13. Interpenetration of the scalloped boundaries will occur, resulting in a microstructure of what appear to be small equiaxed grains, an example of which is shown in figure 11.14a. Measurement of the grain orientations in microstructures such as figure 11.14a (Drury and Humphreys 1986) show that a considerable number of the boundaries are of low angle, although the proportion of high angle boundaries increases with increasing strain (Kassner et al. 1992). An equiaxed microstructure with a large number of high angle boundaries has therefore evolved without the operation of any new microscopic recrystallization mechanism and this process clearly differs from the **discontinuous dynamic recrystallization** discussed in §11.2. Microstructures of this type are commonly formed in aluminium and its alloys deformed to high strains (Perdrix et al. 1981), and were originally interpreted as being due to

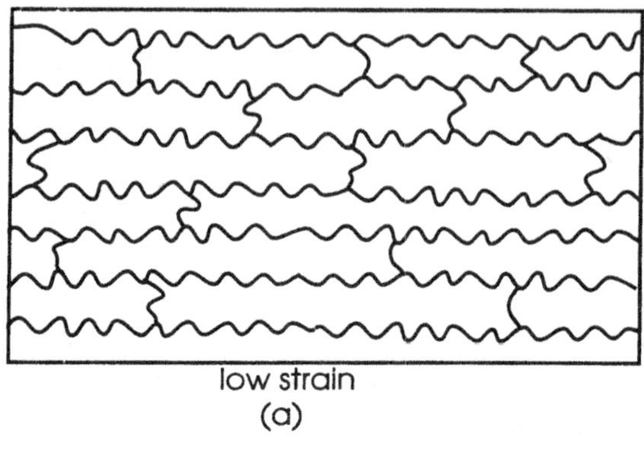

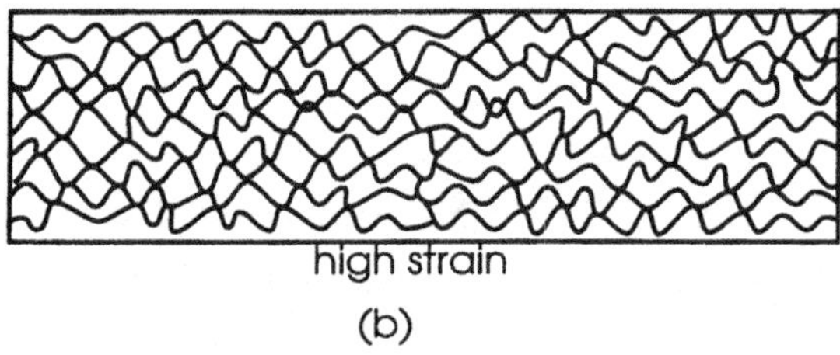

Fig. 11.13. Geometric dynamic recrystallization. a) Low strain. b) High strain.

discontinuous dynamic recrystallization. Humphreys (1982) showed that the origin of such microstructures was a process of grain impingement as discussed above, and this has been confirmed by later more extensive investigations (McQueen et al. 1985, Humphreys and Drury 1986, Solberg et al. 1989, McQueen et al. 1989) and the term **geometric dynamic recrystallization** has been used to describe the phenomenon. The mechanical properties and texture resulting from this process are of interest, and are reviewed by Kassner et al. (1992).

The occurrence of geometric dynamic recrystallization will depend on both the original grain size (D_0) of the material and on the deformation conditions. If we assume (Humphreys 1982) that the condition for geometric dynamic recrystallization is that grain impingement occurs when the subgrain size (D) becomes equal to the grain thickness, then the critical compressive strain (ϵ_c) for the process is

$$\epsilon_c = \ln\left(\frac{K_1 D_0}{D}\right) \qquad (11.20)$$

where K_1 is a constant of the order of unity

The relationship between flow stress and subgrain grain size is given by equation 11.7 and therefore

$$\epsilon_c = \ln(\sigma D_0) + K_2 \qquad (11.21)$$

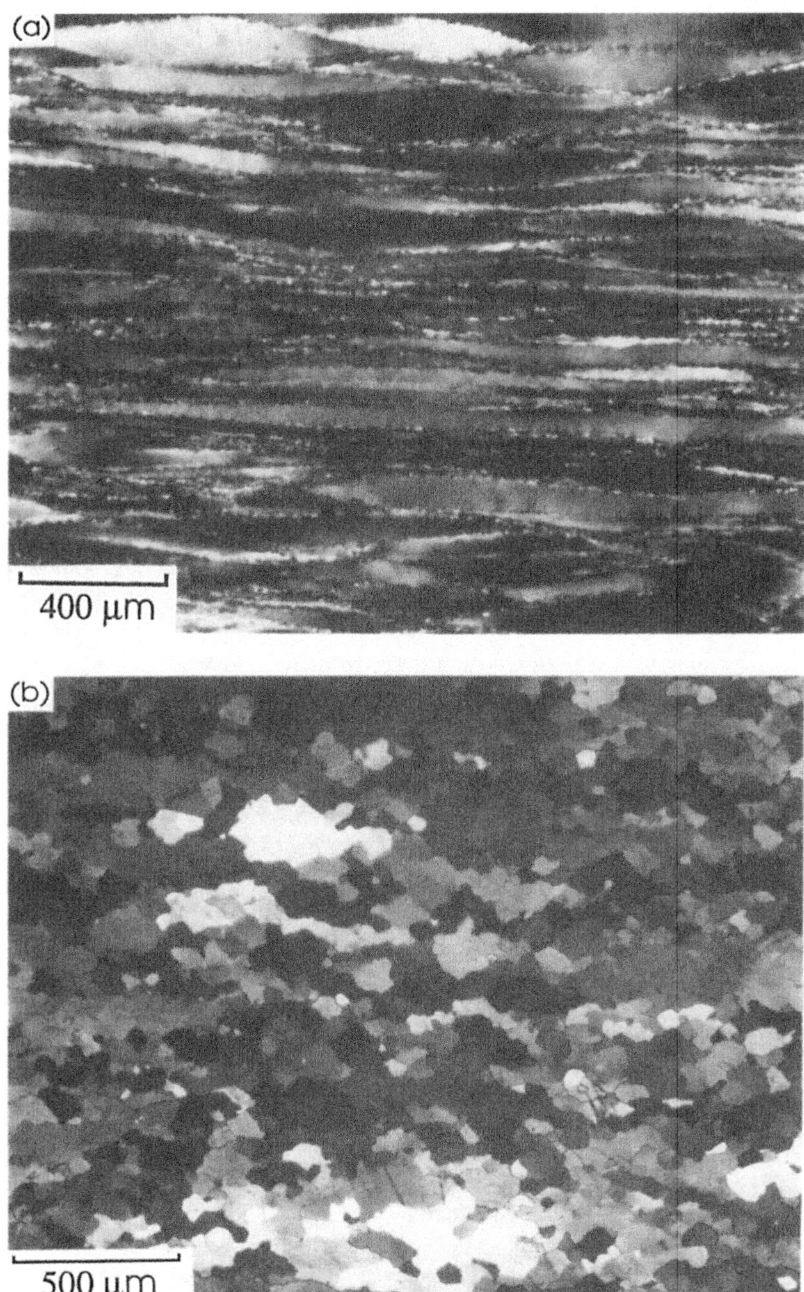

Fig. 11.14. The effect of deformation conditions on geometric dynamic recrystallization in Al-5%Mg. The compression plane is horizontal. a) $\epsilon=0.9$, $\dot{\epsilon}=2\text{x}10^{-3}$, $T=500°C$. b) $\epsilon=1.5$, $\dot{\epsilon}=2\text{x}10^{-3}$, $T=400°C$, (Drury and Humphreys 1986).

or, using equations 11.1 and 11.4,

$$e_c = \ln(Z^{1/m} D_0) + K_3 \tag{11.22}$$

The effect of deformation conditions on the microstructure may be seen in figure 11.14. When the specimen is deformed at high stresses (fig 11.14b), geometric dynamic recrystallization does not occur even at a strain of 1.5 because the subgrain size is small, whereas at low stresses (fig 11.14a), geometric dynamic recrystallization has commenced at the lower strain of 0.9.

Although geometric dynamic recrystallization will occur more readily at higher temperatures or lower Z (equation 11.22), the analysis above shows that even at low deformation temperatures, very large strains, accompanied by some dynamic recovery will lead to boundary impingement. For example, figure 2.7 shows that at high strains at ambient temperatures, the subgrain size reaches a minimum of $\sim 0.5 \mu$m. Therefore, according to equation 11.20, geometric dynamic recrystallization may occur in a metal of initial grain size 50μm, at a strain of ~ 4.6 (99% reduction). In practice, the strain will be considerably lower than this because of the formation of additional high angle boundaries as a result of heterogeneous deformation. In these circumstances, as the deformation temperature is reduced, the distinction between geometric dynamic recrystallization and the continuous recrystallization phenomena discussed in §5.7 disappears.

11.3.2. Dynamic recrystallization by progressive lattice rotation

There is considerable evidence that in certain materials, new grains with high angle boundaries may be formed during straining, by the progressive rotation of subgrains with little accompanying boundary migration. This is a **strain-induced** phenomenon which should not be confused with the subgrain rotation which has been postulated to occur during a static anneal (§5.5.4).

The phenomenon involves subgrains adjacent to pre-existing grain boundaries being progressively rotated as the material is strained. The old grains develop a gradient of misorientation from centre to edge. In the centre of the old grain, subgrains may not be well developed or may have very low misorientations. Towards the grain boundary, the misorientations increase, and at high strains, high angle boundaries may develop. This mechanism was first found in minerals (see §11.5), and has since been found in a variety of non-metallic and metallic materials. In the Geological literature the phenomenon is termed **rotation recrystallization**. There is evidence that it occurs in magnesium (Ion et al. 1982) and in aluminium containing significant solute additions, such as Al-Mg alloys (Gardner and Grimes 1979, Drury and Humphreys 1986) and Al-Zn alloys (Gardner and Grimes 1979).

Figure 11.15 shows the microstructure of an Al-5%Mg alloy in which well developed subgrains have formed along the original grain boundary. In this type of dynamic recrystallization it is generally found that the size of the grains formed at the old boundaries is only slightly larger than the subgrains, and the grain size is therefore given approximately by equation 11.7.

The mechanism by which this progressive subgrain rotation occurs is not yet entirely clear, but as it is most frequently found in materials in which dislocation motion is inhibited by

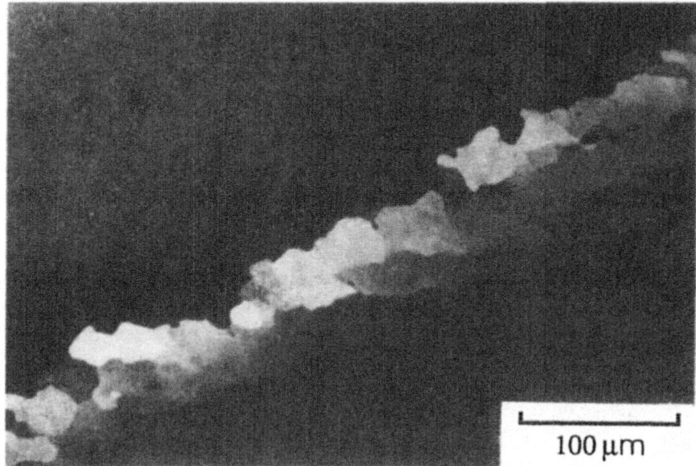

Fig. 11.15. Development of misorientations by lattice rotation adjacent to the grain boundary during deformation of Al-5%Mg, (Drury and Humphreys 1986).

either a lack of slip systems or by solute drag, it is likely that it is associated with inhomogeneous plasticity and accelerated dynamic recovery in the grain boundary regions, and it is possible that grain boundary sliding is also involved. However, the occurrence of a transition in dynamic recrystallization behaviour in minerals between discontinuous and rotation mechanisms (§11.4), raises the possibility that dynamic recrystallization by subgrain rotation in metals may be related to boundary mobility, occurring when boundaries are solute loaded and unable to migrate rapidly (§4.4.2). Although this phenomenon usually results in a partially recrystallized **necklace** microstructure (fig 11.8a-c), at large strains a completely recrystallized structure (fig 11.8d) may be formed (Gardner and Grimes 1979).

11.3.3. Dynamic recrystallization in two-phase alloys

11.3.3.1 Particle-stabilised microstructures

There is some evidence that a process similar to that of progressive lattice rotation (§11.3.2) may occur in alloys containing finely dispersed second-phase particles. The particles pin the subgrains and prevent extensive growth, and there is evidence (Nes 1979, Higashi et al. 1990) that during straining the subgrain/grain misorientation progressively increases. It has been suggested that this mechanism may be responsible for the development of fine grained microstructures during the thermomechanical processing of superplastic aluminium alloys (§12.5). However, as discussed in §12.5, this behaviour is generally observed in alloys which have already been extensively cold or warm worked and at the present time it is unclear as to whether this phenomenon is similar to that discussed in §11.3.2, is yet another type of dynamic recrystallization, or if it is essentially **static recrystallization** being triggered by high temperature deformation of an already deformed material.

11.3.3.2 Dynamic recrystallization by PSN

There is some evidence that a process similar to the particle stimulated nucleation of recrystallization (PSN) which occurs on static annealing of alloys containing large particles (§8.4), may occur during high temperature deformation. In copper, dynamically recrystallized grains have been found at particles of SiO_2 or GeO_2 several microns in diameter (Ardakani and Humphreys 1992). The microstructures developed in these alloys are very different to those in single-phase copper, or in copper containing small particles, in which dynamic recrystallization is associated only with the prior grain boundary regions (fig 11.9).

There is little evidence for particle stimulated dynamic recrystallization in aluminium alloys. Humphreys and Kalu (1987) and Castro-Fernandez and Sellars (1988) found small highly misoriented grains adjacent to large second-phase particles but these grains were generally of a similar size to the subgrains remote from the particles and there was little evidence of their growth. It is likely that such nuclei form by dynamic recovery of the misoriented subgrains formed at particles during deformation (§8.2.4), but that the stored energy of the matrix is too low to allow the nuclei to grow (§8.4.1.2). Although dynamic recrystallization in Al-Mg alloys has been attributed to PSN (McQueen et al. 1984, Sheppard et al. 1983), it is more likely that the recrystallized grains found in these alloys were formed by progressive lattice rotation as discussed in §11.3.2.

Particle stimulated dynamic recrystallization will only be possible if dislocations accumulate at the particles during deformation. This will only occur for larger particles, lower temperatures and higher strain rates (high Z) and therefore, unlike single phase alloys, there will be a particle-dependent lower limit to Z below which the nucleation of PSN will not be possible (fig 11.24). This aspect is considered in more detail in §11.6.4.

11.3.3.3 Phase transformations during hot deformation

In addition to the processes of recovery and recrystallization, there may be concurrent phase transformations during deformation at elevated temperatures. The complex interactions between the processes of deformation, restoration and phase transformation are of great importance in the thermomechanical processing of a number of materials, in particular steels (Gladman 1990, Jonas 1990, Fuentes and Sevillano 1992) and titanium alloys (Williams and Starke 1982, Flower 1990, Weiss et al. 1990). Although we have discussed the underlying principles of many of the individual restoration processes involved, a detailed treatment of this subject is beyond the scope of this book.

11.4. DYNAMIC RECRYSTALLIZATION IN SINGLE CRYSTALS

A considerable research effort has been expended in studying the fundamental aspects of the dynamic recrystallization of single crystals of silver, gold, nickel and copper and its alloys, and this is reviewed by Mecking and Gottstein (1978) and Gottstein and Kocks (1983). It has been shown (Nicklas and Mecking 1979) that dynamic recrystallization does not occur in single crystals of aluminium, although recent research (Yamagata 1993) has questioned this conclusion. As discussed in §11.2, dynamic recrystallization is essentially a grain boundary phenomenon, and therefore its occurrence in single crystals, although scientifically interesting is unlikely to be of direct relevance to the thermomechanical processing of industrial alloys.

As shown in figure 11.16, the stress-strain curve of a copper crystal deformed in tension at high temperatures exhibits a sharp drop, which has been shown to correspond to the occurrence of dynamic recrystallization. There is no clear correlation between the strain and the onset of dynamic recrystallization, but the shear stress for dynamic recrystallization (τ_R) has been shown to be reproducible for given deformation conditions. However, τ_R is a function of the material, the crystal orientation, the deformation temperature and the strain rate (Gottstein et al. 1979, Stuitje and Gottstein 1980, Gottstein and Kocks 1983), and unlike the case for polycrystals, τ_R is not uniquely related to the Zener-Hollomon parameter. It has been found (Gottstein and Kocks 1983) that τ_R is closely related to the steady state flow stress (τ_S), which is determined from extrapolation of the work hardening curves (fig 11.17). The data indicate two regimes with a transition at T = $0.75T_m$. In the low temperature regime $\tau_R = 0.82\tau_S$, and in the high temperature regime $\tau_R = \tau_S$.

The experimental evidence suggests that in the high temperature regime, dynamic recrystallization in a single crystal is triggered by the formation of a single critical nucleus which is a subgrain in the deformed structure. This subgrain then sweeps rapidly through the microstructure. However in the low temperature regime, copious twinning of the nucleus takes place, and rapid growth only occurs when a twin with a high mobility orientation relationship with the matrix (§4.3.2) is formed. It has been suggested (Gottstein and Kocks 1983) that the twins may be generated mechanically in regions of high stress. The onset of dynamic recrystallization in single crystals is therefore seen to be **nucleation controlled** in contrast to polycrystals (§11.2) in which it is **growth controlled**.

In single crystals of copper containing a fine dispersion of oxide particles which inhibit subgrain growth (§5.6), dynamic recrystallization has been found to nucleate at the transition bands which form in the crystals during high temperature deformation (Ardakani and Humphreys 1992).

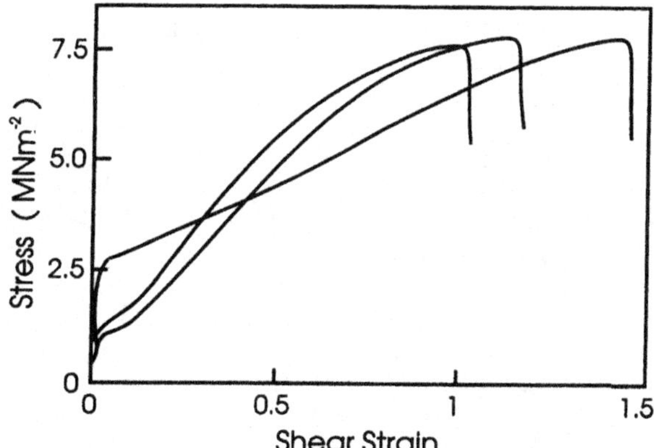

Fig. 11.16. Shear-stress/shear-strain curves of copper crystals of similar orientation deformed at 857°C, (Gottstein et al. 1979).

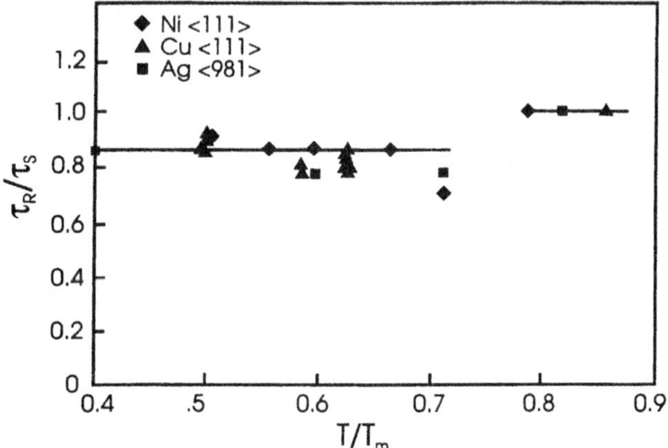

Fig. 11.17. Ratio of the recrystallization stress to the steady state flow stress in single crystals as a function of homologous temperature, (Gottstein and Kocks 1983).

11.5. DYNAMIC RECRYSTALLIZATION IN MINERALS

Dynamic (or **syntectonic**) recrystallization has long been recognised as an important process during the natural deformation under metamorphic conditions, of rock-forming minerals (see e.g. Nicolas and Poirier 1976, Poirier and Guillopé 1979, Poirier 1985, Urai et al. 1986). Although there are substantial differences not only between minerals and metals, but also between the deformation conditions in the Earth's crust and mantle, and in a hot rolling mill, there is a substantial commonality of behaviour between the two classes of materials, and over the past 30 years there has been a very fruitful interchange of scientific ideas between Geology and Materials Science. One of the main interests of the Geologists is in using the microstructure to interpret the deformation history of the mineral. The dynamically recrystallized grain sizes (fig. 11.11), which have been shown to be closely related to stress (equation 11.17) have, together with the subgrain sizes and dislocation densities, been used by Geologists as **paleopiezometers** to determine the stresses to which the mineral has been subjected.

Research in this area has involved the investigation of microstructures in naturally deformed minerals, laboratory deformation of minerals under conditions of high temperature and high pressure, and the in-situ deformation under the optical microscope of low melting point minerals and other transparent analogue crystalline materials. The latter experiments have been able to provide direct observations of the occurrence of dynamic recrystallization albeit at low spatial resolution, that cannot be obtained for metals. In many cases it is also possible to simultaneously obtain local orientation (microtexture) information from minerals in the optical microscope using standard mineralogical methods. Figure 11.18 is an optical micrograph of a naturally deformed quartzite and shows clear evidence of dynamic recrystallization in the regions of the original grain boundaries. Figure 11.19 is an example of an in-situ deformation experiment on camphor, showing the development of a similar microstructure to figures 11.9 and 11.18.

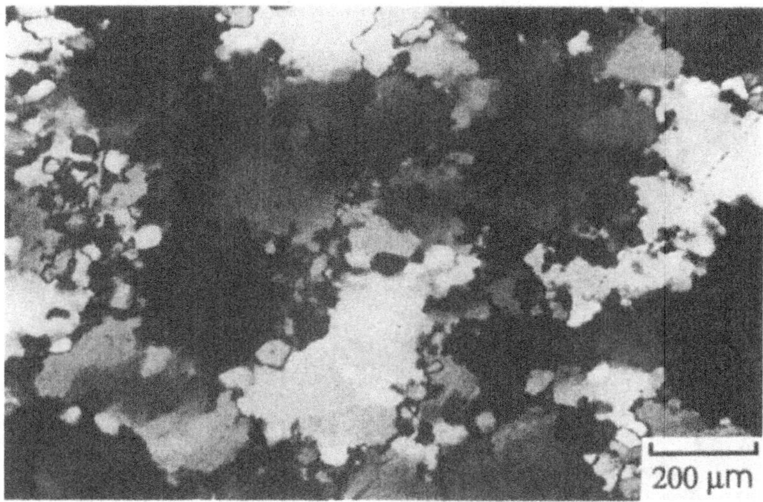

Fig. 11.18. Dynamic recrystallization in naturally deformed quartzite, (Urai et al. 1986).

11.5.1 Boundary migration in minerals

Natural minerals are generally very impure and segregation and precipitation may occur at the grain boundaries leading to a loss of mobility. In addition, bubbles and cavities may also be present. The bonding in minerals is of course different to that in metals, and this will also influence the boundary migration (Kingery 1974, Brook 1976). In many instances, the boundaries in minerals are likely to resemble those of impure ceramics.

It is now well established that in some natural minerals there is a fluid film at the boundary (see Urai et al. 1986) and that this has a large influence on grain boundary migration (Rutter 1983). The structure of a fluid-filled boundary can be described in terms of the two crystal-liquid interfaces and the fluid layer between them, and the behaviour of the boundary will therefore be less constrained by the matching of the two crystals than for a dry boundary. The boundary configuration may then be determined by the energy of the crystal-liquid interfaces, as is the case for crystal growth from the melt, and an orientation dependence of boundary migration rate is therefore expected. Migration mechanisms related to solid-liquid interfaces such as spiral growth involving screw dislocations (cf. Gleiter's boundary migration model discussed in §4.4.1.3) are predicted to be important.

If it is assumed that the migration rate is limited by diffusion in the fluid layer, a rapid increase in migration rate with film thickness is predicted as shown qualitatively in figure 11.20. The increase in mobility may be as much as four orders of magnitude for a 2nm film (Rutter 1983). For thick fluid layers, diffusion across the film is expected to become rate controlling, the mobility then becoming inversely proportional to the film thickness.

11.5.2 Migration and rotation recrystallization

It is well established that in minerals, two types of dynamic recrystallization occur. At high temperatures and high stresses a discontinuous form of dynamic recrystallization similar to

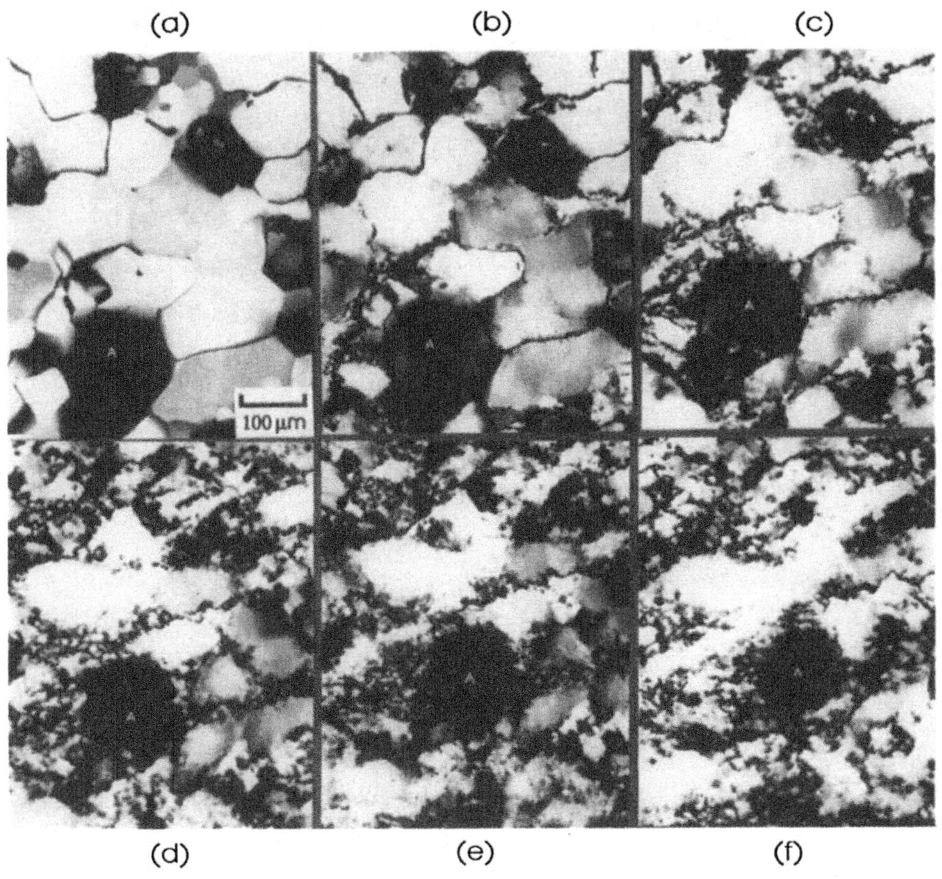

Fig. 11.19. Dynamic recrystallization of camphor at room temperature,
(Tungatt and Humphreys 1981).

that discussed in §11.2 occurs, and this is referred to in the Geological literature as
migration recrystallization. However, at lower temperatures and stresses there is often a
transition to a mechanism similar to that discussed in §11.3.2 (referred to as **rotation
recrystallization**). This transition, which is thought to correspond to the breakaway of
boundaries from their solute atmospheres (fig 4.27), is shown in figure 11.21 for
experimentally deformed halite (NaCl). The driving force for dynamic recrystallization
increases with applied stress and the intrinsic boundary mobility is also a strong function of
temperature, thus accounting qualitatively for the shape of figure 11.21. The boundary
mobility in ionic solids is very sensitive to small solute concentrations, particularly aliovalent
ions, and in figure 11.22, it is seen that the transition temperature in $NaNO_3$ (calcite
structure) between rotation and migration recrystallization is markedly raised by small
amount of Ca^{++} ions.

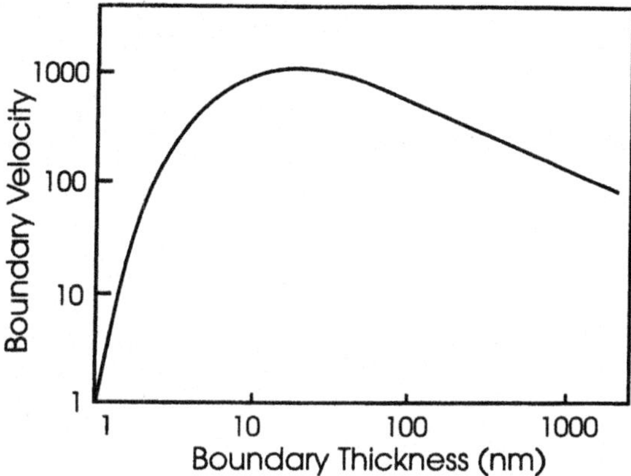

Fig. 11.20. The predicted effect of fluid film thickness on boundary migration rate, (Urai et al. 1986).

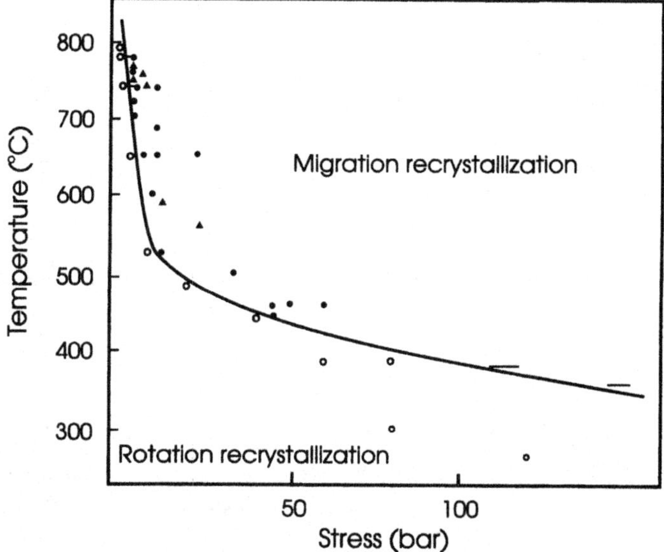

Fig. 11.21. The boundary between migration and rotation recrystallization in halite, (Guillopé and Poirier 1979).

11.6 ANNEALING AFTER HOT DEFORMATION

Recovery and recrystallization following hot deformation are of great technological importance because in many hot working operations such as multi-pass rolling, annealing takes place between the rolling passes. Additionally, the rates of cooling of the material are

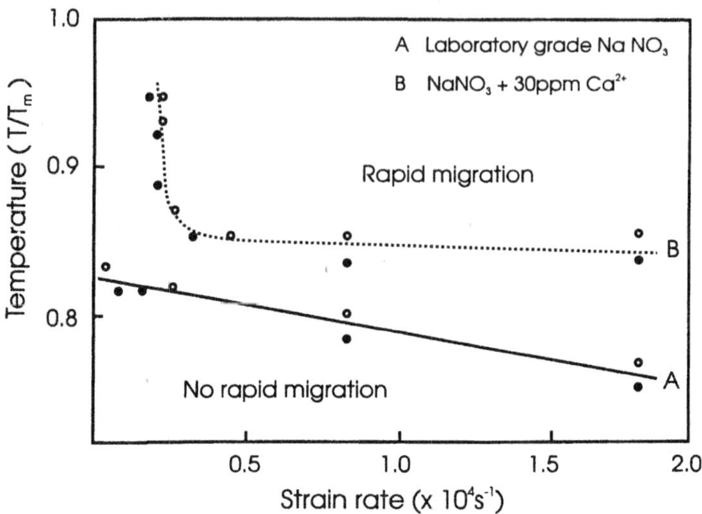

Fig. 11.22. The effect of calcium additions on boundary migration in sodium nitrate, (Tungatt and Humphreys 1984).

generally very low in large-scale metal forming operations, allowing recovery, recrystallization and grain growth to occur immediately after hot deformation.

11.6.1 Static recovery

Because dynamic recovery has already taken place during the deformation, further microstructural changes due to static recovery are generally small. However, some further recovery, including dislocation rearrangement and subgrain growth (Ouchi and Okita 1983) and consequent softening (Sellars et al. 1986) may take place, usually with similar kinetics to those found for static recovery (§5.2.2).

11.6.2 Static recrystallization

Static recrystallization may occur when a hot deformed material is subsequently annealed. This is often very similar to the static recrystallization discussed in earlier chapters, the main difference being that the lower stored energy resulting from hot deformation affects the kinetics of recrystallization. From §11.1.1 we expect the driving pressure and hence the recrystallization behaviour to be strongly dependent on Z, and this effect is well documented (see Jonas et al. 1969). Figure 11.23 shows the effect of Z on the recrystallization kinetics of commercially pure aluminium deformed to a constant strain and annealed at a constant temperature. The effect of strain is also important, and in materials which undergo dynamic recrystallization, the static recrystallization behaviour depends on whether the strain was larger or smaller than that required for dynamic recrystallization (ϵ_c).

Metallographic measurements indicate that static recrystallization can often be described approximately by the JMAK equation (5.17), with the exponent in the range 1.5-2 (Roberts

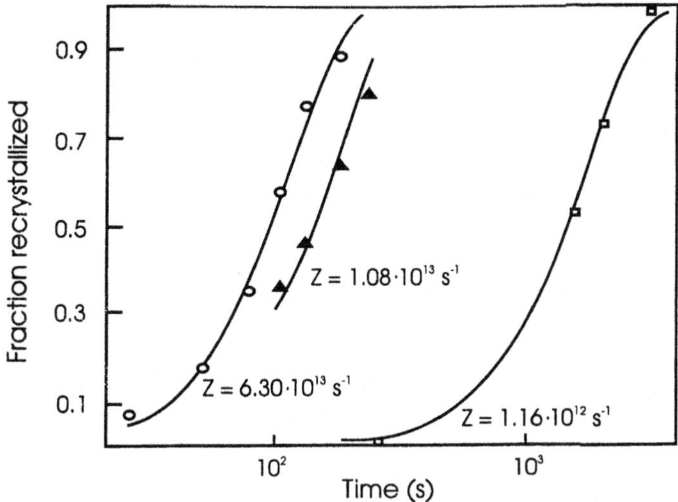

Fig. 11.23. The effect of deformation temperature and strain rate (Zener-Hollomon parameter) on the recrystallization kinetics of commercial purity aluminium deformed to a strain of 3 and annealed at 410°C, (Gutierrez et al. 1988).

1984, 1985). Because of the technological importance of the subject, empirical relationships for the recrystallization behaviour as a function of deformation behaviour have been determined (e.g. Roberts 1985, Sellars 1986, 1992a, Beynon and Sellars 1992), and for C-Mn and HSLA steels of grain size D_0, it is often found that the time for 50% static recrystallization ($t_{0.5}$) at the deformation temperature is given (Sellars and Whiteman 1979) by an equation of the form

$$t_{0.5} = c_1 \, D_0^{\,C} \, \epsilon^{-n} \, Z^{-K} \exp\left(\frac{Q_{REX}}{R \, T}\right) \qquad (11.23)$$

where c_1, C, K and n are constants and Q_{REX} is the activation energy for recrystallization.

The recrystallized grains generally nucleate at the old grain boundaries and are equiaxed. Their size D_R is given by

$$D_R = c_2 \, D_0^{\,C'} \, \epsilon^{-n'} \, Z^{-K'} \qquad (11.24)$$

where c_2, C', K' and n' are constants.

The range of values of the constants in these equations is discussed by Roberts (1985) and Sellars (1986, 1992a). If dynamic recrystallization has occurred prior to static recrystallization, then there is little effect of strain or D_0, i.e. C, C', n and n' tend to zero. Relationships such as these are necessary components of models for industrial hot rolling (§12.3) and the development of such equations from being purely empirical to being more physically-based, remains an important long term scientific objective.

11.6.3 Metadynamic recrystallization

Whenever the critical strain for dynamic recrystallization (ϵ_C) is exceeded, recrystallization nuclei will be present in the material. If straining is stopped, but annealing continued, then these nuclei will grow with no incubation period into the heterogeneous, partly dynamically recrystallized matrix. This phenomenon is known as **metadynamic recrystallization** (Djaic and Jonas 1972, Petkovic et al. 1979). The microstructure of a material which has undergone some dynamic recrystallization is very heterogeneous and may contain

       A - Small dynamically recrystallized grains which are almost dislocation free.
       B - Large dynamically recrystallized grains with a moderate dislocation density.
       C - Unrecrystallized material with a high dislocation density (ρ_m).

Each of these types of region will have a different static annealing behaviour, and the overall annealing kinetics and grain size distributions may be extremely complex. Sakai and colleagues (Sakai et al. 1988, Sakai and Ohashi 1992) have identified several annealing stages in hot-deformed nickel. **Regions A**, may continue to grow during the early stages of post deformation annealing by the mechanism of metadynamic recrystallization. **Regions B** will, if their dislocation density (ρ) is below a critical value (ρ_{RX}), recover, and this has been termed **metadynamic recovery**. If $\rho > \rho_{RX}$ then these regions may subsequently recrystallize statically. **Regions C** will undergo static recovery, followed by static recrystallization. When the material is fully recrystallized then further grain growth may occur (§11.6.5).

11.6.4 PSN after hot deformation

The conditions under which particle stimulated nucleation of recrystallization (PSN) could occur following low temperature deformation were discussed in §8.4. However, if the temperature of deformation is raised, then PSN may become less viable, and we need to consider the effect of deformation temperature on the two criteria for PSN - the **formation** of nuclei within deformation zones, and the **growth** of the nuclei beyond the particle.

11.6.4.1 Deformation zone formation
At high temperatures, dislocations may be able to bypass particles without forming deformation zones. Humphreys and Kalu (1987) have shown that the critical strain rate for the formation of a deformation zone at a particle of diameter **d** is given by :-

$$\dot{e}_c = \frac{K_1 \exp\left(-\dfrac{Q_s}{R\,T}\right)}{T\,d^2} + \frac{K_2 \exp\left(-\dfrac{Q_b}{R\,T}\right)}{T\,d^3} \tag{11.25}$$

where K_1 and K_2 are constants and Q_s and Q_b are the activation energies for volume and boundary diffusion. This relationship has been found to be obeyed in aluminium alloys for a wide range of particle sizes and deformation conditions with $K_1 = 1712 m^2 s^{-1} K$ and $K_2 = 3 \times 10^{-10} m^3 s^{-1} K$.

For particles which are potential PSN sites ($d > 1\mu m$), the second term will generally be negligible, and a simplified version of equation 11.25 may therefore be used. The critical particle diameter for the formation of a zone (d_f) may then, if we assume that the activation energies in equations 11.1 and 11.25 are identical, be expressed in terms of the Zener-Hollomon parameter as

$$d_f = \left(\frac{K_1}{TZ} \right)^{1/2} \tag{11.26}$$

11.6.4.2 Nucleus growth

The growth criterion for PSN (e.g. equation 8.24) is also affected by the deformation conditions, because the stored energy (E_D) is reduced at elevated temperatures, although this effect is more difficult to quantify. As the microstructure following high temperature deformation consists mainly of subgrains, the growth condition (equation 8.24) may be written in terms of the subgrain size (D), using equation 2.7, as

$$d_g = \frac{4\,\gamma_b\,D}{3\,\gamma_s} \tag{11.27}$$

and using equation 11.7 then

$$d_g = \frac{4\,\gamma_b\,K\,G\,b}{3\,\gamma_s\,\sigma} \tag{11.28}$$

The relationship between the flow stress (σ) and Zener-Hollomon parameter is given by equations 11.2 to 11.5, and if, for example we use equations 11.1 and 11.4 then

$$d_g = \frac{4\,\gamma_b\,K'\,G\,b}{3\,\gamma_s\,Z^{1/m}} \tag{11.29}$$

Somewhat different versions of equation 11.29 may be obtained, depending on the particular relationships between subgrain size, flow stress and Z which are used, and at present there is no entirely satisfactory solution.

In figure 11.24 we show the variation of the critical particle diameters for zone formation and growth with Z, according to equations 11.26 and 11.29 (neglecting the small effect of the temperature term in equation 11.26 and using values of the constants appropriate for aluminium). It may be seen that at values of Z less than $\sim 10^{12}$, the condition for PSN is governed by that for **deformation zone formation**, but at higher values it is determined by the criterion for **growth** of the nucleus. Experimental investigations have confirmed this model. Measurements by Kalu and Humphreys (1986) on Al-Si alloys deformed at $10^8 < Z < 10^{12}$ showed that zone formation was the controlling factor, whereas the measurements of Oscarsson et al. (1987) on AA3004 and of Furu et al. (1992) on AA3003, in which specimens were deformed in the range $10^{12} < Z < 10^{16}$, were both consistent with growth control.

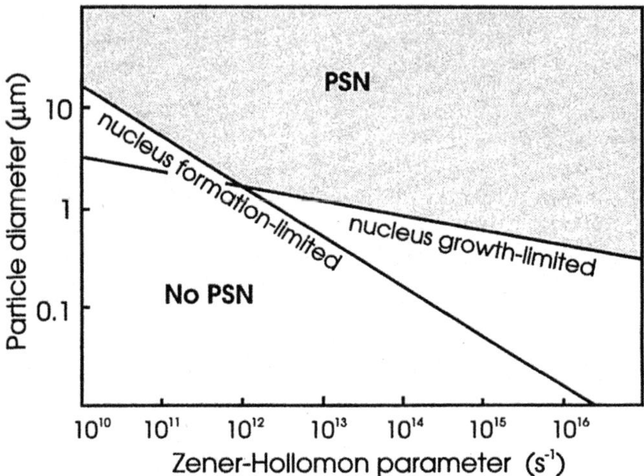

Fig. 11.24. The effect of deformation conditions on PSN.

11.6.5 Grain growth after hot working

When recrystallization is complete, grain growth can take place, and has been described by the empirical equation

$$D^n = D_R^n + c\,t\,\exp\left(\frac{-Q_g}{k\,T}\right) \tag{11.30}$$

where c, n and Q_g are constants.

For steels, very large values of the exponent n (~ 10) have been reported (see Sellars 1986).

Chapter 12

CONTROL OF RECRYSTALLIZATION

12.1. INTRODUCTION.

Although in this book we are primarily concerned with the scientific aspects of annealing phenomena, it is important to remember that this science provides the base for the solid state processing of all metals. Recrystallization plays a vital part in the manufacture of all wrought steel products and contributes enormously to the production of aluminium, copper and other non-ferrous metals. The control of microstructure and texture during recrystallization is therefore of major economic importance.

In this chapter we have selected four industrially important examples to illustrate how control of microstructure and texture during annealing are vital for the optimisation of properties. Although some of our examples may seem common-place, these products are in fact amongst the most highly developed and sophisticated materials available.

12.2. THE PRODUCTION OF ALUMINIUM BEVERAGE CANS.

The modern aluminium beverage can is produced in vast numbers. In the United States alone nearly 80 billion cans are produced each year and about two million tons of rolled can-stock sheet are used worldwide. The production of these cans provides an excellent example of the practical significance that comes from an understanding of the principles of deformation processing, recovery and recrystallization. The basic metallurgical steps involved in the production of the sheet are casting, homogenization, hot rolling, annealing and cold rolling. Understanding of the problems involved requires some knowledge of can making and we begin with a brief account of that process. This is followed by a description of sheet production and a discussion of metallurgical factors such as microstructure and texture.

12.2.1. Can making.

Can making begins with a cupping operation (Hartung 1993). This is done on a multi-stage press which cuts the blank and then forms a shallow, drawn cup with diameter ~90mm.

Modern presses form up to 14 cups in a single stroke at speeds as high as 275 strokes per minute. The cups are then transferred to the can press or "body maker". These are long stroke, mechanical presses that generate a trimmed can in a series of steps, viz:

(i) At the beginning of the stroke the cup is redrawn to the final can diameter.

(ii) The side wall is then thinned and elongated in a series of wall ironing dies: in a typical body maker three ironing operations are involved. The punch is designed to develop a gradual reduction in metal thickness from the dome area to the sidewall, but an annular region of thicker material is maintained near the open end. This region is usually ~50% thicker than the sidewall and its purpose is to provide a sufficient volume of material for subsequent necking and flanging. The sidewall itself is usually a little more than 0.1mm thick.

(iii) At the end of the stroke the bottom of the can is formed into the well known, pressure resistant, stackable dome configuration. The can is removed from the retreating punch by a combination of air pressure and mechanical fingers.

After removal the can is trimmed to remove the ears and roughened edges generated at the open end. The whole process is done at very high speed and a modern body maker/trimmer combination produces in excess of 350 can bodies per minute.

The remaining processes are of less interest metallurgically. They involve washing, preparation of the surface for the inks and varnishes used in labelling and the application of the interior coating. Finally the cans are necked down at the open end and flanged so that the lid can be inserted after filling.

The overall result of this processing is a thin walled pressure vessel with an internal volume of ~375ml. capable of withstanding an internal pressure of 0.62MPa and a longitudinal load of ~135kg in the unpressurised condition; the final wall thickness is ~0.12mm. The reduction in can mass in recent years has been spectacular; the weight of the blank from which the can is drawn has fallen from 22g in 1968, to 19.2g in 1978, to 16.5g in 1988 and to ~12g in 1993.

12.2.2. Can body sheet

As will be explained later the production of can body sheet is highly competitive and explicit production details are not freely available. The details given in this section are basically correct but not explicitly true for any individual manufacturer. The sheet currently used for can making is rolled from ingots of the alloy AA3004. The nominal specification for this alloy is given in Table 12.1 but can manufacturers operate to much tighter schedules than those given in the specification. The easy-open ends that are fitted after the can has been filled are made from a different alloy (AA5182) and are not discussed here.

The sheet is rolled from DC-cast ingots that are 300-760mm thick, 1300-1850mm wide and of length determined by the mill layout and its ingot and coil handling facilities. After scalping the ingots are homogenized at temperatures above ~550°C and soaked for a carefully controlled period. The ingot is then cooled to ~500°C and held for a sufficient time to ensure a uniform temperature for hot rolling in a breakdown mill. During this rolling

Table 12.1.

Nominal composition, alloy AA3004.

Si 0.3% max.	Fe 0.7% max.
Cu 0.25% max.	Mn 1-1.5%.
Mg 0.8-1.3%.	Zn 0.25% max.
All others. Each 0.05% max. Total 0.15% max.	

process which lasts for about 15 minutes the temperature falls to ~300°C and the ingot is reduced to a slab of ~25mm thickness. At this stage the slab is transferred to a hot finishing mill. This may be either a single stand reversing mill or a multiple stand continuous mill that can roll fast enough to prevent recrystallization between passes. Mills of the latter type with 3-4 stands produce coiled metal ~2.5mm thick that recrystallizes without furnace annealing, but in single stand rolling the final coiling temperature is too low for full in-coil recrystallization. A further furnace anneal at ~350°C is required in this case. The current demand is for final sheet ≤ 0.3mm thick with strength corresponding to H19 (UTS 295MPa, YS 285MPa, Elongation 2%). This is achieved by cold rolling to ~87% reduction in thickness. There is a continuing demand for further reduction in the thickness of can stock sheet; in 1978, for example, a typical value was 0.38mm. In the same year the required thickness control ranged from $\pm 7.5\mu$m to $\pm 10\mu$m; the 1993 requirement is $\pm 5\mu$m.

12.2.3. Metallurgical factors

The metallurgical effort in can production is directed towards minimizing the mass of aluminium in each can and eliminating shutdowns. The aim is to use the smallest round blank stamped from the thinnest possible sheet, that will provide sufficient strength after cupping and wall ironing to minimum height. In order to achieve this, strict control is exercised over all aspects of the development of microstructure and texture.

12.2.3.1. Microstructure.

Microstructural control begins at the homogenization stage. The composition of AA3004 is such that a number of different phases, derived from the solidification process and a number of solid state transformations, are in equilibrium at the different temperature ranges involved in the production cycle. Some idea of the complexity is provided by the DSC curve of figure 12.1 and the resistivity results of figure 12.2 (Owen et al. 1991). Three prominent exothermic and four endothermic reactions occur during heating of cast material to 600°C. The weak endotherm (B) at 200-220°C is characteristic of the dissolution of Mg-Si rich G.P. zones while the exothermic doublet (C) at 270-320°C is associated with the precipitation of β'' and β' Mg$_2$Si. The β' phase is dissolved in the temperature range 350-420°C where the endotherm (D) occurs and the exotherm (E) corresponds to the precipitation of platelets of the β phase (Mg$_2$Si). This phase dissolves in turn in the range 470-500°C associated with endotherm (F). The exothermic peak (G) at ~520°C is associated with the precipitation of complex dispersoids of type Al$_6$Mn and complex α phases such as Al$_{15}$(MnFe$_3$)Si$_2$. As heating continues the Al$_6$Mn type phases and the α dispersoids dissolve and coarsen. Finally at 570-600°C the endotherm (H) corresponds to the melting of complex grain boundary eutectic phases and Mg-Si based compounds.

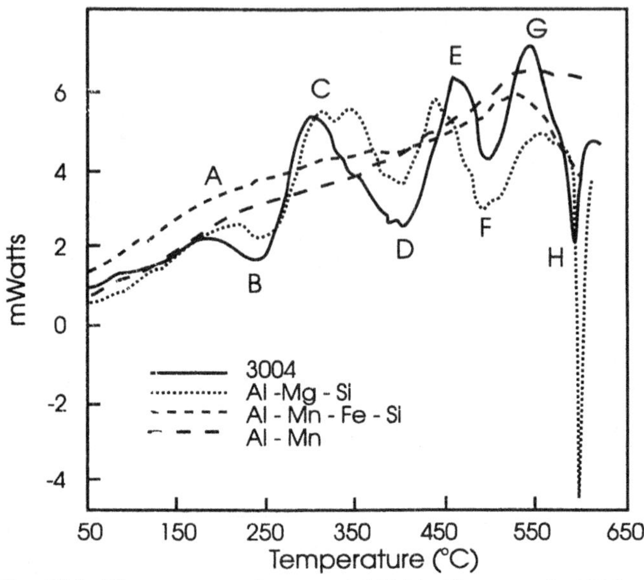

Fig. 12.1. Thermogram of as cast AA3004, (Owen et al. 1991).

Application of these and similar results to the changes occurring during homogenization is not straightforward but some progress has been made. During homogenization at >550°C most of the manganese retained in solid solution after solidification is precipitated (fig 12.2) and two species of second phase particles, $Al_6(FeMn)$ and Mg_2Si, are unstable. The latter dissolves and the silicon released reacts with the $Al_6(FeMn)$ to form the α phase Al(FeMn)Si. The thin interdendritic platelets of $Al_6(FeMn)$ are transformed to coarser α phase particles (5-10μm) and the same phase also nucleates and grows to form a fine dispersion of precipitates that are ~0.1μm in diameter. The duration and temperature(s) of the

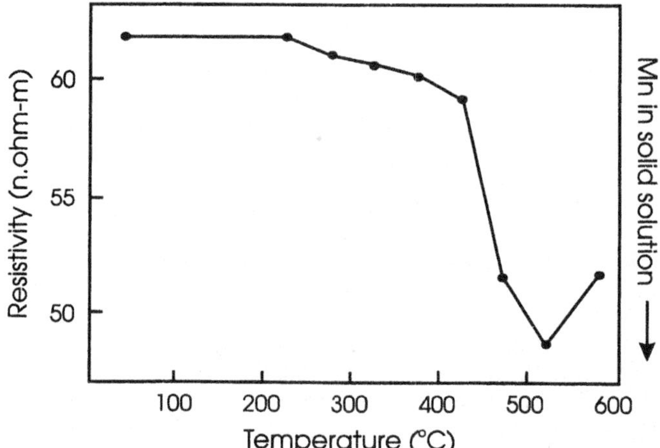

Fig. 12.2. Resistivity changes during slow heating to temperature, AA3004, (Owen et al. 1991).

homogenization determine the mean size of these dispersoids and thereby influence the subsequent recovery and recrystallization behaviour.

Further precipitation of manganese-containing phases occurs during the pre-rolling treatment at 500°C and some precipitation of Mg_2Si occurs at this time. Little precipitation occurs during hot rolling in the break-down mill even though the temperature falls to ~ 300°C over 15 minutes or so. The major metallurgical changes at this stage are the breaking down of interdendritic particles and the establishment of a moderate recrystallized grain size in the matrix. The changes occurring during the final hot and cold rolling stages have already been described.

The continuing demand for thinner sheet has meant that the particle size of non-metallic inclusions must be carefully controlled. These can be introduced at the casting stage as refractory dross and aluminium oxide, and particles as small as $40\mu m$ may lead to splits and pinholes in the $100\mu m$ side walls. The inclusion content is controlled by degassing and casting through ceramic foam filters.

12.2.3.2. Texture.

Efficient use of the blank requires that the cup and can should be drawn and ironed isotropically. Unfortunately the strong textures developed during thermomechanical processing of the sheet lead to the generation of ears and valleys around the top of the can. These may cause trouble during can making and eventually they must be trimmed off. At present, can makers specify that the fractional height of the ears above the valleys of a drawn cup should lie in the range 0.5% to 3% of the can height. The details of the steps taken to meet this requirement and those relative to strength, flatness, surface quality and thickness are not generally available but a generic description will provide some insight into the principles involved.

At the end of the hot rolling and annealing stages the texture consists of a predominant cube component together with residual fcc rolling texture components (Goss, copper and S). The last of these is sometimes called R in the aluminium industry. There is, additionally, a significant volume fraction of grains with random orientations. The relative proportions of the various components are controlled by such factors as the stored energy retained after deformation and the size, number and distribution of precipitates. The aim at this stage is to promote the development of the cube texture and control of this is critical. As indicated in §10.7.6, cube texture grains in rolled fcc metals are often nucleated at heterogeneities in the microstructure such as transition bands and in addition, particularly after high temperature deformation, flattened cube grains retained during the deformation may nucleate cube texture on recrystallization (§10.7.6.2). However, as discussed in §10.6, recrystallization at second-phase particles larger then $\sim 2\mu m$ will promote the development of a random texture at the expense of the cube texture by particle stimulated nucleation (fig 10.16). If a strong cube texture is to be obtained in the hot rolled sheet after recrystallization, the rolling schedule must be such that the favoured cube nucleation sites are developed, i.e. there must be a component of cold working sufficiently large to promote the required inhomogeneities of deformation. At the same time the deformation temperature must be high enough to ensure that particle stimulated nucleation does not develop at the large particles (§11.6.4). The generation of the cube texture during hot rolling and its relationship to the final deep drawing properties has been discussed in detail by Hutchinson et al. (1989b) and Hutchinson and Ekström (1990).

Table 12.2.
Major texture components in cold rolled can body sheet.

Texture	Name	Earing
$\{100\}<001>$	Cube	4-fold 0/90° etc.
$\{110\}<001>$	Goss	2-fold 0/180°
$\{110\}<112>$	Brass	4-fold 45° etc.
$\{112\}<111>$	Copper	4-fold 45° etc.
$\{123\}<412>$	R ($\sim$S)	4-fold 45° etc.

The major texture components after cold rolling are shown in table 12.2. Although the cube texture of the hot-rolled and recrystallized sheet is relatively stable during cold rolling, its strength is reduced as the normal cold-rolled texture components develop (§2.4). Table 12.2 suggests that at this stage strong $\pm45°$ earing would be present, together with minor 0-180° earing. The extent of the $\pm45°$ earing can be minimised if the development of the rolling texture components is kept to a minimum during hot-rolling and annealing, i.e. if the cube texture is maximised.

While the trend is to minimize earing and the consequent trimming and metal loss, a minimum amount is specified (usually 0.5%). There is a very important practical reason for this (Malin 1993). If $\pm45°$ earing is eliminated, the new texture generates two-fold earing at 0° and 180° derived from the Goss component. Although four 45° ears are able to withstand the hold-down pressure supporting the blank during forming, two 0-180° ears cannot do this without some circumferential flow and pinching. The resulting pinched ears tend to jam the body maker during the subsequent re-drawing and ironing.

It is well known that earing depends also on mechanical parameters such as lubrication, blank holder pressure and alignment (Rodrigues and Bate 1984, Bate 1989, Malin and Chen 1993), but a discussion of these is beyond the scope of this book.

It will be clear that texture control is of vital importance during can manufacture, and this is particularly so at the hot-rolled and annealed stage.

12.3 RECENT DEVELOPMENTS IN LOW CARBON, DEEP DRAWING STEELS.

12.3.1 Introduction

The world production of low carbon steel far exceeds that of any other metallic material and while much of that production is used for what may seem simple purposes there is a continuing demand for improvement in processing and properties.
The major steps have been:

      (i) the development of oxygen blown converters,
      (ii) the use of ladle degassing,
      (iii) the introduction of continuous casting and the direct production of slabs,
      (iv) the continuing replacement of batch annealing by continuous annealing lines, and
      (v) the development of low carbon and ultra low carbon steels.

Each of these has influenced the production and properties of deep drawing steels but we are concerned here with only the last two. Although continuous annealing lines have been used for steel sheet products for many years, the best quality drawing steels were traditionally produced by lengthy batch annealing and it was always recognized that development of the required texture and strength were critical problems facing the introduction of continuous annealing. Today these problems have been solved and more than 40 continuous annealing lines are in operation with a total capacity of 22.6 million tonnes per year (Willis 1993).

The first interstitial-free steels appeared in the late 1960's when it was reported that drawability could be improved by small additions of titanium and niobium. In order to achieve carbon contents < 0.01% these steels require an alloy content of 0.05-0.1%. Such amounts lead to loss of surface quality due to the presence of inclusions and also to high alloy costs (Obara et al. 1984). However steel making technology has now reached the point where carbon contents in the range 0.001-0.003% can be produced by short time, ladle-degassing treatments. Such steels require much smaller amounts of alloying elements ($\leq 0.03\%$ Nb or $\leq 0.05\%$ Ti). If additional strength is required this can be achieved by the addition of small amounts of phosphorus ($< 0.1\%$) and by bake hardening.

In the following discussion of these topics we begin with a brief review of the measurement of drawability and this is followed by an account of the effects of changes in annealing practice and steel chemistry. Attention is centred particularly on normal low carbon steels and the relatively new ultra low carbon steels. Only brief mention is made of the conventional interstitial-free steels. Much of the relevant research work has been carried out in the industrial laboratories of Japanese steel producers and much of the relevant literature is to be found in steel industry journals and conference reports.

12.3.2 Background

12.3.2.1 Assessment of formability
It is obvious that the presence of a crystallographic texture in a rolled sheet will affect the strain distribution and plastic flow during forming. This texture-induced anisotropy may take two forms. In the first of these, termed planar anisotropy the flow properties in the sheet plane vary with direction. In the second an appropriate texture will introduce a differential strengthening between the "in-plane" and "through-thickness" directions; such an effect is referred to as **normal anisotropy**.

The criterion for drawability is given by the ratio of the true strains in the width and thickness direction (the **r** value).

$$r = \frac{e_w}{e_t} \tag{12.1}$$

In the presence of planar anisotropy the value of **r** varies with direction in the sheet plane and an average **r** value, $\bar{r}$ is used.

$$\bar{r} = \frac{r_o + 2r_{45} + r_{90}}{4} \tag{12.2}$$

Table 12.3.

Values of r̄ and Δr for major texture components.
(Hutchinson and Bate 1994)

Orientation (10° spread)				Predicted anisotropy (pencil glide)	
{hkl} <uvw>	Euler angles			r̄	Δr
	ϕ_1	Φ	ϕ_2		
{001}<110>	45	0	0	0.45	-0.88
{112}<110>	0	35.3	45	3.33	-5.37
{111}<110>	0	54.7	45	2.80	-0.44
{111}<112>	90	54.7	45	2.80	-0.44
{554}<225>	90	60.5	45	2.87	1.53
{110}<001>	90	90	45	25.43	49.38

r̄ is a convenient measure of normal anisotropy and therefore of drawability. High values of r̄ correlate with good drawability. Planar anisotropy which is given by

$$\Delta r = \frac{r_0 + r_{90} - 2r_{45}}{2} \qquad (12.3)$$

correlates with the extent of earing in deep drawing.

Theoretical values of r̄, calculated by Hutchinson and Bate (1994), on the basis of the Taylor model (§2.3.5) are given in table 12.3. The desired properties of good drawability and minimum earing are seen to be associated with {111}<uvw> type textures; it is for this reason that the strong <111>ND fibre texture found in annealed low carbon steels is so important. In this respect it is to be noted that {554} is only a few degrees removed from {111}. It can also be seen that {001)<110> and {110}<001> textures are undesirable.

Because good drawability is associated with {111} rolling plane textures and bad drawability with {100} rolling plane textures, the ratio of the intensities of the 222 and 200 X-ray reflections from a rolling plane specimen (I_{222}/I_{200}) is often used as a measure of formability and correlates well with r̄.

12.3.2.2 Texture of low carbon steel

As discussed in §2.4.2, the rolling texture of low carbon steel consists of two major orientation spreads. One of these, an almost complete <111> rolling plane normal fibre texture, has {111} parallel to the rolling plane with several prominent directions parallel to the rolling direction; these include <110>, <112> and <123>. The second is a partial fibre texture about the rolling direction with a <110> fibre axis and prominent {001}, {112} and {111} rolling plane components (see figure 2.28). The annealing texture developed depends on the extent of rolling. For low reductions a Goss component occurs together with a {111} rolling plane texture, but as the reduction is increased the Goss component weakens and the {111} texture strengthens with the centre of the spread near

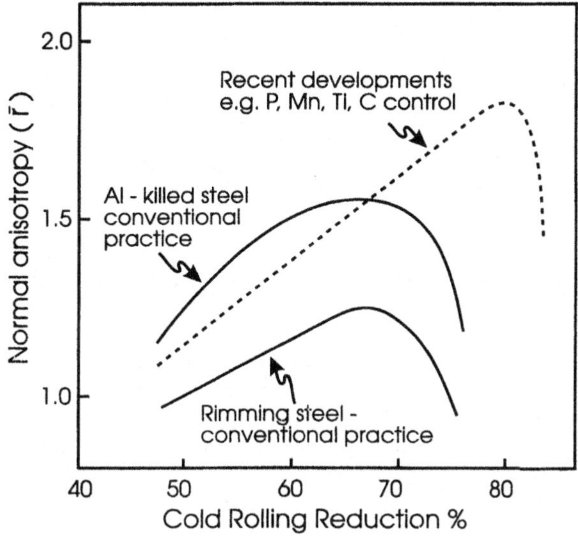

Fig. 12.3. Relationship between cold rolling reduction and $\bar{r}$ for different steel types, (Hutchinson 1984).

$\{111\}<112>$. At 90% reduction the desirable $\{111\}$ texture strengthens further but this is offset by the appearance of deleterious components with near $\{100\}$ orientations. As a consequence there has been an optimum rolling reduction for the traditional low carbon forming steels, viz. rimming and aluminium killed steels (fig 12.3). The recent developments in steel making and processing have enabled the use of higher reductions (up to 90%) and at the same time the achievable $\bar{r}$ value has been raised from the typical values shown in figure 12.3 to ~ 2.5 in ultra low carbon steels containing small amounts of niobium and titanium. This improvement is best understood by considering the effects of processing variables on texture development in aluminium killed steels and the development of microalloyed, ultra low carbon steels.

12.3.2.3 The origin of the texture
The most important single property of a good deep drawing steel is a strong $\{111\}$ rolling plane texture after annealing. As pointed out by Nes and Hutchinson (1989) the limitations of commercial processing are such that modification of rolling textures and/or grain boundary character is not generally possible. This effectively rules out the option of controlling texture development by **growth** processes. It is however, often possible to modify the recrystallization texture by controlling the **nucleation** process.

In general, nucleation is associated with microstructural inhomogeneities and the most significant of these are transition bands, shear bands, grain boundaries and the deformation zones around hard particles. In these very low carbon steels, carbides are few and particle stimulated nucleation is not likely to be significant. Shear band nucleation was investigated by Haratani et al. (1984) who found that shear bands form preferentially in $\{111\}<112>$ volumes and result in the creation of $\{110\}<001>$ Goss oriented nuclei. Shear band formation is favoured by large grain size and high contents of interstitial elements and it follows that a small grain size and low C and N contents are desirable. In these steels transition band nucleation is not significant and the important nuclei are those associated with grain boundaries. Of the two known grain boundary nucleation mechanisms, strain induced grain boundary migration (§6.6.2) would be expected to lead to retention of the rolling texture components. The strengthening of the important {111} texture appears to be due to grain boundary nucleation processes. The bi-crystal experiments of Hutchinson (1989) have shown that the boundaries between rolled grains with orientations within the γ fibre are the source of new grains that are also within the fibre, but rotated ~30° about the rolling plane normal.

12.3.3 Low carbon forming steels.

This category includes the common rimming and aluminium killed steels. In addition to the effects of rolling reduction, r̄ values in these steels are strongly influenced by the coiling temperature after hot rolling and the heating rate during annealing. Figure 12.4, compiled by Hutchinson (1984) from the work of several authors, summarizes the position in 1983. It will be seen that at low coiling temperatures ($\leq 600°C$), the high r̄ values typical of killed steels are found only for the slow heating rates typical of batch annealing. The situation is reversed for high coiling temperatures (~700°C) but the maximum value that can be

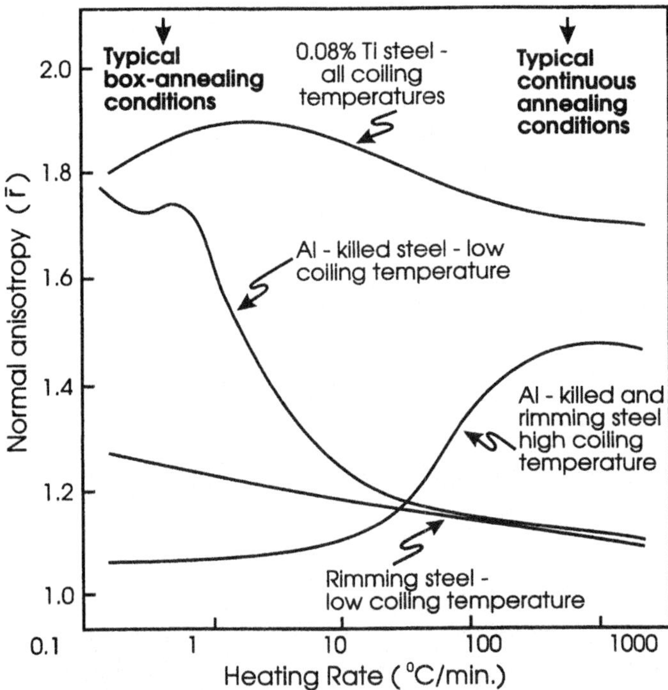

Fig. 12.4. Effect of coiling temperature and heating rate for final anneal on r̄ for different steel types, (Hutchinson 1984).

developed is lower. These effects which provide the basis for the traditional use of batch annealing for the extra deep drawing quality killed steels are discussed below.

12.3.3.1 The aluminium nitride reaction
The origin of the high $\bar{r}$ values associated with aluminium killed steels has been the subject of an immense research effort. The steels have strong $\{111\}$ texture components and very weak $\{100\}$ components and a so-called pancake grain structure in which the recrystallized grains are elongated in the rolling direction.   Understanding of the phenomena involved began with the work of Leslie et al. (1954). An essential requirement is that both aluminium and nitrogen be in solid solution before cold rolling begins and this means that any aluminium nitride (AlN) present must be decomposed prior to hot rolling. It is essential also that the dissolved aluminium and nitrogen should not precipitate during hot rolling or immediately thereafter. Leslie et al. found that precipitation occurred most rapidly at $\sim 800°C$ and established that the extent of precipitation was related to the development of the pancake grain structure. In practice AlN precipitation is prevented by spraying the hot band with water after rolling and before coiling at $\sim 580°C$.

Under slow heating conditions, as in batch annealing, the aluminium atoms diffuse and combine with the nitrogen to form small precipitates which retard recrystallization. The precipitates which occur at sub-boundaries, are readily identified within the recrystallized grains as sheets that are elongated in the rolling direction and dictate the formation of the pancake grain structure. The second-phase particles inhibit both nucleation and growth but the large final grain size indicates that nucleation is the more affected. Dillamore et al. (1967) have related the accompanying enhancement of the desirable $\{111\}$ texture components during recrystallization to the higher stored energy associated with $\{111\}$ oriented volumes in the deformation microstructure (§2.2.3) and the consequent faster recrystallization (fig 6.7).

12.3.3.2 The impact of continuous annealing.
The development of continuous casting opened the way to the large scale introduction of continuous annealing lines. Such lines involve heating rates of the order of several hundred degrees per minute and a reduction in annealing time from several days to ~ 10 min. An excellent summary of the extensive Japanese research work that led to their introduction is given by Abe et al. (1978).

It was found, *inter alia*, that the low hot band coiling temperature ($\sim 580°C$) necessary for batch-annealed, aluminium-killed-steels was no longer appropriate, and the best results were obtained if coiling was carried out at temperatures above 700°C. Under these conditions the eutectoid transformation occurs during the slow cooling of the tightly wound coil rather than during water spraying and the cementite is much coarser and more widely dispersed. The dispersed particles of MnS and AlN also coarsen during the slow cooling and equilibrium is established in the ferrite so that the carbon content in the ferrite is very low. The realization that this almost pure iron is the key parameter controlling the texture was the major factor behind the successful introduction of the modern ultra low carbon drawing steels.

The dramatic effect of carbon on drawability is well illustrated in figure 12.5. Two important factors are immediately obvious. First the value of $\bar{r}$ is dramatically increased at low carbon levels and second there is an optimum value associated with a carbon level near 3ppm. A clear account of the role of carbon in texture control is given by Hutchinson (1984).

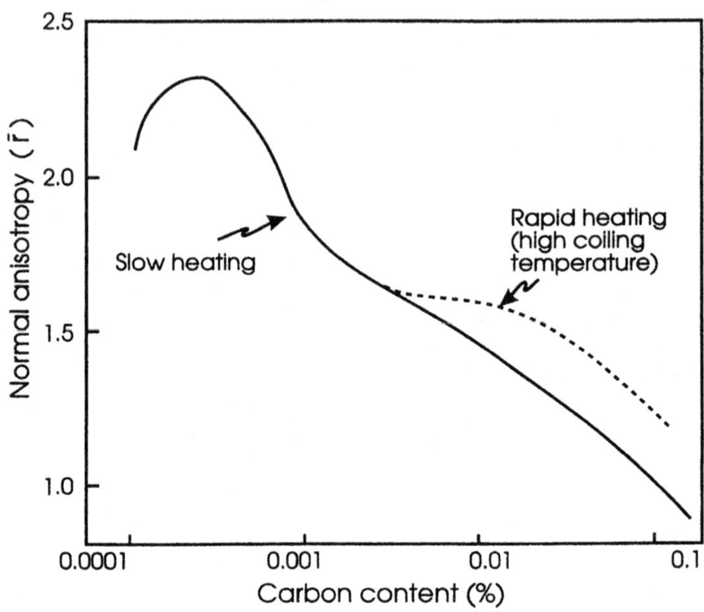

Fig. 12.5. Effect of carbon content on $\bar{r}$ value of steel, (Hutchinson 1984).

Table 12.4 provides a summary of the optimum conditions for batch and continuous annealing of low carbon steels.

Table 12.4.

Optimum conditions for texture and anisotropy development in low carbon steels (Hutchinson 1993).

Parameter	Batch Annealing	Contin. Annealing
Carbon	low (*)	low (***)
Manganese	low (*)	low (**)
Microalloying	Al (***)	( )
Soaking Temp., °C	high (***)	low (*)
Finish Rolling	$> A_3$ (**)	$> A_3$ (**)
Coiling, °C	low, < 600, (***)	high, > 700, (***)
Cold Rolling, %.	~70	~85
Heat. Rate, Anneal	20-50°C/hr. (***)	5-20°C/sec. (**)
Anneal Temp. °C	~720 max.	850 max.

( ), not critical, (*) significant, (**) important, (***) vital.

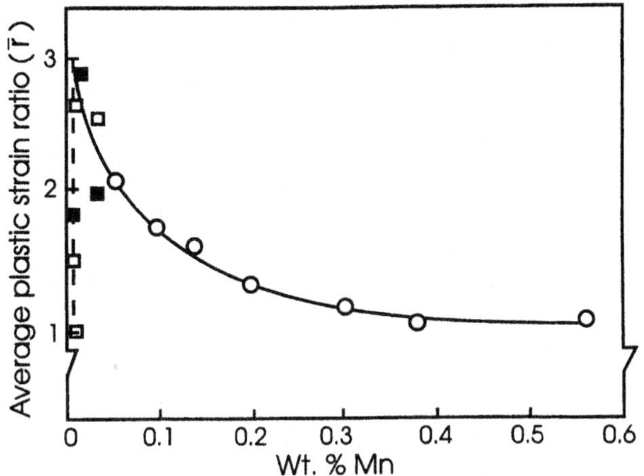

Fig. 12.6. Effect of manganese on r̄ value, (Cline and Hu 1978).

12.3.3.3 The role of manganese

Manganese is the major alloying element present in these steels and its influence is thought to be the result of solid solution effects, phase forming effects and complex interactions with other elements such as carbon, nitrogen, oxygen and phosphorus. The extensive investigations of Hu and colleagues (Hu and Goodman 1970, Hu, 1978, Cline and Hu 1978) showed that manganese adversely affects the drawability (fig 12.6).

It was thought originally that the effect of manganese was due to a solute drag effect that provided the opportunity for nucleation and growth of unfavourably oriented grains. However, it is now known that the effect is due to the simultaneous presence of carbon and manganese, probably in the form of C-Mn dipoles. The importance of this interaction was demonstrated in a number of important papers during the period 1970-1984, and details are given by e.g. Fukuda and Shimazu (1972), Abe et al. (1983), Hutchinson and Ushioda (1984) and Osawa (1984).

12.3.4 The development of ultra low carbon steels

12.3.4.1 Ultra low carbon steels.

Although the very high r̄ values associated with the low carbon steels of figure 12.5 suggest that good drawability is readily available by carbon reduction this is an overly simple view as other property requirements such as ductility and resistance to strain ageing must also be considered. Figure 12.7 gives details of the range of forming steels available from Nippon Steel Corp. (Fudaba et al. 1988). In general the ductility is improved by the low yield strength and high elongation resulting from a decreased carbon content while strain aging can be eliminated by lowering the combined carbon/nitrogen content to <0.001% and light temper rolling (~0.5% reduction). However these effects are limited and the production of the superior grades shown in figure 12.8 requires the additional use of strong carbide forming elements such as niobium and titanium. Little more than the stoichiometric equivalent is needed and these small amounts are considerably less than those used in the earlier interstitial-free steels where the alloy content was often as high as 10x the stoichiometric

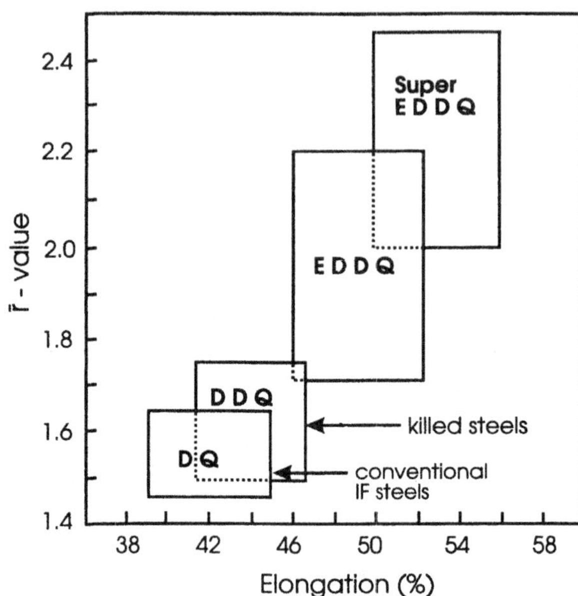

Fig. 12.7. Formability and ductility of various interstitial-free steel grades, (Fudaba et al. 1988).

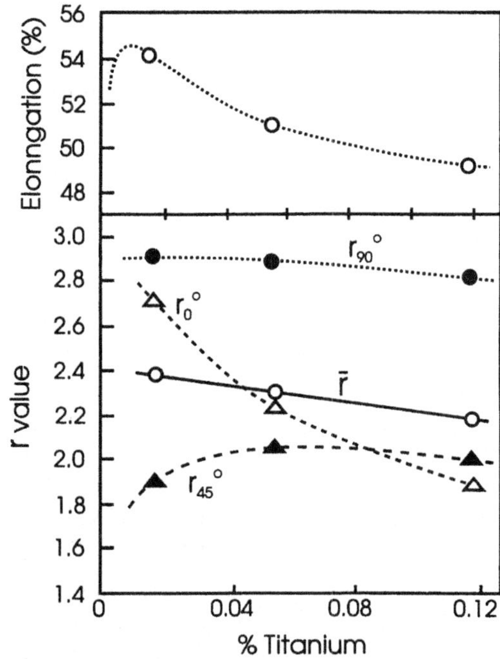

Fig. 12.8. Effect of titanium content on the mechanical properties of 0.8mm, 0.007% Nb steel cold rolled 81%, (Fudaba et al. 1988).

equivalent. Both elements are effective in reducing the planar anisotropy of elongation and Δr. Satoh et al. (1985) found that this was particularly so for niobium.

As indicated in figure 12.8 even better mechanical properties are found in the Super EDDQ grade steels. These easily formed, very low carbon ($\sim 0.0015\%$C), steels contain $\sim 0.005\%$ niobium and $\sim 0.01\%$ titanium and must be annealed at higher temperatures ($\sim 850°$C) in order to develop the required texture. They have $\bar{r}$ values as high as 2.5 with elongation of 53% and have in addition a very low value of planar anisotropy (see the results for $r_{45°}$ in fig 12.8). The absence of planar anisotropy is particularly significant in the production of drawn rectangular shapes.

The outer panels of automobile bodies must be baked to harden the surface coating and this is usually done at $\sim 190°$C. Because of the possibility of strain ageing any carbon in solution should be eliminated or stabilised. In the case of continuously annealed, normal drawing grades of low carbon steel, strain ageing is eliminated by the introduction of an overaging stage. For the higher grade steels containing titanium and niobium this treatment is not necessary. Since the carbides are stable to $\sim 830°$C there is very little carbon in solution during annealing and high cooling rates can be used.

12.3.4.2 Extra low carbon, high strength steels.
In addition to drawability and resistance to strain aging, automobile outer panels should have high strength. This is provided by the addition of solution hardening elements like phosphorus (0.06-0.08%) and manganese (0.5-0.8%). Unfortunately phosphorus in these steels increases the tendency to develop forming cracks but this can be controlled by the further addition of very small amounts of boron (Fudaba et al. 1988).

Further strength can be obtained by taking advantage of the normally deleterious phenomenon of strain aging. Bake-hardenable steels are produced with a small, but controlled, amount of residual solute carbon. Resistance to room temperature aging is obtained but the yield strength is increased when the dislocations become stabilized during baking. As a consequence the yield strength is low during forming but high after baking. The required amount of solute carbon can be obtained by adjustment of the annealing cycle but as has been pointed out earlier, small amounts of solute carbon have a deleterious effect on the development of the desired {111} texture during annealing. The solution is simple and elegant (Obara et al. 1984, Abe and Satoh 1990). As before, the carbon is stabilized as niobium carbide before recrystallization is complete; this ensures a strong {111} texture. The temperature is then raised so that a small amount of carbon goes into solution. As recrystallization is almost complete before this happens the texture is unaffected.

12.4 RECENT DEVELOPMENTS IN GRAIN ORIENTED, SILICON STEEL SHEETS.

12.4.1. Introduction

The production of grain oriented, electrical steels is probably the best known example of the beneficial effects of recrystallization control during processing. The subject has been reviewed many times (see, for example Dillamore 1978b, Luborsky et al. 1983) and we begin with only a brief summary of the requirements and the processes used and a similarly brief account of their significance. This is followed by a review of recent developments relating particularly to changes in composition, the possibility of producing cube oriented sheet and the control of domain size.

12.4.2. Production of silicon steel sheets

The sheets or laminations used as transformer cores are usually made from a 3%Si, carbon-free, sheet steel.   The essential requirements are easy magnetization, low hysteresis loss and low eddy current loss.   The first and second of these are governed principally by composition, orientation and purity.   It would be advantageous to use a somewhat higher silicon content but such alloys are brittle and cannot be cold rolled. Magnetization is easiest in the <100> directions (fig 12.9) and great care is taken to see that a strong preferred orientation is developed with a <100> rolling direction.   In practice the texture that is commonly found is the well known {110}<001> Goss texture.   The state of internal stress and the smoothness of the sheet surface are also important.   Eddy current losses are also affected by these parameters but more importantly by grain size, sheet thickness and stresses introduced by the insulating coatings applied to the sheets.   The properties are optimized by small grain size, reduced thickness and minimum internal stress.   Finally, it would be desirable to have the smallest possible domain size and maximum domain wall mobility (Kramer 1992).

Present day practice stems from the work of Goss (1934a,b) who patented a technique for developing a strong {110}<001> texture.   This process is called the **Armco**, or two-stage cold rolling method, and a typical production schedule is summarized in table 12.5. The data shown are approximately correct, although different manufacturers use slightly modified compositions, temperatures, gas mixtures, coatings etc.

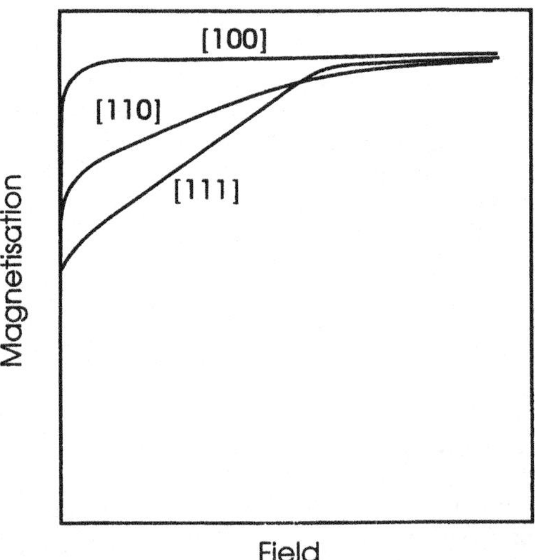

Fig. 12.9. Relationship between magnetisation and field for simple crystallographic directions in iron, (after Honda et al. 1928).

Table 12.5.

Processing of grain oriented steel.

Armco process	Nippon Steel process
a) Soak at ~1340°C; hot roll to 2mm.	a) Soak at ~1350°C; hot roll to ~2mm.
b) Cold roll to 0.5-1mm.	b) Hot band anneal 1125° C, air cool to 900°C, water quench to 100°C
c) Anneal at 950°C in dry H_2/N_2 (80:20)	c) Cold roll
d) Cold roll to finished size.	d) Decarburize at 850°C in wet H_2; dew point 66°C.
e) Decarburize at 800°C in wet H_2; dew point 50°C	e) Coat with MgO + 5% TiO_2.
f) Coat with MgO.	f) Texture anneal at 1200°C in H_2/N_2 (75:25).
g) Texture anneal at 1150°C in pure H_2.	

There are three important requirements in the Armco process:

(i) Provision must be made for the nucleation of $\{110\}<001>$ grains.
(ii) These grains must be able to grow.
(iii) Grains of other orientations must not grow.

The desired $\{110\}<001>$ orientation first appears during the initial hot rolling as a friction-induced shear texture at and near the surface. In the normal course of events it would be largely lost during cold rolling and be replaced by the usual two-component $\{112\}<110>$ + $\{111\}<110>$ texture (§2.4.2). The use of two, light, cold rolling stages ensures that although these components are produced their presence is controlled. In addition, the Goss orientations are not completely lost. They survive at the centres of transition bands separating the above orientations and in this environment nucleation is favoured so that some $\{110\}<001>$ grains appear in the annealing texture after the decarburizing anneal. These grains are somewhat larger than those of the other orientations, and during the final texture anneal they grow by abnormal grain growth to dominate the final texture.

The necessary conditions for their growth are provided by microstructural control. A fine dispersion of manganese sulphide is generated by cooling the slab rapidly before hot rolling. These particles are resistant to rapid coarsening and, by preventing normal grain growth (§9.4) keep the matrix grain size small during the early stages of the final high temperature annealing. As Ostwald ripening and dissolution proceed, abnormal grain growth becomes possible and the desired Goss-oriented grains grow and dominate the microstructure. Abnormal grain growth is also promoted by the presence of a sharp texture (§9.5.3). The possibility of generating undesirable orientations by surface nucleation processes is eliminated by the addition of sulphur to the MgO surface coating. This prevents the growth of surface grains and the sulphide formed is eventually lost by reaction with the hydrogen atmosphere.

A second processing schedule, sufficiently different to avoid direct problems with patent laws, was developed and patented by Nippon Steel Corporation (Taguchi et al. 1966, Sakakura 1969). Details of this **Nippon Steel** process which uses a single cold rolling stage, and aluminium nitride as an additional grain growth inhibitor are also summarized in table 12.5. Typically these steels contain $\sim 0.025\%$ aluminium and $\sim 0.01\%$ nitrogen. The resultant AlN particles are much less stable than MnS and because they coarsen rapidly the processing involves rapid cooling after hot rolling (Step (b) in table 12.5). For the same reason two-stage cold rolling with intermediate annealing is not possible. The resulting large cold rolling reduction results in a decreased $\{110\} <001>$ component and a larger grain size but this is off-set by an increased perfection of the final orientation. The MgO coating used in the final annealing treatment contains metal nitrides as well as sulphur. The nitrides help to control the decomposition of the AlN particles.

There has been continuing controversy about the validity of the patents involved and an investigation by Harase and Shimazu (1988) is significant. These workers investigated two batches of material, one containing only MnS as an inhibitor and the other containing both MnS and AlN. Samples from each batch were processed by both the single, and double cold rolling routes. Abnormal grain growth, with a strong Goss texture, occurred in the presence of AlN particles irrespective of the rolling schedule used. If only MnS particles were present, abnormal grain growth occurred only after two-stage cold rolling. In this material the texture was described as $\{334\} <9,13,3>$ which is similar to the normal $\{111\}$ type textures found in annealed low carbon steels. Both processes are being continually refined but it is generally accepted at this time, that the Nippon Steel process leads to a more perfect Goss texture but to a larger grain size. Some of the changes are referred to in the next section.

12.4.3 The development of the Goss texture

The strong $\{110\} <001>$ texture develops from near-surface grains during the final high temperature anneal. As the grain growth inhibitors coarsen and dissolve, abnormal grain growth occurs. Bölling et al. (1992) point out that the preferred growth of Goss grains requires a suitable texture relationship when coarsening begins. Although very few Goss grains are present at this stage, the more plentiful $\{111\} <112>$ grains have a high coincidence site (low Σ) relationship to the $\{110\} <001>$ grains, and they attribute the emergence of the Goss texture to this factor. Inokuti et al. (1987) found that the Goss grains also had a slight size advantage in the primary microstructure, although this has been disputed by Harase and Shimizu (1988) and Böttcher et al. (1992).

12.4.4. Recent Developments.

Thin gauge, high silicon content, perfect crystal alignment and small domain size are the fundamental features of an ideal electrical steel sheet (Bölling et al. 1992). Recent developments have involved small but significant alterations to composition, improvements in the perfection of the final texture, reductions in sheet thickness, control of the domain structure, a renewed interest in the development of cube-textured material, the production of amorphous sheet by melt spinning techniques and the possibility of direct casting of strip. Some of these are discussed below.

12.4.4.1 Composition.

It has been generally accepted that the magnetic properties of electrical steels would be considerably improved by increasing the silicon content; unfortunately increase beyond the current maximum of $\sim 3.1\%$ leads to embrittlement during cold rolling.

Recently Nakashima et al. (1991a) examined steels produced by the single-stage cold rolling process and containing up to 3.7% Si, and particular attention was paid to the abnormal grain growth aspects. It was found that the effectiveness of grain growth inhibition in the single-stage process decreased as the silicon content increased and that abnormal grain growth did not occur at all in the 3.7% Si sheet (final thickness 0.285mm). However, at the higher silicon contents used in this study, the ratio of the α and γ phases is changed and there is a real possibility that this may have affected the state of the inhibiting precipitates of MnS and AlN. To overcome this, Nakashima et al. adjusted the carbon content of their steels so as to maintain a constant α/γ ratio at a temperature of 1150°C. The problem of silicon-induced cold brittleness could be largely overcome by strip casting of thin sheet or completely overcome by the use of melt spinning to produce amorphous material. Both techniques are being actively investigated and both are likely to be used in the very near future.

Many small changes of composition have been investigated, particularly with respect to the growth inhibitors but details of many of these are to be found only in the voluminous and sketchy patent literature. Selenium and antimony react with manganese to form appropriate inhibiting phases and Fukuda and Sadayori (1989) claim that when used in conjunction with AlN the mean deviation from the $<001>$ for the Goss grains is improved from 7° to 3.5°.

12.4.4.2 Sheet thickness

It is well known that eddy current losses are substantially reduced in thin sheets, but until the early 1980's the thinnest sheets produced were 0.30mm thick. Currently sheets of 0.23, 0.18 and 0.15mm are available but production is not easy. Nakashima et al. (1991b) have shown that while abnormal grain growth produces a nearly perfect Goss texture in thick (0.60mm) sheet, texture loss occurs increasingly in thinner material. This was attributed to a decline in the efficiency of the growth inhibiting process that permitted grains of undesirable orientations to reach the surface and become stabilized. The effect is overcome in current production practice by the addition of nitrogen to the furnace atmosphere during the final anneal (see table 12.5).

12.4.4.3 Cube texture.

If a strong $\{100\}<uvw>$ texture could be induced the resulting steel would have two $<100>$ directions in the plane of the sheet. Obvious advantages would exist for such a product and this would particularly be so for the cube texture, $\{100\}<001>$. The development of such a texture was first reported many years ago by Assmus et al. (1957) and Detert (1959) in thin (<0.6mm) silicon iron sheet. Normal grain growth was retarded in the thin sheets and large grains with a $\{100\}$ surface plane appeared. It was recognised that a process of surface-energy-controlled, abnormal grain growth (sometimes referred to as **tertiary recrystallization**) was occurring and it was found that, in the presence of environmental sulphur, grains with a $\{100\}$ surface plane grew into the matrix structure. If the sulphur was removed $\{110\}$ type grains grew. It was thought at one time that oxygen might be an active species in promoting $\{100\}$ grains (Walter and Dunn 1960a) but this is now considered unlikely. Of the other Group VI elements the surface adsorption of selenium leads to the preferred growth of $\{111\}$ surface oriented grains (Benford and Stanley 1969). The role of surface energy in promoting abnormal grain growth in thin sheets is discussed in §9.5.4.2. Kramer (1992) has reviewed the development and potential uses of cube-oriented sheet. Despite the apparent advantages offered by such sheet, several difficulties remain. The grain size is some four times greater than that of Goss textured sheet and the domain size is larger; both effects are undesirable. As in other products the texture is never perfect and the domain structure in slightly misoriented sheet, where misorientation may occur about each of the surface $\{100\}$ directions, is complicated. Kramer believes that these factors ensure that, in all but a few cases, cube textured sheet cannot compete with the best, current Goss-textured material.

A different approach to the production of cube oriented sheet has been described by Arai and Yamashiro (1989) who reported results for a 3.26% Si-Fe alloy directly cast to 0.37mm thickness before final cold rolling. No inhibitors were used in this work and surface-energy-controlled grain growth occurred. Despite the lack of inhibitors a very strong $\{100\}<uvw>$ texture developed as the final sheet thickness decreased to 0.15mm. In view of the current developments in strip casting this is a very interesting result.

12.4.4.4 Domain structure

As understanding of the physical metallurgy of electrical steels has developed there has been an upsurge of interest in the domain structure. Figure 12.10 from the work of Bölling et al. (1992) shows the domain structure at the surface of a typical, 0.23mm, thick sheet. Points to be noted are the large grain size, the extent of local misorientation and the variation in domain wall spacing from 0.25 - 1.1mm. Ideally the spacing should be small and this can be induced partly by means of the insulating coating applied to the surface, which develops high surface stresses. However, the length of the domains is governed primarily by the grain size which is large in strongly textured material. The problem can be solved simply by lightly scratching the surface in a direction normal to RD; the small local misorientation produced acts as an artificial grain boundary. Domain refining of this type is now carried out in commercial production by pulsed laser irradiation, which produces a very small local strain at the surface without affecting the insulating coatings, and the core losses are reduced by ~10%.

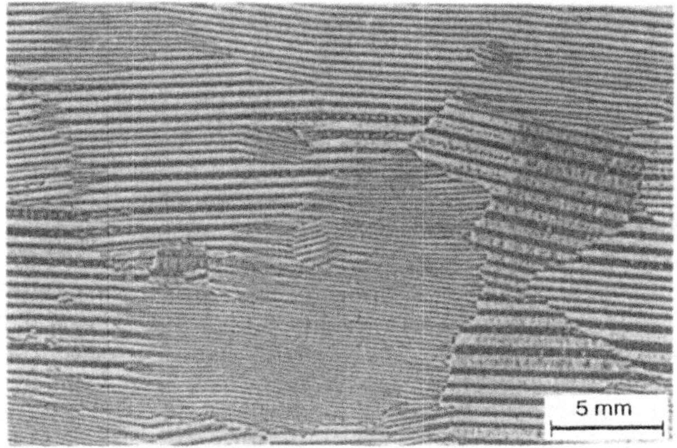

Fig. 12.10. Magnetic domains in high permeability grain oriented steel, (Bölling et al. 1992)

12.5. COMMERCIAL SUPERPLASTIC ALUMINIUM ALLOYS

12.5.1 Superplasticity and microstructure

Superplastic materials are polycrystalline solids which have the ability to exhibit unusually large ductilities. If deformed in uniaxial tension, a normal ductile metal will fail by necking after an elongation of less than 50%. However, superplastic alloys often exhibit elongations of several hundred percent and may exhibit elongation of several thousand percent. In order to resist necking, the material must have a high strain rate sensitivity of the flow stress, or **m** value, where

$$\sigma = K\,\dot{\epsilon}^{m} \qquad (12.4)$$

and σ is the flow stress, $\dot{\epsilon}$ the strain rate and K a constant.

This may be rewritten (Mukherjee et al. 1969) in an expanded form as

$$\dot{\epsilon} = \frac{K'\,D\,G\,b}{k\,T} \cdot \left(\frac{b}{d}\right)^{p} \cdot \left(\frac{\sigma}{G}\right)^{1/m} \qquad (12.5)$$

where **D** is the diffusivity, **d** the grain size, and **p** the grain size exponent is usually between 2 and 3.

For superplasticity, **m** must be greater than 0.3 and for the majority of superplastic alloys **m** lies in the range 0.4 to 0.7. A wide variety of metals, alloys and even ceramics can be made superplastic and there has been extensive development of superplastic aluminium and titanium alloys over the past 30 years. Recent reviews include those by Stowell (1983), Kashyap and Mukherjee (1985), Sherby and Wadsworth (1988), Pilling and Ridley (1989) and Ridley (1990).

The production of superplastic microstructures in **any** crystalline material and the stability of such a microstructure during subsequent high temperature deformation, exploit the principles of recrystallization and grain growth discussed in this book. However, we will restrict the present discussion to superplastic aluminium alloys which provide very good examples of the application of microstructural control by thermomechanical processing to commercial practice. Although the earliest work on superplasticity concentrated on **microduplex** alloys which consist of approximately equal amounts of two phases, the aluminium alloys currently of commercial interest are all **"pseudo-single-phase"** alloys, i.e. they consist of an aluminium matrix with only relatively small amounts of dispersed second phases. Although special alloys have been developed for superplastic applications, it is now known that many conventional alloys may be made superplastic if thermomechanically processed in the appropriate manner.

At high temperatures, a superplastic alloy is characterised by a low flow stress ($<10\text{MPa}$) and a high resistance to non-uniform thinning. This allows the near-net-shape forming of sheet material using techniques similar to those developed for the bulge forming of thermoplastics. Superplastic forming (SPF) is thus an attractive option for the forming of complex shapes from sheet at low stresses, incurring much smaller tooling costs than for conventional cold pressing operations. However, the slow strain rates often necessary for superplastic forming and the higher material costs have so far restricted the commercial exploitation of SPF to a relatively small number of specialised applications.

Superplastic deformation occurs primarily by grain boundary sliding, with the accommodation of strain at grain corners being provided by diffusional and/or dislocation processes. The microstructural requirement is for a small grain size, typically less than 10μm. Equation 12.5 shows that the smaller the grain size, the higher is the strain rate which can be used at a given stress. It is important that the grains are equiaxed, with high angle boundaries. It is also important that dynamic grain growth during SPF is minimised, otherwise, as seen from equation 12.5, the flow stress will increase during forming. The maximum superplastic strain is usually limited by grain boundary cavitation, which again is minimised by a small grain size.

The key to successful SPF is therefore the development of a very small grain size and the inhibition of dynamic grain growth during the forming process.

12.5.2 Development of the microstructure by static recrystallization

This processing route, typified by that developed by Rockwell (Wert et al. 1981), which is illustrated in fig 12.11, is based on the mechanism of **particle stimulated nucleation of recrystallization** (§8.4) and has been used to produce superplastic microstructures in the high strength aerospace 7xxx series alloys (Al-Zn-Mg-Cr) such as AA7075 and AA7475.

Following a solution heat treatment, the alloy is overaged at $\sim 400°$C to produce a dispersion of large ($\sim 1\mu$m) intermetallic particles in addition to a stable dispersion of small ($\sim 0.1\mu$m) chromium-rich dispersoids. The material is then warm-rolled in several passes at 200°C to an 85% reduction. This temperature is low enough (§11.6.4) to ensure that deformation zones develop at the large particles, but high enough to prevent excessive work hardening by the small dispersoids. The alloy is then heated rapidly in a salt bath at 480°C to completely recrystallize. Particle stimulated nucleation occurs at the large particles and grain growth is inhibited by both the large and the small particles. Wert et al. (1981) found that

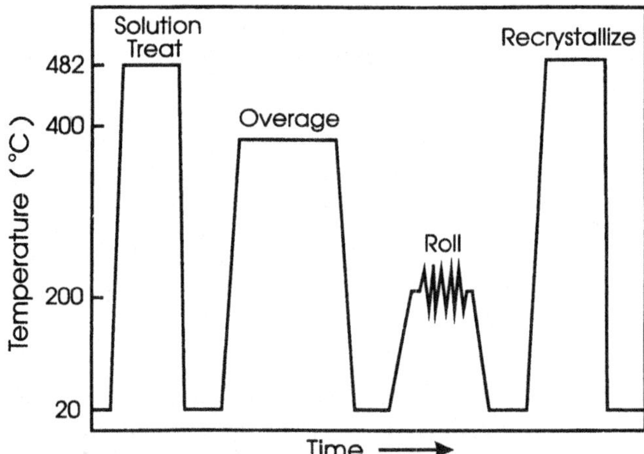

Fig. 12.11. Schematic diagram showing the four step thermomechanical processing treatment used for grain refinement of AA7075, (after Wert et al. 1981).

the number of large particles per unit volume was approximately ten times the number of grains, indicating a very low efficiency of PSN. As the particle size of $\sim 1\mu m$ is on the borderline for PSN (§8.4.1.2) it is likely that recrystallization nucleates first at the larger particles, and the smaller particles are then engulfed before nucleation occurs at them.

AA7475 processed by this route has a grain size of $\sim 10\mu m$ and is capable of tensile elongations of $\sim 1000\%$ if deformed at the relatively slow strain rate of $2\times10^{-4}s^{-1}$ at 520°C.

12.5.3 Development of the microstructure by dynamic recrystallization

An alternative method of producing a fine grain structure in aluminium involves a form of dynamic recrystallization (§11.3.3.1) which is not completely understood. The first recorded example of the use of this processing route was for the **Supral** alloys, such as Supral 100 (Al-6%Cu-0.4%Zr) and Supral 220 (Al-6%Cu-0.4%Zr-0.3%Mg-0.2%Si-0.1%Ge) which were developed specifically for superplastic application (Watts et al. 1976).

The alloys are cast from high temperatures ($\geq 780°C$) and rapidly chilled to avoid the precipitation of primary $ZrAl_3$. The alloy is given an initial heat treatment at $\sim 350°C$ in which most of the Zr is precipitated as a fine dispersion of $ZrAl_3$ particles (volume fraction ~ 0.05 and particle size $\sim 10nm$) which, once formed are relatively stable. A solution treatment may then be given at $\sim 500°C$ which takes $\sim 4\%$ copper into solid solution, leaving the remaining 2% in the form of large ($> 1\mu m$) $CuAl_2$ particles. The alloy is then warm worked at $\sim 300°C$, the small spacing of the $ZrAl_3$ particles preventing recrystallization (§8.5). This may be followed by a final reduction by cold rolling. Finally the heavily rolled sheet is superplastically formed at $\sim 460°$. During the initial stages of this deformation (typically within a strain of 0.4), a microstructure of small ($\sim 5\mu m$) equiaxed grains with high angle boundaries evolves from the deformed microstructure. Tensile elongations of greater than 1000% can readily be obtained for Supral 220 at strain rates of $10^{-3}s^{-1}$ at 460°C. The grain size produced during SPF is strain rate dependent, and in many cases it is beneficial to begin the deformation at a high strain rate which produces a small grain size, and then reduce the strain rate once the recrystallized structure has evolved.

Despite considerable research, the exact nature of this recrystallization process is not clear. The main role of the $ZrAl_3$ particles is to inhibit recovery and prevent static recrystallization of the cold worked material and to prevent grain growth when dynamic recrystallization has occurred. It has been suggested that the role of the large $CuAl_2$ particles is to provide nucleation sites for recrystallization although this has never been clearly demonstrated. It seems certain that in addition to the $ZrAl_3$ particles, the necessary requirements for producing a fine grain structure during the final hot working are substantial solid solution (e.g. Mg, Cu) and prior working, because a binary Al-Zr alloy will not produce the required microstructure (Watts et al. 1976), nor will an alloy which has not been previously worked. The mechanism may be related to that of dynamic recrystallization by progressive grain rotation (§11.3.2), although, as discussed in §11.3.3.1, this is by no means certain.

This method of producing a fine grain size during the superplastic deformation itself is not unique to the Supral alloys, and similar processing routes are used to achieve superplasticity in other aluminium alloys. In particular the newer strong lightweight aluminium-lithium alloys such as AA8090 (Al-2.5%Li-1.2%Cu-0.6%Mg-0.1%Zr) which are also resistant to static recrystallization, may undergo a similar type of dynamic recrystallization and show extensive superplasticity (Ghosh and Ghandi 1986, Grimes et al. 1987).

Chapter 13

COMPUTER MODELLING AND SIMULATION OF ANNEALING

13.1 INTRODUCTION

In order to control the microstructure, texture and properties of an alloy during a complex industrial thermomechanical treatment, there is a need for quantitative models which will accurately predict the effect of the processing parameters on the material which is produced. The empirical approach, which has long been used is now recognised as being of limited value and, in many cases the cost of industrial scale parametric experimental investigations is prohibitively expensive. Many major metal-producing companies, particularly in the aluminum and steel industries, have now recognised that in order to have any predictive value, models are needed which are based on physically sound concepts. Because of the complexity of industrial thermomechanical processing schedules and our lack of a sound understanding of many of the annealing phenomena, this remains a long term objective.

The modelling which has been carried out to date may be divided into two general categories. There are the **micro models** which aim to deal with individual processes such as deformation or annealing, or perhaps only part of these i.e. recovery, recrystallization or grain growth. Then there are the **macro models** which aim to model the total process, and these may involve linking several of the micro models. In this chapter we will examine some of the micro models of annealing and will briefly look at the concept of macro-modelling.

13.2 MICRO MODELS

13.2.1 The role of computer simulation

In discussing recovery, recrystallization and grain growth in earlier chapters, we have tried to show wherever possible, how an understanding of the physical mechanisms can lead to the development of quantitative models of the process. The ideal model would be analytical and based on sound physical principles, which perfectly describe the annealing process, thus allowing prediction of the resultant microstructure, texture and kinetics. It is clear that because of the complexity of the processes involved, we are still far from this goal and in part this is due to the heterogeneous nature of many annealing processes. Whereas it is feasible to accurately describe simple processes such as the annealing of a dislocation dipole with some accuracy by analytical means, the nucleation of primary recrystallization or the onset of secondary recrystallization are catastrophic events which depend not upon the average microstructure, but upon some heterogeneities in the microstructure. As was discussed in chapter 6, analytical methods such as the JMAK approach and its extensions may be useful in giving a broad description of a process, but cannot yet handle the complexities of recrystallization in real materials. For these reasons, much effort has in recent years been put into the development of computer simulations of annealing. There have been several methods of approach, each with its own particular advantages and disadvantages. Some of the models involve simulation close to the atomistic level and once the rules for the motion of the "atoms" are assumed then little further input is necessary. Other computer models use analytical equations to describe parts of the annealing process and allow the computer to deal with the heterogeneity of the annealing.

Computer simulations have in the past been criticised as being incapable of making any significant predictions, and it is of course true that they will not reveal any micromechanisms which occur on a smaller scale than the basic units of the model. However, by dealing with the annealing of a complete microstructure and thus taking into account long range cooperative effects and those due to heterogeneity, the models are capable of revealing hitherto unknown phenomena.

Whilst most of the models produce realistic microstructures or microstructural data, the use of these models is at present limited either by the size of the model or by the lack of detail in the input. Unlike an analytical solution, it is usually impossible for the reader to verify a computer simulation, and there is always a real danger that the output of a model may be the result of faulty, inaccurate or inadequately detailed modelling. For these reasons the results of computer simulation should be treated with great caution. This is illustrated by the problems encountered in using 2-D models to simulate the grain growth of 3-D microstructures as discussed in §9.4.1.4 and §9.5.2.3.

Perhaps at this time, one of the most useful roles of computer simulations is to draw attention to areas where further theoretical or experimental work are needed. A good example of this is the computer simulation of **grain growth**, which has in the past decade, by producing results which have questioned the accepted understanding of the subject, stimulated a large amount of theoretical and experimental work, thus rejuvenating an important field which had become largely dormant. As our understanding of the physical phenomena involved in annealing increases, modelling techniques become more sophisticated, and as the power of computers grows, the role of modelling and computer simulation will become increasingly dominant.

13.2.2 Monte Carlo Simulations

In a series of papers, Anderson and colleagues (e.g. Anderson et al. 1984, Anderson 1986) have developed Monte Carlo methods for the simulation of grain growth in two and three dimensions (§9.2.4.2). More recently, the same approach has been used to study primary recrystallization, abnormal grain growth (§9.5.2.3) and dynamic recrystallization.

13.2.2.1 The method and its application to grain growth
In this method, known as the Potts, or large-Q Potts model, the material is divided into a number of discrete points which represent the centres of small areas or volumes of material

and these points are arranged on a regular lattice. These are the fundamental building blocks of the model. They are regions within which the microstructure is assumed to be homogeneous and structureless. Each block may have an attribute such as an orientation, but no subsidiary microstructure. Each region is given a number corresponding to a grain orientation as shown in figure 13.1, and a grain may comprise one or more blocks. A grain boundary is therefore characterised by the relative numbers (orientation) of the blocks, leading to boundaries of the type 4/6 and 3/9 etc. The grain boundary energy may then be specified in terms of the number pairs. For example, in the simplest case we might take like number pairs (e.g. 3/3) to have a zero energy and all unlike pairs to have the same high energy, which would be a reasonable approximation for high angle grain boundaries. Although, in a 2-D simulation, the boundary energy is satisfactorily simulated by the interaction between nearest neighbour pairs, in 3-D simulations it has been found necessary to consider interactions of up to third nearest neighbours.

The model is run by selecting a block at random, and reorienting it to one of the other grain orientations. The energy of the new state is then determined. If the energy change ΔE is less than or equal to zero then the transition is accepted. However, if ΔE is greater than zero, then the reorientation is accepted with a probability of $\exp(-\Delta E/kT)$. The unit of time in the simulation is the Monte Carlo Step (MCS) which represents N attempted transitions, where N is the number of blocks or lattice sites in the model.

Transitions within a grain will not occur because of the severe energy penalty, but transitions at grain boundaries may occur, giving rise to boundary migration. Figure 13.2 shows how this arises. If the interaction energy between like sites is 0 and that between unlike sites is 1, then the energy of the configuration in 13.2a is greater than that of 13.2b by 2 units, thus providing a driving force for reducing boundary curvature. Consequently, as the model is run, the grains grow, as shown in figure 13.3, exhibiting many of the features of grain growth. Using larger arrays of 200 x 200 points, then realistic microstructures are produced as shown in figure 13.4. The techniques may be extended to three dimensions (Anderson et al. 1985), although the size of the array which can be used is currently limited to around 100 x 100 x 100 points. Comparison of the 2-D and 3-D simulations shows the grain size and size distribution in the 2-D case to be nearly identical to that found from a cross section of the 3-D simulation.

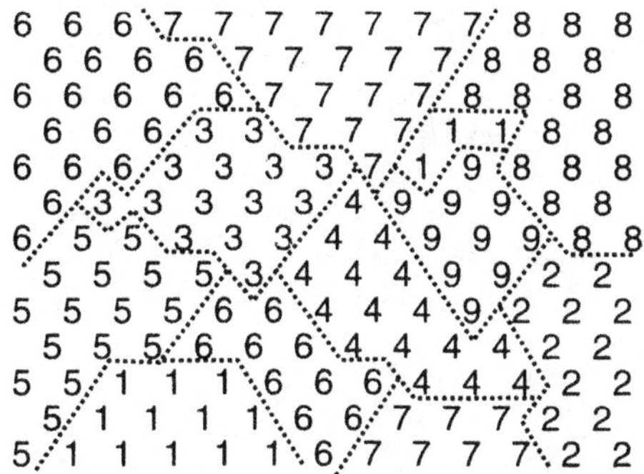

Fig. 13.1. The basis of the Monte Carlo simulation method.

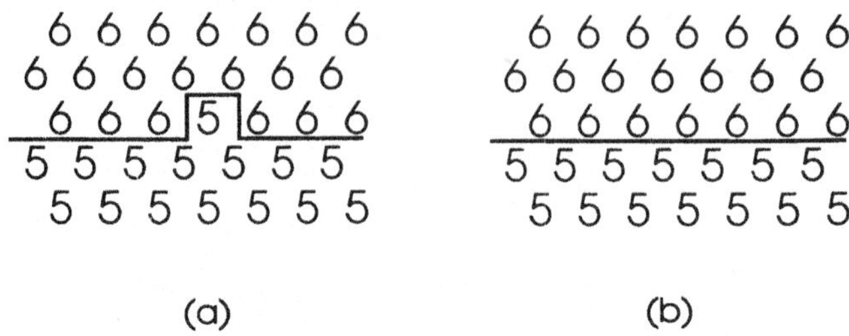

(a)   (b)

Fig. 13.2. Grain boundary migration in the Monte Carlo model.

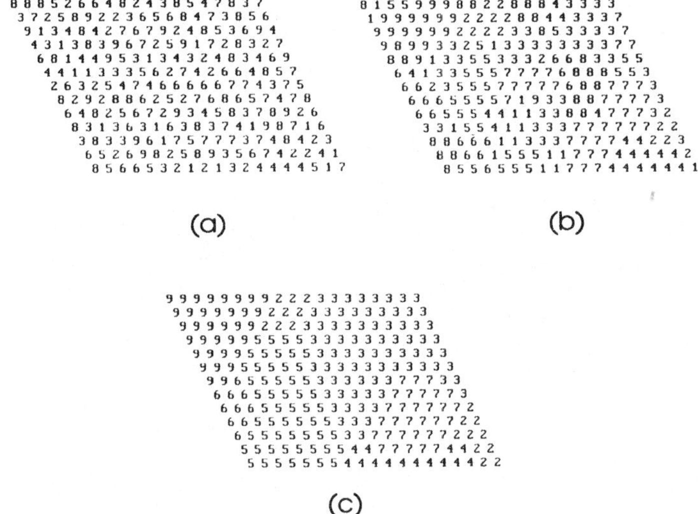

(a)   (b)

(c)

Fig. 13.3. Monte Carlo simulation of grain growth using a small array.

Quantitative measurements of the structures evolved in both 2-D and 3-D simulations exhibit many of the features of grain growth. The grain size distribution $f(R/\bar{R})$ is found to be time invariant, to peak at $R/\bar{R}=1$ and to have an upper cut-off at $R/\bar{R} = 2.7$. Figure 9.6 shows the comparison between the grain size distributions produced in a simulation and those predicted by analytical models (fig 9.6a) and those found experimentally (fig 9.6b). The variation of the mean grain size with time is found to obey equation 9.7, and the more recent results have shown the grain growth exponent to be ~2, although earlier simulations erroneously produced larger values as discussed in §9.2.4.2.

The model has also been used to simulate the effect of second-phase particles on grain growth. This is done by selecting a fraction of individual blocks to be particles. These are assigned a different number to any of the matrix blocks and these sites are not allowed to switch orientation during the simulation. The "particles" therefore have an interfacial energy equal to the grain boundary energy. If we calculate the total energy of the array shown in

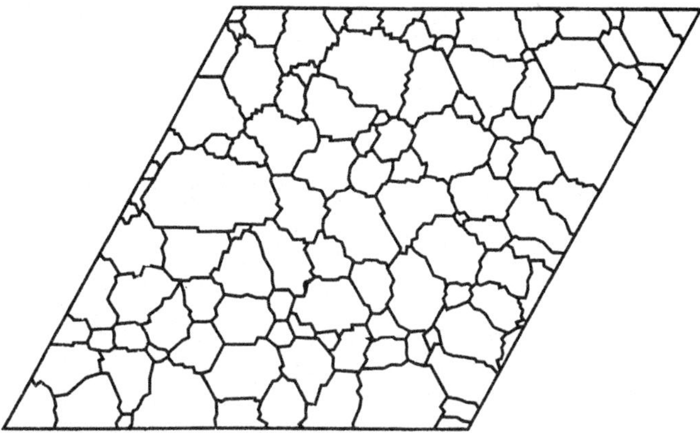

Fig. 13.4. A typical grain structure from a two-dimensional Monte Carlo simulation of grain growth.

figure 13.5a where the "particle is on the grain boundary, we find that it is 2 units lower than that of 13.5b where the particle is inside a grain. Therefore there is an attraction between particles and boundaries leading to a pinning force similar to that discussed in §3.6. The volume fraction of particles may be varied, as it is simply the fraction of sites which are designated as particles. However, the particle size, which is one block, is not varied. Starting with a small initial grain size (one block), it is found that grain growth stops at a grain size which is dependent on the particle volume fraction. In 2-D simulations the limiting grain size (D_{2D}) is given by

$$D_{2D} = c_1 F_V^{-0.5} \qquad (13.1)$$

and for a 3-D simulation

$$D_{3D} = c_2 F_V^{-0.3} \qquad (13.2)$$

where c_1 and c_2 are constants.

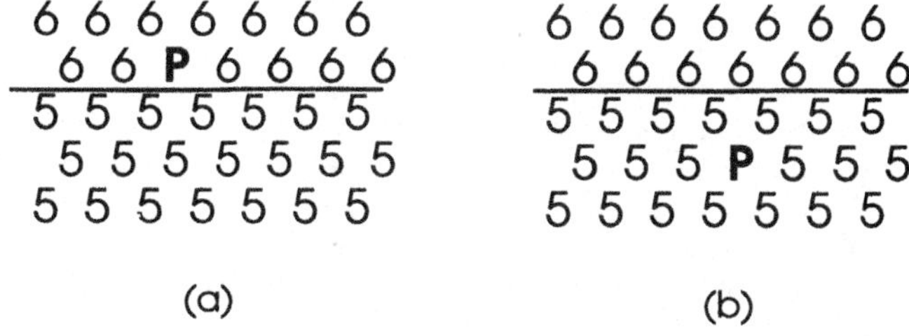

Fig. 13.5. Interaction of a grain boundary and second-phase particle in the Monte Carlo model.

These results are similar to those predicted analytically for alloys containing large volume fractions of particles in the rather unusual case where the initial grain size is much smaller than the interparticle spacing (§9.4.2) and where there is therefore a strong correlation between particles and grain boundaries. The significance of these results is discussed in §9.4.2.4.

Abnormal grain growth (§9.5.2.3) may also be simulated by the Monte Carlo method. This is done by introducing a very large grain, typically of a width equal to that of the array, into a particle-containing microstructure where normal grain growth has stagnated.

13.2.2.2 Application to primary recrystallization

The Monte Carlo simulation technique described above has been extended to the simulation of primary recrystallization (e.g. Srolovitz et al. 1986, Rollett et al. 1989a). A grain structure is first developed by growth as discussed above, and a stored energy H is given to all sites within each grain. In order to simulate heterogeneous deformation, the stored energy within a grain can be varied. Recrystallization "nuclei", which are grains comprising three blocks with $H=0$, are then introduced into the structure. Just as for grain growth, the pinning effect of second-phase particles may be studied.

Although reasonably realistic kinetics and microstructures are obtained, it is not clear that the use of this model for recrystallization is as successful as it is for grain growth. This is because of the size of the model relative to the scale of structure which is of importance in determining the recrystallization of real materials. For example, if the units of the model are equivalent to atoms (which are the units which jump in a real material), then this means that the technique can be criticised for dealing only with unrealistically small volumes of material (a cube of side 20nm). Although it is probably not necessary to go down to the atomic scale, it is clear that a model in which a particle is represented by a single lattice point is inadequate, because the effect of particles on recrystallization is strongly dependent on particle size as well as spacing (§8.3). Similarly, by failing to deal in detail with the nucleation stage, which is dependent on variations in the microstructure over distances of the order of 100nm, the model is clearly limited. Using empirical rules for nucleation, recovery and stored energy, and taking into account coarse spatial heterogeneities in the microstructure, realistic microstructures, kinetics and grain size distributions can be determined using the Monte Carlo simulation method. However, there are difficulties with some aspects of the simulation. For example, grain growth of the **unrecrystallized** grains is allowed to occur in the model (Srolovitz et al. 1986), whereas in real materials this does not occur.

At the present time, the computer Avrami models discussed in §13.2.4, which produce comparable types of data, i.e. kinetics and grain size distributions, but which can deal with much larger numbers of nuclei, appear to have an advantage in modelling primary recrystallization. The Monte Carlo model is a powerful technique, but if it were to adequately simulate recrystallization without reliance on empirical nucleation rules, then it would probably need to have a linear dimension of less than 10nm as the base unit, and would need to encompass a linear distance of at least 100μm in order to produce a statistically significant microstructure. This would require a lattice of 10^4 points in one dimension or 10^{12} points in three-dimensions, i.e. considerably larger than can be dealt with at the present time.

13.2.2.3 Application to dynamic recrystallization

The Monte-Carlo simulation method has recently been applied to dynamic recrystallization (Rollett et al. 1992, Peczac and Luton 1993). The 2-D simulations, which employ a 200x200 lattice, are similar to the static recrystallization simulations discussed above, except that the stored energy is allowed to increase with strain (time). The rate of increase of energy is calculated by assuming a work hardening relationship for the material, and in the later simulation, dynamic recovery has been taken into account.

In order to model dynamic recrystallization, "nuclei", consisting of a group of three units are introduced at random into the lattice. These nuclei have no internal energy when they are introduced, but subsequently acquire energy during the "deformation". Nuclei grow or shrink according to the usual energy criteria used in these simulations (§13.2.1). There is no explicit criterion for nucleation such as a critical strain, and the growth of the nuclei is determined by energetic criteria. The flow stress of the material is taken to be proportional to the square root of the total internal energy of the lattice points (i.e. equation 2.2) and thus stress-strain curves may be obtained. There are a number of parameters which need to be given to the model. These include the rates of work hardening and recovery and the rate of nucleation and its dependence on the other parameters. The stress-strain curves produced by the simulations are similar to the experimental ones shown in figure 11.6. The results show that stress-strain behaviour is not very sensitive to the hardening or nucleation laws, but is strongly affected by the recovery rate.

Such models provide a means of examining the effect of variables on dynamic recrystallization. The outputs, which may be compared with experiment or theory are the stress-strain data and the grain size. The simulations are still in their infancy and the advantages of the Monte-Carlo approach over the well established methods which use semi-empirical equations to model hot working (§13.3) have yet to be demonstrated.

13.2.2.4 Success and limitations of the model

The merit of the Monte Carlo model lies in its inherent simplicity. In its simplest form it can develop a realistic microstructure given little more than boundary energies. In its ultimate form, the blocks would be atoms, and if the correct interatomic potentials were used, then in principle this approach, which is a type of **molecular modelling** such as is currently of great interest in Physics and Chemistry for simulating reactions, would give a true atomic-scale simulation of annealing processes. The model develops the complete microstructure and is essentially deterministic apart from the local random fluctuations at the boundaries. It therefore automatically includes the topographic factors which have been a continuing problem in developing theories of grain growth (§9.2).

The Monte Carlo method is most successful when used in its most simple form, e.g. for grain growth in single phase materials, as in this situation the model requires very little input. As the method is applied to more complex problems such as static and dynamic recrystallization, more assumptions need to be incorporated, more relationships, either empirical or analytical, need to be fed into the model, and the unit blocks of the model need to have a larger number of attributes. In these situations the main role of the model is to provide a framework which will allow the effects of microstructural heterogeneity to be included, and as such, the model competes with the computer Avrami models (§13.2.4) which start with analytical/empirical predictions and use the computer to allow a study of the effects of heterogeneity.

13.2.3 Cellular models

These models are based on the assumption that the smallest microstructural unit of importance during annealing is a grain or subgrain, and they represent the microstructure as a cellular structure. The earliest simulations in this field were the use of bubble rafts by Bragg and Nye (1947) and Smith (1952) to study grain growth. The coarsening of soap froths between glass plates has also been used as a grain growth simulation (see Weaire and Glazier 1992) and has also been studied in its own right. Although the mechanisms of grain growth and froth coarsening are clearly different as the latter involve gas diffusion, it is interesting that detailed measurements of the kinetics of soap froth coarsening show a remarkable similarity to those for grain growth (Ling et al. 1992). The computer models discussed below are currently two-dimensional, but are being extended to 3-D (Nagai et al. 1992). It should again be emphasised that 2-D simulations of the annealing of 3-D microstructures may be misleading, and the results should be interpreted with caution.

13.2.3.1 Modelling orientation-independent grain growth.
If we consider a 2-D array of grains, then the grains themselves may be represented only by their vertices and this reduces the amount of data needed to be held in the computer. For example, figure 13.4 was obtained using a Monte Carlo simulation in two-dimensions with an array of 150 x 150 i.e. 22500 lattice points. There are approximately 100 grains in the structure, with an average of six sides, and thus there are some 200 vertices. Representing this structure by the node points therefore gives significant compression of the data and allows larger numbers of grains to be used in such a model. In the type of simulation, which was first proposed by Fullman (1952), the boundary energy is assumed to be isotropic. The forces due to the boundaries on the vertices at which the cells meet are then calculated, a mobility is assumed, and the vertices are moved accordingly (e.g. Soares et al. 1985). This approach, which neglects the curvature of the boundaries can be justified if the boundary mobilities are greater than those of the vertices, so that vertex motion is rate controlling. Alternatively, from a knowledge of the vertex positions, the local boundary curvatures may be calculated and the boundaries and hence the vertices moved accordingly (Weaire and Kermode 1983, Ceppi and Nasello 1984, Frost et al. 1986, Kawasaki et al. 1989). A detailed comparison of the various models is given by Atkinson (1988).

Once the initial microstructure has been constructed, often by a Voronoi network, subsequent events are determined by the equations governing the motion of the boundaries and vertices. As no random paths are chosen, the models are truly deterministic. During annealing, nodes may be driven into contact and in such cases appropriate action must be incorporated into the computer code. For example, nodes moving in the directions shown in figure 13.6a will switch partners as shown in figure 13.6b. Similarly, three-sided grains will generally tend to shrink and to vanish as shown in figure 13.6c and 13.6d.

Frost and colleagues have made extensive use of their model to simulate grain growth in thin metallic films, an area of importance in the technology of integrated circuits, and in which the limitation of the model to two dimensions does not pose a serious problem. They have incorporated grain boundary grooving into the model and an example of their simulation is shown in figure 13.7.

13.2.3.2 An orientation-dependent model
In the models discussed above, the grain boundary energies were taken to be isotropic. This may be a reasonable assumption for grain growth, although, as discussed in §9.3, grain

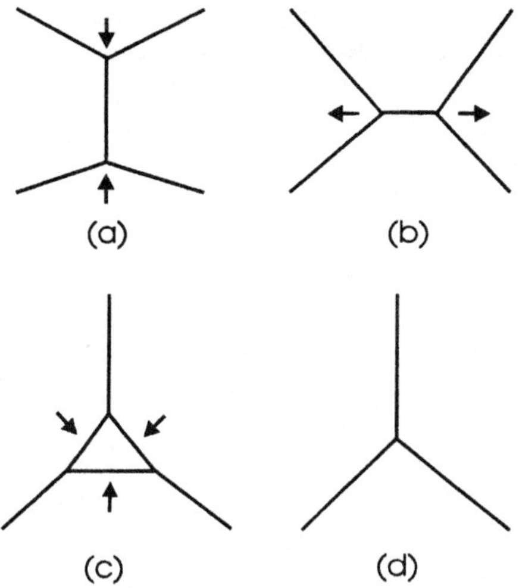

Fig. 13.6. a) b) node switching events. c) d) shrinking and disappearance of a three-sided grain.

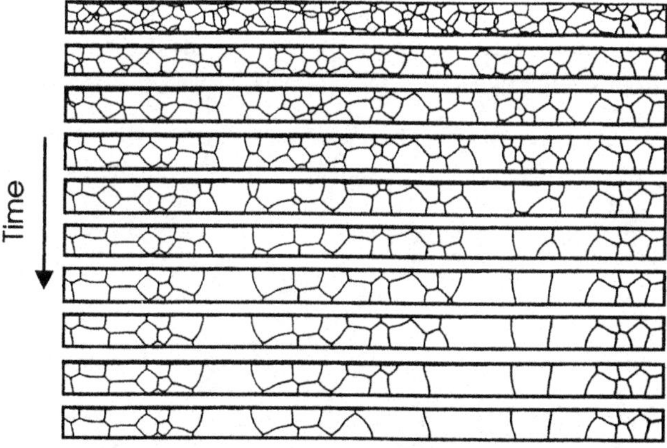

Fig. 13.7. Simulation of grain growth in thin metallic strip. Thermal grooving is included in the model, and the grain structure stagnates before a complete bamboo structure is formed, (Walton et al. 1992).

orientation effects are important even in this case. In order to simulate recovery or the nucleation of recrystallization, the model has to be modified so as to take account of the grain and subgrain orientations, because the energy and mobility of low angle boundaries are strongly dependent on orientation, and a two-dimensional version of such a model has been discussed by Humphreys (1992a,b). A model such as this may of course only be used to simulate the recovery of a material in which the dislocations are arranged into low angle boundaries (e.g. a high stacking fault energy material such as Al or Fe).

A microstructure consisting of a two-dimensional network of grains or subgrains is constructed. The (sub)grains are represented by the vertices N_j as shown in figure 13.8, and the positions of these, together with the identification of their neighbours is stored in the computer. Apart from the constraint that each vertex should connect three boundaries, there are no constraints on their spatial distribution. Each subgrain is given a number O_j, representing its crystallographic orientation, and the distribution of orientations can be varied, so as to represent any required orientation gradients. The three orientations associated with each vertex are also stored in the computer.

This is a flexible framework into which grains or subgrains of any spatial or angular distribution may be introduced, and it is possible to produce microstructures which are reasonably realistic representations of deformed or recovered structures. Running on a desktop PC, such a model can accommodate up to 4000 grains or subgrains. The model requires the input of data for the boundary energies and mobilities. As the orientations of adjacent grains are known, the boundary misorientation angle is also determined, and the boundary energy may be calculated if the relationship between boundary angle and misorientation is known. For low angle boundaries (up to 15°) it is reasonable to use the Read-Shockley relationship (equation 3.5). For higher angles, the boundary energy is usually taken to be constant, although any desired relationship may be incorporated. The mobility of a low angle boundary, is known to be dependent on the boundary misorientation, although there is little guidance from either experiment or theory as to the form of such a relationship (§4.2.2), and various relationships have been investigated.

The model has been tested by comparing the growth kinetics and grain size distributions in a high angle boundary array with that produced by 2-D Monte Carlo simulations (Humphreys 1992a) and has been found to produce similar results to those described in §13.2.2.

13.2.3.3 Simulations of recovery and recrystallization
A model of this type may be used to investigate recovery. Preliminary experiments have suggested that the subgrain growth rate exponent (**n** in equation 5.27) for recovery in the absence of a long range orientation gradient is >3 and that the average boundary misorientation decreases during annealing as shown in figure 5.22. However in the presence

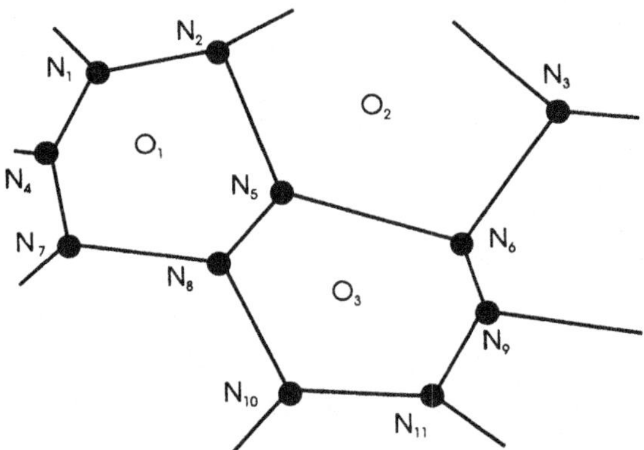

Fig. 13.8. The basis of an orientation-dependent network model, (Humphreys 1992a).

of a long range orientation gradient, **n** is closer to 2 and misorientations tend to increase during recovery (Humphreys 1992b). It is also possible to investigate separately the effects of boundary energy and mobility using this type of model as is shown in figure 9.9.

The model has been used to simulate some of the possible mechanisms of recrystallization nucleation, including the annealing of a transition band in a deformed material as shown in figure 13.9. It should be emphasised that recrystallization nucleation, which is no more than heterogeneous recovery, follows as a natural consequence of the spatial and angular heterogeneities in the starting microstructure and is not specifically programmed into the model. Pinning by particles may also be introduced into such a model, and by producing a suitable orientation gradient at a large particle, particle stimulated nucleation of recrystallization may be simulated (Humphreys 1992a).

This type of modelling is at present in its early stage of development, and as it has not yet been extended to three dimensions, any results obtained should be regarded as tentative. However, this type of approach is likely to be increasingly used, particularly fo investigating the effect of deformation heterogeneities on annealing. As in all simulations of recovery and recrystallization, probably the greatest problem is to define the deformed microstructure which is the starting point for the annealing simulations.

13.2.4 Computer Avrami models

As discussed in §6.4, one of the main problems in accurately predicting the kinetics of recrystallization and the development of grain structures is the spatial inhomogeneity of the nucleation and growth processes. By using analytical relationships to describe nucleation and growth rates etc., but using a computer to deal with the spatial distribution effects, significant progress has been made in developing more realistic models for recrystallization. Such an approach has the merit of being readily able to assimilate new developments in nucleation and growth theories within the framework of the model.

13.2.4.1 The basic method

An early two-dimensional simulation of this type was demonstrated by Mahin et al. (1980), and the Trondheim group (Saetre et al. 1986a, Marthinsen et al. 1989, Furu et al. 1990) have extended the simulation to three dimensions and extensively refined the model. In its most general form, nuclei are distributed within a cube at a given rate, and these then grow according to a specified growth law, the transformation being complete when grains impinge on one other. Microstructures are then obtained from two-dimensional sections of the volume, which are analyzed by a binary tree construction. A coarse grid is constructed and each grid point is examined to see whether

Case A - No grain has grown into this region (unrecrystallized)
Case B - One grain has grown into this region (grain interior)
Case C - More than one grain has (hypothetically) grown into this area (polygranular)

For cases A and B no further analysis is necessary. For case C, this area of the grid is subdivided into two and re-examined only for the grains which had intruded into the grid point, the process being repeated until either the grid point becomes a grain interior or the ultimate resolution of the grid is reached (typically 512x512 points), in which case this point is marked as being a boundary (fig 13.10). Such a recursive algorithm in which the number

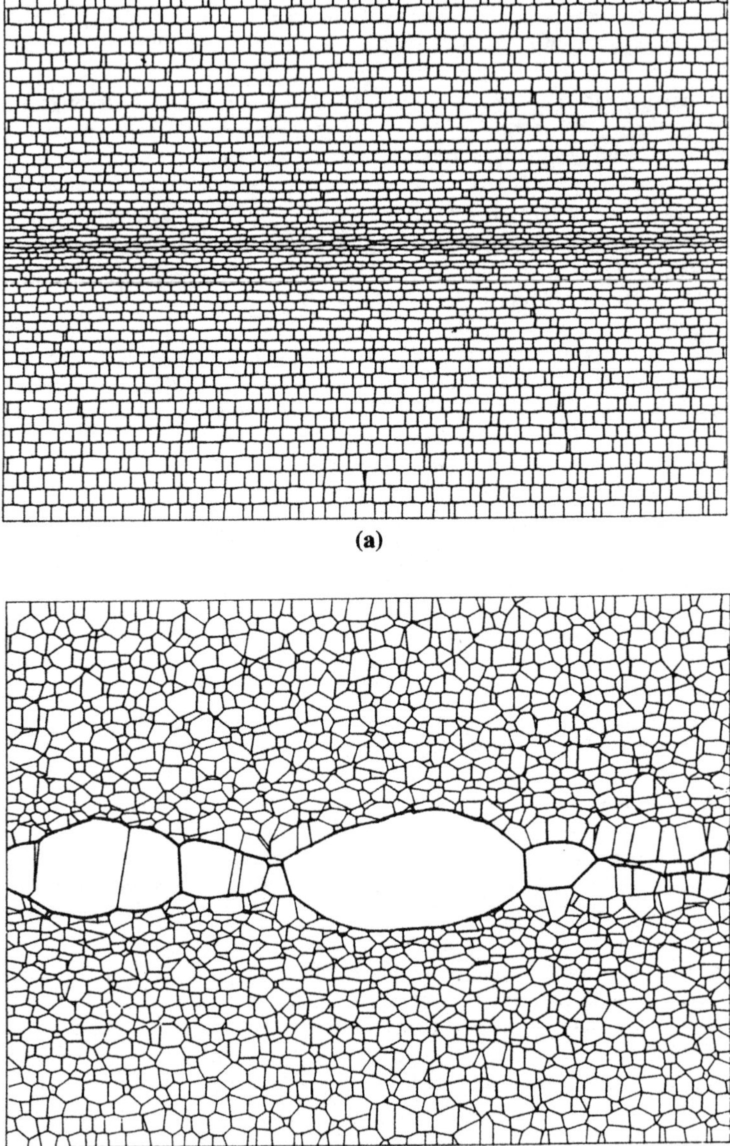

Fig. 13.9. Simulation of recrystallization nucleation in a transition band. a) The initial microstructure consists of subgrains which are smaller and more elongated in the transition band. There is also an imposed vertical orientation gradient of 5° per subgrain over the ten central subgrains. b) The microstructure evolved on "annealing" shows the development of large elongated grains of similar orientations but with high angle boundaries (bold lines) to the recovered structure, (Humphreys 1992a).

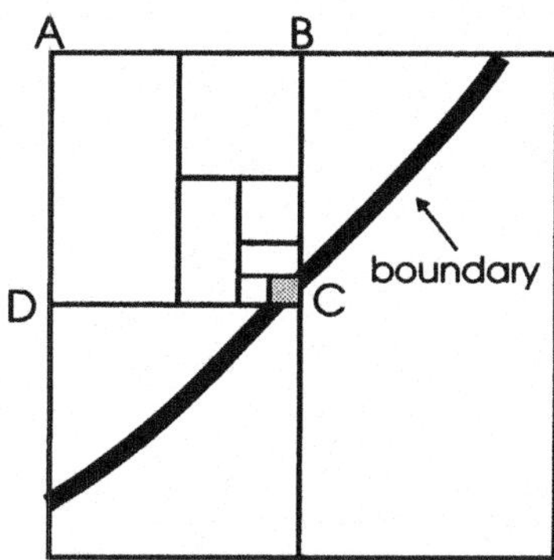

Fig. 13.10. The method of computer analysis of a microstructure, showing how the use of successively finer grids, enables the grain boundary which was initially located in the area ABCD, to be accurately placed.

of grains to be tested is reduced at each iterative step is very efficient. Typical microstructures produced from this type of model are shown in figure 13.11. The Trondheim simulation will accommodate more than 10^4 nuclei and is thus capable of producing sufficient grains to enable statistically meaningful grain size distribution data to be obtained.

13.2.4.2 Application to recrystallization

For randomly distributed nuclei, the kinetics are, as expected, similar to those predicted on the analytical models discussed §6.3. Figure 13.12 shows the recrystallization kinetics for a constant nucleation rate (n=4), and figure 13.13 for site saturated nucleation (n=3). The models allow the spatial distribution of nuclei to be varied, and figure 13.14 shows how an inhomogeneous distribution of nuclei affects the JMAK exponents, which decrease as recrystallization proceeds. As is discussed in §6.4.2.3, this is an important result. The flexibility of this type of simulation allows many parameters to be varied, and the effects of concurrent recovery, variations of growth rate and the effects of particle and solute pinning of boundaries have been investigated.

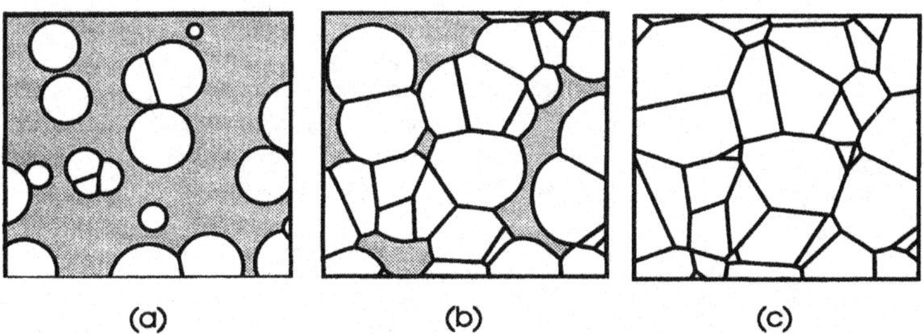

Fig. 13.11. Microstructures developing during a 3-D Avrami simulation with site-saturated nucleation.

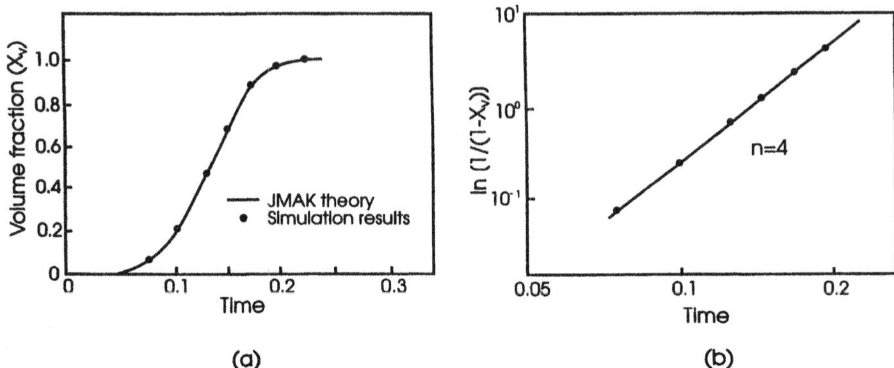

Fig. 13.12. An Avrami simulation with a constant nucleation rate, (Marthinsen et al. 1989).
a) Comparison of fraction recrystallized with JMAK analysis.
b) Log plot of data showing Avrami exponent equal to 4.

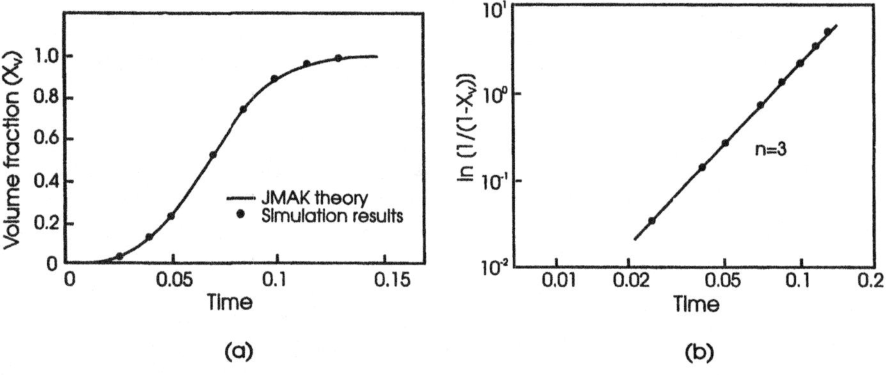

Fig. 13.13. An Avrami simulation with site saturated nucleation, (Marthinsen et al. 1989).
a) Comparison of fraction recrystallized with JMAK equation.
b) Log plot of data showing Avrami exponent equal to 3.

A sensitive test of a simulation is the grain structure which it produces, and grain size distributions obtained from these simulations have been compared with those measured experimentally. In general it has been found that the experimental distribution is significantly broader than the calculated distribution and the symmetry of the experimental and calculated distributions is not the same. Even with an inhomogeneous distribution of the nucleation sites, it has not yet been possible to obtain very good agreement between simulation and experiment (Marthinsen et al. 1989, Furu et al. 1990). In addition, comparison of the microstructures produced from simulations with those found experimentally shows some differences, possibly because the simulations do not allow the grain boundary vertex angles to move towards equilibrium by local grain growth.

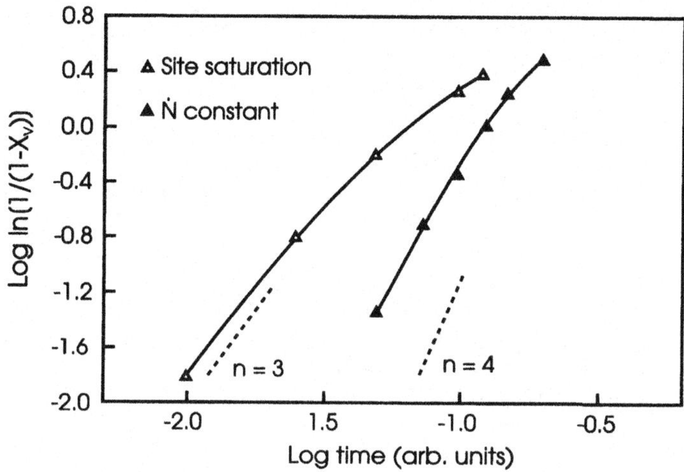

Fig. 13.14. The effect of inhomogeneous nucleation on recrystallization kinetics. The Avrami exponent decreases as recrystallization proceeds, (Furu et al. 1990).

This type of simulation is a very powerful means of dealing with inhomogeneities in recrystallization. It provides a framework within which smaller-scale micro-models dealing with recovery and the nucleation of recrystallization can readily be incorporated, and is likely to become increasingly useful.

13.3 MACRO MODELS

The development of quantitative physically-based models of thermomechanical processing is already established in the steel industry and is being developed for the aluminium industry. Although detailed consideration of such models is beyond the scope of this book, it is instructive to see how some of the basic concepts discussed earlier in this book fit into the larger perspective of industrial practice.

13.3.1 Overall model

The outline of such a model for industrial multi-pass hot rolling, due to Sellars (1992a) is shown in figure 13.15a. where five sub-models are shown within the overall model and the interaction between the sub-models is shown.

(i) Microstructure
The **structure** sub-model which is of particular relevance and which will be discussed further is seen to be a key part of the model, interacting directly with the **deformation, temperature, mechanics and behaviour** sub-models. In figure 13.15, the microstructure is represented by a single parameter **S**, although in practice several parameters incorporating grain, subgrain and dislocation structure, together with texture are actually required.

(ii) Deformation
The microstructure and deformation parameters coupled with appropriate constitutive equations determine the flow stress of the material during hot deformation.

(iii) Mechanics
The **Mechanics** sub-model takes the high temperature flow stress, and with the operating conditions of the equipment, material dimensions and required reduction, then calculates the working forces. There is an output to the **temperature** sub-model through the heat generated during working.

(iv) Temperature
Temperature is non-uniform through the material and changes continuously with time and stock geometry. The surface oxide film and lubricants will also affect heat transfer to the environment and to the rolling mill. Finite difference computing methods are used to calculate instantaneous values of temperature.

13.3.2 Structure sub-model

The **structure** sub-model takes the external variables such as strain, strain rate, temperature and time from other sub-models, together with the microstructure, and using the appropriate equations, it describes the microstructural changes, including dislocation content, grain and subgrain structure, texture and phase transformation during the process. **Micro-models**, such as the recrystallization models discussed above, are needed for these processes, although as yet, many of these models are largely empirical.

The model is used to predict the **dynamic** microstructural changes which take place during the rolling and also the **static** changes which may take place between passes and after rolling

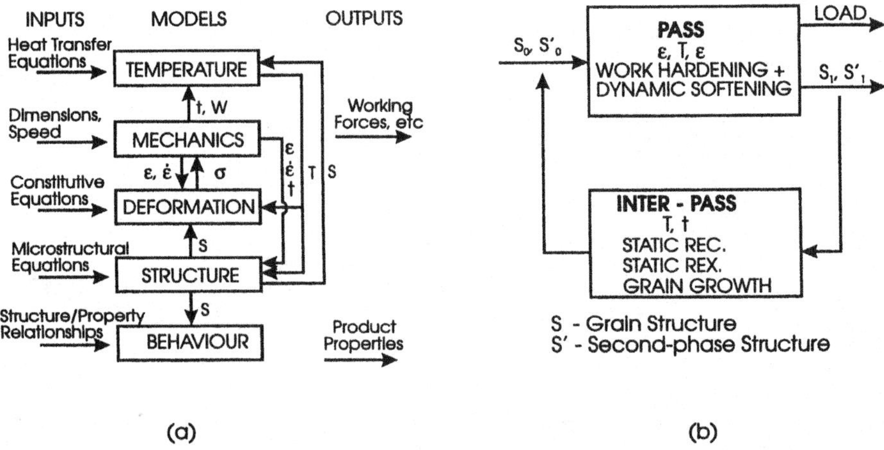

Fig. 13.15. a) Schematic diagram of a model for thermomechanical processing showing the interaction between five sub-models. b) Schematic diagram of the structure sub-model, (Sellars 1992a).

is complete, as shown in figure 13.15b. In this schematic diagram, two structural parameters S_M relating to the microstructure and texture of the matrix alloy and S_P relating to the size and distribution of the second-phase particles are shown.

The initial values of these parameters are set by the microstructure of the heated ingot and during a rolling pass they are modified by dynamic hardening and softening. In the interval between passes the microstructure is modified by static annealing, giving the parameters for entry to the second pass etc. These parameters are passed to the **deformation** sub-model, and together with the constitutive equations and external variables they determine the flow stress. After the required number of passes, the final microstructure after cooling to room temperature is computed, thus providing data from which the microstructure and properties of the finished product may be determined.

13.3.3 Application of the model

13.3.3.1 Steels

Models such as that discussed above have been successfully used by the steel industry for some time (see Jonas 1990, Sellars 1990). The predicted change in grain size during multiple rolling of C-Mn steel is shown in figure 13.16. The steel is reduced from 250mm slab in equal passes of 15% reduction, with 20 seconds between passes. The broken line, which represents the predicted recrystallization kinetics shows that after each roughing pass (**R**) the grain size is refined by complete recrystallization and when the grain size is below 100μm, some grain growth occurs between passes. Complete recrystallization also occurs during the first four finishing passes (**F**), but at the lower temperatures recrystallization is incomplete between passes. For comparison, data has been computed (solid line) using a recrystallization rate which is 5 times slower. Although recrystallization is incomplete between all finishing passes, the overall changes in grain size remain similar.

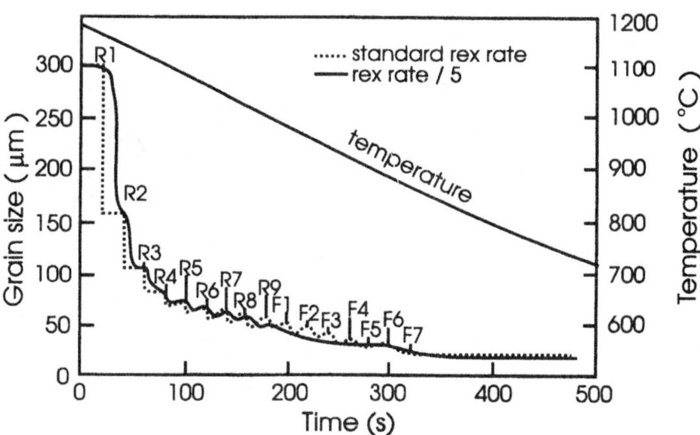

Fig 13.16. The predicted evolution of microstructure during the hot rolling of 20mm C-Mn steel plate, (Sellars and Whiteman 1979).

The relative insensitivity of the output to differences in the input microstructural equations and the good quantitative agreement between predicted and observed microstructures has made modelling an accepted tool for the industrial thermomechanical processing of steel.

13.3.3.2 Aluminium alloys

For aluminium alloys, the models are less well developed. There are several reasons for this including:

(i)   Accumulation of strain over several passes occurs before recrystallization takes place

(ii)   The microstructure and texture are very dependent on the history of the specimen, more so than in steels, where the austenite/ferrite transformation may wipe out much of the previous processing history by changing the grain size and orientation.

(iii)   The development of microstructure and texture is critically dependent on the influence of second-phase particles, whose size and distribution may alter during the processing.

An example of the output of a model for the industrial rolling schedule of 5.5mm Al-1%Mg strip is shown in figure 13.17. The temperature history of the material is shown in figure 13.17a. It is seen that the chilling effect of the rolls causes a large drop in surface temperature during each pass. However, the work of deformation causes a rise of temperature in the centre of the strip. In the interpass period the temperature gradients diminish, and are seen to become negligible as the sheet becomes thin.

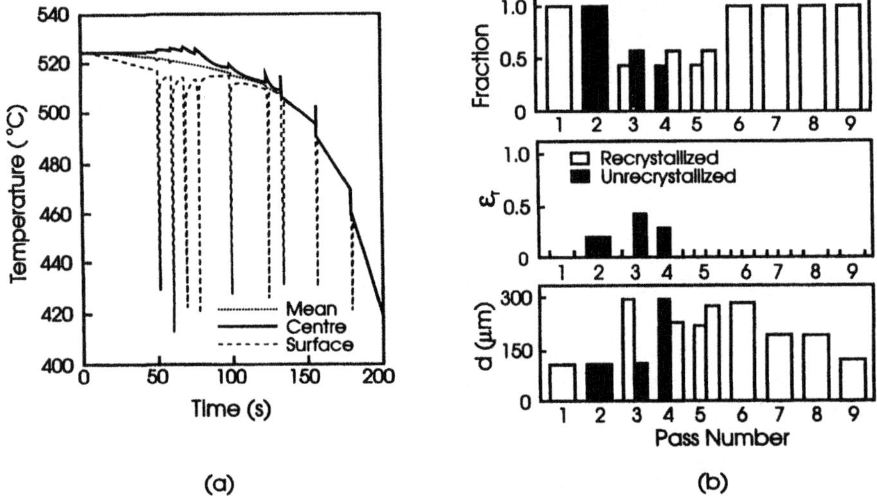

Fig. 13.17. Predictions for the hot rolling of Al-1%Mg 5.5mm strip. a) Temperature changes in centre, mean and surface. b) The predicted microstructures, assuming no dependence on initial grain size, (Sellars 1992a).

Figure 13.17b shows the predicted fraction recrystallized and the grain size. It is seen that there is no recrystallization after pass 1 and the strain accumulates with that of pass 2. After passes 3 and 4 there is partial recrystallization, and thereafter, recrystallization occurs after each pass.

APPENDIX

The crystallographic orientation or texture is an important parameter describing the microstructure of a crystalline material. Traditionally, textures have been determined by **x-ray diffraction** and represented by **pole figures**, but in recent years, new methods of texture representation and determination have also become widely used. The purpose of this appendix is to provide the non-specialist with sufficient information to understand the discussions of texture in the book.

A1. REPRESENTATION OF TEXTURES

A simple treatment of texture representation is given in this section, and further introductory accounts may be found in Hatherly and Hutchinson (1979), Cahn (1991b) and Humphreys (1993). Detailed accounts of texture representation with particular reference to orientation distribution functions (**ODF**) may be found in the works of Bunge (e.g. Bunge 1982), and Randle (1992) discusses texture representation with particular reference to microtextures.

A1.1. Pole figures

A pole figure is a stereographic projection which shows the distribution of a particular crystallographic direction in the assembly of grains that constitutes the specimen. If it is to have meaning it must also contain some reference directions that relate to the material itself. Traditionally these directions refer to the forming process, e.g. the drawing direction in wires or the rolling direction etc. in rolled sheets. Idealized pole figures for a drawn wire and a rolled sheet are given in figure A1.

In figure A1a, the drawing direction of a wire specimen is shown at the top and the distribution of $<100>$ directions indicates that the grains have these directions at 45° and 90° to the wire axis. This axis, is therefore, parallel to a $<110>$ direction and the texture is described as a $<110>$ fibre texture. Figure A1b refers to a rolled sheet. The orthogonal

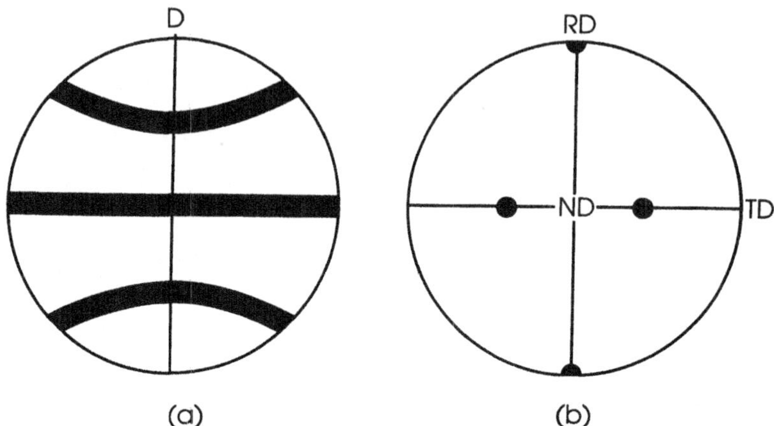

(a)   (b)

Fig. A1. Idealized 100 pole figures for: (a) drawn wire showing <110> fibre texture; (b) rolled sheet showing {110}<001> rolling texture.

specimen axes, rolling direction (RD), transverse direction (TD) and sheet normal direction (ND) are plotted with ND at the centre and RD at the top. Once again the distribution of the <100> directions is shown. There is a concentration of these directions at RD and in the plane defined by ND and TD at 45° to ND. Such a texture is described by the notation {110}<001> which states that planes of the form {110} are parallel to the surface of the sheet and directions of the form <001> are parallel to the rolling direction.

The distribution of intensity in real pole figures is much more diffuse than shown above. The intensity distribution is usually represented by contour lines with values 1 to n times **R** where **R** is the value associated with a specimen of completely random orientation. The texture of heavily rolled copper (fig 2.20a) shows the typical spread observed.

A1.2 Inverse pole figures

The inverse pole figure is particularly useful for deformation processes such as wire drawing or extrusion, which require the specification of only a single axis. The frequency with which a particular crystallographic direction coincides with the specimen axis is plotted in a single triangle of a stereographic projection (fig 2.33). Rolling textures can also be described by inverse pole figures but in this case two or sometimes three separate plots are presented, one being used for each of the principal strain axes ND, RD (and TD if required). This method is used much more frequently for bcc steels than for fcc materials.

A1.3 Orientation distribution functions and Euler space

The description of texture by pole figures is incomplete. The information provided refers only to the statistical distribution of a single direction and there is no way of using this to obtain the complete orientation of individual grains or volume elements. A better description

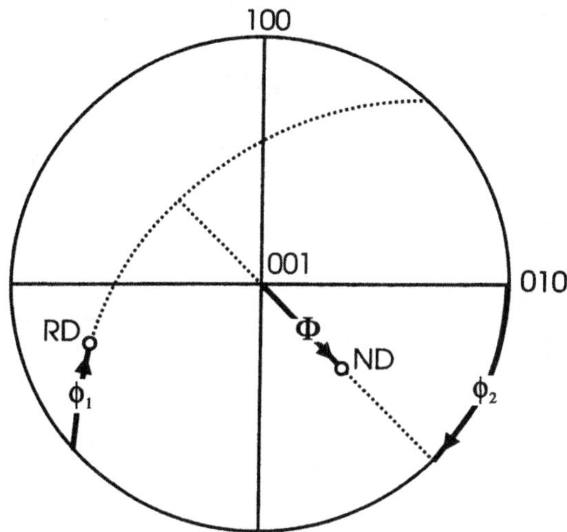

Fig. A2. Definition of Euler angles used for rolled sheet.

is given by the ODF which describes the orientation of all the discrete volumes in the aggregate. ODF analysis was developed originally for materials with cubic crystallography and orthorhombic sample symmetry, i.e. for sheet products. There have been a few studies of hexagonal metals but most of the literature and most of what follows refer to rolled materials with fcc or bcc structures. Some explanation of the formalism used to describe the ODF is necessary but no description of the mathematics involved in generating the ODF will be given. The interested reader is referred instead to the definitive work of Bunge (1982).

Consider the case of a rolled sheet in which a particular volume element has the orientation (hkl)[uvw]. The orientation of this element can be described in terms of three Euler angles. Several different notations have been used to define these angles, but that of Bunge is most common and will be used here. The crystallographic axes are represented in the normal way in a standard projection (fig A2) and the specimen orientation is specified by the reference directions ND and RD. The angles Φ and ϕ_2 completely specify the direction ND. RD lies in the plane normal to ND and the angle ϕ_1 completely specifies the direction RD. Because three variables have been used to define (hkl)[uvw], the ODF can only be displayed as a three dimensional plot with the three Euler angles as axes as shown in figure A3a. For rolled fcc materials the data are normally shown as a series of slices taken through the three dimensional ODF space at $\phi_2 = 0, 5, 10.....90°$; as shown in figure A3b.

Equations A1-A9 define the relationship between Euler angles and Miller indices for cubic materials. In order to obtain consistent results, equations A1 to A6 should be used to obtain Miller indices from Euler angles, and equations A7 to A9 to obtain Euler angles from Miller indices.

$$h = \sin(\Phi)\,\sin(\phi_2) \qquad\qquad\qquad (A1)$$

$$k = \sin(\Phi)\,\cos(\phi_2) \qquad\qquad\qquad (A2)$$

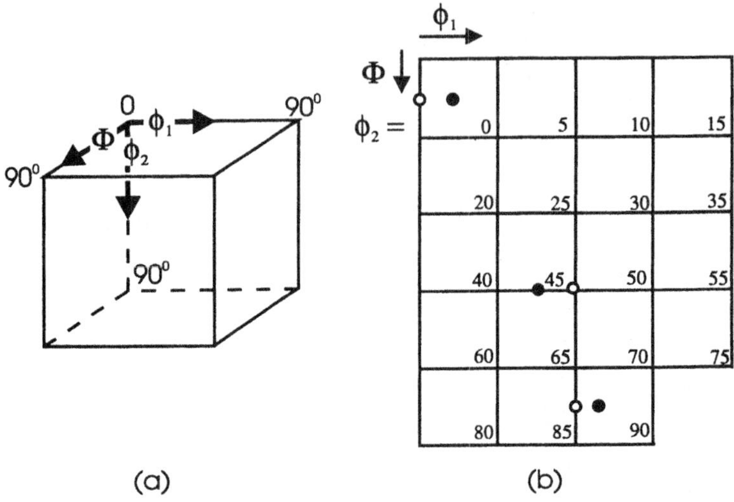

Fig. A3. a) Location of Euler angles in ODF space; b) ODF sections showing location of $\{011\}<2\bar{1}1>$ (filled circles), and $\{110\}<001>$ (open circles) orientations .

$$1 = \cos(\Phi) \tag{A3}$$

$$u = \cos(\phi_1)\cos(\phi_2) - \sin(\phi_1)\sin(\phi_2)\cos(\Phi) \tag{A4}$$

$$v = -\cos(\phi_1)\sin(\phi_2) - \sin(\phi_1)\cos(\phi_2)\cos(\Phi) \tag{A5}$$

$$w = \sin(\phi_1)\sin(\Phi) \tag{A6}$$

$$\tan(\Phi)\cos(\phi_2) = \frac{k}{l} \tag{A7}$$

$$\tan(\phi_2) = \frac{h}{k} \tag{A8}$$

$$\cos(\Phi)\tan(\phi_1) = \frac{l\,w}{k\,u - h\,v} \tag{A9}$$

Table A1 gives the Euler angles for $\{110\}<112>$ and $\{110\}<001>$, two of the orientations that are commonly used to describe the textures of fcc metals, and figure A3b indicates where these occur in ODF space. It should be noted that the general orientation $\{hkl\}<uvw>$ appears more than once in the customary 90x90x90° cube of Euler space.

The use of ODFs allows a more quantitative description of textures than is possible with pole figures. Although the interpretation of a full ODF (e.g. fig 2.22) is not immediately apparent to the non-specialist, there are, in any material only a relatively few important orientations (e.g. tables 2.7 and 2.9) and these are readily identified in the ODF sections (e.g. fig A3b).

Table A1

Euler angles for some texture components.

Component	ϕ_1	Φ	ϕ_2
$\{011\}<2\bar{1}1>$	35	45	0
	55	90	45
	35	45	90
$\{110\}<001>$	90	90	45
	0	45	0
	0	45	90

More importantly, the ODF allows the identification of texture **fibres** as shown in figure 2.23, and quantitative plots of intensity along these fibres (e.g. fig 2.24) provide very detailed information. The volume fractions of any texture components (often defined as orientations within $10°$ of the ideal) may also be readily calculated from the ODF data. Such simple yet quantitative representations of the data (e.g. table 2.8, figure 10.16) may be compared directly with theoretical predictions or may form part of the specification of an industrial product.

Despite the benefits derived from the use of ODFs and their general acceptance, there are a number of disadvantages associated with the use of Euler space (Randle 1992).

(i) Each orientation appears three times in the conventional $90°x90°x90°$ cube (table A1 and fig A3b).

(ii) The population of Euler space by a random array of orientations is very distorted. If $\Phi = 0$ the orientation is determined by $(\phi_1+\phi_2)$ and all points in the plane $\Phi = 0$ and having the same value of $(\phi_1+\phi_2)$ represent the same orientation. This is particularly confusing with respect to orientations of the form $\{001\}<hk0>$.

(iii) Significant fibres in the texture often lie on curves (fig 2.23) and may be difficult to recognise.

(iv) There is no direct association between Euler space and the specimen coordinate system.

A1.4 Rodrigues-Frank space

Some of the problems discussed above can be overcome by the use of other three-dimensional representations and of these the most suitable appears to be that advocated by Frank (1988), and based on the analysis of Rodrigues (1840). A brief description follows, but for more detailed accounts the reader is referred to the text by Randle (1992) or to Frank's paper.

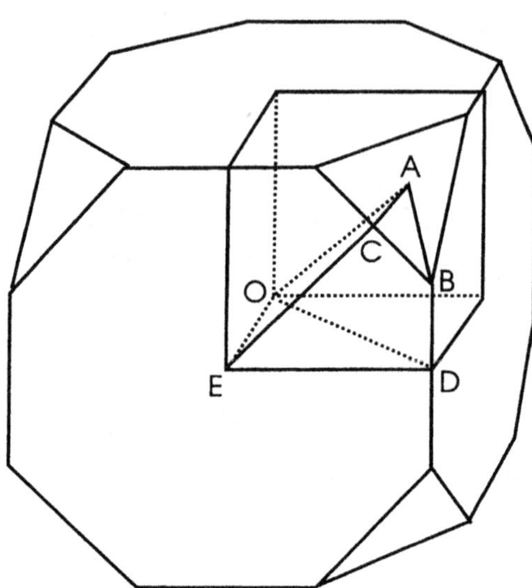

Fig. A4. The fundamental zone of Rodrigues-Frank space for holosymmetric cubic crystals.
Internal volumes define 1/8th and 1/48th of the fundamental zone, (after Randle 1992).

The concept of an angle/axis of rotation is widely used to describe the misorientation
relationship between neighbouring grains (§3.2). In order to express the absolute orientation
of a crystal, the reference crystal is taken to be the standard cube crystal orientation. If the
axis is defined by a vector L and the rotation angle by θ the required relationship is given by
the so-called **Rodrigues vector**

$$\underline{R} = L \tan(\theta/2) \qquad\qquad (A10)$$

The three orthogonal axes $\mathbf{R_1, R_2, R_3}$ define a **Rodrigues-Frank (R-F)** space in which all
possible angle/axis combinations are found. There are 24 possible R vectors and in practice
that with the smallest rotation angle is used. By definition this has the smallest R vector and
it follows that all such R vectors lie close to the origin of **R-F** space. A full set of equivalent
orientations lies in a fundamental zone of **R-F** space but in the case of high symmetry
crystals only a small part of this zone is needed. Figure A4 shows the form of the
fundamental zone for cubic crystals and details of the reduced volumes that suffice for high
symmetry. In figure A5 some common texture components in cubic crystals are plotted in
the required reduced volume; details of the parameters associated with these orientations are
given in Table A2.

The major advantages of this method of representation, which is still in its infancy have been
summarized by Randle (1992):

(i)   Each orientation appears only once in the fundamental zone.

(ii)   Rotations about a common axis fall on a straight line. This means that
         the identification of fibre components is simple.

(iii)   The volume element of **R-F** space is much more homogeneous than is the case for Euler space.

(iv)   The axes of **R-F** space coincide with those of the specimen.

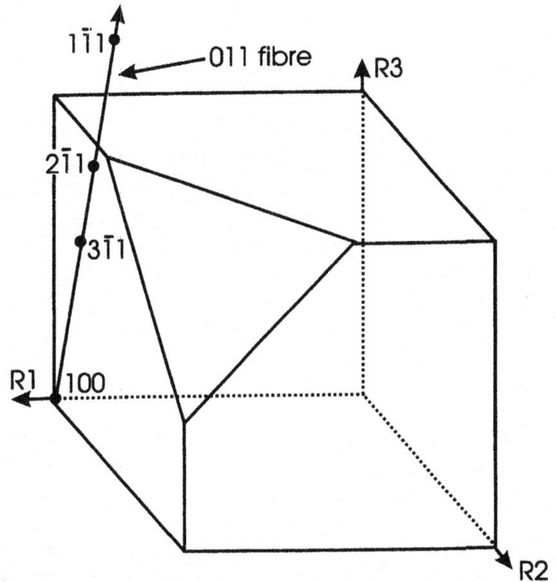

Fig. A5. Some common texture components plotted in one octant of the fundamental zone of Rodrigues-Frank space. Note that the [011] fibre texture components lie on a straight line, (Randle 1992).

Table A2

Orientation Parameters for Some Texture Components.

Orientation	Angle/Axis Rodrigues vectors $(R_1; R_2; R_3)$	Euler Angles $(\phi_1; \Phi; \phi_2)$
(011)[100]	45°/1; 0; 0	0; 45; 0
(011)[2̄11]	56.6°/0.77; 0.25; 0.59 (0.41; 0.13; 0.32)	35.3; 45; 0
(011)[3̄11]	51.3°/0.86; 0.19; 0.47 (0.41; 0.09; 0.22)	25.2; 45; 0
(011)[11̄1]	69.7°/0.59; 0.31; 0.74 (0.41; 0.21; 0.52)	54.7; 45; 0

It is important to remember that **R-F** space is three dimensional and although there has been some use of sections taken through the reduced zones, these are not easy to comprehend, and this is likely to limit the use of R-F space for the representation of bulk textures.   In addition, unlike the methods discussed above the data cannot be directly extracted from bulk x-ray data (Becker and Panchanadeeswaran 1989).   However, it does have advantages for the representation of **misorientation** data as discussed below.

A1.5 Misorientations

The increasing availability and sophistication of equipment for measuring single orientations (§A3) has led to a considerable interest in the misorientations that exist across grain boundaries and the association of these misorientations with textures. The difference between a normal texture orientation and a misorientation is simply that in the former case the external axes of the specimen provide the frame of reference, whilst in the latter the axes of one of the grains serves this purpose. The misorientation parameters can be expressed in a number of ways.

(i)   Misorientations may be expressed as Euler angles and displayed in Euler space as misorientation distribution functions (MODF).

(ii)   The axes of misorientation may be represented on inverse pole figures. In this case the angle of misorientation (θ) is conveniently plotted on an axis orthogonal to the inverse pole figure, and the data presented as sections of constant θ.

(iii)   The misorientations may be represented in R-F space. As this method has some particular advantages and is not as widely known as the first two, it is discussed in more detail below.

If the crystal axes are used to describe the axes of **R-F** space and if the direction cosines of the rotation axis are written so that $L_1 > L_2 > L_3$ then the Rodrigues vector lies within 1/48th of the fundamental zone. The properties of this space (fig A6), which is the volume OABCDE of figure A4, are very simple. The R vectors for the simple low index, crystallographic axes lie along its edges and low angle boundaries appear near the origin (O) and are easily recognised. More importantly, coincident site lattice misorientations appear on straight lines through the origin.

A2 MEASUREMENT OF MACROTEXTURES

A2.1 X-ray diffraction

The most commonly used X-ray techniques are those developed by Schulz (1949). Most measurements involve materials that have been rolled or annealed and originally two separate methods of examination, involving back reflection and transmission techniques were required to obtain a complete pole figure. Nowadays the transmission method is rarely used and useful pole figures (covering an area of up to 85° from the centre) are obtained by the back reflection technique. If complete pole figures are required these can be calculated from an ODF which has been obtained from a number of partial pole figures.

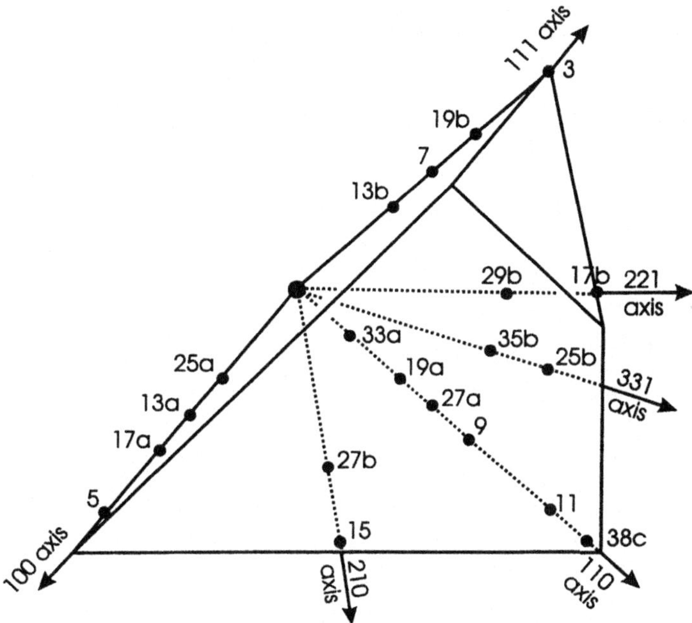

Fig. A6. Subvolume (1/48th) of the fundamental zone of Rodrigues-Frank space showing location of the [100], [110] and [111] axes and the location of some coincident site lattices, (after Randle 1992).

A typical specimen is some 25mm square with a flat surface, and the specimen must be thick enough (>0.2mm) to prevent penetration of the incident X-ray beam. The specimen is mounted in a two-circle goniometer (fig A7) which permits simultaneously, a rotation through an angle, δ, about its normal and a rotation, α, about an orthogonal axis that lies in the plane defined by that normal and the incident and diffracted beams. These beams which are restricted by a series of slits are set at the appropriate Bragg angles for diffraction from the required plane. The intensity of the diffracted beam is measured by normal counting methods and normalised to that obtained from a randomly oriented standard specimen. Because of the absorption effects that develop as α approaches 90° this technique provides a partial pole figure that extends only some 70-85° from the centre.

Nowadays the level of instrumentation is such that the gathering of data and the preparation of the final pole figure are controlled by computers. For a full account of a modern, computer controlled goniometer see Hirsch et al. (1984).

Although most texture determinations are made with X-ray equipment and the Schulz back reflection method there are some severe limitations involved. The most important of these is the small volume of material actually examined. The depth to which the incident beam penetrates (and from which the diffracted beams emerge) is governed principally by the wavelength of the X-rays used and the absorption coefficient of the specimen material and is rarely greater than 0.1mm. It has been pointed out many times in this book that the most prominent feature of a deformed metal is the heterogeneity of the microstructure and it will

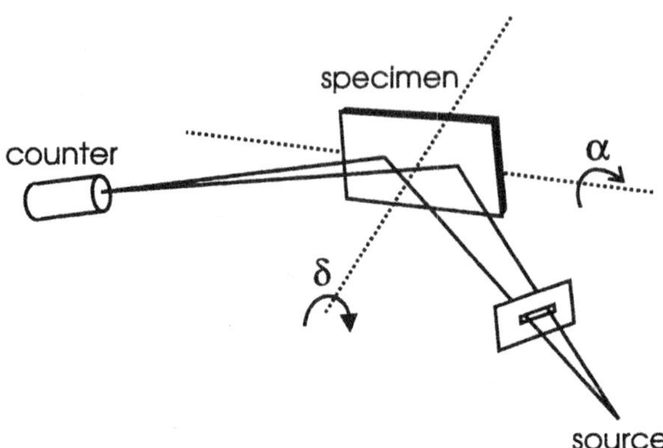

Fig. A7. The reflection method for pole figure determination.

be clear that there must always be some doubt as to whether or not an X-ray based texture result is truly representative of a rolled specimen. In many rolled products the texture varies through the thickness of the sheet, and in most cases texture studies are made on mid-plane sections. Because recrystallization does not necessarily occur homogeneously in such a material similar doubts must also apply to annealed specimens.

If an ODF is required, then 3-4 separate pole figures are measured from a sample. These need not be complete and modern practice uses only the Schulz back reflection method. The time to acquire the data depends on the material, the resolution required, and the strength of the texture, but typically a single partial pole figure is collected in ~ 1hr. The separate pole figures are determined sequentially and in some cases the goniometer head is capable of holding several specimens so that continuous overnight data collection is possible. The raw data are used to derive an orientation distribution function, **f**, for a particular orientation, usually by a series expansion method, and the combination of all possible **f** values gives the ODF.

There are still a number of problems arising in the calculation of ODFs from pole figures. One of these is the so-called "ghost" problem which affected all of the early ODFs and manifests itself by the generation of components that are known not to be present in the texture (Matthies 1979, Lücke et al. 1981). In addition the peak intensities may be reduced by as much as 10-30%. This difficulty has its origin in the use of the series expansion technique to calculate the ODF from the experimental pole figures. Such expansions have both odd and even terms but assumptions made about symmetry led initially to the use of only the even coefficients and the appearance of the non-existent ghost peaks in the ODF. Ghosts in bcc textures are less significant than in fcc and the nature of the texture components is such that the ghosts appear only in high intensity regions. Because of this they cannot usually be recognised. Exceptions occur only for the $\{112\}<110>$ and $\{001\}<110>$ components. These problems do not exist if the ODFs are constructed from individual orientation measurements (§A3.4)

The very high intensities of X-rays emitted by synchrotrons enable rapid data collection (Szpunar and Davies 1984), and it is possible to investigate changes taking place during the recrystallization of deformed materials. However this technique has not yet been widely exploited.

A2.3 Neutron diffraction

The availability of thermal neutrons with a wavelength of ~1Å provides an opportunity for the use of neutron diffraction in texture studies. Because the absorption of neutrons in most metals is low it is then possible to use large specimens. A steel specimen of thickness 10mm will absorb about 20% of a typical neutron beam whereas a 0.1mm specimen will absorb >90% of a similar X-ray beam. Neutron beam techniques are therefore useful for the examination of coarse grained materials and for gathering information from the full thickness of an inhomogeneous material. In some cases it is possible to examine directly the changes occurring during annealing (see, for example, Juul Jensen et al. 1984).

A.3. MEASUREMENT OF MICROTEXTURES

There are many occasions when it is desirable to obtain data about the local array of orientations present in particular parts of the specimen, and the textures within small specified volumes are generally referred to as microtextures. Although microtexture data are often more difficult to collect than macrotexture data, by linking the orientation and spatial parameters this approach provides a more complete description of the specimen, and can supply information which is capable of resolving many of the uncertainties highlighted in this book. Most of the techniques used for obtaining microtextures have been developed in the past 10 years, and it is therefore a relatively new and developing area.

There is a range of methods available, each with its own particular application, and further details may be found in the reviews of Humphreys (1988b), Randle (1992) and Schwarzer (1993). Table A3 summarises many of the techniques, and tabulates their spatial and angular resolutions.

A3.1 Optical methods

The optical techniques used by geologists and mineralogists and applied to transparent non-cubic minerals are described in standard mineralogical texts. Using plane polarised light in a transmission optical microscope equipped with a universal stage (goniometer), the specimen is manipulated until extinction of a grain or subgrain is achieved, and hence the crystallographic direction parallel to the optic axis of the microscope is determined.

Table A3
Electron beam methods for microtexture determination

Pattern	Instrument	Highest spatial resolution	Angular resolution (°)
Spot	TEM (SAD)	$0.5\text{-}1.5\mu m$	2
	TEM (microbeam)	10nm	2
	FEG-STEM	1nm	5
Kikuchi	TEM (SAD)	$0.5\text{-}1.5\mu m$	0.2
	TEM (microbeam)	10nm	0.2
	FEG-STEM	2nm	0.2
SACP	SEM	$10\mu m$	0.5
	TEM/STEM	$1\text{-}2\mu m$	0.5
EBSP	SEM	$0.5\text{-}1\mu m$	0.5

Surface films such as anodic films, whose thickness or surface topography are dependent on the orientation of the underlying crystalline material may sometimes be used to obtain information about orientations. In the case of cubic metals, such as aluminium, the high symmetry of the material makes it difficult to obtain unambiguous data from an anodised specimen (Saetre et al. 1986b). However for hexagonal metals such as magnesium or titanium, the orientation of the basal plane may be obtained with the use of a polarising reflection microscope (Couling and Pearsall 1957).

A3.2 Deep etching (Köhlhoff) technique.

This etching technique (Köhlhoff et al. 1988b) has been used with considerable success by Duggan, Köhlhoff and their collaborators (e.g. Duggan et al. 1993) in recent years to study copper and copper alloys. The heavily etched surface is examined at a magnification in the range 500-2000X in an optical or scanning electron microscope, and a typical example is seen in figure 10.23. The basis of the technique lies in the fact that the {111} planes are attacked more slowly than others so that a relief structure of tilted {111} planes is developed. The lines of the internal structure within the grains define the intersections of the {111} planes and are therefore <110> directions. The various etched patterns are characteristic of the crystallographic orientation and may be readily identified to an accuracy of ~5° and with a spatial resolution of ~10μm.

A3.3 Single orientations by transmission electron microscopy

Techniques for the determination of the orientation of small regions of a specimen in the transmission electron microscope from **spot** or **Kikuchi line** diffraction patterns have been well established for many years and are fully discussed in textbooks on electron microscopy. However, their use for the determination of local textures, which requires the acquisition and solution of many diffraction patterns has increased with the availability of computer based on-line techniques for the rapid solution of the patterns (e.g. Schwarzer and Weiland 1984).

Transmission electron diffraction is particularly suited to applications requiring high spatial resolution, as electron microscopes can produce beams of less than 10nm diameter. Thus the technique is suitable for the examination of deformed materials in which the cell or subgrain size is of the order of 0.5μm. The crystallographic orientation may be determined from the diffraction spots, but because of relaxation of the Bragg diffracting conditions in thin specimens, the angular resolution is usually in the range 2-5° (Duggan and Jones 1977). The orientation may be determined to a much higher accuracy (<0.1°) using the Kikuchi line patterns in a convergent beam diffraction pattern.

Such a pattern may be measured in the microscope relatively simply. There are two methods for doing this. The pattern may be acquired on a CCTV camera attached to the microscope, digitised and then stored. A computer generated cursor is then superimposed on the pattern and used to measure 9 points which define the width and direction of three intersecting Kikuchi bands. Alternatively, the same information is obtained by using a computer to deflect the pattern in the microscope until the required point is coincident with a reference mark on the microscope screen. The data are normally analyzed according to the method of Heimendahl et al. (1964).

A3.4 Single orientations by scanning electron microscopy

Diffraction patterns may be obtained using backscattered electrons in the SEM. The specimen surface must be carefully prepared (e.g. electropolished), and satisfactory results will only be obtained if the pattern comes from a coherently diffracting volume, i.e. the area from which the pattern is obtained must have a single orientation.

Earlier methods (selected area channelling patterns, **SACP**) employed a rocking beam and a standard electron detector. However, because of their poor angular resolution and small angular pattern range, this technique has now been superseded by the use of electron back scatter patterns and will not be discussed further.

Electron backscatter patterns (**EBSP**), are being increasingly used to determine microtextures (the technique is also termed Electron Backscatter Diffraction **EBSD** or Backscattered Kikuchi Diffraction **BKD**). In this method, a specimen is tilted, typically by 70° in an SEM, and a stationary beam produces a diffraction pattern, again similar to a TEM Kikuchi pattern, on a phosphor-coated glass screen as shown in figure A8 (Dingley 1984, Dingley and Randle 1992). The pattern is digitised by a very sensitive CCTV camera and in most cases is analyzed semi-automatically by selecting either three low-index poles or three intersecting bands in the pattern, with a computer cursor. The data are compared with look-up tables to identify the poles, and may be analyzed using the method of Heimendahl et al. (1964), the acquisition and analysis procedure typically taking ~ 1 minute per pattern. The spatial resolution of the technique is $\sim 1\mu m$, making it unsuitable for highly deformed material, but ideal for recovered, recrystallized or hot-worked samples.

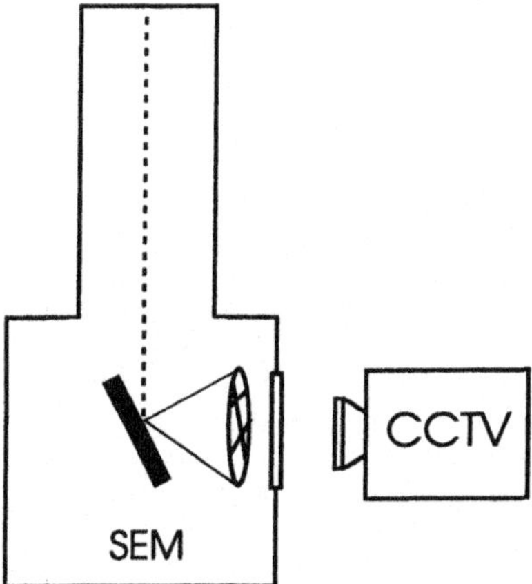

Fig. A8. Electron backscatter patterns in the SEM.

Methods are now available (e.g. Kunze et al. 1993, Juul Jensen 1993) for fully automated analysis of the patterns, and with computer-control of either the specimen stage or the beam, it is possible to scan a large area of a specimen and acquire and analyze some 10,000 points in a period of a few hours. These data may then be used to construct orientation maps on which absolute or relative orientations may be displayed. The data may also be used to construct ODFs, and this has the advantage over the x-ray diffraction method (§A2.1) that the problem of "ghosts" is eliminated.

It is often required to obtain a backscattered electron image of the area being analyzed, and this may be difficult because of the large specimen tilt, particularly if the specimen surface is not smooth. In order to overcome this problem, a special side-mounted electron detector, placed close to the transmission phosphor, is normally used to collect the electrons for a backscattered image.

A3.5 Pole figures by transmission electron microscopy

A microtexture method which does not require the measurement of individual diffraction patterns has been developed by Humphreys (1983) and Weiland and Schwarzer (1984). Using this method, which is implemented on a TEM, a pole figure is obtained from a selected area, typically 5-10μm in diameter, of a thin specimen. A eucentrically-mounted specimen is tilted through an angle of $\pm 50°$ in steps of $\sim 2°$, and at each tilt the intensity around a low index Debye ring is measured by scanning the beam over a transmission electron detector. The technique is fully automated and has a similar geometry to the transmission x-ray method.

Using this method, partial pole figures are obtained from the selected areas in a few minutes. The technique is ideally suited to the study of highly deformed materials in which the number of cells or subgrains is too large for the acquisition of individual TEM patterns to be feasible. It has been used to study the orientations near second-phase particles and the texture of shear bands.

REFERENCES

Aaronson, H.I., Laird, C. and Kinsman, K.R. (1962), in **Phase Transformations**, ASM, Metals Park, Ohio, 313.

Abbruzzese, G. and Brozzo, P. (1992), (eds) **Grain Growth in Polycrystalline Materials**. Trans Tech Publns. Switzerland.

Abbruzzese, G. and Lücke, K. (1986), Acta Metall. 34, 905.

Abbruzzese, G., and Lücke, K. (1992), in **Grain Growth in Polycrystalline Materials**. ed. Abbruzzese and Brozzo. Trans Tech Publns. 597.

Abbruzzese, G., Heckelmann, I. and Lücke, K. (1992), Acta Metall. 40, 519.

Abbruzzese, G., Lücke, K. and Eichelkraut, H. (1988), in **ICOTOM 8**, eds. Kallend and Gottstein, TMS, Warrendale, 693.

Abe, H. and Satoh, S. (1990), Kawasaki Steel Tech. Rep. 22, 48.

Abe, H., Suzuki, T. and Okada, S. (1983), Tetsu-to-Hagané, 69, S1415.

Abe, H., Nagashima, S, Hayami, S. and Nakaoka, K. (1978), in **ICOTOM 5**, eds. Gottstein and Lücke, Springer-Verlag, 2, 21.

Abe, H., Suzuki, T. and Takagi, K. (1980), Trans. ISIJ, 20, 100.

Aboav, D.A. and Langdon, T.G. (1969), Metallography, 2, 171.

Abrahamson, E.P. and Blakeny, B.S. (1960), Trans. Metall. Soc. A.I.M.E. 218, 1101.

Adam, C.M. and Wolfenden, A. (1978), Acta Metall. 26, 1307.

Adcock, F. (1922), J. Inst.Met. 27 73.

450 References

Ahlborn, H., Hornbogen, E. and Köster, U. (1969), J.Mats. Sci. 4, 944.

Akef, A. and Driver, J.H. (1991), Mat. Sci. and Eng. A132, 245.

Alexander, B. and Balluffi, R.W. (1957), Acta Metall. 5, 666.

Altherthum, H. (1922). Z. Metallk. 14, 417.

Amelinckx, S. and Strumane, R. (1960), Acta Metall. 8, 312.

Anand, L. and Gurland, J. (1975), Metall. Trans. 6A, 928.

Anand, L. and Gurland, J. (1976), Acta Metall. 24, 901.

Anderson, M.P. (1986), Proc. 7th Int. Risø Symposium. Risø, Denmark. ed. Hansen, 15.

Anderson, M.P., Grest, G.S. and Srolovitz, D.J. (1985), Scripta Metall. 19, 225.

Anderson, M.P., Grest, G.S. and Srolovitz, D.J. (1989a), Phil. Mag. B59, 293.

Anderson, M.P., Grest, G.S., Doherty, R.D., Li, K. and Srolovitz, D.J. (1989b), Scripta Metall. 23, 753.

Anderson, M.P., Srolovitz, D.J., Grest, G.A. and Sahni, P.S. (1984), Acta Metall. 32, 783.

Anderson, W.A. and Mehl, R.F. (1945), Trans. Metall. Soc. A.I.M.E. 161, 140.

Ando, H., Sugita, J. Onaka, S. and Miura, S. (1990), J. Mats Sci 9, 314.

Arai, K.J. and Yamashiro, Y. (1989), in **MRS Int. Mtg. on Advanced Materials**, 11, 187.

Ardakani, M.G. and Humphreys, F.J. (1992), in **Recrystallization'92.** eds. Fuentes and Gil Sevillano. Trans Tech Publications. 213.

Ardakani, M.G. and Humphreys, F.J. (1994), Acta Metall. 42, 763.

Ardell, A.J. (1972), Acta Metall. 20, 601.

Ardell, A.J. (1985), Metall. Trans. 16A, 2131.

Aretz, W., Ponger, D. and Gottstein, G. (1992), Scripta Metall. 27, 1593.

Argon, A.S. and Moffatt, W.C. (1981), Acta Metall. 29, 293.

Argon, A.S., Im, J. and Safoglu, R. (1975), Metall. Trans. 6A, 825.

Ashby, M.F. (1966), Phil. Mag. 14, 1157.

Ashby, M.F. (1970), Phil. Mag. 21, 399.

Ashby, M.F. (1980), Proc. 1st Int. Risø Symp. Risø, Denmark. ed. N. Hansen, 325.

Ashby, M.F. and Centamore, R.M.A. (1968), Acta Metall. 16, 1081.

Ashby, M.F. and Palmer, I.G. (1967), Acta Metall. 15, 420.

Ashby, M.F., Harper, J. and Lewis, J. (1969), Trans. Metall. Soc. A.I.M.E. 245, 413.

Assmus, F., Boll, R., Ganz, D. and Pfeifer, F. (1957), Z. Metallk. 48, 341.

Atkinson, H.V. (1988) Acta Metall. 36, 469.

Atwater, H.A., Thompson, C.V. and Smith, H.I. (1988), J. Appl. Phys. 64, 2337.

Aust, K.T. (1969), Can. Met. Quart. 8, 173.

Aust, K.T. and Rutter, J.W. (1959a), Trans. Metall. Soc. A.I.M.E. 215, 119.

Aust, K.T. and Rutter, J.W. (1959b), Trans. Metall. Soc. A.I.M.E. 215, 820.

Aust, K.T. and Rutter, J.W. (1963), in **Recovery and Recrystallization of metals**. ed. Himmel. Interscience, 131.

Aust, K.T., Ferran, G. and Cizeron, G. (1963), C.R. Acad. Sci. Paris, 257, 3595.

Avrami, M. (1939), J. Chem. Phys. 7, 1103.

Babcock, S.E. and Balluffi, R.W. (1989a), Acta Metall. 37, 2357.

Babcock, S.E. and Balluffi, R.W. (1989b), Acta Metall. 37, 2367.

Bacroix, B. and Jonas J.J. (1988), Text. Microstruct. 8/9, 267.

Bahk, S. and Ashby, M.F. (1975), Scr. Met. 9, 129.

Bailey, J.E. and Hirsch, P.B. (1960), Phil. Mag. 5, 485

Bailey, J.E. and Hirsch, P.B. (1962), Proc. R. Soc. Lond. A267, 11.

Bainbridge, D.W., Li, C.H. and Edwards, E.H. (1954), Acta Metall. 2, 322.

Baker, I. (1991), in **Structure and Property Relationships for Interfaces**, eds. Walter et al, ASM, 67.

Baker, I. and Gaydosh, D.J. (1987), Metallography, 20, 347.

Baker, I. and Martin, J.W. (1980), J. Mater. Sci. 15, 1533.

Baker, I. and Martin, J.W. (1983a), Metal Sci. 17, 459.

Baker, I. and Martin, J.W. (1983b), Metal Sci. 17, 469.

Baker, I. and Munroe, P.R. (1990), in **High Temperature Aluminides and Intermetallics,** eds. Whang et al, TMMMS, 425.

Baker, I., Viens, D.V. and Schulson, E.M. (1984), J. Mats. Sci. 19, 1799.

Ball, J. and Gottstein, G. (1993a), Intermetallics, 1, 171.

Ball, J. and Gottstein, G. (1993b), Intermetallics, 1, 191.

Ball, J., Mitteau, J. and Gottstein G. (1992), in **Ordering and Disordering in Alloys**, ed. Yavari, 138.

Balluffi, R.W. (1980), (ed.), **Grain Boundary Structure and Kinetics**. ASM Ohio.

Balluffi, R.W. Metall. Trans. A. (1982), 13, 2069.

Balluffi, R.W., Koehler, J.S. and Simmons, R.O. (1963), in **Recovery and Recrystallization of metals**. ed. Himmel. Interscience. 1.

Barioz, C., Brechet, Y., Legresy, J.M., Cheynet, M.C., Courbon, J., Guyot, P. and Ratnaud, G.M. (1992), Proc. 3rd Int. Conf on Aluminium, Trondheim. 347.

Barrett, C.R., Nix, W.D. and Sherby, O.D. (1966), Trans. ASM. 59, 3.

Barrett, C.S. (1939), Trans. Metall. Soc. A.I.M.E. 135, 296.

Barrett, C.S. (1940), Trans. Metall. Soc. A.I.M.E. 137, 128.

Barrett, C.S. and Massalski, T. (1980), **Structure of Metals,** 3rd ed., Pergamon Press, Oxford.

Barto, R.L. and Ebert, L.J. (1971), Metall. Trans. 2, 1643.

Bate, P. (1989), Technical Report, CRC/TR/89/4C, Comalco Research Centre.

Bauer, C.L. (1974), Canad. Met. Q. 13, 303.

Bauer, C.L. and Lanxner, M. (1986), in **Grain boundary structure and related phenomena. Proc. JIMIS-4.** Suppl. to Trans. Jap. Inst. Met. 27, 411.

Bay, B. and Hansen, N. (1979), Metall. Trans. A10, 279.

Bay, B., Hansen, N., Hughes, D.A. and Kuhlmann-Wilsdorf, D. (1992), Acta Metall. 40, 205

Beck, P.A., (1953), Acta Metall. 1, 230.

Beck, P.A. (1954), Adv. Phys. 3, 245.

Beck, P.A. (1963). In **Sorby Centennial Symposium on the History of Metallurgy**. ed. Smith. Met. Soc. Conf. no. 27. Gordon and Breach, New York. 313.

Beck, P.A. and Hu, H. (1952), Trans. Metall. Soc. A.I.M.E. **194**, 83.

Beck, P.A. and Hu, H. (1966), in **Recrystallization, Grain Growth and Textures**, ed. Margolin, ASM, 393.

Beck, P.A. and Sperry, P.R. (1949), Trans. Metall. Soc. A.I.M.E. **180**, 240.

Beck, P.A. and Sperry, P.R. (1950), J. Appl. Phys. **21**, 150.

Beck, P.A., Holdsworth, M.L. and Sperry, P.R. (1949), Trans. Metall. Soc. A.I.M.E. **180**, 163.

Beck, P.A. Ricketts, B.G. and Kelly, A. (1959), Trans. Metall. Soc. A.I.M.E. **215**, 949.

Beck, P.A., Sperry, P.R. and Hu, H. (1950), J. Appl. Phys. **21**, 420.

Becker, R. and Panchanadeeswaran, S. (1989), Text. Microstruct. **10**, 167.

Bellier, S.P. and Doherty, R.D. (1977), Acta Metall. **25**, 521.

Benford, J.G. and Stanley, E.B. (1969), J. App. Phys. **40**, 1583.

Benjamin, J.S. (1970), Metall. Trans. **1**, 2943.

Berger, A., Wilbrandt, P.J. and Haasen, P. (1983), Acta Metall. **31**, 1433.

Berger, A., Wilbrandt, P.J., Ernst, F., Klement, U. and Haasen, P. (1988), Prog. Mater. Sci. **32**, 1.

Bever, M.B. (1957), in **Creep and Recovery**, ASM, Cleveland. 14.

Bever, M.B., Holt, D.L. and Titchener, A.L. (1973), Prog. Mater. Sci. **17**. 1.

Beynon, J.H. and Sellars, C.M. (1992), ISIJ **32**, 359.

Bhatia, M.L. and Cahn, R.W. (1978), Proc. R. Soc. Lond. A. **362**, 341.

Bishop, G.H., Harrison, R.J., Kwok, T. and Yip, S. (1980), in **Grain boundary Structure and Kinetics**. ed. Balluffi. ASM Metals Park, Ohio. 373.

Blade, J.C. and Morris, P.L. (1975), Proc. 4th Int. Conf. on Textures. Cambridge. 171.

Blicharski, M., Nourbaksh, S. and Nutting, J. (1979), Met.Sci. **13**, 516

Bockstein, B.S., Kopetsky, C.V. and Shvindlerman, L.S. (1986), Metallurgia. Moscow, 224.

Bolingbroke, R.K., Creed, E., Marshall, G.J. and Ricks, R.A. (1993), In **Aluminium alloys for packaging.** eds. Morris et al. TMS, Warrendale, USA, 215.

Bolingbroke, R.K., Marshall, G.J. and Ricks, R.A. (1994). Materials Science Forum, <u>157-162</u>, 1145.

Bolingbroke, R.K., Marshall, G.J. and Ricks, R.A. (1995). Proc. 16th Int. Risø Symp. eds Hansen et al. 281.

Bolling, G.F. and Winegard, W.C. (1958), Acta Metall. <u>6</u>, 283.

Bölling, F., Günther, K., Böttcher, A. and Hammer, B. (1992), Steel, <u>63</u>, 405.

Bollmann, W. (1970), in **Crystal Defects and Crystalline Interfaces.** Springer-Verlag, Berlin.

Böttcher, A., Gerber, T. and Lücke, K. (1992), Mats. Sci. Tech. <u>8</u>, 16.

Bourelier, F. and Le Hericy, J. (1963), in **Ecrouissage, Restauration, Recristallisation,** Presses Univ. de France, Paris, 33.

Bowen, A.W. (1990). Mats. Sci. and Tech. <u>6</u>, 1058

Bowen, A.W. and Humphreys, F.J. (1991). Proc. **ICOTOM** 9. ed. Bunge. Avignon. 715.

Bowen, A.W., Ardakani, M. and Humphreys, F.J. (1991). Proc. 12th Risø Int. Symp. eds. Hansen et al. Risø, Denmark. 241.

Bowen, A.W., Ardakani, M. and Humphreys, F.J. (1993), in Proc. **ICOTOM 10,** ed. Bunge, Clausthal, Trans Tech pubs. 919.

Bowles, J.S. and Boas, W. (1948), J. Inst. Met. <u>74</u>, 501.

Bragg, L. and Nye, J.F. (1947). Proc. R. Soc. Lond. <u>A190</u>, 474.

Brandon, D,G., Ralph, B., Ranganathan, S. and Wald, M.S. (1964), Acta Metall. <u>12</u>, 813.

Brimhall, J.L., Klein, M.J. and Huggins, R.A. (1966), Acta Metall. <u>14</u>, 459.

Brook, R.J. (1976), in **Ceramic Fabrication Processes.** ed. Wang. Academic Press, New York, 331.

Brown, A.F. (1952), Advan. Phys. <u>1</u>, 427.

Brown, K. (1972), J.Inst.Met. <u>100</u>, 341.

Brown, K. and Hatherly, M. (1970), J. Inst. Met. <u>98</u>, 317.

Brown, L.M. (1985), Proc. 5th Int. Conf. on Strength of Metals and Alloys. <u>3</u>, 1551.

Buckley, R.A. (1979), Met. Sci. 13, 67.

Bunge, H. (1982), **Texture Analysis in Materials Science,** Butterworth, London.

Bunge, H.-J., and Dahlem-Klein, E. (1988), in **ICOTOM 8,** eds. Kallend and Gottstein, TMS, Warrendale, 705.

Burgers, J.M. (1940), Proc. Phys. Soc. (London), 52, 23.

Burgers, W.G. (1941). **Rekristallisation, Verformter Zustand und Erholung.** Leipzig.

Burgers, W.G. and Louwerse, P.C. (1931), Z. Phys. 67, 605.

Burgers, W.G. and Snoek, J.L. (1935), Z. Metallk. 27, 158.

Burke, J.E. (1949), Trans. Metall. Soc. A.I.M.E. 180, 73.

Burke, J.E. (1950), Trans. Metall. Soc. A.I.M.E. 188, 1324.

Burke, J.E. and Turnbull, D. (1952), Prog. Metal Phys. 3, 220.

Byrne, J.G. (1965), **Recovery, Recrystallization and Grain Growth.** McMillan, New York.

Cahn, J.W. (1956), Acta Metall. 4, 449.

Cahn, J.W. (1962), Acta Metall. 10, 789.

Cahn, J.W. and Hagel, W. (1960), in **Decomposition of Austenite by Diffusional Processes,** ed. Zackey and Aaronson, Interscience. Publ. New York. p131.

Cahn, R.W. (1949), J.Inst. Metals, 76, 121.

Cahn, R.W. (1983), in **Physical Metallurgy.** ed. Cahn and Haasen. Elsevier Science Publishers, 3rd edition. p1595.

Cahn, R.W. (1990), in **High Temperature Aluminides and Intermetallics,** eds. Whang et al, TMS, 245.

Cahn, R.W. (1991a), in **Intermetallic Compounds,** ed. Izumi, Jap. Inst. Met. 771.

Cahn, R.W (1991b), in **Processing of Metals and Alloys,** ed. Cahn, VCH, Heinheim, 429.

Cahn, R.W. and Westmacott, K.H. (1990), Cited in Cahn, (1991a),

Cahn, R.W., Takeyama, M., Horton, J.A. and Liu, C.T. (1991), J. Mat. Res. 6, 57.

Carmichael, C. Malin, A.S. and Hatherly, M. (1982), Proc. 6th Int. Conf. on Strength of Metals and Alloys. ed. Gifkins. Melbourne. p.381.

Carpenter, H.C.H. and Elam, C.F. (1920). J. Inst. Met. 24.2, 83.

Carrington, W., Hale, K.F. and McLean, D (1960), Proc. R. Soc. Lond. 259A , 303.

Castro-Fernandez, F.R. and Sellars, C.M. (1988), Mats. Sci. and Tech. 4, 621.

Castro-Fernandes, F.R., Sellars, C.M. and Whiteman, J.A. (1990), Mats. Sci. and Tech. 6, 453.

Ceppi, E.A. and Nasello, O.B. (1984), Scr. Met. 18, 1221.

Chadwick, G.A. and Smith, D.A. (1976), **Grain boundary structure and properties.** Academic Press, N. York.

Chan, H.M and Humphreys, F.J. (1984a), Proc El. Mic. Soc. America. 476.

Chan, H.M. and Humphreys, F.J. (1984b), Acta Metall. 32, 235.

Chan, H.M. and Humphreys, F.J. (1984c), Metal Science, 18, 527.

Charpy, G. (1910). Rev. Met. 7.1, 655.

Chin, G.Y. (1969), in **Textures in Research and Practice**, eds. Grewen and Wassermann, Berlin, 236.

Chin, L.I.J. and Grant, N.J. (1967), Powder Metall. 10, 344.

Christian, J.W. (1965), **The Theory of Transformations in Metals and Alloys**. Pergamon, Oxford.

Chung, C.Y., Duggan, B.J., Bingley, M.S. and Hutchinson, W.B. (1988), Proc. 8th Int. Conf. on Strength of Metals and Alloys, 1 319.

Clareborough, L.M. (1950), Aust. J. Sci. Res. 3A, 72.

Clareborough, L.M., Hargreaves, M.E. and West, G.W. (1955), Proc. R. Soc. Lond. A232, 252.

Clareborough, L.M., Hargreaves, M.E. and West, G.W. (1956), Phil. Mag. 1, 528.

Clareborough, L.M., Hargreaves, M.E. and Loretto, M.H. (1963), in **Recovery and Recrystallization of metals**. ed. Himmel. Interscience. 43.

Cline, R.S. and Hu, H. (1978), unpublished research; quoted by Hu (1978),

Coble, R.L. and Burke, J.E. (1963), in **Progress in Ceramic Science**. Pergamon Press.

Cooke, B.A. and Ralph, B. (1980), Proc. 1st Int. Risø Symp. Risø, Denmark, eds Hansen et al. 211.

Cooke, B.A., Jones, A.R. and Ralph, B. (1979), Met. Sci. 13. 179.

Cook, M. and Richards, T.L. (1940), J. Inst. Met. 66, 1.

Cook, M. and Richards, T.L. (1946), J. Inst. Met. 73, 1.

Cotterill, P. and Mould, P.R. (1976), **Recrystallization and Grain Growth in Metals**. Surrey Univ. Press. London.

Cottrell, A.H. (1953), **Dislocations and Plastic Flow in Crystals**. OUP, Oxford.

Cottrell, A.H. and Aytekin, V. (1950), J. Inst. Metals. 77, 389.

Couling, S.L. and Pearsall, G.W. (1957), Trans. Metall. Soc. A.I.M.E. 209, 939.

Dadras, M.M. and Morris, D.G. (1993), Scr. Met. et Mat. 28, 1245.

Dahl, O. and Pawlek, F. (1936), Z. Metallk. 28, 320.

Davies, R.G. and Stoloff, N.S. (1966), Trans. Metall. Soc. A.I.M.E. 236, 1905.

Demianczuc, D.W. and Aust, K.T. (1975), Acta Metall. 23, 1149.

Derby, B. (1991), Acta Metall. 39, 955.

Derby, B. and Ashby, M.F. (1987), Scripta Metall. 21, 879.

Detert, K. (1959), Acta Metall. 7, 589.

Detert, K. (1978.), in **Recrystallization of Metallic Materials**, ed. F. Haessner, Dr. Riederer Verlag GmbH, Stuttgart, 97.

Dillamore, I.L. (1978a), in Proc. ICOTOM 5, eds. Gottstein and Lücke, Springer-Verlag, Berlin, 67

Dillamore, I.L. (1978b), in **Recrystallization of Metallic Materials,** ed. F. Haessner, Springer, 223.

Dillamore, I.L. and Katoh, H. (1974), Met.Sci. 8, 73.

Dillamore, I.L., Katoh, H. and Haslam, K. (1974), Texture, 1, 151.

Dillamore, I.L. and Roberts, W.T. (1965), Met. Rev. 10 (39), 271.

Dillamore, I.L., Roberts, J.G. and Busch, A.C. (1979), Met.Sci. 13, 73

Dillamore, I.L., Smith, C.J.E. and Watson, T.W. (1967), Met. Sci. J. 1, 49.

Dillamore, I.L., Morris, P.L., Smith, C.J.E. and Hutchinson, W.B., (1972), Proc. R. Soc. Lond. A. 329, 405

Dimitrov, O. Fromageau, R. and Dimitrov, C. (1978), in **Recrystallization of Metallic Materials.** ed. Haessner. Dr. Riederer-Verlag GMBH, Stuttgart, 137.

Dingley, D.J. (1984), Scan. Elec. Mic. 11, 74.

Dingley, D.J. and Pond, R.C. (1979), Acta Metall. 28, 667.

Dingley, D.J. and Randle, V. (1992), J. Mats. Sci. 27, 4545.

Distl. J.S., Welch, P.I. and Bunge, H.J. (1982), Scripta Metall. 17, 975.

Djaic, R.A.P. and Jonas, J.J. (1972), J. Iron and Steel Inst. 210, 256.

Doherty, R.D. (1975), Metall. Trans. 6A, 588.

Doherty, R.D. (1978), in **Recrystallization of Metallic Materials,** ed. F. Haessner, Dr. Riederer Verlag GmbH, Stuttgart, 23.

Doherty, R.D. (1982), Metal Sci. 16, 1.

Doherty, R.D. and Cahn, R.W. (1972), J. Less Common Metals. 28, 279.

Doherty, R.D., Kashyap, K. and Panchanadeeswaran, S. (1993). Acta Met. 41, 3029.

Doherty, R.D., Li, K., Anderson, M.P., Rollett, A.R. and Srolovitz, D.J. (1990), In **Recrystallization'90**, ed. Chandra. TMS, 122.

Doherty, R.D., Li, K., Kashyap, K., Rollett, A.R. and Srolovitz, D.J. (1989), Proc 10th Risø Symp. eds. Bilde-Sorensen et al. Risø, Denmark. 31.

Doherty, R.D. and Martin, J.W. (1962), J. Inst. Metals. 91, 332.

Doherty, R.D. and Martin, J.W. (1964), Trans. ASM. 57, 874.

Doherty, R.D., Rollett, A.R. and Srolovitz, D.J. (1986), Proc 7th Risø Symp. eds. Hansen et al. Risø, Denmark. 53.

Doherty, R.D. and Szpunar, J.A. (1984), Acta Metall. 32, 1789.

Drolet, J.P. and Galibois, A. (1968), Acta Metall. 16, 1387.

Drouard, R., Washburn, J. and Parker, E.R. (1953) Trans. Metall. Soc. A.I.M.E. 197, 1226.

Drury, M.D. and Humphreys, F.J. (1986), Acta Metall. 34, 2259.

Drury, M.D., Humphreys, F.J. and White, S.H. (1989), J. Mater. Sci. 24, 154.

Duggan, B.J., Hatherly, M., Hutchinson, W.B. and Wakefield, P.T. (1978a), Met.Sci. 12, 343

Duggan B.J., Hutchinson, W.B. and Hatherly, M. (1978b), Scr. Met. 12, 1293.

Duggan, B.J. and Jones, I.P. (1977), Texture. 2, 205.

Duggan B.J. and Lee, W.B. (1988), in **ICOTOM 8,** eds. Kallend and Gottstein, TMS, Warrendale, 625.

Duggan, B.J., Lücke, K., Köhlhoff, G. and Lee, C.S. (1993), Acta Metall. 41, 1921.

Dunn, C.G. (1966), Acta Metall. 14, 221.

Dunn, C.G. and Koh, P.K. (1956), Trans. Metall. Soc. A.I.M.E. 206, 1017.

Dunn, C.G. and Walter, J.L. (1966), in **Recrystallization, Grain Growth and Textures**. ASM, Ohio. 461.

Dutta, S.K. and Spriggs, R.M. (1970), J. Am. Ceram. Soc. 53, 61.

Eastwood, L.W. Bousu, A.E. and Eddy, C.T. (1935), Trans. Metall. Soc. A.I.M.E. 117, 246.

Edward, G.H., Etheridge, M.A. and Hobbs, B.E. (1988), Text. Microstruct. 5, 127.

Eichelkraut, H., Abbruzzese, G. and Lücke, K. (1988), Acta Metall. 36, 55.

Ells, C.E. (1963), Acta Metall. 11, 87.

Embury, J.D., Poole, W.J. and Koken, E. (1992), Scripta Metall. 27, 1465.

Emren, F., von Schlippenback, U. and Lücke, K. (1986), Acta. Met. 34, 2105.

English, A.T. Backofen, W.A. (1964), Trans. Metall. Soc. A.I.M.E. 230, 396

Every, R.L. and Hatherly, M. (1974), Texture, 1, 183

Ewing, J.A. and Rosenhain, W. (1900). Phil. Trans. Royal Soc. 193A, 353.

Exell, S.F. and Warrington, D. (1972), Phil. Mag. 26, 1121.

Faivre, P. and Doherty, R.D. (1979), J. Mats. Sci. 14, 897.

Farag, M.M., Sellars, C.M. and Tegart, W.McG (1968), in **Deformation under Hot Working Conditions**. Iron & Steel Inst. London, ed. Moore. p103.

Farrell, K. Schauffhauser, A.C. and Houston, J.T. (1970), Metall. Trans. 1, 2899.

Feltham, P. (1957), Acta Metall. 5, 97.

Feltner, P.K. and Loughhunn, D.J. (1962), Acta Metall. 10, 685

460 References

Ferran, G., Cizeron, G. and Aust, K.T. (1967), Mem. Sci. 64, 1064.

Ferry, M., Munroe, P., Crosky, A. and Chandra, T. (1992), Mats. Sci. and Tech. 8, 43.

Flower, H.M. (1990), Mats. Sci. and Tech. 6, 1082.

Form, W., Gindreaux, G. and Mlyncar, V. (1980), Metal Sci. 14, 16.

Foster, K., Kramer, J.J. and Weiner, G.W. (1963), Trans. Metall. Soc. A.I.M.E. 227, 185.

Frank, F.C. (1988), Metall. Trans. 19A, 403.

Fridman, E.M., Kopezky, C.V. and Shvindlerman, L.S. (1975), Z. Metallk. 66, 533.

Fridy, J.M., Marthinsen, K., Rouns, T.N., Lippert, K.B., Nes, E. and Richmond, O. (1992), Proc. 3rd Int. Conf on Aluminium, Trondheim. 333.

Friedel, J. (1964), **Dislocations**. Addison-Wesley, London.

Frois, C. and Dimitrov, O. (1966), Ann. Chim. Paris. 1, 113.

Frost, H.J. (1992), in **Grain Growth in Polycrystalline Materials**. ed. Abbruzzese and Brozzo. Trans Tech Publns. 903.

Frost, H.J. and Ashby, M.F. (1982), **Deformation-Mechanism Maps**. Pergamon Press.

Frost, H.J. and Thompson, C.V. (1988), J. Electronic Mats. 17, 447.

Frost, H.J., Thompson, C.V. and Walton, D.T. (1990), Acta Metall. 38, 1455.

Frost, H.J., Thompson, C.V. and Walton, D.T. (1992) in **Grain Growth in Polycrystalline Materials**. ed. Abbruzzese and Brozzo. Trans Tech Publns. 543

Frost, H.J, Whang, J. and Thompson, C.V. (1986), Proc 7th Risø Int. Symp., eds. Hansen et al. Risø, Denmark. 315.

Fudaba, K., Akisue, O. and Tokunaga Y. (1988), in **27th CIM Conf.** (Montreal), 290.

Fuentes, M. and Sevillano, J.G. (eds), (1992), **Recrystallization'92**. San Sebastian, Spain, Trans Tech Publications.

Fukuda, F. and Sadayori, T. (1989), Steel Times Int. July., 38.

Fukuda, N. Shimizu, M. (1972), Sosei-to-Kako, 13, 841.

Fullman, R.E. and Fisher, J.C. (1951), J. Appl. Phys. 22, 1350.

Fullman, R.L. (1952), in **Metal Interfaces**. ASM, Cleveland, Ohio, 179.

Furley, J. and Randle, V. (1991), Mats. Sci. and Tech. 7, 12.

Furu, T. (1992) Doctoral Thesis. NTH, Trondheim.

Furu, T. and Nes, E. (1992), in **Recrystallization'92**. ed. Fuentes and Sevillano. San Sebastian, Spain. 311.

Furu, T., Marthinsen, K. and Nes, E. (1990), Mats. Sci. and Tech. $\underline{6}$, 1093.

Furu, T., Marthinsen, K. and Nes, E. (1992), in **Recrystallization'92**. ed. Fuentes and Sevillano. San Sebastian, Spain. 41.

Galina, A.V., Fradkov, V.Y. and Shvindlerman, L.S. (1987) Fiz. Metal. Metalloved. $\underline{63.6}$, 1220.

Gardner, K.J. and Grimes, R. (1979), Metal Sci. $\underline{13}$, 216.

Gawne, D.T. and R.A. Higgins, R.A. (1969), In **Textures in Research and Practice.** eds Grewen and Wasserman. Springer, New York. 319.

Gawne, D.T. and R.A. Higgins, R.A. (1971), J. Mats. Sci. $\underline{6}$, 403.

Ghosh, A.K. and Ghandi,C. (1986),in **Strength of Metals and Alloys. Proc. 7th Int. Conf. on Strength of Metals and Alloys.** ed. McQueen et al. Pergamon Press. $\underline{3}$, 2065.

Gialanella, S., Cahn R.W., Malagelada, J., Surinach, S. and Barò, M.D. (1992), in **Kinetics of Ordering Transformations in Metals**, eds. Chang et al, 161.

Gifkins, R.C. (1952), J. Inst. Metals. $\underline{81}$, 417.

Gil Sevillano, J., van Houtte, P. and Aernoudt, E. (1980), Prog. Mater. Science. $\underline{25}$, 69.

Gilman, J.J. (1955), Acta Metall. $\underline{3}$, 277

Gjostein, N.A. and Rhines, F.N. (1959), Acta Metall. $\underline{7}$, 319

Gladman, T. (1966), Proc. R. Soc. Lond. $\underline{A294}$, 298.

Gladman, T. (1980), Proc. 1st Int. Risø Symposium. Risø, Denmark. ed. Hansen. p. 183.

Gladman, T. (1990), Mats. Sci. and Tech. $\underline{6}$, 1131.

Gladman, T. (1992) in **Grain Growth in Polycrystalline Materials**. ed. Abbruzzese and Brozzo. Trans Tech Publns. 113.

Gleiter, H. (1969a), Acta Metall. $\underline{17}$, 565.

Gleiter, H. (1969b), Acta Metall. $\underline{17}$, 853.

Gleiter, H. (1969c), Acta Metall. $\underline{17}$, 1421

Gleiter, H. (1970a), Acta Metall. $\underline{18}$, 117.

Gleiter, H. (1970b), Acta Metall. $\underline{18}$, 23.

Gleiter, H. (1970c), Z. Metallk. $\underline{61}$, 282.

Gleiter, H. (1971), Phys. Stat. Solidi (b), $\underline{45}$, 9.

Gleiter, H. (1980), in **Grain Boundary Structure and Kinetics**, ASM, Ohio, 427.

Gleiter, H. and Chalmers, B. (1972), Prog. Mater. Sci. $\underline{16}$, 1.

Gokhale, A.M. and DeHoff, R.T. (1985), Metall. Trans. $\underline{16A}$, 559.

Gondi, P., Montanari, R. and Sili, A. (1992), in **Grain Growth in Polycrystalline Materials.** eds. Abbruzzeze and Brozzo. Trans Tech Pubs. Switzerland. 591.

Goodenow, R.H. (1966), Trans. ASM. $\underline{59}$, 804.

Goodhew, P.J. (1979), Metal Sci. $\underline{13}$, 108.

Goodhew, P.J. (1980), in **Grain Boundary Structure and Kinetics.** ed. Balluffi, ASM, Ohio, 155.

Gordon, P. (1955), Trans. Metall. Soc. A.I.M.E. $\underline{203}$, 1043.

Gordon, P. and El Bassyoumi, T.A. (1965), Trans. Metall. Soc. A.I.M.E. $\underline{223}$, 391.

Gordon, P. and Vandermeer, R.A. (1962), Trans. Metall. Soc. A.I.M.E. $\underline{224}$, 917.

Gordon, P. and Vandermeer, R. (1966), in **Recrystallization, Grain Growth and Textures.** ASM, Metals Park, Ohio. 205.

Gordon, R.S., Marchant, D.D. and Hollenburg, G.W. (1970), J. Am. Ceram. Soc. $\underline{53}$, 399.

Goss, N.P. (1934a), U.S. Patent 1,965,559.

Goss, N.P. (1934b), Trans. ASM, $\underline{23}$, 511.

Gottstein, G. (1984), Acta Metall. $\underline{32}$, 1117.

Gottstein, G., Zabardjadi, D. and Mecking, H. (1979), Metal Sci. $\underline{13}$, 223.

Gottstein, G. and Kocks, U.F. (1983), Acta Metall. $\underline{31}$, 175.

Gottstein, G. and Schwarzer, F. (1992), in **Grain Growth in Polycrystalline Materials.** eds. Abbruzzeze and Brozzo. Trans Tech Pubs. Switzerland. 197.

Gottstein, G. and Shvindlerman, L.S. (1992), Scripta Metall. $\underline{27}$, 1521.

Gottstein, G. and Shvindlerman, L.S. (1993), Acta Metall. $\underline{41}$, 3267.

Gottstein, G., Nagpal, P. and Kim, W. (1989), Mat. Sci. and Eng. A108, 165.

Gottstein, G., Zabardjadi, D. and Mecking, H. (1976), Proc. 4th Int. Conf. on Strength of Metals and Alloys. Nancy, 1126.

Gottstein, G., Murmann, H.C., Renner, G., Simpson, C.J. and Lücke, K. (1978) in **ICOTOM 5**, eds. Gottstein and Lücke, Aachen, 521.

Gould, D., Hirsch, P.B. and Humphreys, F.J. (1974), Phil. Mag. 30, 1353.

Graham, C.D. and Cahn, R.W. (1956) Trans. Metall. Soc. A.I.M.E. 206, 517.

Grant, E., Porter, A. and Ralph, B. (1984), J. Mats. Sci. 19, 3554.

Green, R.E., Liebmann, G.B. and Yoshida, H. (1959), Trans. Metall. Soc. A.I.M.E. 215, 610.

Grest, G.S., Srolovitz, D.J. and Anderson, M.P. (1985), Acta Metall. 33, 509.

Grewe, H.G., Schmidt, P.F. and Schur, K. (1973), Z. Metallk. 64, 502.

Grewel, G. and Ankem, A. (1989), Metall. Trans. 20A, 39.

Grewel, G. and Ankem, A. (1990), Metall. Trans. 21A, 1645.

Grewen, J. (1973), Proc. 3rd Coll. Eur. sur Textures, Pont-a-Mousson, 195.

Grewen, J. and Huber, J. (1978), in **Recrystallization of Metallic Materials,** ed. F. Haessner, Springer, 111.

Grey, E.A. and Higgins, G.T. (1973), Acta Metall. 21, 309.

Grimes, R., Miller, W.S. and Butler, R.G. (1987), J. de Physique. Colloque C3. 48. 239.

Grimmer, H., Bollmann, W. and Warrington, D.H. (1974), Acta Cryst. 30A, 197.

Gronski, R. (1980), in **Grain Boundary Structure and Kinetics**, ed. Baluffi, ASM, Ohio, 45.

Grovenor, C.R.M., Smith, D.A. and Goringe, M.J. (1980), Thin Solid Films, 74, 269.

Grunwald, W. and Haessner, F. (1970), Acta Metall. 18, 217.

Gryzliecki, J., Truszkowski, W., Pospiech, J. and Jura, J. (1988), Text. Microstruct. 14-18, 1061.

Guillopé, M. and Poirier, J.P. (1979), J. Geophys. Res. 84, 5557.

Gutierrez, I., Castro, F.R., Urcola, J. and Fuentes, M. (1988), Mats Sci & Eng. A 102, 77.

Haase, O. and Schmid, E. (1925), J. Phys. 33, 413.

Haasen, P. (1993), Metall. Trans. 24A, 1001.

Habiby F. and Humphreys, F.J. (1993), Text. Microstruct. 20, 125.

Haessner, F. (ed.) (1978), **Recrystallization of Metallic Materials**. Dr. Riederer-Verlag. GMBH Stuttgart.

Haessner, F. (1990), in **Recrystallisation 90**, ed. Chandra, TMS, Warrendale, 511.

Haessner, F. and Hofmann, S. (1978), in **Recrystallization of Metallic Materials**. ed. Haessner. DR Riederer Verlag, Stuttgart. 63.

Haessner, F. and Holzer, H.P. (1974), Acta Metall. 22, 695.

Hansen, N. (1975) Mem. Sci. Rev. Met. 72, 189.

Hansen, N. (1990), Mats. Sci. and Tech. 6, 1039.

Hansen, N. and Bay, B. (1972), J. Mats. Sci. 7, 1351.

Hansen, N. and Bay, B. (1981), Acta Metall. 29, 65.

Hansen, N. and Juul Jensen, D. (1991), in **Hot Deformation of Aluminium Alloys**, ed. Langdon et al, TMS, Warrendale, 3.

Hansen N. and Kuhlmann-Wilsdorf, D. (1986), Mat. Sci. and Eng. 81, 141.

Hansen, N., Leffers, T. and Kjems, J.K. (1981), Acta Metall. 29, 1523.

Harase, J. and Shimazu, R. (1988), Trans. JIM, 29, 388.

Haratani, T. Hutchinson, W.B. Dillamore, I.L. and Bate, P. (1984), Metal Sci. 18, 57.

Hartung, W. (1993), private communication.

Hasegawa, T. and Kocks, U.F. (1979), Acta Metall. 27, 1705.

Hasson, G.C. and Goux, C. (1971), Scripta Metall. 5, 889.

Hatherly, M. (1959), J.Inst.Met. 88, 60.

Hatherly, M. (1982), in **Proc. 6th Int. Conf. on Strength of Metals and Alloys,** ed. Gifkins, Pergamon, Oxford 1181.

Hatherly, M. and Dillamore, I.L. (1975), J. Aus. Inst. Met. 20, 71.

Hatherly M. and Hutchinson, W.B. (1979), **An Introduction to Textures in Metals**. Institution of Metallurgists, Monograph 5.

Hatherly M. and Malin, A.S. (1979), Met. Tech. $\underline{6}$, 308.

Hatherly, M. and Malin, A.S. (1984), Scripta. Met. $\underline{18}$, 449

Hatherly, M., Malin, A.S. and Carmichael, C.M. (1984), Proc ICOTOM 7, Holland, 245.

Hatherly, M., Malin, A.S., Carmichael, C.M, Humphreys, F.J. and Hirsch, J. (1986), Acta Metall. $\underline{34}$, 2247.

Hazzledine, P.M. and Oldershaw, R.D.J. (1990), Phil. Mag. $\underline{61A}$, 579.

Heckelmann, I, Abbruzzese, G. and Lücke, K. (1992), in **Grain Growth in Polycrystalline Materials**. ed. Abbruzzese and Brozzo. Trans Tech Publns. 391.

Heimendahl, M., Bell, W. and Thomas, G. (1964), J. Appl. Phys. $\underline{35}$, 3614.

Hellman, P. and Hillert, M. (1975), Scand. J. Metall. $\underline{4}$, 211.

Herbst, P. and Huber, J. (1978), in **Texture of Metallic Materials**, eds. Gottstein and Lücke, (Berlin, Springer Verlag), $\underline{1}$, 452.

Hibbard, W.R. and Dunn, C.G. (1956), Acta Metall. $\underline{4}$, 306.

Hibbard, W.R. and Tully, W.R. (1961), Trans. Metall. Soc. A.I.M.E. $\underline{221}$, 336.

Higashi, K., Uno, M., Matsuda, S., Ito, T. and Tanimura, S. (1990). In **Recrystallization'90**, ed. Chandra. TMS, 711.

Higgins, G.T. (1974), Met Sci. J. $\underline{8}$, 143.

Higgins, G.T., Wiryolukito, S, and Nash. P. (1992), in. **Grain Growth in Polycrystalline Materials**. ed.Abbruzzese and Brozzo. Trans Tech Publns. 671.

Hillert, M. (1965), Acta Metall. $\underline{13}$, 227.

Hillert, M. (1979), Met. Sci. $\underline{13}$, 118.

Hillert, M. (1988), Acta Metall. $\underline{36}$, 3177.

Hillert. M. and Purdy, G.R. (1978), Acta Metall. $\underline{26}$, 333.

Hillert, M. and Sundman, B. (1976), Acta Metall. $\underline{24}$, 731.

Himmel, L. (ed.), **Recovery and Recrystallization of Metals**. Interscience, N.York. (1963)

Hirsch, J. (1986), Proc. 7th Risø Int. Symp., eds. Hansen et al. Risø, Denmark. 349.

Hirsch, J. (1990), in **Recrystallization 90**, ed. Chandra, TMS, Warrendale, 759.

Hirsch, J. and Engler, O. (1995). Proc. 16th Int. Risø Symp. eds Hansen et al. 49.

Hirsch, J., Loeck, M., Loof, L. and Lücke, K. (1984), Proc. ICOTOM 7, ed. Brakman et al, Noordwijkerhout, 765.

Hirsch, J. and Lücke, K. (1985), Acta Metall. 33, 1927.

Hirsch, J. and Lücke, K. (1988a), Acta Metall. 36, 2863.

Hirsch, J. and Lücke, K. (1988b), Acta Metall. 36, 2883.

Hirsch, P.B. and Humphreys, F.J. (1969), in **Physics of Strength and Plasticity**. ed. A. Argon. MIT Press. 189.

Hirth, J.P. (1972), Metall. Trans. 3, 3047.

Hirth, J.P. and Lothe, (1968), **Theory of Dislocations**, Wiley.

Hjelen, J, Ørsund, R. and Nes, E. (1991), Acta Metall. 39, 1377.

Hjelen, J., Weiland, H., Butler, J., Liu, J., Hu, H. and Nes, E. (1990). Proc. **ICOTOM 9**. ed. Bunge. Avignon. 983.

Holmes, E.L. and Winegard, W.C. (1959), Acta Metall. 7, 411.

Holmes, E.L. and Winegard, W.C. (1961), Can. J. Phys. 39, 1223.

Holt, D.L. (1970), J. Appl. Phys. 41, 3197.

Honda, K., Masunoto, H. and Kaya, S. (1928), Sci. Rep. Tohoku. Imp. Univ. 17, 11.

Hondros, E.D. and Seah, M.P. (1977), Int. Met. Reviews. 222. 1.

Honeff, H. and Mecking, H. (1978), Proc. ICOTOM 5, eds. Gottstein and Lücke, Springer-Verlag, Berlin, 265

Honeycombe, R.W.K. (1985), **The Plastic Deformation of Metals**. Edward Arnold.

Honeycombe, R.W.K. and Boas, W. (1948), Aust. J. Sci. Res. 1A, 70.

Hornbogen, E. (1970), Practische Metallographie, 9, 349.

Hornbogen, E. (1977), J. Mats. Science. 12, 1565.

Hornbogen, E. (1980), Proc. 1st Int. Risø Symp. Risø, Denmark. eds. Hansen et al. 199.

Hornbogen, K. and Köster, U. (1978), in **Recrystallisation of Metallic Materials**, ed. F. Haessner, Dr. Riederer Verlag GmbH, Stuttgart, 159.

Hornbogen, E. and Kreye, H. (1969), in **Texturen in Forschung und Praxis,** eds. Grewen and Wassermann, Springer-Verlag, Berlin, 274.

Hornbogen, E. and Lütjering, G. (1975), Proc. 6th Int. Conf. on Light Metals. Leoban, Vienna. Al-Verlag, Düsseldorf.

Howell, P.R. and Bee, J.V. (1980), Proc. 1st Int. Risø Symp. Risø, Denmark, eds Hansen et al. 171.

Hu, H. (1962), Trans. Metall. Soc. A.I.M.E. 224, 75.

Hu, H. (1963), in **Recovery and Recrystallization of metals**. ed. Himmel. Interscience. 311.

Hu, H. (1974), Can. Met. Quart. 13, 275.

Hu, H. (1978), in **ICOTOM 5**, eds. Gottstein and Lücke, Springer-Verlag, 2, 3.

Hu, H. and Goodman, S.R. (1970), Metall. Trans. 1, 3057.

Hu, H. and Smith, C.S. (1956), Acta Metall. 4, 638.

Huber J. and Hatherly, M. (1979), Met. Sci. 13, 665.

Huber J. and Hatherly, M. (1980), Z. Metallk. 71, 15

Hughes, D. (1993), Acta Metall. Mater. 41, 1421

Hull, D. and Bacon, D. J. (1984), **Introduction to Dislocations**. 3rd Edn. Pergamon.

Humfrey, J.C.W. (1902). Phil. Trans. Royal Soc. 200, 225.

Humphreys, F.J. (1977) Acta Metall. 25, 1323

Humphreys, F.J. (1979a), Acta Metall. 27, 1801.

Humphreys, F.J. (1979b), Metal Sci. 13 , 136.

Humphreys, F.J. (1980), Proc. 1st Int. Risø Symp. Risø, Denmark, eds Hansen et al. 35.

Humphreys, F.J. (1982), Proc. 6th Int. Conf. on Strength of Metals and Alloys. ed. Gifkins. Melbourne. 1, 625.

Humphreys, F.J. (1983), Text. Microstruct. 6, 45.

Humphreys, F.J. (1985), in **Dislocations and Properties of Real Materials**, ed. Loretto. Inst. Metals, London, 175.

Humphreys, F.J. (1988a), Proc. 9th Int. Risø Symp. Risø, Denmark, eds Anderson et al. 51.

Humphreys, F.J. (1988b), in **ICOTOM 8**, eds Kallend and Gottstein, TMS, Warrendale, USA. 171.

Humphreys, F.J. (1990), in **Recrystallization 90**, ed. Chandra, TMS, Warrendale, 113.

Humphreys, F.J. (1991), in. **Processing of Metals and Alloys**. 9, ed. R.W. Cahn. VCH, Germany. 373

Humphreys, F.J. (1992a), Mats. Sci. and Tech. 8, 135.

Humphreys, F.J. (1992b), Scripta Metall. 27, 1557.

Humphreys, F.J. (1993), **Crystallographic textures**. Institute of Materials, London, Engineering materials software series, No. PD570.

Humphreys, F.J. and Ardakani, M.G. (1994), Acta Metall. 42, 749.

Humphreys, F.J. and Chan, H.M. (1996), Mats. Sci. and Tech. 12, 143.

Humphreys, F.J. and Hirsch, P.B. (1976), Phil. Mag. 34, 373.

Humphreys, A.O. and Humphreys, F.J. (1994), Proc. 4th Int. Conf on Aluminium, Atlanta, 1, 211.

Humphreys, F.J. and Juul Jensen, D. (1986). Proc. 7th Int. Risø Symp. Risø, Denmark. 93.

Humphreys, F.J. and Kalu, P.N. (1987), Acta Metall. 35, 2815.

Humphreys, F.J. and Kalu, P.N. (1990), Acta Metall. 38, 917.

Humphreys, F.J. and Martin, J.W. (1967), Phil. Mag. 16, 927.

Humphreys, F.J. and Martin, J.W. (1968), Phil. Mag. 17, 365.

Humphreys, F.J. Miller, W.S. and Djazeb, R. (1990), Mats. Sci. and Tech. 6, 1157.

Humphreys, F.J. and Ramaswami, V. (1973), Proc. 3rd Int Conf. on High Voltage Electron Microscopy. Oxford. 268-272.

Hunderi, O. and Ryum, N. (1992a), Acta Metall. 40, 543.

Hunderi, O. and Ryum, N. (1992b), in **Grain growth in polycrystalline materials**. ed. Abbruzzese and Brozzo. Trans Tech Publns. 89.

Hunderi, O., Ryum, N. and Westengen, H. (1979), Acta Metall. 27, 161.

Hutchinson, W.B. (1974), Metal Sci. J. 8, 185.

Hutchinson, W.B. (1984), Int. Met. Rev, 29, 25

Hutchinson, W.B. (1989), Acta Metall. 37, 1047.

Hutchinson, W.B. (1993), in Proc. **ICOTOM 10,** ed. Bunge, Clausthal, Trans Tech pubs. 1917.

Hutchinson, W.B. and Bate, P. (1994), Private communication.

Hutchinson, W.B. and Duggan, B.J. (1978), Metal Sci. $\underline{12}$, 372.

Hutchinson, W.B. and Ekström, H.E. (1990), Mats. Sci. and Tech. $\underline{6}$, 1103.

Hutchinson, W.B. and Nes, E. (1992), in **Grain Growth in Polycrystalline Materials**. ed. Abbruzzese and Brozzo. Trans. Tech. Publns., Rome, 385.

Hutchinson, W.B. and Ryde, L. (1995), Proc. 16th Int. Risø Symposium. Risø, Denmark. ed. Hansen et al., 105.

Hutchinson, W.B. and Ushioda, K. (1984), Scand. J. Met. $\underline{13}$, 269.

Hutchinson, W.B., Besag, F.M.C. and Honess, C.V. (1973), Acta Metall. $\underline{21}$, 1685.

Hutchinson, W.B., Duggan, B.J. and Hatherly, M. (1979), Met. Tech. $\underline{6}$, 398.

Hutchinson, W.B., Jonsson, S. and Ryde, L. (1989a), Scripta Metall. $\underline{23}$, 671.

Hutchinson, W.B., Oscarsson, A. and Karlsson, Å. (1989b), Mat. Sci. and Tech., $\underline{5}$, 1118.

Ibe, G. and Lücke, K. (1966), in **Recrystallization, Grain Growth and Textures,** ed Margolin, ASM, Metals Park, 434.

In der Schmitten, W.P. Haasen, P. and Haessner, F. (1960), Z. Metallk. $\underline{51}$, 101.

Inakazu,N., Kaneno, Y. and Inoue, H. (1991), Text. Microstruct. $\underline{14\text{-}18}$, 847.

Inokuti. Y. and Doherty, R.D. (1978), Acta Metall. $\underline{26}$, 61.

Inokuti, Y., Maeda, C. and Ito, Y. (1987), Trans. ISIJ, $\underline{27}$, 140.

Inoue, H. and Inakazu, N. (1988), in **ICOTOM 8,** eds. Kallend and Gottstein, TMS, Warrendale, 997.

Ion, S.E., Humphreys, F.J. and White, S. (1982), Acta Metall. $\underline{30}$, 1909.

Ito, K. (1988). In **Homogenising and Annealing of Aluminium and Copper Alloys.** eds Merchant et al. TMS, Warendale, 169.

Jack, D.B., Koken, E. and Underhill, R. (1989), Proc. 10th Risø Int. Symp. eds. Bilde-Sørensen et al. Risø, Denmark. 403.

Jackson, P.A. (1983) Scripta Metall. $\underline{17}$, 199.

Jang, J.S. and Koch, C.C. (1990), J. Mat. Res. $\underline{5}$, 498.

Jefferies, Z. (1916). Trans. Metall. Soc. A.I.M.E. 56, 571.

Johnson, W.A. and Mehl, R.F. (1939), Trans. Metall. Soc. A.I.M.E. 135, 416.

Jonas, J.J. (1990), In **Recrystallization'90**, ed. Chandra. TMS, 27.

Jonas, J.J., Sellars, C.M. and Tegart, W.J. McG. (1969), Metallurgical Reviews. Review 130, 1.

Jones, A.R. and Hansen, N. (1981), Acta Metall. 29, 589.

Jones, A.R., Ralph, B. and Hansen, N. (1979), Proc. R. Soc. Lond. A368, 345.

Juul Jensen, D. (1993), Text. Microstruct. 20, 55.

Juul Jensen, D. (1995), Proc. 16th Int. Risø Symp. eds Hansen et al. 119.

Juul Jensen, D. and Hansen, N. (1986), Proc. 7th Risø Int Symposium. Risø, Denmark. eds Hansen et al. 379.

Juul Jensen, D., Hansen, N. and Humphreys, F.J. (1985), Acta Metall. 33, 2155.

Juul Jensen, D., Hansen, N. and Humphreys, F.J. (1988), in **ICOTOM 8**, eds. Kallend and Gottstein, TMS, Warrendale, 431.

Juul Jensen, D., Hansen, N., Kjems, J.K. and Leffers, T. (1984). Proc. ICOTOM 7, eds. Brakman et, al, Noordwijkerhout, 777.

Kaibyshev, O., Kaibyshev, R. and Salishchev, G. (1992), in **Recrystallization'92.** eds. Fuentes and Gil Sevillano. Trans Tech Publications. 423.

Kalischer, S. (1881). Ber. 14, 2747.

Kalu, P.N. and Humphreys, F.J. (1986), Proc. 7th Risø Int. Symposium. Risø, Denmark. eds Hansen et al. 385.

Kalu, P.N., and Humphreys, F.J. (1988), in **ICOTOM 8,** eds. Kallend and Gottstein, TMS, Warrendale, 511.

Kamma, C. and Hornbogen, E. (1976), J. Mat. Sci. Letters. 11, 2340.

Kapadia, C.M. and Leipold, M.H. (1974), J. Am. Ceram. Soc. 57, 41.

Kashyap, B.P. and Mukherjee, A.K. (1985), in **Superplasticity**, ed. Baudelet and Suery. Grenoble, CNRS. 4,

Kassner, M.E., McQueen, H.J. and Evangelista, E. (1992), in **Recrystallization'92.** eds. Fuentes and Gil Sevillano. Trans Tech Publications. 151.

Kawahara, K. (1983), J. Mat. Sci. 19, 949.

Kawasaki, K. Nagai, T. and Nakashima, K. (1989), Phil. Mag. B, 60, 399.

Kim, Y-W, and Griffith, W.M. (1988), eds. **Dispersion Strengthened Aluminium Alloys**. TMS, Warrendale, USA.

King, A.H. and Smith, D.A. (1980), Acta Cryst. A36, 335.

Kingery, W.D. (1974), J. Am. Ceram. Soc. 57-2, 74.

Kingery, W.D. and Francois, B. (1965), J. Am. Ceram. Soc. 48, 546.

Kivilahti, J.K., Lindroos, J.K., and Lehtinen, B. (1974), in **High Voltage Electron Microscopy**. ed. Swann. Academic Press. 195.

Koch, C.C. (1991), in **Processing of Metals and Alloys**, ed. Cahn, Weinheim, VCH, 193.

Kocks, U.F. and Canova, G. R. (1981), Proc. 2nd Int. Risø Symp. Risø, Denmark. ed. Hansen et al. 35.

Koehler, J.S., Henderson, J.W. Bredt, J.H. (1957), in **Creep and Recovery**, ASM, Cleveland, 1.

Kohara, S. Parthasarathi, M.N. and Beck, P.A. (1958), Trans. Metall. Soc. A.I.M.E. 212, 875.

Köhlhoff, G.D., Hirsch, J., von Schlippenbach, U. and Lücke, (1981), in **ICOTOM 6**, Tokyo, 489.

Köhlhoff, G.D., Malin, A.S., Lücke, K. and Hatherly, M. (1988a), Acta Metall. 36, 2841

Köhlhoff, G.D., Sun, X. and Lücke, K. (1988b), in **ICOTOM 8**, eds. Kallend and Gottstein, TMS, Warrendale, 183.

Koken, E, Chandrasekaran, N., Embury, J.D. and Burger, G. (1988), Mats. Sci. & Eng. A104, 163.

Kolmogorov, A.N. (1937), Izv. Akad. Nauk. USSR-Ser-Matemat. 1(3), 355.

Kopetski, C.V., Sursaeva, V.G. and Shvindlerman, L.S. (1979), Sov. Phys. Solid State. 21(2), 238.

Korbel, A., Embury, J.D. Hatherly, M., Martin, P.L. and Erbsloh, H.W. (1986), Acta Metall. 34, 1999

Köster, U. and Hornbogen, E. (1968), Z. Metallk. 59, 792.

Koul, A.K. and Pickering, B. (1982), Acta Metall. 30, 1303.

Krakow, W and Smith, D.A. (1987), Ultramicroscopy, 22, 47.

Kramer, J.J. (1992), Metall. Trans. 23A, 1987.

Kronberg, M.L. and Wilson, F.H. (1949), Trans. Metall. Soc. A.I.M.E. 185, 501.

Kuhlmann (now Kuhlmann-Wilsdorf), (1948), Z. fur Phys. 124, 468.

Kuhlmann-Wilsdorf, D. (1970), Metall. Trans. 1, 3173.

Kuhlmann-Wilsdorf. D. (1989), Mat. Sci. Eng. A113, 1.

Kuhlmann (now Kuhlmann-Wilsdorf), D., Masing, G. and Raffelsiefer, J. (1949), Z. Metallk. 40, 241.

Kunze, K., Wright, S.I., Adams, B.L. and Dingley, D.J. (1993), Text. Microstruct. 20, 41.

Kurtz, S.K and Carpay, F.M.A. (1980), J. Appl. Phys. 51, 5745.

Lawley, A., Vidoz, E.A. and Cahn, R.W. (1961), Acta Metall. 9, 287.

Lee, C.S., Smallman, R.S. and Duggan, B.J. (1993), Script. Met. et Mat. 29, 43.

Leffers, T. (1981), Proc. 2nd Int. Risø Symp. ed. Hansen et al. Risø, Denmark. 55.

Leffers, T. and Juul Jensen, D. (1988), Text. Microstruct. 8/9, 467.

Lerf, R. and Morris, D.G. (1991), Acta Metall. 39, 2430.

Leslie, W.C., Michalak, J.T. and Aul. F.W. (1963), in **Iron and its Dilute Solid Solutions**. ed. Spencer and Werner. Interscience. New York. 119.

Leslie, W.C., Rickett, R.L., Dotson, C.L. and Watson C.S. (1954), Trans. ASM, 46, 1470.

Lewis, M.H. and Martin, J.W. (1963), Acta Metall. 11, 1207.

Li, C.H., Edwards, E.H., Washburn, J. and Parker, E.R. (1953), Acta Metall. 1, 223.

Li, J.C.M. (1960), Acta Metall. 8, 563.

Li, J.C.M. (1962) J. Appl. Phys. 33, 2958.

Li, J.C.M. (1966) in **Recrystallization, Grain Growth and Textures**. ASM, Ohio, 45.

Liebmann, B. and Lücke, K. (1956), Trans. Metall. Soc. A.I.M.E. 206, 1443.

Liebmann, B. Lücke, K. and Masing, G. (1956), Z. Metallk. 47, 57.

Lindh, E., Hutchinson, W.B. and Ueyama, S. (1993), Scripta Metall. 29, 347.

Ling, S., Anderson, M.P., Grest, G.S. and Glazier, J.A. (1992), **In Grain Growth in Polycrystalline Materials**. ed. Abbruzzese and Brozzo. Trans Tech Pubs. Zurich. p39.

Liu, C.T. (1984), Int. Met. Rev. $\underline{29}$, 168.

Liu, J. and Doherty, R.D. (1986), in **Aluminium Technology'86**. Institute of Metals, London. p347.

Liu, Y.L, Hansen, N and Juul Jensen, D. (1989), Metall. Trans. $\underline{20A}$, 1743.

Liu, Y.L, Hansen, N and Juul Jensen, D. (1991), Proc. 12th Int. Risø Symp. Risø, Denmark. eds. Hansen et al. 67.

Lloyd, D.J. and Kenny, D. (1980), Acta Metall. $\underline{28}$, 639.

Lojkowski, W., Gleiter, H. and Maurer, R. (1988), Acta Metall. $\underline{36}$, 69.

Louat, N.P. (1974), Acta Metall. $\underline{22}$, 721.

Louat, N.P. (1982), Acta Metall. $\underline{30}$, 1291.

Louat, N.P., Duesbery, M.S. and Sadananda, K. (1992), in **Grain Growth in Polycrystalline Materials**. ed. Abbruzzese and Brozzo. Trans Tech Publns. 67.

Luborsky, F.E., Livingston, J.D. and Chin, J.Y. (1983), in **Physical Metallurgy**. eds. Cahn and Haasen. North-Holland. 1698.

Lücke, K. (1984), in **ICOTOM 7**, eds. Brakman et al, Noordwijkerhout, 195.

Lücke, K. and Detert, K. (1957), Acta Metall. $\underline{5}$, 628.

Lücke, K. and Engler, O. (1992). Proc. 3rd Int Conf. on Aluminium Alloys. Trondheim. 439.

Lücke K. and Hölscher, M. (1991), Text. Microstruct. $\underline{14\text{-}18}$, 585.

Lücke, K. and Stüwe, H.P. (1963), in **Recovery and Recrystallization in Metals**. ed. L. Himmel. Interscience Publications. p171.

Lücke, K. and Stüwe, H.P. (1971), Acta Metall. $\underline{19}$, 1087.

Lücke, K., Abbruzzese, G. and Heckelmann, I. (1990), in **Recrystallisation'90**, ed. Chandra. TMS, 37.

Lücke, K., Heckelmann, I. and Abbruzzese, G. (1992), Acta Metall. $\underline{40}$, 533.

Lücke, K., Rixen, R. and Senna, M. (1976), Acta Metall. $\underline{24}$, 103.

Lücke, K., Pospiech, J., Virnich, K.H. and Jura, J. (1981), Acta Metall. $\underline{29}$, 167

Luton, M.J. and Sellars, C.M. (1969), Acta Metall. $\underline{17}$, 1033.

474 References

Lytton, J.L., Westmacott, K.H. and Potter, L.C. (1965), Trans. Metall. Soc. A.I.M.E. 233, 1757.

Mackenzie, J.K. (1958), Biometrika, 45, 229.

Mäder, K. and Hornbogen, E. (1974), Scripta Met 8, 979.

Mahin, K.W., Hanson, K. and Morris, J.W. (1980), Acta Metall. 28, 443.

Majid, I. and Bristowe, P.D. (1987), Scripta Metall. 21, 1153.

Maksimova, E.L., Shvindlerman, L.S. and Straumal, B.B. (1988), Acta Metall. 36, 1573.

Malin, A.S. (1978), Ph.D. thesis, University of New South Wales.

Malin, A.S. (1993), private communication.

Malin, A.S. and Chen, B.K. (1992), In **Aluminium alloys for packaging.** eds. Morris et al. TMS, Warrendale, USA, 251.

Malin, A.S. and Hatherly, M. (1979), Met.Sci. 13, 463

Malin, A.S., Huber, J. and Hatherly, M. (1981), Z. Metallk. 72, 310.

Malin, A.S. Hatherly, M., Huber, J. and Welch, P. (1982a), Z. Metallk. 73, 489

Malin, A.S., Hatherly, M. and Piegerova, V. (1982b), Proc. 6th Int. Conf. on Strength of Metals and Alloys, ed. Gifkins, Pergamon, Oxford, 523.

Margolin, H. (ed.), (1966), **Recrystallization, Grain growth and Textures.** ASM, Ohio, USA, (1966),

Marshal, G.J. and Ricks, R.A. (1992), in Proc. **Recrystallization'92.** ed. Fuentes and Gil Sevillano. San Sebastian, Spain. 245.

Marthinsen, K., Lohne, O and Nes, E. (1989), Acta Metall. 37, 135

Martin, J.W. (1980), **Micromechanisms in Particle Hardened Alloys.** Cambridge Univ. Press.

Martin, J.W. and Doherty, R.D. (1976), **The Stability of Microstructure in Metals.** Cambridge University Press.

Masing, G. and Raffelsieper, J. (1950), Z. Metallk. 41, 65.

Masteller, M.S. and Bauer, C.L. (1978), in **Recrystallization of Metallic Materials**, ed. F. Haessner, Dr. Riederer Verlag, GmbH, Stuttgart, 251.

Marthinsen, K., Lohne, O. and Nes, E. (1989), Acta Metall. 37, 135.

Matsura, K. and Itoh, Y. (1992), in **Grain Growth in Polycrystalline Materials**. ed. Abbruzzese and Brozzo. Trans Tech Publns. 331.

Matthies, S. (1979), Phys. Stat. Sol. (b), <u>92</u>, K135

Mathur, P.S. and Backofen, W.A. (1973), Met.Trans. <u>1</u>, 1105.

Maurice, C. and Driver, J.H. (1993), Acta Metall. <u>41</u>, 1653.

May, J.E. and Turnbull, D. (1958), Trans. Metall. Soc. A.I.M.E. <u>212</u>, 769.

May, M. and Erdmann-Jesnizer, F. (1959), Z. Metallk. <u>50</u>, 434.

McCarney, C.G. and Pierce, D.H. (1992), Script. Met. et Mat. <u>28</u>, 1173.

McElroy, R.J. and Szkopiak, Z.C. (1972), Int. Met. Reviews. <u>17</u>. 175.

McQueen, H.J. (1981), Metal Forum. <u>4</u>, 81.

McQueen, H.J. and Jonas, J.J. (1975), in **Treatise on Materials Science & Technology**. <u>6</u>. ed. Arsenault. Academic Press, N.York. p393.

McQueen, H.J., Evangelista, E., Bowles, J. and Crawford, G. (1984), Metal Sci. <u>18</u>, 395.

McQueen, H.J., Knustad, O., Ryum, N. and Solberg, J.K. (1985), Scr. Met. <u>19</u>, 73.

McQueen, H.J., Solberg, J.K., Ryum, N. and Nes, E. (1989), Phil. Mag. <u>A60</u>, 473.

Mecking, H. and Gottstein, G. (1978), in **Recrystallization of metallic materials**, ed. F. Haessner, Dr. Riederer Verlag, GmbH, Stuttgart, 195.

Mehl, R.F. (1948), in **ASM Metals Handbook**, ASM, Metals Park, Ohio. p259.

Meyers, M.A. and Murr, L.E. (1978), Acta Metall. <u>26</u>, 951.

Michalak, J.T. and Hibbard, W.R. (1957), Trans. Metall. Soc. A.I.M.E. <u>209</u>, 101.

Michalak, J.T. and Hibbard, W.R. (1961), Trans. ASM. <u>53</u>, 331.

Michalak, J.D. and Hu, H. (1978), quoted in Hu, in **ICOTOM 5(2),** eds. Gottstein and Lücke, Springer-Verlag, Berlin, 3.

Michalak, J.T. and Paxton, H.W. (1961), Trans. Metall. Soc. A.I.M.E. <u>221</u>, 850.

Middleton, A.B., Pfeil, L.B. and Rhodes, E.C. (1949), J. Inst. Met. <u>75</u>, 595.

Miura, H., Kato, M. and Mori, T. (1990), Coloque de Phys. <u>C1-51</u>, 263.

Montariol, F. (1963), Metaux et Corrosion, <u>38</u>, 223.

Morral, J.E. and Ashby, M.F. (1974), Acta Metall. 22, 567.

Morris, D.G. and Morris, M.A. (1991), J. Mat. Sci., 26, 1734.

Morris, L. (1976), Proc. 4th Int Conf. on Strength of Metal and Alloys. Nancy, France. p649.

Morris, P.L. and Duggan, B.J. (1978), Metal Sci. 12, 1.

Mott, N.F. (1948), Proc. Phys. Soc. 60, 391

Mukherjee, A.K., Bird, J.E. and Dorn, J.E. (1969), Trans. ASM. 62, 155.

Mullins, W.W. (1958), Acta Metall. 6, 414

Mullins, W.W. and Vinals, J. (1989), Acta Metall. 37, 991.

Murr, L.E. (1975), **Interfacial Phenomena in Metals and Alloys**. Addison-Wesley, Reading, p131.

Mykura, H. (1980), in **Grain Boundary Structure and Kinetics**. ed. Balluffi, ASM Ohio, 445.

Nagai, T., Ohta, S and Kawasaki, K. (1992), in **Grain Growth in Polycrystalline Materials**. ed. Abbruzzese and Brozzo. Trans Tech Publns. 313.

Nakashima, S., Takashima, K. and Harase, J. (1991a), ISIJ, 13, 1007.

Nakashima, S., Takashima, K. and Harase, J. (1991b), ISIJ, 13, 1013.

Nauer-Gerhardt, C.U. and Bunge, H.J. (1988), in **ICOTOM 8**, eds. Kallend and Gottstein, TMS, Warrendale, 505.

Nes, E. (1976), Acta Metall. 24, 391.

Nes, E. (1979), Metal Sci. 13, 211.

Nes, E. (1985), Proc. Symp on **Microstructural Control During Processing of Aluminium Alloys**. New York, 95.

Nes, E. (1995), Acta Metall. Mater. 43, 2189.

Nes, E. and Hutchinson, W.B. (1989), Proc. 10th Int. Risø Symp. Risø, Denmark. ed. Bilde-Sørensen, 233.

Nes, E., Hutchinson, W.B. and Ridha, A.A. (1986), Proc. 7th Int. Conf. on Strength of Metals and Alloys, ed. McQueen, 1, 57.

Nes, E., Ryum, N and Hunderi, O. (1985), Acta Metall. 33, 11.

Nes, E. and Solberg, J.K. (1986), Mat. Sci. and Tech., 2, 19.

Nicklas, B. and Mecking, H. (1979), in **Strength of Metals and Alloys**. eds. Haasen, Gerold and Kostorz. Pergamon Press, Oxford. 391.

Nicolas, A. and Poirier, J.P. (1976), **Crystalline Plasticity and Solid State Flow in Metamorphic Rocks.** Wiley-Interscience.

Nourbackhsh, S. and Nutting, J. (1980), Acta Metall. 28, 357.

Novikov, V.Y. (1979), Acta Metall. 27, 1461.

Obara, T., Satoh, S., Nishida, N. and Irie, T. (1984), Scand. J. Met. 13, 201.

Olgaard, D.L. and Evans, B. (1986), J. Am. Ceram. Soc. 69, C272.

Orowan, E. (1942), Nature, 149, 643.

Ørsund, R, and Nes, E. (1988), Scripta Metall., 22, 671.

Ørsund, R. and Nes, E. (1989), Scripta Metall. 23, 1187.

Ørsund, R., Hjelen, J. and Nes, E. (1989), Scripta Metall. 23, 1193.

Osawa, K. (1984), Tetsu-to-Hagane, 70, S552.

Oscarsson, A., Ekstrom, H-E. and Hutchinson, W.B. (1992), in **Recrystallization'92**. eds. Fuentes and Gil Sevillano. Trans Tech Publications. 177.

Oscarsson, A., Hutchinson, W.B. and Ekstrom, H-E. (1991), Mats. Sci. and Tech. 7, 554.

Oscarsson, A., Hutchinson, W.B. and Karlsson, A., (1987), 8th ILMT, (Leoban, Vienna), 531.

Ouchi, C. and Okita, T. (1983), Trans. ISIJ. 23, 128.

Owen, N.J., Lykins, M.L., Stanton, G. and Malin, A.S. (1991), in **Recrystallization '90**, ed. Chandra, TMS, Warrendale, 649.

Palmer, J.E., Thompson, C.V. and Smith, H.I. (1987), J. Appl. Phys. 62, 2492.

Palumbo, G. and Aust, K.T. (1990), in **Recrystallization'90**. ed. Chandra. TMS. 101.

Palumbo, G. and Aust, K.T. (1992), in **Materials Interfaces**. eds. Wolf and Yip. Chapman and Hall, London. 190.

Pan, Y. and Adams, B.L. (1994), Scripta Metall. 30, 1055.

Pande, C.S. (1987), Acta Metall. 35, 2671.

Parker, E.R. and Washburn (1952), J. Trans. Met. Soc. AIME. $\underline{194}$, 1076.

Parthasarathi, and Beck, P.A. (1961), Trans. Metall. Soc. A.I.M.E. $\underline{221}$, 831.

Peczak, P. and Luton, M.J. (1993), Acta Metall. $\underline{41}$, 59.

Percy, J. (1864). **Metallurgy - Iron and Steel**. London.

Perdrix, C., Perrin, M.Y. and Montheillet, F. (1981), Mem. Sci. Rev. Met. $\underline{78}$, 309.

Perryman, E.C.W. (1955), Trans. Aime. $\underline{203}$, 1053.

Petkovic, R.A., Luton, M.J. and Jonas, J.J. (1975), Can, Metall. Q. $\underline{14}$, 137.

Petkovic, R.A., Luton, M.J. and Jonas, J.J. (1979), Acta Metall. $\underline{27}$, 1633.

Petrovic, V. and Ristik, M.M. (1980), Metallography, $\underline{13}$, 219.

Phillips, V.A. (1966), Trans. Metall. Soc. A.I.M.E. $\underline{236}$, 1302

Pilling, J. and Ridley, N. (1989), **Superplasticity in Crystalline Solids**. Institute of Metals, London.

Poirier, J.P. (1985), **Creep of Crystals**. Cambridge University Press, Cambridge.

Poirier, J.P. and Guillopé, M. (1979), Bull. Mineral. $\underline{102}$, 67.

Polmear, I.J. (1989), **Light Alloys**, Edward Arnold, London.

Pond, R.C. (1980) in **Grain Boundary Structure and Kinetics**. ed. Balluffi, ASM Ohio, 13.

Porter, A. and Ralph, B. (1981), J. Mats. Sci. $\underline{16}$, 707.

Porter, J. and Humphreys, F.J. (1979), Metal Sci. $\underline{13}$, 83.

Preston, O. and Grant, N.J. (1961), Trans. Metall. Soc. A.I.M.E. $\underline{221}$, 164.

Prinz, F. Argon, A.S. and Moffatt, W.C. (1982), Acta Metall. $\underline{30}$, 821

Puchi, E.S., Beynon, J.H. and Sellars, C.M. (1988), in **Int. Conf. on Thermomechanical Processing of Steels, Thermec-88**. Tokyo. ISI Japan. $\underline{2}$, 572.

Rabkin, E.I., Shvindlerman, L.S. and Straumal, B.B. (1991), Int. J. Mod. Phys. $\underline{5-19}$, 2989.

Rae, C.M.F. (1981), Phil. Mag. $\underline{A44}$, 1395.

Rae, C.M.F. and Smith, D.A. (1981), Proc. ICOTOM 6, Tokyo. 528.

Rae, C.M.F., Grovenor, C.R.M. and Knowles, K.M. (1981), Z. Metallk. $\underline{72}$, 798.

Rajković, M. and Buckley, R.A. (1981), Met. Sci. 15, 21.

Ralph, B., Shim, K.B., Huda, Z. Furley, J. and Edirisinghe, M. (1992), in **Grain Growth in Polycrystalline Materials**. ed. Abbruzzese and Brozzo. Trans Tech Publns. 129

Randle, V. (1992), **Microtextures**. The Institute of Materials, London.

Randle, V. and Brown, A. (1989), Phil. Mag. 59A, 1075.

Randle V. and Ralph, B. (1986), Acta Metall. 34, 891.

Randle, V. and Ralph, B. (1987), J. Mats. Sci. 22, 2535.

Rath, B.B., Ledrerich, R.J., Yolton, C.F. and Froes, F.H. (1979), Metall. Trans. 10A, 1013.

Rath, B.B. and Hu, H. (1972), in **The Nature and Behaviour of Grain Boundaries**. ed H. Hu. Plenum Press, NY. p405.

Ray, R.K. Hutchinson, W.B. and Duggan, B. J. (1975), Acta Metall. 23, 831.

Read, W.T. and Shockley, W. (1950), Phys. Rev.78, 275.

Read, W.T. (1953), **Dislocations in Crystals**. McGraw Hill.

Reid, C.N. (1973), **Deformation Geometry for Materials Scientists**. Pergamon Press, Oxford.

Reiter, S.F. (1952), Trans. Metall. Soc. A.I.M.E. 194, 972

Rhines, F.N. and Craig, K.R. (1974), Metall. Trans. 5, 413.

Ridha, A.A. and Hutchinson, W.B. (1982), Acta Metall. 30, 1929

Ridley, N. (1990), Mats. Sci. and Tech. 6, 1145.

Ringer, S.P., Li, W.B. and Easterling, K.E. (1989), Acta Metall. 37, 831.

Riontino, G., Antonione, C., Battezzati, L., Marino, F. and Tabasso, M. (1979), J. Mater. Science. 14, 86

Rivier, N. (1983), Phil. Mag. B47, L45.

Roberts, W. (1984), in ASM Seminar **Deformation, Processing and Structure**. ed. Krauss. St Louis, Missouri. ASM. p109

Roberts, W. (1985), Proc. 7th Int. Conf. on Strength of Metals and Alloys, ed. McQueen, 3, 1859.

Roberts, W. and Ahlblom, B. (1978), Acta Metall. 26, 801.

Rodrigues, C. (1840). J. des Mathematiques, $\underline{5}$, 380.

Rodrigues, P. and Bate, P. (1984), in **Textures in Non-Ferrous Metals and Alloys,** eds. Marchant and Morris, AIME.

Roessler, B., Novik, D.T. and Bever, M.B. (1963), Trans. Metall. Soc. A.I.M.E. $\underline{227}$, 985.

Rollett, A.J., Luton, M.J. and Srolovitz, D.J. (1992), Acta Metall. $\underline{40}$, 43.

Rollett, A.D., Srolovitz, D.J., Doherty, R.D. and Anderson, M.P. (1989a), Acta Metall. $\underline{37}$, 627.

Rollett, A.D., Srolovitz, D.J. and Anderson, M.P. (1989b), Acta Metall. $\underline{37}$, 1227.

Rosen, A. Burton, M.S. and Smith, G.V. (1964), Trans. Metall. Soc. A.I.M.E. $\underline{230}$, 205.

Rosi, F.D., Alexander, B.H. and Dube, C.A. (1952), Trans. Metall. Soc. A.I.M.E. $\underline{194}$, 189.

Rutter, E.H. (1983), J. of Geol. Soc. $\underline{140}$, 725.

Rutter, J.W. and Aust, K.T. (1960), Trans. Metall. Soc. A.I.M.E. $\underline{218}$, 682.

Rutter, J.W. and Aust, K.T. (1965), Acta Metall. $\underline{13}$, 181

Ryde, L., Hutchinson, W.B. and Jonsson, S. (1990), in **Recrystallization'90**, ed. Chandra. TMS. p313.

Ryum, N. (1969), Acta Metall. $\underline{17}$, 831.

Ryum, N. and Hunderi, O. (1989), Acta Metall. $\underline{37}$, 1375.

Ryum, N., Hunderi, O. and Nes, E. (1983), Scripta Metall. $\underline{17}$, 1281.

Sachs, G. (1928) Z. Verein, Deut.Ing. $\underline{72}$, 734.

Saetre, T.O. and Ryum, N. (1992), in **Grain Growth in Polycrystalline Materials**. eds. Abbruzzese and Brozzo. Rome. Trans Tech Publns. 373.

Saetre, T., Hunderi, O. and Nes, E. (1986a), Acta Metall. $\underline{34}$, 981.

Saetre, T.O., Solberg, J.K. and Ryum, N. (1986b), Metallography, $\underline{19}$, 347.

Sah, J.P., Richardson, G.J. and Sellars, C.M. (1974), Metal Sci. $\underline{8}$, 325.

Sakai, T., Akben, M.G. and Jonas, J.J. (1983), Acta Metall. $\underline{31}$, 631.

Sakai, T. and Jonas, J.J. (1984), Acta Metall. $\underline{32}$, 189.

Sakai, T. and Ohashi, M. (1992), in **Recrystallization'92.** eds. Fuentes and Gil Sevillano. Trans Tech Publications. 521.

Sakai, T., Ohashi, M. and Chiba, K (1988), Acta Metall. $\underline{36}$, 1781.

Sakakura, A. (1969), J. App. Phys. $\underline{40}$, 1539.

Samajdar, I. and Doherty, R.D. (1995). Scripta Metall. Mater. $\underline{32}$, 845.

Samuels, L.E. (1954), J. Inst. Met. $\underline{83}$, 359.

Sanders, R.E., Baumann, S.F. and Stump, H.C. (1986). Proc. 1st Int. Conf. on Aluminium Alloys, Charlottsville, USA. $\underline{3}$, 1441.

Sanders, T.H. and Starke, E.A. (1982), Acta Metall. $\underline{36}$, 927.

Sandström, R. (1977a), Acta Metall. $\underline{25}$, 897.

Sandström, R. (1977b), Acta Metall. $\underline{25}$, 905.

Sandström, R. (1980), Z. Metallk. $\underline{71}$, 681.

Sandström, R. and Lagneborg, R. (1975), Scripta Metall. $\underline{9}$, 59.

Sandström, R. Lehtinen, E. Hedman, B., Groza, I and Karlsson, J. (1978), J. Mats. Sci. $\underline{13}$, 1229.

Sass, S.L. and Bristowe, P.D. (1980) in **Grain Boundary Structure and Kinetics.** ed. Balluffi, ASM Ohio, 71.

Satoh, S., Obara, T., Takasaki, J., Yasuda, A. and Nishida, M. (1985), Kawasaki Steel Tech. Rep., $\underline{12}$, 36.

Sauter, H., Gleiter, H. and Baro, G. (1977), Acta Metall. $\underline{25}$, 457.

Sauveur, A. (1912). Int. Assoc. Testing Mater. Proc. VI Cong. New York. $\underline{2.6}$, 1.

Savart, F. (1829). Ann. Chim. Phys. $\underline{41}$, 61.

Schlaffer, and Bunge, H. (1974), Texture, $\underline{3}$, 157

Schmidt, E. and Wassermann, G. (1927), Z. Physik, $\underline{42}$, 779

Schmidt, J. (1989), Thermochim. Acta $\underline{151}$, 333.

Schmidt, J. and Haessner, F. (1990), Z.f. Phys B. Condensed matter. $\underline{81}$, 215.

Schmidt, U., Lücke, K. and Pospeich, J. (1975), in **Fourth European Texture Conference,** Met. Soc., London, 147.

Schulz, L.G. (1949). J. Appl. Phys. **20**, 1030.

Schwarzer, R.A. (1993), Text. Microstruct. **20**, 7.

Schwarzer, R. and Weiland, H. (1984), in **ICOTOM 7,** eds. Brakman et al., Noordwijkerhout, 839.

Seeger, A. and Haasen, P. (1956), Phil. Mag. **3**, 470.

Seidman, D.N. (1992), in **Materials Interfaces.** (eds. Wolf and Yip), Chapman and Hall, London. 58.

Sellars, C.M. (1978), Philos. Trans. Royal Soc. **288**, 147

Sellars, C.M. (1986), in Proc. 7th Int. Risø Symposium. Risø, Denmark. ed. Hansen et al. 167.

Sellars, C.M. (1990), Mats. Sci. and Tech. **6**, 1072.

Sellars, C.M. (1992a), Proc. 3rd Int. Conf on Aluminium, Trondheim. **3**, 89.

Sellars, C.M. (1992b), in **Recrystallization'92.** eds. Fuentes and Gil Sevillano. Trans Tech Publications. 29.

Sellars, C.M., Irisarri, A.M. and Puchi, E.S. (1986) in **Microstructural Control** in **Aluminium Alloys**. eds Chia and McQueen. Met. Soc. AIME. Warrendale, USA. 179.

Sellars, C.M. and Whiteman, J.A. (1979), Met. Sci. **13**, 187.

Sheppard, T., Parson, N. and Zaidi, M.A. (1983), Metal Sci. **17**, 481.

Sherby, O.D. and Burke, P.M. (1967), Prog. Mater. Sci. **13**, 325.

Sherby, O.D. and Wadsworth, J. (1988), in **Superplasticity in Aerospace**. eds. Heikennen & McNelly. TMS.

Shvindlerman, L.S. and Straumal, B.B. (1985), Acta Metall. **33**, 1735.

Simpson, C.J., Aust, K.T. and Wineguard, W.C. (1970), Metall. Trans. **1**, 1482.

Sinha, P.P and Beck, P.A. (1961), J. App. Phys. **32**, 1222.

Sircar, S. and Humphreys, F.J. (1994), Proc. **4th Int. Conf on Aluminium**. eds. Sanders and Starke,Chan Atlanta. **1**, 170.

Smidoda, K., Gottschalk, W. and Gleiter, H. (1978), Acta Metall. **26**, 1833

Smirnova, N.A., Levit,V.I., Pilyugin, V.I., Kusnetzov, R.I., Davydova, L.S. and Sazonov, V.A. (1986), Fiz. Met Metalloved. **61**, 1170.

Smith, C.J.E. and Dillamore I.L. (1970), Met. Science J. $\underline{4}$, 161

Smith, C.S. (1948), Trans. Metall. Soc. A.I.M.E. $\underline{175}$, 15.

Smith, C.S. (1952), in **Metal Interfaces**, ASM, Cleveland, 65.

Smith, D.A. (1992), in **Grain Growth in Polycrystalline Materials.** eds. Abbruzzeze and Brozzo. Trans Tech Pubs. Switzerland. 221.

Smith, D.A., Rae, C.M.F. and Grovenor, C.R.M. (1980), in **Grain boundary Structure and Kinetics.** ed. Balluffi. ASM Metals Park, Ohio. 337.

Soares, A, Ferro, A. and Fortes, M. (1985). Scripta Metall. $\underline{19}$, 1491

Sorby, H.C. (1887). J. Iron and Steel Inst. $\underline{31.1}$, 253.

Solberg, J.K., McQueen, H.J., Ryum, N. and Nes, E. (1989), Phil. Mag. $\underline{A60}$, 447.

Sparks, C.N. and Sellars, C.M. In **Recrystallization'92**. eds. Fuentes and Sevillano. Trans Tech Pubs. 557.

Speich, G.R. and Fisher, R.M. (1966), in **Recrystallization, Grain Growth and Textures.** ASM, Ohio, 563.

Srolovitz, D.J., Grest, G.S. and Anderson, M.P. (1985), Acta Metall. $\underline{33}$, 2233.

Srolovitz, D.J., Grest, G.S. and Anderson, M.P. (1986), Acta Metall. $\underline{34}$, 1833.

Srolovitz, D.J., Anderson, M.P., Sahni, P.S. and Grest, G.S. (1984a), Acta Metall. $\underline{32}$, 793.

Srolovitz, D.J., Anderson, M.P., Grest, G.S. and Sahni, P.S. (1984b), Acta Metall. $\underline{32}$, 1429.

Stead, J.E. (1898). J. Iron and Steel Inst. $\underline{53.1}$, 145.

Stewart, A.T. and Martin, J.W. (1975), Acta Metall. $\underline{23}$, 1.

Stiegler, J.O., Dubose, C.K.H., Reed, R.E. and McHargue, C.J. (1963), Acta Metall. $\underline{11}$, 851

Stoloff, N.S. and Davies, R.G. (1964), Acta Metall. $\underline{12}$, 473.

Stowell, M.J. (1983), Proc. 4th Int. Risø Symposium. eds Bilde-Sørensen et al. Risø, Denmark. 119.

Straker, H.R. and Holt, D.L. (1972), Acta Metall. $\underline{20}$, 569.

Stuitje, P.J.T. and Gottstein, G. (1980), Z. Metallk. $\underline{71}$, 279.

Stüwe, H.P. (1978) In **Recrystallization of Metallic Materials**. ed. Haessner. Dr Riederer Verlag GMBH, Stuttgart. 11.

Sun, R.C. and Bauer, C.L. Acta Metall. (1970), $\underline{18}$, 639.

Sursaeva, V.G., Andreeva, A.V., Kopezky, C.V. and Shvindlerman, L.S. (1976), Solid State Physics, $\underline{41}$, 1013.

Sutton, A.P. and Balluffi, R.W. (1987), Acta Metall. $\underline{35}$, 2177.

Sutton, A.P. and Balluffi, R.W. (1995), **Interfaces in Crystalline Materials**. Oxford University Press, Oxford.

Szpunar, J.A. and Davies, S.T. (1984), Proc. ICOTOM 7, eds. Brakman et, al, Noordwijkerhout, 845.

Taguchi, S., Sakakura, A. and Takashima, H. (1966), U.S. Patent 3,287,183.

Takayama, Y., Tozawa, T., Kato, H.,Furushiro, N. and Hori,S. (1992), in **Grain Growth in Polycrystalline Materials**. ed. Abbruzzese and Brozzo. Trans Tech Publns. 325.

Takeuchi, A. and Argon, A.S (1976), J. Mats. Sci. $\underline{11}$, 1547.

Taylor, G.I. (1938), J.Inst.Met. $\underline{62}$, 307

Tegart, W.J. McG. (1992), In **Recrystallization'92**. eds. Fuentes and Gil Sevillano. San Sebastian, Trans Tech Publications. 1.

Thompson, A.W. (1977), Metall. Trans. $\underline{8A}$, 833.

Thompson, C.V. (1992), in **Grain growth in polycrystalline materials**. ed. Abbruzzese and Brozzo. Trans Tech Publns. 245

Thompson, C.V., Frost, H.J. and Spaepen, F. (1987), Acta Metall. $\underline{35}$, 887.

Thomson-Russell, K.C. (1974), Planseeber. Pulvermetall. $\underline{22}$, 264.

Tiedema, T.J., May, W. and Burgers, W.G. (1949), Acta Cryst. $\underline{2}$, 151.

Tien, T.Y. and Subbaro, E.C. (1963), J. Am. Ceram. Soc. $\underline{46}$, 489.

Titchener, A.L. and Bever, M.B. (1958), Prog. Metal Phys. $\underline{7}$, 247.

Tungatt, P.D. and Humphreys, F.J, (1981), Proc. 2nd Int. Risø Symp. Risø, Denmark. ed. Hansen et al. 393.

Tungatt, P.D. and Humphreys, F.J. (1984), Acta Metall. $\underline{32}$, 1625.

Turnbull, D. (1951), Trans. Metall. Soc. A.I.M.E. $\underline{191}$, 661.

Tweed, C., Hansen, N. and Ralph, B. (1982), Metall. Trans. 14A, 2235.

Twiss, R.J. (1977), Pure Appl. Geophys. 115, 227.

Urai, J.L., Means, W.D. and Lister, G.S. (1986), in **The Patterson Volume**, Geophysical Monograph 36. The American Geophysical Union. 161.

Urcola, J.J. and Sellars, C.M. (1987), Acta Metall. 35, 2649.

Ushioda, K. Ohsone, H. and Abe, M. (1981), in **ICOTOM 6,** (Tokyo), 829.

Ushioda, K., Hutchinson, W.B., Agren, J. and von Schlippenbach, U. (1989), Mats. Sci. and Tech. 2, 807.

van Drunen, G. and Saimoto, S. (1971), Acta Metall. 19, 213.

Vandermeer, R.A. and Gordon, P. (1959), Trans. Metall. Soc. A.I.M.E. 215, 577.

Vandermeer, R.A. and Gordon, P. (1963), in **Recovery and Recrystallization of Metals.** ed. Himmel. Interscience. p211.

Vandermeer, R.A. and Hu, H. (1994), Scripta Metall. 42, 3071.

Vandermeer, R.A. and Rath, B.B. (1989a) Metall. Trans. A, 20A, 391.

Vandermeer, R.A. and Rath, B.B. (1989b), Proc. 10th Risø Int. Symposium. eds Bilde-Sørensen et al. Risø, Denmark. 589.

Vandermeer, R.A. and Rath, B.B. (1990), in **Recrystallization'90**, ed. Chandra. TMS. p49.

Varma, S.K. (1986), Mats. Sci. & Eng. 82, 19.

Varma, S.K and Wesstrom, B.C (1988) J. Mats Sci Let. 7 1092.

Varma, S.K. and Willetts, B.L. (1984), Metall. Trans. 15A, 1502.

Vasudevan, A.K., Petrovic, J.J. and Roberson, J.A. (1974), Scripta Metall. 8, 861.

Vatne, H.E. and Nes, E. (1994), Scripta Metall. Mater. 30, 309.

Vatne, H.E., Daaland, O. and Nes, E. (1994), Materials Science Forum, 157-162, 1087.

Verbraak, C.A. (1958), Acta Metall. 6, 580.

Verbraak, C.A. (1975), in **ICOTOM 5**, eds. Gottstein and Lücke, Springer-Verlag, 111.

Verbraak, C.A. and Burgers, W.G. (1957), Acta Metall. 5, 765.

Vidoz, A.E., Lazarevic, D.P. and Cahn, R.W. (1963), Acta Metall. 11, 17.

Virnich, K.H. and Lücke, K. (1978) in **ICOTOM 5,(2),** eds. Gottstein and Lücke, Springer-Verlag, 3.

Viswanathan. R. and Bauer, C.L. (1973), Acta Metall. <u>21</u>, 1099.

Vitek, V., Sutton, A.P., Smith, D.A. and Pond, R.C. (1980), in **Grain Boundary Structure and Kinetics**. ed. Balluffi, ASM Ohio, 115.

von Neumann, J. (1952), in **Metal Interfaces**. ASM, Cleveland, Ohio. p108

von Schlippenbach, U., Emren, F. and Lücke, K. (1986), Acta Metall. <u>34</u>, 1289

von Schlippenbach, U. and Lücke, K. (1984), in **ICOTOM 7**, eds. Brakman et al, Noordwijkerkout, 159.

Wakefield, P.T. and Hatherly, M. (1981), Met.Sci. <u>15</u>, 109

Walter, J.L. (1965), J. Appl. Phys. <u>36</u>, 1213.

Walter, J.L. and Dunn, G.C. (1959), Acta Metall. <u>7</u>, 424.

Walter, J.L. and Dunn, C.G. (1960a), Trans. Metall. Soc. A.I.M.E. <u>218</u>, 914.

Walter, J.L. and Dunn, G.C. (1960b), Acta Metall. <u>8</u>, 497.

Walter, J.L. and Koch, E.F. (1963), Acta Metall. <u>11</u>, 923.

Walton, D.T., Frost, H.J. and Thompson, C.V. (1992), in **Grain Growth in Polycrystalline Materials**. ed. Abbruzzese and Brozzo. Trans Tech Publns. 531.

Wang, J., Horita, Z., Furukawa, M., Nemoto, M., Tsenev, N., Valiev, R., Ma, Y. and Langdon, T. G. (1993), J. Mater. Res. <u>8</u>, 2810.

Warrington, D. H. (1980), in **Grain Boundary Structure and Kinetics**. ed. Balluffi, ASM Ohio, 1.

Washburn, J. and Murty, G. (1967), Can. J. Physics, <u>45</u>, 523

Washburn, J. and Parker, E.R. (1952), Trans. Metall. Soc. A.I.M.E. <u>194</u>, 1076.

Wassermann, G. (1963), Z. Metallk. <u>54</u>, 61.

Wassermann, G. and Grewen, J. (1962), **Texturen Metallischer Werkstoffe**, Springer Verlag, Berlin.

Wassermann, G., Bergmann, H.W. and Fromeyer, G. (1978), in **ICOTOM 5,(2)**, eds. Gottstein and Lücke, Springer-Verlag, Berlin, 37.

Watanabe, T. (1992a), Scripta Metall. <u>27</u>, 1497.

Watanabe, T. (1992b), in **Grain Growth in Polycrystalline Materials**. ed. Abbruzzese and Brozzo. Trans Tech Publns. 209.

Watanabe, T., Fujii, H., Oikawa, H. and Arai, K.I. (1989), Acta Metall. 37, 941.

Watts, B.M., Stowell, M.J., Baikie, B.L. and Owen, D.G.E. (1976), Met. Sci. 10, 189, 198.

Weaire, D. and Glazier, J.A. (1992), in **Grain Growth in Polycrystalline Materials**. ed. Abbruzzese and Brozzo. Trans Tech Publns. 27.

Weaire, D. and Kermode, J. (1983), Phil. Mag. B48, 245.

Weaire, D. and Rivier, N. (1984), Contemporary. Phys. 25, 59.

Weiland, H. (1992), Acta Metall. 40, 1083.

Weiland, H., Dahlem-Klein, E., Fiszer, A. and Bunge, H. (1988), in ICOTOM 8, eds. Kallend and Gottstein, TMS, Warrendale, 717.

Weiland, H. and Hirsch, J. (1991). Proc. ICOTOM 9. ed. Bunge. Avignon. 647.

Weiland, H. and Schwarzer, R.A. (1984), Proc. ICOTOM 7, ed. Brakman et al, Noordwijkerhout, 765.

Weins, M.J. (1972), in **Grain Boundaries and Interfaces,** eds. Chaudhri and Matthews. North Holland. 138.

Weiss, I., Srinivasan, R. and Froes, F.H. (1990), in **Recrystallization'90**. ed. Chandra. TMS, Ohio. 609.

Wert, J.A. and Austin, L.K. (1985), Metall. Trans. 19A, 617.

Wert, J.A., Paton, N.E., Hamilton, C.H. and Mahoney, M.W. (1981), Metall. Trans. 12A, 1267.

Wilbrandt, P-J. (1988), in ICOTOM 8, Santa Fe, USA. 573.

Williams, J.C. and Starke, E.A. (1982), in **Deformation Processing and Structure**. ed. Krauss. ASM. Ohio. 279.

Willis, D.J. (1982), unpublished work (quoted by Hatherly 1982),

Willis, D.J. (1993), private communication.

Willis, D.J. and Hatherly, M. (1978), in ICOTOM 5, eds. Gottstein and Lücke, Springer-Verlag, Berlin, 465.

Wilsdorf, H. and Kuhlman-Wilsdorf, D. (1953), Acta Metall. 1, 394.

Wolf, D. and Merkle, K.L. (1992), in **Materials Interfaces**. (eds. Wolf and Yip), Chapman and Hall, London. 87.

Wolf, D. and Yip, S. (1992) (eds.) **Materials Interfaces**. Chapman and Hall, London.

Yamagata, H. (1993), Scripta Metall. $\underline{27}$, 727.

Yeung, W.Y. (1990), Acta Metall. Mater. $\underline{38}$, 1109.

Yeung, W., Hirsch, J. and Hatherly, M. (1988), in **ICOTOM 8**, eds. Kallend and Gottstein, TMS, Warrendale, 631.

Yoshida, H., Liebmann, B. and Lücke, K. (1959), Acta Metall. $\underline{7}$, 51.

Zener, C. (1948), see Smith (1948),

SUBJECT INDEX

Printed in Dunstable, United Kingdom

83755687R00291